Treetops at Risk

Margaret Lowman • Soubadra Devy • T. Ganesh
Editors

Treetops at Risk

Challenges of Global Canopy Ecology and Conservation

Springer

Editors
Margaret Lowman
North Carolina Museum
of Natural Sciences
North Carolina State University
Raleigh, NC, USA

Soubadra Devy
Ashoka Trust for Research in Ecology
and the Environment (ATREE)
Bangalore, KA, India

T. Ganesh
Ashoka Trust for Research in Ecology
and the Environment (ATREE)
Bangalore, KA, India

ISBN 978-1-4614-7160-8 ISBN 978-1-4614-7161-5 (eBook)
DOI 10.1007/978-1-4614-7161-5
Springer New York Heidelberg Dordrecht London

Library of Congress Control Number: 2013939349

Printed on acid-free paper

Springer is part of Springer Science+Business Media (www.springer.com)

Foreword

E.O. Wilson has said that we are letting "nature slip through our fingers." Depleting Earth's fabric of life means we could slip through hers. In a few decades, I have personally witnessed the loss or serious decline of about half of the coral reefs, kelp forests, seagrass meadows and mangroves, elimination of 90 % of many kinds of fish and other ocean wildlife, creation of hundreds of coastal dead zones, a sharp reduction in oxygen-generating carbon dioxide-absorbing phytoplankton, acidification of the ocean, acceleration of global warming, a swift reduction of polar ice, and introduction of tons of toxins and plastic into the blue heart of the planet.

The good news is that we have arrived at the "sweet spot" in our history. Never before could we so clearly grasp our dependence on – and loss of – intact, living systems that underpin everything we care about. Never again will there be a chance as good as the present time to restore and protect the natural systems that keep us alive. Now we know what no one knew when I was a child. Making peace with nature is the best hope for having an enduring future for humankind.

In this volume, international forest scientists describe the challenges, methods, and their hope for conservation of these terrestrial systems. Like the oceans, forests are critical hot spots for biodiversity, and their future health is intertwined with that of humans.

Raleigh, NC, USA Sylvia Earle (aka "Her Deepness")

Contents

Part II Climate Change

Part III New Approaches

Part IV Education and Outreach

Part V Ecosystem, Services and Sustainability

Contributors

L.A. Ashton Environmental Futures Centre, Griffith School of the Environment, Griffith University, Nathan, QLD, Australia

Gregory P. Asner Department of Global Ecology, Carnegie Institution for Science, Stanford, CA, USA

M. Baimas-George Department of Biology, Colgate University, Hamilton, NY, USA

Rohini Balakrishnan Centre for Ecological Sciences, Indian Institute of Science, Bangalore, India

N. Barbier IRD, UMR AMAP, Montpellier, France

James Barsimantov Department of Environmental Studies, University of California Santa Cruz, Santa Cruz, CA, USA

Kamaljit S. Bawa Ashoka Trust for Research in Ecology and the Environment (ATREE), Bangalore, KA, India

S.L. Boulter Environmental Futures Centre, Griffith School of the Environment, Griffith University, Nathan, QLD, Australia

T. Bourguignon Department of Biological Sciences, National University of Singapore, Singapore

Anna E. Burns Department of Zoology, La Trobe University, Bundoora, VIC, Australia

C.J. Burwell Environmental Futures Centre, Griffith School of the Environment, Griffith University, Nathan, QLD, Australia

Biodiversity Program, Queensland Museum, South Brisbane, QLD, Australia

Catherine L. Cardelús Department of Biology, Colgate University, Hamilton, NY, USA

Rose Caspa Agronomic and Development Research Institute, Yaoundé, Cameroon

Lynne Cherry Young Voices on Climate Change, Lynne Cherry, Washington, DC, USA

Tiffany Clark Department of Biology, Baker University, Baldwin City, KS, USA

North Carolina Museum of Natural Sciences, Raleigh, NC, USA

P. Couteron IRD, UMR AMAP, Montpellier, France

Mohan P. Devkota Department of Botany, Tribhuvan University, Amrit Campus, Kathmandu, Nepal

Soubadra Devy Ashoka Trust for Research in Ecology and the Environment (ATREE), Bangalore, KA, India

Daniela Dutra Elliott Botany Department, University of Hawaii at Manoa, Honolulu, HI, USA

Alexandra Fischer-Pardow Department of Plant Ecology and Systematics, University of Kaiserslautern, Kaiserslautern, Germany

Evariste Fongnzossie Fedoung University of Douala, Douala, Cameroon

Logan Gallardo Department of Biology, Baker University, Baldwin City, KS, USA

North Carolina Museum of Natural Sciences, Raleigh, NC, USA

T. Ganesh Ashoka Trust for Research in Ecology and the Environment (ATREE), Bangalore, KA, India

Frances Gatz Frances Gatz, Environmental Expeditions, Silver Spring, MD, USA

David B. Greenberg Environmental Change Institute, Oxford University Centre for the Environment, School of Geography and the Environment, University of Oxford, Oxford, United Kingdom

School of Geography, University of Leeds, Leeds, United Kingdom

Penelope Greenslade Centre for Environmental Management, School of Science and Engineering, Mt Helen, University of Ballarat, Ballarat, VA, Australia

Australian National University, ACT, Australia

Nimal Gunatilleke University of Peradeniya, Peradeniya, Sri Lanka

Savitri Gunatilleke University of Peradeniya, Peradeniya, Sri Lanka

Andrew J. Hamilton Department of Agriculture and Food Systems, The University of Melbourne, Dookie Campus, Dookie College, VIC, Australia

Harold Heatwole Department of Zoology, North Carolina State University, Raleigh, NC, USA

Rebecca C.-C. Hsu Taiwan Forestry Research Institute, Taipei, Taiwan

Universiteit van Amsterdam, Institute for Biodiversity and Ecosystem Dynamics (IBED), Amsterdam, GE, The Netherlands

Manjari Jain Institute of Evolutionary Biology and Environmental Studies, University of Zurich, Zurich, Switzerland

Centre for Ecological Sciences, Indian Institute of Science, Bangalore, India

Leon Kaganovskiy Touro College, New York, NY, USA

Roland Kays North Carolina Museum of Natural Sciences, Raleigh, NC, USA

Smithsonian Tropical Research Institute, Panama, República de Panamá North Carolina State University, Raleigh, NC, USA

R.L. Kitching Environmental Futures Centre, Griffith School of the Environment, Griffith University, Nathan, QLD, Australia

Ch. Körner Institute of Botany, University of Basel, Basel, Switzerland

Bhaskar Krishnamurthy TREE Foundation, Sarasota, FL, USA

Jagdish Krishnaswamy Ashoka Trust for Research in Ecology and the Environment (ATREE), Bangalore, KA, India

M.J. Laidlaw Queensland Department of Science, Information Technology, Innovation and the Arts, Queensland Herbarium, Toowong, QLD, Australia

Michael Lakatos Department of Plant Ecology and Systematics, University of Kaiserslautern, Kaiserslautern, Germany

C.L. Lambkin Biodiversity Program, Queensland Museum, South Brisbane, QLD, Australia

William F. Laurance Centre for Tropical Environmental and Sustainability Science (TESS) and School of Marine and Tropical Biology, James Cook University, Cairns, QLD, Australia

Maurice Leponce Biological Assessment Section, Royal Belgian Institute of Natural Sciences, Brussels, Belgium

Simon L. Lewis School of Geography, University of Leeds, United Kingdom

Department of Geography, University College London, London, WC1E, United Kingdom

Zoë Lindo Department of Biology, University of Western Ontario, London, ON, Canada

Wen-Yao Liu Key Laboratory of Tropical Forest Ecology, Xishuangbanna Tropical Botanical Garden, Chinese Academy of Sciences, Yunnan, China

Thomas Lovejoy George Mason University, VA, USA

The Heinz Center for Science, Economics and the Environment, Washington, DC, USA

Margaret Lowman North Carolina Museum of Natural Sciences, North Carolina State University, Raleigh, NC, USA

S.C. Maunsell Environmental Futures Centre, Griffith School of the Environment, Griffith University, Nathan, QLD, Australia

E. McCord Department of Biology and Environmental Studies, New College of Florida, Sarasota, FL, USA

William R. Miller Department of Biology, Baker University, Baldwin City, KS, USA

North Carolina Museum of Natural Sciences, Raleigh, NC, USA

Andrew W. Mitchell Global Canopy Programme (GCP), Oxford, UK

Mark W. Moffett National Museum of Natural History, Smithsonian Institution, Washington, DC, USA

Demetria Mondragón Centro Interdisciplinario de Investigación para el Desarrollo Integral Regional (CIIDIR) unidad Oaxaca, Santa Cruz Xoxocotlán, Oaxaca Mexico

Nalini M. Nadkarni Center for Science and Mathematics Education, University of Utah, Salt Lake City, UT, USA

A. Nakamura Environmental Futures Centre, Griffith School of the Environment, Griffith University, Nathan, QLD, Australia

Bernard-Aloys Nkongmeneck University of Yaoundé I, Yaoundé, Cameroon

F. Ødegaard Norwegian Institute for Nature Research, Trondheim, Norway

Claire M.P. Ozanne Department of Life Sciences, Centre for Research in Ecology, University of Roehampton, Holybourne Avenue, London, UK

Raphaël Pélissier Ecology Department, French Institute of Pondicherry, Pondicherry, India

IRD, UMR AMAP, Montpellier, France

Pierre Ploton Ecology Department, French Institute of Pondicherry, Pondicherry, India

IRD, UMR AMAP, University of Yaoundé I, Yaoundé, Cameroon

Christophe Proisy IRD, UMR AMAP, Montpellier, France

Alex Racelis Department of Environmental Studies, University of California Santa Cruz, Santa Cruz, CA, USA

Department of Biology, University of Texas Pan American, Edinburg, TX, USA

Nitin D. Rai Ashoka Trust for Research in Ecology and the Environment (ATREE), Bangalore, KA, India

B.R. Ramesh Ecology Department, French Institute of Pondicherry, Pondicherry, India

D.C. Randle St. Francis High School, St. Francis, MN, USA

Yves Roisin Evolutionary Biology and Ecology, Université Libre de Bruxelles, Brussels, Belgium

Stefan A. Schnitzer Department of Biological Sciences, University of Wisconsin – Milwaukee, WI, USA

Markus Seibel Geographisches Institut der Humboldt-Universität zu Berlin, Berlin, Germany

Reinmar Seidler University of Massachusetts, Boston, MA, USA

K.S. Seshadri Ashoka Trust for Research in Ecology and the Environment (ATREE), Bangalore, KA, India

Siddappa Setty Ashoka Trust for Research in Ecology and the Environment (ATREE), Bangalore, KA, India

S.P. Singh Centre for Ecology, Development and Research (CEDAR), Dehradun, Uttarakhand, India

Liang Song Key Laboratory of Tropical Forest Ecology, Xishuangbanna Tropical Botanical Garden, Chinese Academy of Sciences, Yunnan, China

Nigel E. Stork Environmental Futures Centre, Griffith School of Environment, Griffith University, Nathan, QLD, Australia

Rajesh Thadani Centre for Ecology, Development and Research (CEDAR), Dehradun, Uttarakhand, India

Tamara Ticktin Botany Department, University of Hawaii at Manoa, Honolulu, HI, USA

Stephen M. Turton Centre for Tropical Environmental & Sustainability Sciences, School of Earth and Environmental Sciences, James Cook University, Cairns, QLD, Australia

Sybille Unsicker Institute of Ecology, University of Jena, Jena, Germany

Vivek Ramachandran Ashoka Trust for Research in Ecology and the Environment (ATREE), Bangalore, KA, India

Manipal University, Manipal.edu, Madhav Nagar, Manipal, India

Bryson Voirin Max Plank Institute for Ornithology, Radolfzell, Germany

Smithsonian Tropical Research Institute, Panamá, República de Panamá

Carl W. Wardhaugh School of Marine and Tropical Biology, James Cook University, Cairns, QLD, Australia

Alemayu Wassie Eshete Department of Forestry, Bahir Dar University, Bahir Dar, Ethiopia

David M. Watson School of Environmental Sciences, and Institute for Land, Water and Society, Charles Sturt University, NSW, Australia

Martin Wikelski Max Planck Institute for Ornithology, Radolfzell, Germany

North Carolina Museum of Natural Sciences, Raleigh, NC, USA

Neville N. Winchester Department of Geography, University of Victoria, British Columbia, VIC, Canada

Jan H.D. Wolf Universiteit van Amsterdam, Institute for Biodiversity and Ecosystem Dynamics (IBED), Amsterdam, GE, The Netherlands

Stephen P. Yanoviak Department of Biology, University of Louisville, Louisville, KY, USA

FONDATION elf

Part I
Emerging Issues

Introduction

The notion of "treetops at risk" is an understatement. According to Laurance (this section), over half of the world's original forests have been destroyed. Although tree-planting remedies often occur, many countries restore nonnative forests or plantations, both of which have significantly different impacts on planetary health. In this first section, called "Emerging Issues," authors comment on challenges facing our global forests—from logging to climate change to insidious shrinkage of forest boundaries, the human race is intent upon reducing the world's forest resources. Canopies are part of this critical package—forest canopies provide unique ecosystem services to all life on Earth, and so the notion of "treetops at risk" indicates our current plight. The authors in this section represent synthesizers, leaders, and consensus builders in the world of canopy biology. They all spoke at the 5th International Canopy Conference in Bangalore, India, during 2010. From that collaborative conference, this book had its inspiration. In this introductory section, authors provide commentary about solutions, threats, and the future of global forests and their canopies.

Chapter 1
The Role of Scientific Conferences to Foster Conservation Solutions for Global Forests

Margaret Lowman, Soubadra Devy, and T. Ganesh

Keywords Canopy conference • Science education • India • Forest conservation

The professional group of biologists who prioritize studies of the forest canopy hosted their first international conference in Sarasota, Florida, in 1994, and the field's first textbook was published the same year. Subsequently, canopy scientists have convened a dedicated conference every 4 years, with a mission of advancing the scientific field through collaborative exchange of ideas. For its first time ever, the fifth International Canopy Conference (ICC) convened in an emerging country, hosted by Ashoka Trust for Research in Ecology and the Environment (ATREE) in Bangalore, India. Also, for the first time ever, education outreach was added as a new session to the conference agenda, including a highly popular canopy education workshop that drew Indian students, teachers, scientists, and stakeholders from all regions of the country. This session was advertised as an open forum where laypersons and teachers could interface with canopy scientists, who enthusiastically shared their experiences, thereby taking science outside the conventional walls of academia. In a country like India, this was groundbreaking. In subsequent feedback, we were inspired by the reports of burgeoning education outreach activities in India spawned by this conference session.

Like many developing countries, India needs success in advancing its forest conservation. The Western Ghats forest region in southern India is one of India's

M. Lowman (✉)
North Carolina Museum of Natural Sciences, North Carolina State University,
121 W. Jones Street, Raleigh, NC 27603, USA
e-mail: canopymeg@gmail.com

S. Devy • T. Ganesh
Ashoka Trust for Research in Ecology and the Environment (ATREE), Royal Enclave,
Srirampura, Jakkur Post, Bangalore, KA 560064, India
e-mail: soubadra@atree.org; tganesh@atree.org

M. Lowman et al. (eds.), *Treetops at Risk: Challenges of Global Canopy Ecology and Conservation*, DOI 10.1007/978-1-4614-7161-5_1,

Fig. 1.1 Teaching Indian students about canopy access

four designated biodiversity hotspots, and the other three also contain critical forest habitat. India is losing 2 % of its forests annually (see Seidler et al.'s chapter), and charismatic megafauna such as tigers, leopards, and Indian elephants remain poster children for India's forest conservation, as well as the potential economy of ecotourism for future generations. The promotion of education outreach at the ICC fostered new enthusiasm in forest canopy conservation using canopies as a "hook" to inspire rapid action. New outreach programs promoted by canopy scientists throughout India after this conference included (1) tree planting programs in schoolyards, (2) citizen surveys of butterflies, (3) Fulbright funding for canopy education outreach, (4) participation by canopy scientists in Earthwatch field courses for HSBC Bank employees, (5) graduate training programs in canopy access (Fig. 1.1), (6) publication of a canopy science book for the public, and (7) a certificate course in conservation for young professionals. All these activities illustrate the genesis of a new

Fig. 1.2 Future women scientists of Assam, India

culture among Indian canopy scientists to engage in education outreach. One year later, hundreds of Indian students and thousands of citizens benefited directly from this one session at the canopy conference.

The notion of including science education and outreach as part of a technical scientific conference agenda was not part of the first ICC, nor of many other scientific meetings several decades ago. Almost 100 years old, the Ecological Society of America (ESA) has witnessed an exponential increase in education sessions at its annual meetings, from less than five in 1990 to over 25 in 2010. The emerging priorities of communicating science and fostering education of diverse stakeholders is an increasingly important platform for scientific conferences, especially given the urgent priorities for STEM (science, technology, engineering, and mathematics) education initiatives in many countries. In India, the priority of education outreach at the canopy conference provided a new model for scientific meetings. This session fostered discussion of a critical question: Will canopy scientists pursue business-as-usual in the face of significant forest degradation, or challenge themselves to do things differently, such as prioritizing education of youth, policy makers, and citizens? Keynote speaker Dr. Thomas Lovejoy (former President, Heinz Center for Economics and Environment, Washington, DC) remarked to the international attendees, "By any measure, tropical forests are in big trouble." But to date, none of the four prior forest canopy conferences celebrated clear links between conventional scientific data collection and forest conservation. The Bangalore, India, conference, for the first time ever, inspired conversation about the role of women in science (especially for developing countries) (Fig. 1.2) and the notion of actions to educate

Fig. 1.3 Environmental education in rural India – bird spotting

many diverse stakeholders about the importance of their environment. Will innovative approaches for both research and education aimed at engaging citizens and policy makers to achieve better metrics of success for forest canopy conservation?

In India, the notion of "treetops at risk" is a clear threat to human quality of life and also to ecosystem health. Over one hundred teachers of India's southern states representing diverse cultures, ages, gender, and religions eagerly attended the conference and listened to important take-home messages: (1) develop short courses in canopy science for teachers, citizens, and K-12 (Fig. 1.3); (2) create climate change awareness in schools and for regional government; (3) share resources; and (4) plant trees to "green India" as a strategy to offset climate change and sustainable services.

The ICC fostered a dialogue where a combination of economics and environment, via more effective science communication and education outreach, may hold the key to producing a scientifically literate generation of citizens and policy makers. If every practicing scientist were to give 10 % of his or her professional time to education outreach, not exclusively limited to technical publications, then perhaps the notion of a scientifically literate public could be attained. Currently in India and elsewhere, many science professionals still think that education outreach does not fall into their purview. The expanding footprint of education integrated with science at technical conferences provides one metric of hope for improving science literacy among diverse stakeholders, not just scientists.

This volume was inspired by the 5th International Canopy Conference held in Bangalore, India, but the authors and issues have subsequently burgeoned, as has the mission of this book. By sharing the story of Bangalore, we hope to set the stage for this book as a "wake-up call" for scientists in many fields of expertise – we must not only uncover the secrets of how ecosystems operate, but we must also disseminate our findings in a manner that inspires conservation and solutions by the diverse stakeholders that comprise the seven billion people of planet Earth. The stakes for conservation have never been higher. And the ultimate height of the forest canopies represents an epicenter for many global solutions – carbon storage to offset global warming, biodiversity of millions of species, productivity of billions of leaves, shade for humans, freshwater cycling, and essential spiritual sanctuaries.

Chapter 2
Greening the Planet?

Thomas Lovejoy

Keywords Climate change • Carbon sequestration • Ecosystem restoration

In 1896 Swedish scientist Svante Arrhenius addressed a very important question: Why is the Earth a habitable temperature for humans and other forms of life? Why isn't it too cold? The answer (building on the work of the British scientist Tyndall earlier in the century) was the greenhouse effect. He even did a quite accurate projection of what double preindustrial levels of CO_2 would do to planetary temperature. What he would not have known was the detailed history of planetary temperature for the previous hundreds of thousands of years – in particular that the last 10,000 years were an unusually stable period of global climate and temperature. During those ten millennia, much of modern civilization (agriculture, human settlements) developed, and all ecosystems adjusted to a stable climate.

That of course is changing because of human activity and rising levels of carbon dioxide in particular. Global temperature is currently approaching 0.9° increase over preindustrial levels. CO_2 concentrations are approaching 400 parts per million (ppm) versus 280 ppm in preindustrial time. The lag between reaching a concentration level and the accumulation of heat means that the planet will reach another 1.0° increase even if CO_2 concentrations rise no further. Most of the response to climate change to date is relatively minor. Nonetheless, there is substantial reduction in glaciers in the tropics and sea ice in the Arctic Ocean. Sea level is rising in part because of thermal expansion of water but increasingly because of glacial melting. While the sea level end point is fairly clear from understanding past climates, the rate at which it will be reached is not.

T. Lovejoy (✉)
George Mason University, VA, USA

The Heinz Center for Science, Economics and the Environment, Washington, DC, USA
e-mail: lovejoy@heinzcenter.org

M. Lowman et al. (eds.), *Treetops at Risk: Challenges of Global Canopy Ecology and Conservation*, DOI 10.1007/978-1-4614-7161-5_2,

There is definite increase in the frequency of wildfires from warmer summers, earlier snowmelt, and in certain instances drought associated with climate change. In Costa Rica's Monteverde Cloud Forest, an increasing number of dry days are recorded, since clouds tend to form at higher altitudes more frequently than in the past. This could lead to major change if the trend continues because those forests depend primarily on condensation from clouds as a source of moisture.

Already we are seeing responses in nature. Species are changing their annual cycles: flowering species blooming earlier, birds nesting earlier, and the like. Species are already changing where they occur, e.g., the Edith's Checkerspot Butterfly is clearly moving northward and upward in altitude in western North America. With double preindustrial levels, it has been projected that sugar maple will occur in Canada but not New England. Mostly seen so far in temperate and boreal habitats are decoupling events in which two coincident features of nature begin to disassociate because one uses temperature as a cue and the other day length. These are all relatively minor changes, essentially ripples in nature. Nonetheless, this kind of change is being detected in almost every place that scientists look. So this is now statistically robust.

More serious abrupt change is also beginning to be detected. Tropical coral reefs are particularly sensitive to short periods of elevated water temperature and undergo bleaching events. These are happening with increasing frequency every year with grim implications for the one-twelfth of humanity that depends on the reefs for well-being. Similar abrupt change is occurring in the coniferous forests of western North America. There warmer summers and milder winters have tipped the balance in favor of native bark beetles leading to massive defoliation and subsequent tree mortality (up to 70 % of stands in some instances). Even larger-scale change is being detected in terms of ocean chemistry where elevated CO_2 absorbed by the oceans also produces carbonic acid. Ocean pH is already 0.1 pH unit more acidic than in preindustrial times. This carries enormous implications for animal species that build shells or skeletons of calcium carbonate.

In the Amazon the hydrological cycle that produces much of the rainfall and works synergistically with the rain forest (in addition to providing rain further south in agricultural regions) may be approaching a tipping point. A World Bank model of the combined effects of fire, deforestation, and climate change suggests a possible tipping point leading to Amazon dieback in the southern and eastern Amazon at around 20 % deforestation. It is currently at about 19 %. This has significant implications to the health of forest canopies. Looking ahead, we can anticipate that additional climate change would disproportionately affect island species and species at high altitudes. And from past climate change, we know that ecosystems do not move but rather that individual species move individually: almost an ecosystem disassembly with surviving species forming novel ecosystems that are hard to manage in advance.

All of that argues strongly that the 2° of increase that is the target of negotiations is really too much for most ecosystems. Yes, one can make species less vulnerable by restoring natural connections in landscapes. In the end, however, it becomes imperative to seek ways to reduce the amount of climate change that would otherwise take place. Geo-engineering schemes which purport to do so mostly do not because they address temperature (the symptom) rather than the cause (greenhouse gas concentrations). The biology of the living planet can be particularly helpful. A substantial amount of the excess CO_2 in the atmosphere actually comes from three centuries of destruction and degradation of ecosystems. Ecosystem restoration at a planetary scale could easily draw down 50 ppm from the atmosphere by converting it into thriving forests, grasslands, and agroecosystems. It is not enough, so we need to find economic ways to do so non-biologically.

It is time to recognize that this planet works as a linked biological and physical system and we should manage it as such. Re-greening the Emerald Planet will make Earth more habitable for ourselves and other forms of life. Efforts to plant trees and restore forest canopy cover represent an important stewardship action.

Chapter 3
Comparative Canopy Biology and the Structure of Ecosystems

Mark W. Moffett

Keywords Architecture • Biofilm • Biodiversity • Biomechanics • Coral reef • Epiphyte • Kelp • Periphyton • Rhizosphere • Stratification

1 Summary

- The way ecologists think about canopy biology as a scientific discipline could lead them to overlook different communities of spatially fixed organisms that may have properties usefully compared to or contrasted with forest canopies. This chapter represents a series of discussions and reviews on the possible nature and limits of canopy biology and introduces the prospect of a general comparative science of biological canopies.
- Rather than restricting canopy biology to plants in terrestrial systems, I argue that *canopy* can be defined in terms of the parts of any community of sessile organisms that emerge from a substratum, the structural products derived from them included. This opens the field to diverse communities that could share many properties with forest (or plant) canopies. I overview the canopy literature on kelp forests, algal turfs, periphyton, bacterial films, and coral reefs. The word *canopy* has already been applied to each of these ecosystems. Periphyton and biofilms in particular have great potential as model systems for studying assembly rules for the physical structure and dynamics of canopies.
- Among the similarities and differences between these types of canopy are the distribution of resources, such as light and nutrient gradients; factors affecting these distributions, such as the flow of air or water; and the resulting disposition of

M.W. Moffett (✉)
National Museum of Natural History, Smithsonian Institution, 10th Street and Constitution Avenue, Washington, DC 20013-7012, USA
e-mail: naturalist@erols.com

M. Lowman et al. (eds.), *Treetops at Risk: Challenges of Global Canopy Ecology and Conservation*, DOI 10.1007/978-1-4614-7161-5_3,

the species living within this matrix. Prioritizing a search for general principles must be a primary goal of comparative canopy biology because the microhabitats generated by stratification and other complex distribution patterns within communities are critical to sustaining global biodiversity.

- I consider five rationales for the practice, the norm prior to this article, of limiting discussions of canopy biology to forests: if people have unique interactions with or concerns about forest canopies, if a substantive basis exists for treating trees as a distinct category of plant, if a substantive basis exists for treating trees as a distinct category in forests, if a substantive basis exists for treating forests as a distinct category of terrestrial community, or if attributes of the tree-crown residents and ground-rooted forest plants prove distinct. In no case is there unequivocal evidence for the usefulness of separating the study of forest canopies from the study of the aerial parts of other terrestrial plant communities.
- In mainstream ecology, the organisms of a community are typically studied in two dimensions, or as isolated points on the earth. While canopy biology encompasses all aspects of the study of the portion of a community that projects into a medium, the discipline can in large part be distinguished as the science of treating plants (or other sessile hosts) as three dimensional.
- In many studies of canopy species, the inhabiting organisms' relationship to the canopy is treated as incidental. "Putting the canopy into canopy biology"—that is, contributing fundamentally to canopy biology as an independent field of study—requires that the attributes of the canopy form an integral part of the research. This would include issues of community ecospace; properties emerging from community residents in aggregate, such as stratification of microclimate; host distributions; host architectures; properties of a canopy's structural elements; and characteristics of the open spaces within a community.
- Applying the principles of canopy biology to communities as a whole, including those parts of sessile organisms and their associated species living on or in the substratum, suggests the value of developing a more comprehensive science, referred to here as structural ecology.

Words compartmentalize information, and that has many repercussions (Lakoff 1987). This affects how we identify subjects for academic study (Bates 1960; Hull 1988). The word *canopy* has been applied to vegetation in varied (Moffett 2000) and often inconsistent (Moffett 2002) ways, leading to varied interpretations of the domain of canopy biology. To select one example, if by canopy we mean the highest plant surfaces in a forest, as many authors do (e.g., Kricher 1997; the "outer canopy"), it is likely that, through habit, canopy biologists will develop a search image effective at picking out information only on the highest parts of the forest. In what philosopher W.T. Jones (in Bateson 1972) describes as the "topography of ignorance," our knowledge of parallel and potentially useful studies on, say, shrubs and herbs, whether in forests or shorter ecosystems, or studies on the lower parts of forest trees, is likely to be marginalized. Such intellectual fragmentation is borne out by the literature. For example, citations on stratification in herbaceous communities, such as Monteith (1975/1976) or Liira et al. (2002), tend to be scarce in forest-canopy publications. One consequence can be independent discoveries by different academics, such as those working in agriculture, economic entomology, or landscape ecology, along parallel research tracks, thus wasting time and effort.

How, then, to delineate a discipline? While any "conceptual framing" can be arbitrary and subjective (Bohr 1955), the most compelling criteria are those of widest general importance to the experts. Consider the shifts in content and perspective between a biochemist's focus on molecules and a cell biologist's concern with structural integration, or a psychologist's focus on mind and a sociologist's interest in communities. In this respect, it makes little sense to grant canopy biology an independent status as a discipline if by canopy, we specifically mean "outer canopy." After all, most aspects of the biology of the outer canopy change gradually from those features found among the plant organs lying beneath the topmost foliage. This is not to deny that some biologists must concentrate on the outer canopy because of its relatively pronounced features, such as the disproportionate significance of upper leaves to forest productivity.

What is the most fruitful basis for delineating canopy biology as a discipline? I will show that there is no clear evidence, for example, for trees or forests being distinct from other terrestrial vegetation. Then I consider modifications of the definition of canopy that would encourage canopy biologists to compare environments previously ignored by forest ecologists, such as coral reefs and bacterial films. Following this, I present some themes for an expanded canopy biology and consider the utility of establishing a field of study more comprehensive than canopy biology—one that applies the principles of canopy biology to communities of sessile organisms treated as a whole, including those with a portion of their anatomies positioned on or beneath a substratum. I conclude by addressing the central role of comparative canopy biology, as well as addressing this more expansive discipline, referred to here as "structural ecology," as a means of interpreting patterns of biodiversity at a global level.

2 Seeing the Forest for the Herbs

The word *canopy* is often applied to the upper parts of forest ecosystems. Moffett (2001) argued for incorporating the literature on all aerial parts of any terrestrial plant community as canopy biology. Because attributes of this vegetation and the species residing within it are likely to change gradually, in a relatively continuous way from one community to the next, and within a community from one height to the next, I propose that we adhere to this approach except when dealing with "concepts or situations necessarily restricted to trees" (Moffett 2000). In fact, no one has specified any rationale for the common practice of restricting the scope of canopy discussions to trees or to forests (as in the useful distinction made between

Fig. 3.1 Forest cross section revealed where the bank of the Rio Napo has fallen away. One can identify a number of strata, including an emergent tree towering above its neighbors (*at left*); the overstory trees in direct sunlight; the understory trees in the shade of the overstory; and a shrub layer below those. Between the overstory and understory is visible a dark open space commonly used as a flyway by birds and bats. Because plant growth is dynamic with the height of each plant dependent on its age, species, health, and situation, including the distance and density of plant crowns overhead, each stratum in a community is unlikely to occur as exact layers at consistent heights, but instead will be "a product of localized conditions (e.g., varying jaggedly in a patchwork of vegetation types or successional stages), and, even in a uniform environment, it need not exist at one height above ground but rather may occur relative to the distance from the outer canopy" (Moffett 2000) (Photograph by Mark W. Moffett, Minden Pictures. All rights reserved)

"tree canopies" and "forest canopies" by Shaw 1996). Consider five criteria by which forest canopies could merit this kind of separate attention (Fig. 3.1):

1. *Humans have unique interactions with and concerns about forest canopies.* Many people have grave concerns about the extinction of canopy species and great interest in the value of canopy products to societies past, present, and future. Because conserving the uppermost level of a forest is not possible without conserving its bottom, I would argue that conservation issues are most sensibly considered not specifically as canopy biology, but under the general rubric of ecology, which encompasses numerous aspects of economics, ethnography, and environmentalism.
2. *There is a substantive basis for treating trees as a distinct category.* The diverse suite of characteristics associated with trees serves "as an example of the molding of the entire phenotype by selection pressures" that have come about convergently in numerous lineages (Niklas 1997). The ecological impetus discussed most often in reference to plant height is competition, especially for light (Tilman 1982; Givnish 1995; Leigh 1999; Sterck and Schieving 2007). The fundamental basis of "treeness" per se, however, may be biomechanical (Niklas and Kerchner 1984).

Any self-supporting terrestrial plant growing beyond a certain height appears to be developmentally channeled into evolving a main vertical trunk built of the stiffest available structural elements (especially around its perimeter), surmounted by a branched crown—in short, it becomes a tree. The transformation seems to occur in a similar way under diverse environmental conditions, a result of gradual shifts from small herbaceous structural designs to an architecture that allows large upright plants to cope with bending or torsion (Niklas 2000). If many of these changes occur at some critical juncture during the evolution of increasing mass or stature (e.g., from 3 to 5 m in height: Givnish 1983), the tree-growth form could be sufficiently distinct to treat the study of the canopies in ecosystems in which the uppermost stratum is dominated by trees (i.e., forests) as an independent research discipline. Although this transformation process appears central to our very conception of "tree," it remains poorly understood (Givnish 1984). Consider that trees allocate a large portion of their photosynthate to supportive tissues, and that, if they have high crowns, they pay a high price in aerodynamic drag, friction during fluid transport, and increased potential for structural failure (see Vogel 1996). Because of such costs, we would expect that plants would readily lose the tree-growth form when doing so would increase fitness, especially given that the character of woodiness, and, apparently, treeness, has been labile in plant evolution (Judd et al. 1994; Dodd et al. 1999; Kim et al. 2004). Nevertheless, trees in deserts, savannas, and other open ecosystems grow relatively very tall, even though they occur widely separated from their neighbors and so by ecological criteria seem conspicuously overbuilt. Tree species adapted to these habitats may be large because they use their trunks to store water (Holbrook 1995); depend on height to avoid herbivory, as arborescent cacti do from tortoises (Dawson 1966) and acacias from giraffes (Brooks and Owen-Smith 1994); are maximizing reproductive dispersal (Richards 1986); are avoiding damage from fast-moving ground fires (Givnish 1995); or are shading out grasses that compete with them for water (Walter 1973). Yet such factors appear neither pervasive nor severe enough to explain the almost ubiquitous occurrence of widely-spaced trees radically taller than the other plants in their communities. For example, giraffes always forage below 5 m, while savanna acacia trees often exceed 20 m in height (Young and Isbell 1991, T.P. Young, pers. comm.). Plant evolutionary mechanics could hold the key to this apparent mystery; further investigations in this area might shed light on the nature of "treeness" itself.

3. *There is a substantive basis for treating trees as a distinct category within forests.* Although adult trees are usually assigned to strata separate from other plants in a forest, the question of whether mature specimens are distinct as a group or are part of a continuum that contains other, smaller forest plants has not been clearly resolved, in part because of inconsistencies among the research approaches to stratification (Parker and Brown 2000). In a frequency distribution of the size of mature vascular plant individuals, excluding vines, in a forest, is there a distinct peak (or peaks) corresponding with trees? Size-frequency distributions are common in studies of animal diversity, while, apparently, are absent from studies of complete plant communities (although some size-frequency distributions have

been included in studies of only the trees: Poorter et al. 2008). This is presumably because modular construction and indeterminate growth can make plant size difficult to assess (e.g., Sterck et al. 2003).

4. *There is a substantive basis for treating forests as a distinct category.* Forests could be considered a distinct category of vegetation if, by some parameter of community physiognomy, they can be separated from other terrestrial ecosystems in a nonarbitrary way. Suppose we graph some measure of overall community height—say, for each major community type in a classification of ecosystems, the modal height reached by the vegetation averaged over randomly chosen points on the ground. Suppose the distribution indeed turns out to be bi- or multimodal, such that forest systems represent a distinct peak. This would suggest that forests are more than an arbitrary construct that humans have split off from a continuum of natural communities. Perhaps on that basis, forest canopies can be distinguished as a separate research discipline. But so far, no information seems to exist on patterns of overall height across communities. If the available classifications are biased with respect to height (e.g., if ecosystem taxonomists have been "splitters" with respect to forest communities), or if community categories are largely artificial, at best representing opportunistic associations of species (Brown 1995), that could make attaining such information difficult.
5. *Attributes of tree-crown residents or ground-rooted plants in a forest prove distinct.* If future studies demonstrate that forests harbor canopy communities distinct in some fundamental and reasonably abrupt way from the organisms dwelling on progressively lower kinds of vegetable communities, that might be taken as evidence for distinguishing forest canopies as a distinct biological entity. This seems unlikely, however, given that most resident canopy organisms respond not to height but to environmental factors that happen to correlate with height (Moffett 2000). Thus, epiphytes seemingly associated with high forest-canopy situations occur closer to the ground when conditions allow (McCune 1993; Benzing 2012). Biodiversity in forest canopies can be extreme, but since most inventories of biodiversity to date have been made in tree crowns (e.g., Stork et al. 1997; Foottit and Adler 2009), the relation between species diversity and community scaling is open to question. For any given latitude, how much of the high diversity of forest-canopy-dwelling species can be attributed to the canopies offering a relatively large overall mass, surface area, productivity, or microhabitat richness? There is also little basis to date for asserting that the organizational principles manifested by ground-rooted plants in forests (such as in the way the trees distribute horizontally or vertically) may be distinct from those operating in other communities, beyond matters of scaling that might be expected to vary in a reasonably continuous manner with successively shorter vegetative types (Moffett 2000).

In summary, there seems no unequivocal basis for the common practice of treating forest canopies independent of the study of the canopies of other terrestrial plant communities. Further general investigations into the nature of trees and forests, of course, may prove me wrong. As I have written elsewhere (Moffett 2000):

> Furthermore, I encourage the application of "canopy" to all flora, reserving the phrase "forest canopy" to concepts or situations necessarily limited to trees. ...Broadening our perspective on

canopies encourages us to pursue the reasonable hypothesis that most or all ecological processes scale up from a meadow to a redwood grove, so that problems considered intractable in the latter can be addressed by looking at shorter [and faster growing] systems.

3 More to Pond Scum than Meets the Eye

The Essence of knowledge is generalization. —

Hans Reichenbach (1951)

Incorporating into canopy science all the studies of all aboveground (aerial) plant organs and their occupants within any community (Moffett 2000) is a given, reflecting a history of the use of the word canopy with both cultivated and natural forb and grass communities (e.g., Monteith 1975–1976, Campbell and Norman 1989, Burrows 1990, Roxburgh et al. 1993, Hirose and Werger 1995). But this represents only a first step in the development of a truly comparative discipline. Indeed, though widely unappreciated by terrestrial "macrobiologists," the word *canopy* is used by aquatic and microbial scientists to describe ecosystems that share many properties with terrestrial plant canopies. In contrast to terrestrial systems, where the hosts are usually vascular plants, these sessile organisms are typically algal species, zooxanthellae-bearing animals, or erect animal species such as hydroids and bryozoans. For an exception, see the information on seagrass communities, including issues bearing on canopy stratification, architecture, and biodiversity by Heck and Wetstone (1977), Kikuchi and Peres (1977), Williams and Heck (2001), and Pogoreutz et al. (2012).

The existence of plant-dominated aquatic systems and the remarkable parallels between these and other aquatic systems and terrestrial plant communities (described below) are among the reasons we should consider the common focus of canopy biology on terrestrial systems to be arbitrary and antiquated. To encompass these various kinds of hosts, canopy can reasonably be redefined as *the parts of any community of sessile organisms that emerge from a substratum*. Canopy biology (or canopy science) is by this criterion *the study of the canopy of a community, including the organs of the sessile individuals and any affixed products of those organisms, and anything in, on, or between those organs and products*. The affixed "products" can range from dead organisms, such as tree snags, to the secreted skeletons of corals or algal mucilage.

In this analysis, *sessile* describes an organism that emerges from or adheres to a substratum at positions fixed over a large part of its life history. *Substratum* refers to any surface or structural matrix that provides points of attachment for a sessile species, fixing the location of individuals or colonies. (I prefer this word to substrate, which is confusing, especially for microbial ecosystems, because it is also used to describe enzyme reactions.) The substratum thereby establishes the spatial relations between sessile individuals, including, to some degree, the organs that project into the fluid medium (the "canopy structure"). In turn, the sessile communities variously transform and stabilize the substratum (e.g., Stevenson 1996). (Many authors have applied the words structure and architecture to communities less

Fig. 3.2 The two upper strata in a kelp forest off the Aleutian Islands of Alaska. An understory of *Cymathera triplicata* kelp grows under an overstory of another kelp species, *Alaria fistulosa* (Photograph by David Duggins. All rights reserved. Reproduced with permission)

literally than I have here—e.g., simply to denote biodiversity or nonphysical attributes of organization, such as niche spaces and trophic hierarchies; e.g., Connell 1975.) The substratum typically is a solid, but the air–water interface may give some level of stability to the relative position of organisms, such as in floating algal mats (metaphyton). I exclude from the canopy communities, or portions of communities, distributed entirely within the substratum matrix, as in the terrestrial soil community or its aquatic equivalent, the epipelic or bottom-sediment community (for instance, the microphytobenthos; MacIntyre et al. 1996). In some situations, the canopies of "different" ecosystems can be studied as one. In forest shade, stream-dwelling algae may show some of the same physiological adaptations as understory terrestrial plants (Robinson and Minshall 1986; Hill 1996).

The following sections summarize the canopy biology of diverse ecosystems.

Kelp Canopies. Describing kelp communities off South America, Darwin (1839) wrote, "I can only compare those great aquatic forests of the southern hemisphere with the terrestrial ones in the inter-tropical regions." The term "kelp forest" has been common in the literature ever since. Application of the term "canopy" to kelp began with Jack Kitching, who, using a milk can with a window made from an old glass cookie box, was the first scientist to successfully dive into this ecosystem (Kitching et al. 1934). In these and other algal communities, there is a relationship between canopy height and algal growth form (Neushul 1972; Hay 1986; Steneck and Dethier 1994). All kelp forests convergently accommodate guilds of species that fall into five distinct "canopies," or strata (including coralline crust as a stratum; Dayton 1985) (Fig. 3.2), the same number typically described for tropical forests (e.g., Richards 1996). The composition of the understory community depends on the distribution and diversity of the species above them (Wernberg et al. 2005; Irving and Connell 2006). The largest and most complex canopies occur in shallow, productive sites (Vadas and Steneck 1988), as might be predicted given that water attenuates light sharply as compared to air in terrestrial communities.

Many findings from kelp forests parallel those for terrestrial communities, such as light attenuation through strata in relation to frond coverage (Gerard 1984), which at the benthos beneath kelp communities often declines to ca. 1 % of surface light, similar to tropical rainforests (Richards 1996); the importance of disturbance and gaps to succession (Neushul 1971; Foster 1975; Dayton 1975a; Hurby 1976; Pearse and Hines 1979; Reed and Foster 1984; Dayton et al. 1999); the value of sunflecks to understory growth and survivorship (Wing et al. 1993); differential survivorship and growth rates that result from the shading of benthic plants (Kastendiek 1982; Santelices and Ojeda 1984; Dean et al. 1989; Miller et al. 2011) and phytoplankton (Borchers and Field 1981); and other diverse competitive effects leading to specialized shade-tolerant (understory) communities (Dayton 1975b; Dayton et al. 1999). Unless predation on them is severe, sessile animals may outcompete kelp in low-light conditions in deeper parts of the benthos (Foster 1975), a pattern that holds to some extent in understory shade in shallower waters, though low-light adapted algae also occur there. Dayton (1971) distinguishes competition for space on the substratum ("primary space") from competition within the volume above that surface, that is, within the canopy ("secondary space"), a concept worthy of widespread application. Because of their flexible tissues, kelp and other (Carpenter 1986) algal communities could in some ways be more ecologically comparable to grassland than to terrestrial forest (but see Holbrook et al. 1991), even though kelp can rise 50 m or more in height because of their reliance on the opportunities for flotation offered by water. The capacity for upward growth in kelp is by this means greatly enhanced over nonwoody terrestrial plants.

There have been general studies on the relation between canopy residents and kelp forest structure. Many fish stratify in a kelp forests, although this generally becomes less pronounced as the fish mature (Anderson 1994), and structural features of the forests influence fish diversity (Russell 1977). Shading by the kelp overstory can reduce algal growth rates in lower strata, thereby altering the abundance of some fish relative to canopy gaps (Carr 1989; Schmitt and Holbrook 1990; Jones 1992). Manipulations of physical structure are common in the study of kelp communities, showing, for example, that simplifying canopy structure can increase fish mortality by removing refuges (Anderson 2001). Predators can be so efficient at feasting on prey that pass through kelp canopies that recruitment of these prey species to ecosystems closer to shore is strikingly reduced (Gaines and Roughgarden 1987).

In macroalgal mats on boulders and within tide pools, canopy-resident diversity relates to algal architectural complexity (Dean and Connell 1987; Hacker and Steneck 1990; for a successional study of this kind of ecosystem, see Sousa 1979). Williams and Seed (1992) review the positive and negative effects of epiphytic animals on large algae.

Periphyton and Algal Turf Canopies. Periphyton (aufwuchs) constitute a "complex community of microbiota (algae, bacteria, fungi, animals, inorganic and organic detritus) that is attached to substrata" (Wetzel 1983; for further terms, see Weitzel 1979)—the microbial equivalent of an epiphyte mat. Indeed, periphyton can be epiphytic (Ruinen 1961, 1975; Morris et al. 1997; Claflin 1968; Whipps et al. 2008; for the rhizosphere equivalent, typically involving more bacteria than algae, see Pearce et al. 1995), albeit "periphyton" also applies to growth on nonliving or deceased substrata, such as submerged rock or leaf litter. Periphyton "have extensive vertical development on a small scale, and cells within the community matrix are tightly packed" (Boston and Hill 1991). They show a repeatable pattern of succession (Lowe et al. 1996; Sekar et al. 2004; Passy and Larson 2011), a pattern that can be disrupted by a high disturbance regime as occurs with communities on the surface of sand grains (Miller et al. 1987) except when algal mucilage binds the grains together, allowing further community development (Hoagland et al. 1982). Succession proceeds from a monolayer community to a stratification of species and chemistry within a matrix of cells and their secretions (Jørgensen et al. 1979; Jørgensen and Revsbech 1983; Kuenen et al. 1986; Lassen et al. 1994; Johnson et al. 1997; Fierer et al. 2010). Stalked microalgae can contribute to the greater depth of late-successional communities, resulting in "an upperstory of growth perhaps functionally analogous to the canopies characteristic of terrestrial forests" (Hoagland et al. 1982). Mature periphyton communities often have a dense understory of small cells overgrown by filamentous algae, though some interesting variants on this are shown in Fig. 3.3. The upper stratum provides attachment points for diatoms that live as epiphytes (Roos 1979; Roemer et al. 1984; Marks and Power 2001), described as "dependent organisms" or pseudoperiphyton, and treated as part of the same community as their hosts (Sládecková 1962). In addition, there appear to be algae that are the functional equivalent of vines (or more accurately, hemiepiphytes; Fig. 3.3b). While some sessile algae cannot adhere to other algae and so require a direct connection with the substratum, other species may preferentially attach to algae in the layer below and thereby epiphytically form a canopy stratum of their own (Fig. 3.3a). This strategy is unknown and probably biomechanically impossible within terrestrial canopies. Vines positioning their foliage uniformly above that of their hosts (Putz 1995) may approach it, although being rooted to the ground, these canopy plants are not epiphytes.

Canopy complexity of periphyton may be greater at sites with higher light intensities (Hudon and Bourget 1983). Furthermore, the internal physiognomy depends on local flow regime, in part because greater turbulence increases the penetration of nutrients and light (Peterson 1996). Wetzel (1993) argues, however, that such penetration is difficult (unless aided by grazers: Hillebrand et al. 2008) and that the high productivity of periphyton is a result of efficient recycling of nutrients within their canopies. Growth of the outer layer can shade the strata below (Johnson et al. 1997; Dodds et al. 1999) and block nutrient inputs to the understory (McCormick and Stevenson 1991; Peterson and Grimm 1992). Shading can lead to understory deterioration (Stock and Ward 1991) in time causing a community to slough from its substratum (Meulemans and Roos 1985). Substrata are colonized both by these detached communities (detrital microcosms; Korte and Blinn 1983) and by isolated

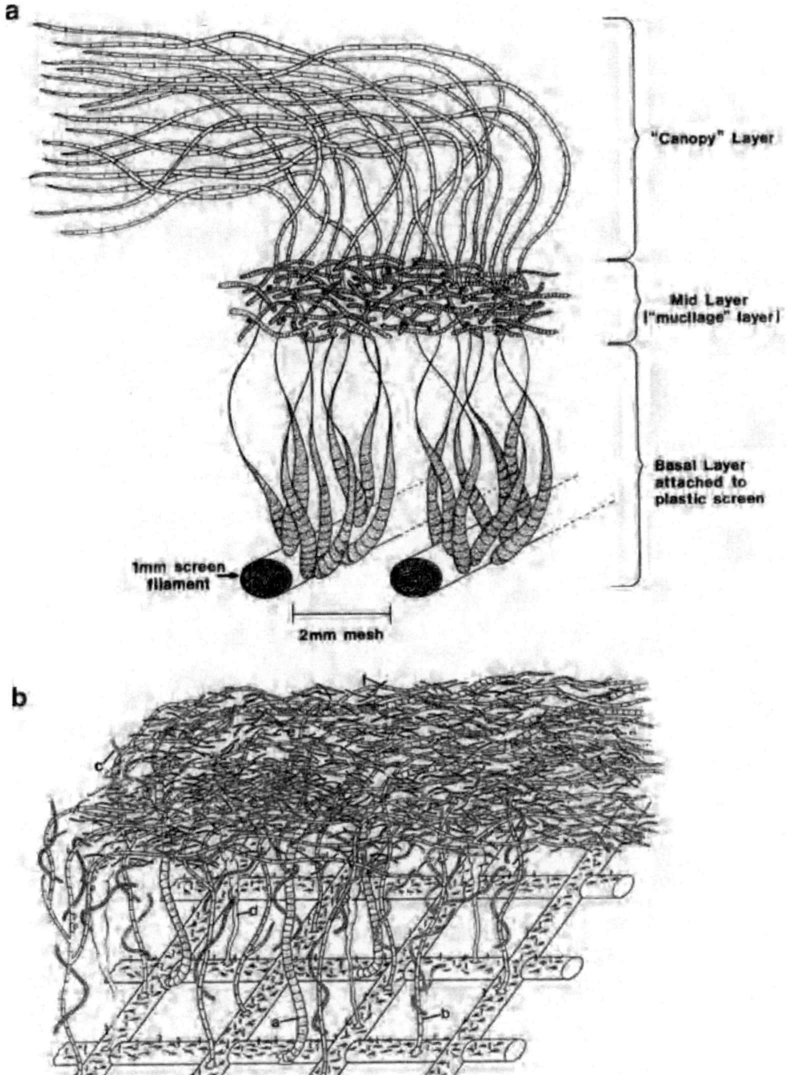

Fig. 3.3 Multilayered periphyton attached to a screen in a freshwater aquarium (**a**) and to a screen in an Algal Turf Scrubber (**b**). Typically the layer directly attached to the screen (the hard rock or carbonate substratum in the wild) is dominated by either blue green (cyanobacteria) or a diminutive green (*Stigeoclonium*) or diatom algae (W. Adey, pers. com.). In (**a**), the upper layers are attached to the layer below them and not the substratum, and so these entire strata are epiphytic on lower strata. The middle layer consists of diatoms and small filamentous algae suspended in mucilage generated by the community. In contrast, when the overstory is made up of large, typically branched filaments (**b**), these are also attached to the substratum, along with the blue-green or small green and red "understory" algae. Much like vines (especially nomadic ones: Moffett 2000), the *Spirogyra* and *Dichotomosiphon* in 3b spiral around their supporting hosts like vines, apparently moving along them by rotation and circumnutation sensu Darwin (Yeh and Gibor 1970); in monoculture, the *Spirogyra* will even coil around one another like plant tendrils with no alternative place to go. These two illustrations show what I would call "monodominant periphyton," because only one species is affixed to the substratum, but more biodiverse communities clearly exist (Both illustrations are from Adey and Loveland (2007). Reproduced with the permission of the authors. All rights reserved)

Fig. 3.4 Algal turf from St. Croix dominated by *Herposiphonia*. Width of the image is about 1 mm (SEM photography by Robert S. Steneck. All rights reserved. Reproduced with permission)

cells in suspension (Stevenson 1983). Sloughing can be reduced where understory algae species are able to produce more photopigments or become increasingly heterotrophic in dim conditions (Tuchman 1996; Peterson 1996), in which case the periphyton can last longer and achieve higher biomasses. Because of the cycle of growth, death, and sloughing, the community that establishes after a site is grazed by herbivores may depend on the prior successional status at the site (Peterson 1996), among other factors (Tuchman and Stevenson 1991).

Turfs are filamentous periphyton communities, typically a few millimeters high (Fig. 3.4) that occur widely on coral reef surfaces and produce the bulk of reef primary productivity (Adey and Steneck 1985). Disturbances (say, by herbivores or abrasion from the movements of taller algae overhead; Russell 2007) reduce turf height and increase light penetration and turbulence through their canopy (Carpenter 1986; Williams and Carpenter 1990; Carpenter and Williams 1993; Cheroske et al. 2000). Stratification can occur within turfs, but is limited (Hackney et al. 1989; R.C. Carpenter, pers. comm.).

Bacterial Films. Almost all bacteria live packed within surface-bound multispecies communities called biofilms (Watnick and Kolter 2000). Until the 1980s, bacteria were studied mainly by traditional sampling and culture methods. Extrapolations from monospecies planktonic laboratory cultures led to serious misunderstandings about bacterial ecosystems (Costerton et al. 1995). Especially in light-exposed habitats, cyanobacteria and their accumulated remains can induce thick perennial accretions called microbial mats or stromatolites (Stal 2000), but non-photosynthetic mats thrive in dark habitats where there are few grazers (Teske and Stahl 2001).

Biofilms often intergrade with eukaryote-dominated periphyton, and increasingly the term has been used broadly to include periphyton and even fungi (Reynolds and Fink 2001; Ramage et al. 2009). In the algal-dominated systems discussed in the previous section, for example, either bacteria or algae can colonize early in succession (Hoagland et al. 1982; Roeselers et al. 2007), and pioneer species of bacteria, in combination with certain diatoms and fungi, may "precondition" the substratum for adherence of subsequent algae, or provide the foundation they require for attachment (Korte and Blinn 1983; Burkholder and Wetzel 1989; Hodoki 2005). In a developed periphyton community, the bacteria can be nutritionally dependent on excreted algal products (Haack and McFeters 1982; Sobczak 1996), but a reverse dependency can arise when nutrients are scarce (Scott et al. 2008), suggesting facultative mutualisms can develop (cf. Carr et al. 2005). The remainder of this section will focus on communities dominated by bacteria, which have become the most extensively studied canopy system.

For macroscopic canopies, interest typically falls into two arenas: studies of the substratum-bound species themselves (the hosts) and studies on species that live within the canopy generated by the hosts (canopy residents). This distinction is apparently of limited utility for bacterial biofilms and mats, which are formed of cells en masse, with a very strong coupling of structure and function. Light penetration and chemical gradients are critical to the makeup of such communities, where the concentration of any nutrient consumed in the biofilm typically decreases with depth (Jørgensen and Revsbech 1983; Kühl et al. 1996; Wimpenny and Kinniment 1995; Stal 2000; Ward et al. 2006; Stewart and Franklin 2008), such that anaerobic or anoxic species often occupy the depths of a film (Jørgensen et al. 1986; Sagan and Margulis 1988, p. 53–54; Ramsing et al. 1993; van Gemerden 1993). More complex aspects of spatial heterogeneity in these communities are becoming evident (Kühl and Polerecky 2008). The internal organization of photosynthetic biofilms and mats often includes stratification in relation to light levels (Kühl and Fenchel 2000; Ward et al. 2006; Ramsing et al. 2000; Al-Najjar et al. 2012).

While most biofilm residents are sessile or at least relatively immobile within the cell matrix (Costerton et al. 1995), the integrity of biofilms may be as much a product of a secreted matrix of polymers, which are complexly structured (Lawrence et al. 2007) and essential to biofilm formation, as of the fixed location of organisms (Flemming and Wingender 2010). Thus, while some species are held in space by intercellular connections (Schaudinn et al. 2007), motile species can shift position within the film—for example, moving within the matrix to different "microzones" (strata)—in response to light or chemical cycles (Doemel and Brock 1977; Garcia-Pichel et al. 1994; Stal 2000); this is also true of some periphyton-dwelling diatoms (Johnson et al. 1997). Such movements can bring about distinct biofilm morphologies (Klausen et al. 2003). Biofilm residents are phenotypically distinct from conspecific planktonic forms, which function as a dispersal stage. The sloughing of bacteria from films, adhesion of the colonists to a surface or within the matrix, and other aspects of establishment and development of films have been documented or modeled in three dimensions (Bryers 2000; Kierek-Pearson and Karatan 2005), as has the ecological succession of biofilm residents through time (Lawrence et al. 1995; Korber et al. 1995; Jackson 2003; Fierer et al. 2010).

The matrix of bacteria in a biofilm was originally described as developing channels or pores through which nutrients enter and wastes leave the community (Costerton et al. 1994; Massol-Deyá et al. 1995; Stoodley et al. 1999a). The structure of many biofilms is so partitioned into independent units by these "waterways" that on close inspection, they do not resemble a "film" at all but rather a landscape of cone-shaped and mushroom-shaped bodies called "microcolonies" distributed intriguingly like trees in a forest (Fig. 3.5). When mature microcolonies grow into contact, they do not fuse but rather they wave past each other when shear forces operate (Stoodley et al. 1999a; J.W. Costerton pers. com.). Microcolony formation is widespread, but not universal (Wimpenny and Colasanti 1997), in natural and artificial systems, both for monocultures and for mixed-species communities, and is presumed to result from nutrient limitation and niche exploitation (Costerton et al. 1995). The distribution of microcolonies (including both the cells and the exopolymeric materials they secrete) and the channels and other spaces between them is referred to as biofilm architecture (Lawrence et al. 1991). Both the species composition and the nature of the substratum affect this architecture (Costerton et al. 1995; Wimpenny and Colasanti 1997). So does water flow around the films (Beyenal and Lewandowski 2002), which may cause entire microcolonies to drift across a surface (Stoodley et al. 1999b; Rupp et al. 2005; Venugopalan et al. 2005), straining the definitions of "sessile" and "canopy." There is a close coupling between such structural patterns, water flow, and the chemical landscape within such biofilm canopies (Staal et al. 2011).

The structure within and between microcolonies "provides very diverse habitats on a small scale, favoring biodiversity" (Flemming and Wingender 2010). Microbiologists have described biofilms as coordinated communities with "primitive homeostasis, a primitive circulatory system and metabolic cooperativity" that can "resemble the tissues formed by eukaryotic cells" (Costerton et al. 1995; Costerton and Lappin-Scott 1995). As Foster (2010) points out, such analogies can be overdone in that there is still little evidence of coordination in biofilms, particularly between species. Nevertheless, there appears to be a stronger interdependency between different taxa in a biofilm than is typically described between most plants in a forest, such that certain bacteria depend on the metabolic products of other microbes (Kühl et al. 1996; van Gemerden 1993; Molin et al. 2000; Paerl et al. 2000) and show other forms of cooperation (Crespi 2001; Mitri et al. 2011) and communication (Davies et al. 1998; Irie and Parsek 2008). Coordination can arise through rapid evolution (e.g., of parasitisms; Hansen et al. 2007) in part through gene transfer within the films (Molin and Tolker-Nielsen 2003; Madsen et al. 2012). In addition to these mutualistic "consortia," biofilms and microbial mats are home to diverse competitive and predator–prey interactions (Lawrence et al. 1995; Allison et al. 2000; Foster and Bell 2012; van Gemerden 1993). A question

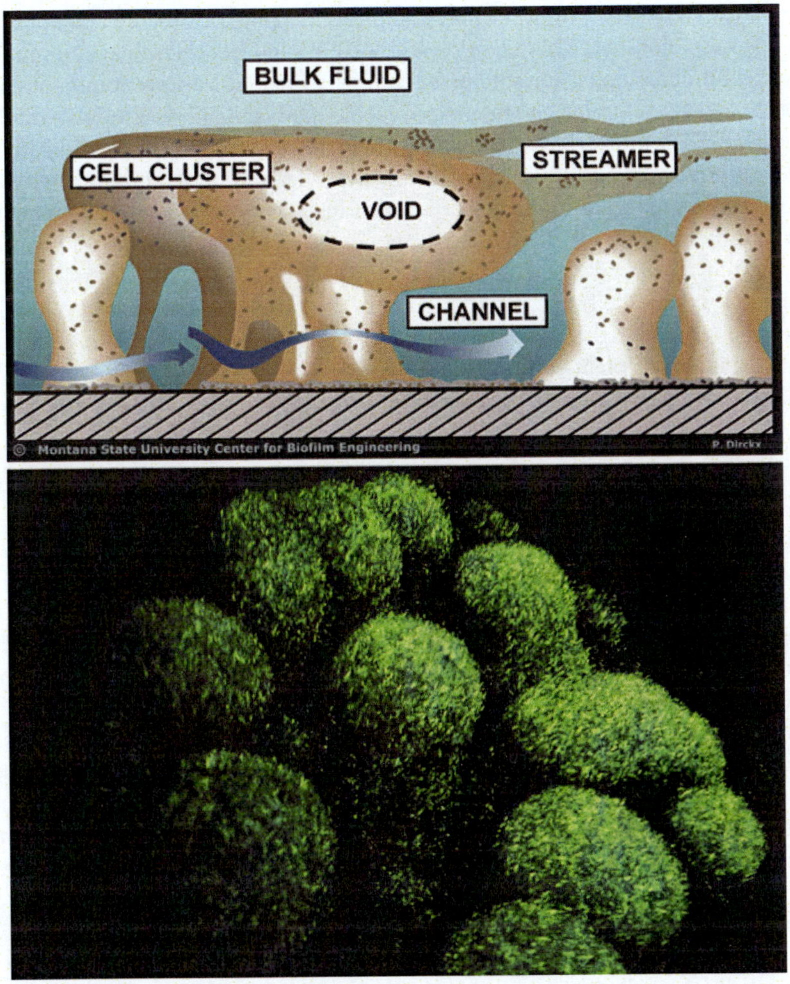

Fig. 3.5 Bacterial films. The schematic drawing of a generalized biofilm shows microcolonies separated by channels. Streamers develop as water flow increases, with the arrow indicating the water flow penetrating the film (drawing by Peg Dirckx, reproduced with permission). Below is a confocal scanning laser microscopic image of microcolonies composed of the single bacterial species *Pseudomonas aeruginosa* (E.S. Lauchnor and B. Pitts, Center for Biofilm Engineering, Montana State University. Reproduced with permission)

wide open for investigation is how selective forces and the potential for intense competition among different strains and species can lead to the seemingly cohesive systems that appear stable, functionally integrated, and structurally convergent (Nadell et al. 2008).

Stolzenbach (1989) and subsequent authors have applied the word "canopy" to biofilms, and microbiologists have been increasingly interested in employing more integrative approaches (e.g., from landscape ecology: Battin et al. 2007). At the same time, new technological advances allow combined mapping of structure, composition, and function in biofilms (Kühl and Polerecky 2008). Indeed, microbial communities can be easily manipulated for studies of canopy structure: Various combinations of microbial species or strains can be mixed, centrifuged, and allowed to form biofilms to investigate assembly rules for canopies under specified nutrient or substratum regimes. While most biofilm studies rely on well-defined communities of one or a few species, formation of biodiverse microbial mats can be induced by treatment of natural sediments (Kühl et al. 2003). All canopy biologists can benefit from knowing about biofilm research (e.g., Guerrero et al. 2002). However, the central role of a polymer matrix and the frequency of facultative symbioses in bacterial communities suggest that some of the algae-based periphyton could be more straightforward microbial analogs of plant communities.

Coral Reefs as Canopies. Dahl (1973) writes of coral reefs that "organisms often occur in many layers and the substratum itself is organism generated." Not surprisingly, overarching corals have been described as producing a "canopy" overtopping an "understory" community (Baird and Hughes 2000). While a coral's supportive structure is not living, it is an immediate by-product of living things and therefore can be treated as a part of a canopy in much the same way as snags are treated as part of a terrestrial canopy. Actually, there is a veneer of living tissue on live coral, much as there is a small zone of living phloem surrounding the mostly "dead" xylem of trees, so that in fact in both ecosystems, a large part of the structural foundation of the canopy is dead.

Somewhat like trees (Horn 1971), the architectures of photosynthetic corals change with light regime (Porter 1976). Death or suppression of the growth of corals from shading by other corals results in an understory that can include shade-tolerant phototropic species (Stimson 1985; Anthony and Hoegh-Guldberg 2003) and that incorporates abundant sessile heterotrophs (Karlson 1999; Baird and Hughes 2000). Other shade-tolerant (or, in the case of sessile animals, shade-indifferent) species can densely occupy the undersurfaces of the corals themselves (Jackson et al. 1971; Maida et al. 1994). Colonization of these habitats can depend on active larval choice for "cryptic" (shaded) microsites (Maida et al. 1994; Mundy and Babcock 1998). The change of species composition with shading can parallel community changes resulting from light falloff with depth in the water column, such that understory shade permits certain deepwater species to extend their distribution into shallow water (an effect for which there is no terrestrial equivalent). In general, however, these understory corals are distinct from reef-building corals, as the latter, when adapted to dim conditions, tend to occur in deeper or more turbid water. Nutrient input to the understory is strongly dependent on wave action (Lowe et al. 2008). Other species prosper in the gaps formed by the death of overstory coral colonies, yielding a habitat mosaic (Stimson 1985). The coral reef community includes algae (among them the turfs discussed previously) that show a complex pattern of competition depending on their heights and interfrond densities (Steneck 1997). The structural complexity of coral reefs (including that of the algae within them: Levin and Hay 1996; see review by Graham and Nash 2012) provides for a high abundance and diversity of resident organisms, such as fishes (Dahl 1973; Steele 1999; Holbrook et al. 2003), clinging invertebrates (Vytopil and Willis 2001; Fraser and Sedberry 2008), and zooplankton (Porter 1974). Species richness declines when this complexity degrades (e.g., as a result of environmental disturbances: Wilson et al. 2007b; Graham et al. 2009; Alvarez-Filip et al. 2009, 2011).

Other Canopies. Various other sessile animals form dense aggregations that could be studied as canopies including both clonal (e.g., anemones, hydroids, bryozoans, ascidians) and nonclonal species such as mussels and barnacles (Paine and Suchanek 1983), for example, in fouling communities (Boyle et al. 2007). Further, if we allow that communities of these organisms have canopies, we can apply the idea of an extended phenotype (i.e., extending the definition of phenotype to include nonliving products of an organism such as nests or retreats: Dawkins 1982; Turner 2000) to enlarge the concept of canopy almost indefinitely, depending on our interests. Least controversial would be static structures like coral skeletons whose architectures and spatial relations are produced directly by living things that serve as "ecosystem engineers" (Jones et al. 1997). Tubes of polychaetes (Bell and Coen 1982) and stream fly larvae (Pringle 1985) attract assemblages of plants and animals. These structures recolonize rapidly after defaunation, forming communities organized around tube architecture (Bell and Coen 1982). On land, patches of earthworm castings (Maraun et al. 1999; Aira et al. 2009) or of fungal fruiting bodies (O'Connell and Bolger 1997) are possible analogs of plant canopies.

Systems that are not canopies by any definition could be useful models for examining certain features of canopy life. Suspended bacteria can stratify under conditions of low turbulence (Guerrero and Mas 1989). Studies of the planktonic cells show that "increased productivity produces a physical scaffold to support biological heterogeneity (as, for example, in the spatial complexity of forest canopies) on which other species can build" (Morin 2000).

4 The Geometry of Canopy Biology

If we expand our definition of *canopy* beyond what I suggested in Moffett (2000), to encompass all parts of any community of sessile organisms that project into a medium, on what basis might the discipline of canopy science rest on firmer (and more fruitful) ground?

Ecologists traditionally treat plants (or other sessile organisms) in two dimensions, or as points on the earth. While such topics as orientation in canopy ants, the conservation of orchids, and the foliage uptake of pollutants bear on canopy biology (i.e., sensu Moffett 2000), much of the research on these topics falls squarely within this tradition. For example, most studies of orientation in canopy ants ignore 3-D spatial issues bearing on plant topographies, as well as other issues that could potentially be unique to canopy substrata, such as the properties of pheromone diffusion from bark as opposed to that from leaves. In a typical experiment, ants are not studied in their canopy environment, as when a species normally found on foliage is housed on a flat laboratory surface. While we can learn a great deal about canopy organisms with this approach, none of it has to do specifically with their origins in the canopy.

By contrast, the core of canopy biology as an independent discipline can be characterized in large part as the science of treating communities of plants (or other sessile organisms) as three dimensional. This is becoming increasingly tractable even for whole forests (Shugart et al. 2010). More generally, creating a robust canopy science requires us to "put the canopy into canopy biology" through research that directly contributes to understanding the aspects of life associated with sessile organisms. This can be achieved by introducing the "z" axis or other canopy attributes (as has been done in the studies on ant orientation by Jander and Voss (1963), Beugnon and Fourcassie (1988), Fourcassie and Beugnon (1988), Jander (1990), Klotz and Reid (1992), and Wohlgemuth et al. (2001)). To take one example, the book *Geometry of Ecological Interactions*, by Dieckmann et al. (2000), conforms to the 2-D "mainstream tradition" except for some material on three-dimensional gap structure, which is the only part of their coverage of ecological geometry that, by my criterion, represents canopy research. The same philosophy applies to aquatic systems, which offer unique experimental opportunities: Consider the value of artificial reefs of varied architecture in understanding the productivity and diversity of canopy residents (Carr and Hixon 1997).

Moffett (1999) outlined attributes falling into six general categories that put the canopy into canopy biology, calling these the discipline's core issues. By framing questions with respect to one or more of the categories (reviewed below), researchers can fundamentally contribute to canopy biology as a discipline. To keep my treatment brief, I have chosen examples from the forest literature alone, although information on each category is available for other kinds of canopies as well. While most scientific results can be partitioned along these categories, they clearly are not independent issues; many studies make important contributions when looking at two or more of the six. Ultimately, much of canopy science may codify according to how these attributes have contributed, over ecological and evolutionary time, to the diversity of both the host organisms and their occupants.

1. *Community Ecospace*. For any canopy, the quantity and quality of space available to the canopy dwellers depend on the host structure. How much so is a matter of conjecture. Any small, nimble bird or agile climbing animal, such as a gibbon, seems to experience canopies as a volume, although even here, not all points in the volume may be accessible: Vegetation may be too dense for a flying bird (Cuthill and Guilford 1990) or too widely spaced for primates to cross (Cant 1992; Cannon and Leighton 1994). Small, flightless arthropods, such as ants and earthworms, are unlikely to register community ecospace in this way. Ants are restricted to mere millimeters of every surface within their territory, such that—despite models to the contrary (Hölldobler and Lumsden 1980)—ants experience canopies as something between a 2-D and 3-D space (Moffett 1994, 2010). In essence, a canopy represents for them a highly warped surface. Like a science-fiction ship using a wormhole to bridge points normally experienced as quite distant from one another, weaver ants create shortcuts through this space by linking their bodies into chains, thus enabling them to quickly access new branches (Hölldobler and Wilson 1977). In this way, they can bridge entire tree crowns that otherwise

could be reached only by way of a long march down to, across, and up from the ground.

2. *Aggregate Properties of the Community*. Here, I include the nonuniform vertical distribution of canopy species and their architectural parts (Fig. 3.1) and the concomitant stratification of other elements of the canopy environment, such as microclimate (other gradients can exist, such as radially from the interior to the exterior of a tree crown [Freiberg 1997] or a bacterial microcolony [Stewart and Franklin 2008]). Parker and Brown (2000) (cf. McElhinny et al. 2005) criticize studies of stratification for their lack of reproducibility, inconsistent terminology, and other weaknesses. Regardless of the difficulties, understanding stratification is at the core canopy science, and to accommodate a variety of research interests, the word is best applied flexibly (Moffett 2000). While many studies of terrestrial nutrient interception treat the canopy as a single "black box" with overall inputs and outputs (e.g., Coxson and Nadkarni 1995), some researchers have uncovered complex internal patterns within canopies that act as atmospheric filters (Wiman et al. 1985; Meyers et al. 1989; Lovett and Lindberg 1992) and nutrient transfer systems (Pike 1978; Reiners and Olson 1984; Coxson et al. 1992; Lindo and Whiteley 2011). Defining strata or gradients requires broad spatial averaging, while explaining emerging patterns necessitates research at a finer spatial scale. For example, bark *p*H can vary with height in the trees (Hyvärinen et al. 1992). If the height distribution of an epiphyte corresponds generally to that of a certain bark *p*H, we could propose that these plants prefer that *p*H. Testing this hypothesis requires determining the detailed distribution of bark *p*H at the locations occupied by the epiphytes (e.g., Gauslaa 1995) and then manipulating the *p*H in the field or a laboratory (e.g., Hallingbäck 1990). In short, as I wrote about stratification elsewhere (Moffett 2000): A common finding is that short distances traversed vertically in canopies are equivalent in effect to changes likely to occur over much greater horizontal distances (e.g., Geiger 1965; Russell et al. 1989), demonstrating the critical importance for the height dimension as an environmental determinant. The vertical richness in microhabitat may be the primary reason so much diversity packs into structurally complex ecosystems. This could explain the high alpha diversity and low beta diversity of epiphytes in relation to other plants (McCune and Antos 1981), intimating the utility of quantifying a vertical component to beta diversity (DeVries et al. 2012).
3. *Host Distribution*. Many canopy species may be specialists at residing on one or a few plant taxa (e.g., Erwin 1982) suggesting that hosts can be considered islands over evolutionary time (Janzen 1968, 1973) (Fig. 3.6), which could help in modeling the processes that occur within and between plants that generate patterns of species distribution and diversity. Rather than treating hosts as islands in a uniform ocean to conform to the perspective of MacArthur and Wilson (1967), "patchwork" biogeography models could treat communities as a continuum of host islands of varied acceptability as sources of food, retreats, or transit routes for each canopy-dwelling species. Diversity is further organized on scales both smaller and larger than that of a host. A distinct community can develop on any stable canopy element that qualifies as an island sensu Haila (1990), such as a

Fig. 3.6 The patchwork of different tree species in a forest canopy is shown here by the distribution of gaudy flowering trees (*Vochysia ferruginea*) in the Osa Peninsula of Costa Rica (Mark W. Moffett/Minden Pictures. All rights reserved)

flower (Seifert 1975), a phytotelmatum (Jenkins and Kitching 1990; Richardson 1999), a leaf (to a microbe: Andrews 2006), or even certain microclimatic features (Herwitz and Slye 1992). Even ant territories could form habitat islands within tropical canopies (Moffett 2010, p. 132–133). The territories of different ant species are distributed as a mosaic that overlays, but are partially independent of, the mosaic of tree crowns (e.g., Dejean et al. 1999). Ants scour their territories to drive off intruders and kill prey while promoting the survival of species-specific assemblages of associates (Hölldobler and Wilson 1990). As with other island-like canopy features, the persistent domination of different aggressive ants with large colonies across the canopy volume may add to the potential for other species to pack into vegetable space.

4. *Host Architecture*. All canopies, from redwood forests to biofilms (Lawrence et al. 1991), have varied architectural parts. In forest ecology, a burgeoning literature on this topic covers the size, angles, distributions, development, and spatial relations of aerial plant parts. Classically, the models of Hallé et al. (1978) have been used to describe the architecture of trees (Fig. 3.7). One key practitioner considers the "Hallé-Oldeman architectural model" classification "comparable to the development of the binary system of nomenclature by Linnaeus" (Tomlinson 1983). Nonetheless, the system has been little used by non-morphologists, arguably in part because of the overall neglect of the potential importance of substratum architecture on canopy organisms (but see Hallé 1990). There are studies of the effects of simple architectural attributes, especially branch angle and width, on the growth of epiphytes (e.g., Rasmussen 1975; Ingram and Nadkarni 1993; Moe and Botnen 1997; Zotz 2006) and on animal locomotion,

Fig. 3.7 The architecture of canopy species is exemplified by an immature *Ceiba pentandra* tree from Cameroon, showing the "bicycle spoke" branching pattern of Massart's architectural model (Mark W. Moffett/Minden Pictures. All rights reserved)

particularly in reptiles (Losos and Greene 2009; Byrnes and Jayne 2012) and primates (Demes et al. 1995; Povinelli and Cant 1995; Dagosto and Yamashita 1998; Hamrick 1998; McGraw 1998; Young 2009, just to mention a few citations on positional behavior). Yet few broadly scaled studies of how community-level aspects of plant architecture influence canopy life have been done, excluding some examples for insects (e.g., Lawton 1983; Sinoquet et al. 2009). Consider that many tropical canopies have well-beaten vertebrate highways extending from tree to tree, which can be detected by the epiphytes that spread to each side of a branch like hair from a part (Perry 1978; Sillett et al. 1995). Perry (1978) found evidence of multispecies use and active pruning of these trails. But as yet, no one has mapped such a trail in relation to the tree architectures available locally, or documented how the trail originates, how long it lasts, or how usage shifts with changes in canopy structure and resource availability.

5. *Open Space*. Not even biofilms are a continuous matrix of organisms: All canopies consist of a framework occupying a dynamic fluid matrix that has open communication and exchange with the adjacent atmosphere or hydrosphere, which typically includes the bulk of canopy volume (Chiarucci et al. 2002). Free space (air or water) within canopies merits special consideration, because of its potential effects on microclimate and on the locomotion or dispersal of organisms. Many

Fig. 3.8 Open spaces within forests include shyness between tree crowns, as shown in a grove of *Dryobalanops lanceolata* trees in Peninsular Malaysia (*above*). Tree spacing is a challenge for the mobility of canopy-dwelling species, including for weaver ants (*Oecophylla smaragdina)*, which bridge gaps by forming living chains of workers (*below*) (Both images are from Mark W. Moffett/ Minden Pictures. All rights reserved)

aspects of this topic remain virtually ignored. Space between forest trees is commonly distinguished through the use of two categories: *gaps*, the result of plant death, and the openings resulting from *shyness*, which is often the result of diminished plant growth and reconfiguration—that is, plant foraging (Hutchings and de Kroon 1994), although physical abrasion can also be involved in shyness patterns (Franco 1986) (Fig. 3.8). Gaps are commonly studied spaces, because of their role in forest succession and species-diversity patterns (Lieberman et al. 1989). Spaces represent barriers to some species and pathways (flyways) or resource (light) access and growth opportunities to others (e.g., Aluja et al. 1989; Brady et al. 1989; Cuthill and Guilford 1990; Cannon and Leighton 1994; Brigham et al. 1997; Aylor 1999; Svenning 2000; Montgomery and Chazdon 2001; Dial et al. 2004; Randlkofer et al. 2010; Byrnes and Jayne 2012). Little information exists, however, on how the distribution of open space might be involved in structuring the populations of canopy residents. There is a tendency to think in terms of canopy structures such as trunks and branches when the space between structures could be the resource used, as might be the case among gliding animals (Moffett 2000), though competition for space per se is seldom likely (Wilson et al. 2007a). Within the open spaces, boundary layers—the regions of lowered fluid velocity that exist around any surface in a turbulent medium—are a general feature of attached communities. Their presence partially isolates canopies from the surrounding medium and thereby can increase community reliance on efficient and potentially autogenically controlled internal (within-canopy) nutrient cycling. This isolation may be particularly important in flowing water (Mulholland 1996), where canopy physiognomy can substantially ameliorate the downstream displacement of chemicals or of any organisms that are moving within a canopy or that have a poorly developed capacity to attach to a substratum.

6. *Physical and Chemical Properties of Structural Elements*. In all canopies, the sessile hosts present associated species with a variety of surfaces, both between host individuals or species and within each host (such as wood versus leaves in a tree). These structural elements vary in their properties; examples include the capacity for insulation or water absorption, tendency to leach nutrients, efficiency at transmitting vibration, and their texture (Fig. 3.9), stability, density, hardness, compliance, stiffness, strength, pH, and so on. How do such variables affect life on or in a host? One of the oldest areas of canopy investigation in terrestrial biology is the question of substratum choice by epiphytes, especially cryptogams (e.g., Barkman 1958). Another area of intensive study has been herbivory as it relates to secondary compounds, nutrient content, and the mechanical difficulties of feeding (Schowalter 2011). Outside of these focal points, the literature is widely scattered, leaving many potential research avenues of enormous prospects.

5 Getting to the Root of the Matter

Much of the language and thinking of forest canopy biology has been predicated on notions of plants as supports for the organisms residing in them (epiphytes, vines, and so on). Structural support between individuals occurs as well in the rhizosphere (Moffett 2001), reminding us that distinctions between above- and belowground plant parts can be arbitrary. In many ways, it would be logical to define words such as "epiphyte" so that they apply to the host in its entirety (e.g., a tree from root to

Fig. 3.9 The physical and structural properties of trees are important to canopy residents. Here a jararaca snake (*Bothrops jararaca*) is able to climb the trunk of an *Araucaria* tree in Brazil by gripping irregularities the rough bark (Mark W. Moffett/ Minden Pictures. All rights reserved)

crown: in marine biology, a more inclusive term, epibiont, describes any nonparasitic species living on any part of the surface of another organism; Wahl 1997). For many research purposes, this idea suggests the value of expanding beyond the realm of canopy biology to fashion a comprehensive science of sessile communities, from crowns to roots (or their equivalents). This nascent subject, which I refer to as *structural ecology*, would apply the principles of canopy biology to communities as a whole.

This would be no big leap for communities structured around sessile taxa lacking elaborate organs that penetrate the substrate, such as algae or corals. In such species, nutrient inputs tend to be greatest in the upper (outer) canopy rather than at or in the substratum, such that traits adaptive for light and nutrient procurement function almost entirely in synchrony (McCormick 1996). I focus instead on plants, with their extensive root systems. It is true that roots evolved from shoots prior to the evolution of leaves (Barlow 1994; Langdale 2008) and have remained developmentally distinct from leaves. In separating canopy from rhizosphere, however, it is more significant that roots and shoots are not necessarily distinguishable in either function or location relative to the ground surface: Shoots can absorb nutrients and water (Parker 1983; Schaefer and Reiners 1989) and can occur belowground, where they are referred to as rhizomes, and roots can occur aboveground, where they can be photosynthetic (Benzing 1991) or have ventilation and aeration functions, as in mangroves. The distinction made by botanists between belowground and above ground (and often between root and shoot) has been largely methodological, a matter of choice between using a climbing rope or a shovel, for instance. The resulting academic fragmentation can be transcended, as in research on root stratification (or lack of clear strata) by several groups of researchers (Mommer et al. 2010; Frank et al. 2010; Kesanakurti et al. 2011; de Kroon et al. 2012). Of the published studies in recent years on the rhizosphere, efforts of this kind most closely follow the paradigm for canopy studies developed in this review.

The field of structural ecology will by necessity emerge in slow increments. While it is true that ecology took a long time to enter the treetops (Moffett and Lowman 1995), in many ways, it is not the aerial world but the subterranean one that remains most alien to us today, given the rhizosphere's complexity (compare Beare et al. 1995 with Freiberg 1997), inaccessibility barring in most cases wholesale destructive intrusion (Smit et al. 2000), and difficulties in identifying plants from their roots (Mommer et al. 2011). Root systems and the species associated with their rhizosphere therefore could merit consideration equal to the rainforest canopy as the last—though assuredly not the highest (sensu Moffett 1994)—biotic frontier (André et al. 1994, 2002; Decaëns 2010). Compared with their crown architecture, the architecture of plant root systems in natural settings is poorly known (but see, e.g., Jeník 1978; Atger and Edelin 1993). This includes the links between the above- and belowground communities, which thus far have largely been framed in terms either of lone plants or of the above- versus belowground ecologies as gross-level compartments, rather than in terms of detailed architecture and stratification (Coleman et al. 1983; van der Putten et al. 2001; Moore et al. 2003; de Deyn and van der Putten 2005; Wardle et al. 2004; Poorter et al. 2012). Despite confounding effects from competition (e.g., Mahall and Callaway 1992; Burgess et al. 1998; Schenk 2006), root systems in nature are seldom isolated: Consider the drawings in Weaver and Clements (1929) of herb and grass communities (the distribution of tree roots is not known to this detail, but see Chilvers (1972), Lyford (1975), and the figures in Külla and Lõhmus (1999)) (Fig. 3.10). Still, 3-D studies of roots that extend beyond simple depth measurements of the kind reviewed by Jackson et al. (1996) and Schenk and Jackson (2002) are scarce (Tsegaye et al. 1995 a, b; Lynch et al. 1997; Ge et al. 2000; Pages 2000; Pages et al. 2000; Danjon et al. 2008; Fang et al. 2012), and most examples have been considered at the level of a single plant rather than of a community, although see the descriptions of the architecture and stratification of root systems by Caldwell and Richards (1986), Mommer et al. (2010), and Postma and Lynch (2012).

How would parasitology stand as a coherent discipline if its practitioners chose to look at the head of their subject while ignoring what its feet were doing? By amalgamating the findings from canopy biology with those from soil sciences, terrestrial biologists could fashion a comprehensive science of plant associates.

Fig. 3.10 Rooting pattern of a group of Norway spruce trees, where root growth and distribution reflects competitive interactions (Gebauer and Martinková 2005) that are most intense near the soil surface, as they are for most plants (Photo by Roman Gebauer. All rights reserved. Reproduced with permission)

6 Conclusions Vis-à-vis Biodiversity

There are species from every kingdom of life that attach to or grow from substrata, and, when aggregated, these organisms form "a three-dimensional complex of structures," as Spies (1998) described forests, whose parts that emerge from the substratum in aggregate can be (and often have been) described as canopies. Often, terrestrial studies appear to be categorized as "canopy biology" either on the basis of inaccessibility, as when specialized gear is required to access trees (Moffett and Lowman 1995), or on the basis of the communities' cover-like properties. (In the latter case, *canopy* is treated as synonymous with overstory or used more broadly as any stratum in which sessile host organs are distributed so as to appreciably shade the layers below, as the word can be used in marine science; Dayton 1975a; Baird and Hughes 2000.) Because such criteria are arbitrary and of limited general import,

I apply the word canopy to all above-substratum parts of sessile communities (of plants, when I originally described this idea in Moffett 2000). The common feature distinguishing studies as canopy biology is the treatment of sessile communities in three spatial dimensions, along with certain other attributes that are unique to life within this "canopy space."

Why are the three-dimensional attributes of communities important enough to serve as the basis for a field of study? Taken in aggregate, the architecture of affixed life forms such as plants serves as scaffolding for much of the biological richness of this planet. By projecting into fluid media, canopies can augment productivity through increasing the biotic mass and the live surface area available for capturing and processing nutrients and energy and can enhance α-diversity through transforming in a heterogeneous way the climatic and chemical properties of the space they occupy, potentially multiplying the available niche space associated with a given surface area of the earth (e.g., Morin 2000; Moffett 2000; Hill and Hill 2001; Walla et al. 2004; DeVries et al. 2012). Moreover, canopies can provide retreats from predators, competitors, and adverse conditions, as well as additional surface area to attract or accumulate new species and the nutrients required to support a rich community (e.g., Crowder and Cooper 1982; Dean and Connell 1987; Lovett and Lindberg 1992; Jones et al. 1997; Johansen et al. 2008). We have yet to explain the differences and even more intriguing similarities in the physical structure and dynamics of canopies that develop in air as contrasted with water, and across orders of magnitude in host size: For example, compare Edred Corner's views on trees with Timothy Allen's on microalgae (Corner 1967; Allen 1977). Surely these size extremes are of special fascination; that may be the primary reason (beyond fulfilling our sense of adventure) that many of us climb trees to collect our data. Will scaling functions prove sufficiently linear to permit straightforward extrapolations across the full expanse of the different ecosystems?

Finally, canopy biology will have to be integrated into a more coherent understanding of sessile communities as a whole—that is, by taking similar account of the role of within-substratum organs such as roots and holdfasts. A central challenge of this holistic approach to communities, which I have called structural ecology, will be to determine the rules by which ecosystems, including both the sessile species and the residents of those species, assemble in all their three-dimensional glory, thereby creating opportunities for increased production and diversity—an idea that can be traced back to MacArthur and MacArthur (1961).

Acknowledgments I published Moffett (2001) in an up-and-coming journal for canopy research that unfortunately folded. I thank Margaret ("Meg") Lowman, who, after I talked about this subject for the 2009 International Canopy Conference in Bangalore, gave me the chance to update the article here, so that it can come before a new audience. Luckily, I was able to contact many of the people who advised me on the first version. Although doubtless I missed a few things, I found that little has been done in the last decade on the issues I raise, except for the burgeoning research on biofilms. For assistance with the philosophy of discipline formation, I thank Mary Catherine Bateson, David L. Hull, Peter Harries-Jones, George Lakoff, Paul Ryan, and Carol Wilder; for thoughts on trees, forests, and the dimensionality of canopies, David Ackerly, Timothy F.H. Allen, Robert G. Bailey, Dennis Baldocchi, Frans Bongers, James H. Brown, Jerome Chave, Joel

Clement, Raphael Didham, Brian J. Enquist, Stephen P. Ellner, Thomas J. Givnish, Juan Gouda, Paul G. Jarvis, David King, Steve Lindberg, Richard Law, Orie Loucks, Margaret Lowman, Gary Lovett, Jonathan Majer, Lauri Oksanen, Karl J. Niklas, John M. Norman, Geoffrey Parker, Serguei Ponomarenko, Hank H. Shugart, Jonathan Silvertown, Anthony R.E. Sinclair, Frank Sterck, Steven Sutton, Bastow Wilson, Neville Winchester, and Truman Young; for information on roots, Peter W. Barlow, Jan Čermák, Hans de Kroon, A. Roland Ennos, Lewis Feldman, Roman Gebauer, Robert B. Jackson, Donald R. Kaplan, Takashi Kohyama, Krista Lõhmus, Jonathan Lynch, Liesje Mommer, John Moore, James H. Richards, Wim van der Putten, Fernando Tuya, and Yoav Waisel; and for views on algal, bacterial, coral reef, and other canopies, Mohammad A Al-Najjar, Yves Basset, David Benzing, James D. Bryers, Robert C. Carpenter, the late and great J. William Costerton, Paul Dayton, Dirk de Beer, Phil DeVries, Peg Dirckx, Michael Dolan, David Duggins, Michael Franklin, Elizabeth Gladfelter, Ken Heck, Brian Helmuth, Helmut Hillebrand, Walter R. Hill, Mark Hixon, Bo Barker Jørgensen, Ronald H. Karlson, Mimi Koehl, Michael Kühl, Scott Larned, Steven E. Lindow, Rex L. Lowe, my recently deceased friend Lynn Margulis, Susan Merkel, Peter J. Morin, Cindy Morris, James W. Porter, Jennifer H. Richards, Kenneth P. Sebens, Vaclav Smil, Robert S. Steneck, Valerie Behan-Pelletier, Alice Tangerini, Chantal Vis, Martin Wahl, and Susan L. Williams. Lapses in coverage and logic can be attributed to me.

References

Adey WH, Loveland K (2007) Dynamic aquaria: building living ecosystems, 3rd edn. Academic, New York

Adey WH, Steneck RS (1985) Highly productive eastern Caribbean reefs: synergistic effects of biological, chemical, physical, and geological factors. In: Reaka ML (ed) The ecology of coral reefs, 3. Office of Undersea Research, Rockville

Aira M, McNamara NP, Piearce TG, Domínguez J (2009) Microbial communities of *Lumbricus terrestris* L. middens: structure, activity, and changes through time in relation to earthworm presence. J Soils Sediments 9:54–61

Allen TFH (1977) Scale in microscopic algal ecology: a neglected dimension. Phycologia 16: 253–257

Allison DG, Gilbert P, Lappin-Scott HM, Wilson M (eds) (2000) Community structure and co–operation in biofilms. Cambridge University Press, Cambridge, UK

Al-Najjar MAA, de Beer D, Kühl M, Polerecky L (2012) Light utilization efficiency in photosynthetic microbial mats. Environ Microbiol 14:982–992

Aluja M, Prokopy RJ, Elkinton JS, Laurence WF (1989) Novel approach for tracking and quantifying the movement patterns of insects in three dimensions under seminatural conditions. Environ Entomol 18:1–7

Alvarez-Filip L, Dulvy NK, Gill JA, Côté IM, Watkinson AR (2009) Flattening of Caribbean coral reefs: region–wide declines in architectural complexity. Proc R Soc Lond B Biol 276:3019–3025

Alvarez-Filip L, Dulvy NK, Gill JA, Côté IM, Watkinson AR, Gill JA (2011) Coral identity underpins architectural complexity on Caribbean reefs. Ecol App. 21:2223–2231

Anderson TW (1994) Role of macroalgal structure in the distribution and abundance of a temperate reef fish. Mar Ecol Prog Ser 113:279–290

Anderson TW (2001) Predator responses, prey refuges, and density–dependent mortality of a marine fish. Ecology 82:245–257

André HM, Noti M-I, Lebrun P (1994) The soil fauna: the other last biotic frontier. Biodivers Conserv 3:45–56

André HM, Ducarme X, Lebrun P (2002) Soil biodiversity: myth, reality, or conning? Oikos 96:3–24

Andrews JH (2006) Population growth and the landscape ecology of microbes on leaf surfaces. In: Bailey MJ, Lilley AK, Timms-Wilson TM, Spencer-Phillips PTN (eds) Microbial ecology of aerial plant surfaces. CAB International, Wallingford

Anthony KR, Hoegh-Guldberg O (2003) Variation in coral photosynthesis, respiration and growth characteristics in contrasting light microhabitats: an analogue to plants in forest gaps and understoreys? Funct Ecol 17:246–259

Atger C, Edelin C (1993) Premières données sur l'architecture comparée des systèmes racinaires et caulinaires. Can J Bot 72:963–975

Aylor DE (1999) Biophysical scaling and the passive dispersal of fungus spores: relationship to integrated pest management strategies. Agric For Meteorol 97:275–292

Baird AH, Hughes TP (2000) Competitive dominance by tabular corals: an experimental analysis of recruitment and survival of understorey assemblages. J Exp Mar Biol Ecol 251:117–132

Barkman JJ (1958) Phytosociology and ecology of cryptogamic epiphytes. Van Gorcum, Assen

Barlow PW (1994) Rhythm, periodicity and polarity as bases for morphogenesis in plants. Biol Rev 69:475–525

Bates M (1960) The forest and the sea. Random House, New York

Bateson MC (1972) Our own metaphor. Knopf, New York

Battin TJ, Sloan WT, Kjelleberg S, Daims H, Head IM, Curtis TP, Eberl L (2007) Microbial landscapes: new paths to biofilm research. Nat Rev Microbiol 5:76–81

Beare MH, Coleman DC, Crossley DA Jr, Hendrix PF, Odum EP (1995) A hierarchical approach to evaluating the significance of soil biodiversity to biogeochemical cycling. Plant Soil 170:5–22

Bell SS, Coen LD (1982) Investigations on epibenthic meiofauna. I. Abundances on and repopulation of the tube–caps of *Diopatra cuprea* (Polychaeta: Onuphidae). Mar Biol 67:303–309

Benzing DH (1991) Aerial roots and their environments. In: Waisel Y, Eshel A, Kafkafi U (eds) Plant roots: the hidden half. Marcel Dekker, New York

Benzing DH (2012) Air plants: epiphytes and aerial gardens. Cornell University Press, Ithaca

Beugnon G, Fourcassié V (1988) How do red wood ants orient when foraging in a three dimensional system? II. Field experiments. Ins Soc 35:106–124

Beyenal H, Lewandowski Z (2002) Internal and external mass transfer in biofilms grown at various flow velocities. Biotechnol Prog 18:55–61

Bohr N (1955) Science and the unity of Knowledge. In: Leary L (ed) The unity of knowledge. Doubleday, Garden City

Borchers P, Field JG (1981) The effect of kelp shading on phytoplankton production. Bot Mar 24:89–91

Boston HL, Hill WR (1991) Photosynthesis–light relations of stream periphyton communities. Limnol Oceanogr 36:644–656

Boyle M, Janiak D, Craig S (2007) Succession in a Humboldt Bay marine fouling community: the role of exotic species, larval settlement and winter storms. In: Proceedings 2004 Humboldt Bay symposium, California Sea Grant, San Diego

Brady J, Gibson G, Packer MJ (1989) Odour movement, wind direction, and the problem of host–finding by tsetse flies. Physiol Entomol 14:369–380

Brigham RM, Grindal SD, Firman MC, Morissette JL (1997) The influence of structural clutter on activity patterns of insectivorous bats. Can J Zool 75:131–136

Brooks R, Owen-Smith N (1994) Plant defenses against mammalian herbivores: are juvenile *Acacia* more heavily defended than mature trees? Bothalia 24:211–215

Brown JH (1995) Macroecology. University of Chicago Press, Chicago

Bryers JD (2000) Biofilm formation and persistence. In: Bryers JD (ed) Biofilms II: process and applications. Wiley-Liss, New York

Burgess SSO, Adams MA, Turner NC, Ong CK (1998) The redistribution of soil water by tree root systems. Oecologia 115:306–311

Burkholder JM, Wetzel RG (1989) Epiphytic microalgae on a natural substratum in a phosphorus–limited hardwater lake: seasonal dynamics of community structure, biomass and ATP content. Arch Hydrobiol Suppl 83:1–56

Burrows CJ (1990) Processes of vegetation change. Unwin Hyman, London

Byrnes G, Jayne BC (2012) The effects of three-dimensional gap orientation on bridging performance and behavior of brown tree snakes (*Boiga irregularis*). J Exp Biol 215:2611–2620

Caldwell MM, Richards JM (1986) Competing root systems: morphology and models of absorption. In: Givnish TJ (ed) On the economy of plant form and function. Cambridge University Press, Cambridge, UK
Campbell GS, Norman JM (1989) The description and measurement of plant canopy structure. In: Russell G, Marshall B, Jarvis PG (eds) Plant canopies: their growth, form and function. Cambridge University Press, Cambridge, UK
Cannon CH, Leighton M (1994) Comparative locomotor ecology of gibbons and macaques: selection of canopy elements for crossing gaps. Am J Phys Anthropol 93:505–524
Cant JGH (1992) Positional behavior and body size of arboreal primates: a theoretical framework for field studies and an illustration of its application. Am J Phys Anthropol 88:273–283
Carpenter RC (1986) Partitioning herbivory and its effects on coral reef algal communities. Ecol Monogr 56:345–363
Carpenter RC, Williams SL (1993) Effects of algal turf canopy height and microscale substratum topography on profiles of flow speed in a coral forereef environment. Limnol Oceanogr 38:687–694
Carr MH (1989) Effects of macroalgal assemblages on the recruitment of temperate zone reef fishes. J Exp Mar Biol Ecol 126:59–76
Carr MH, Hixon MA (1997) Artificial reefs: the importance of comparisons with natural reefs. Fisheries 22(4):28–33
Carr GM, Morin A, Chambers PA (2005) Bacteria and algae in stream periphyton along a nutrient gradient. Freshwater Biol 50:1337–1350
Cheroske AG, Williams SL, Carpenter RC (2000) Effects of physical and biological disturbances on algal turfs in Kaneohe Bay, Hawaii. J Exp Mar Biol Ecol 248:1–34
Chiarucci A, Mistral M, Bonini I, Anderson BJ, Wilson JB (2002) Canopy occupancy: how much of the space in plant communities is filled? Folia Geobot 37:333–338
Chilvers GA (1972) Tree root pattern in a mixed eucalypt forest. Aust J Bot 20:229–234
Claflin TO (1968) Reservoir aufwuchs on inundated trees. Trans Am Microsc Soc 87:97–104
Coleman DC, Reid CPP, Cole CV (1983) Biological strategies of nutrient cycling in soil systems. Adv Ecol Res 13:1–55
Connell JH (1975) Some mechanisms producing structure in natural communities: a model and evidence from field experiments. In: Cody ML, Diamond JM (eds) Ecology and evolution of communities. Harvard University Press, Cambridge, MA
Corner EJH (1967) On thinking big. Phytomorphology 17:24–28
Costerton JW, Lappin-Scott HM (1995) Introduction to microbial biofilms. In: Lappin-Scott HM, Costerton JW (eds) Microbial biofilms. Cambridge University Press, Cambridge, UK
Costerton JW, Lewandowski Z, Caldwell DE, Korber DR, Lappin-Scott HM (1995) Microbial biofilms. Ann Rev Microbiol 49:711–745
Costerton JW, Lewandowski Z, DeBeer D, Caldwell D, Korber D, James G (1994) Biofilms: the customized microniche. J Bacteriol 176:2137–2142
Coxson DS, Nadkarni NM (1995) Ecological roles of epiphytes in nutrient cycles of forest ecosystems. In: Lowman MD, Nadkarni NM (eds) Forest canopies. Academic, New York
Coxson DS, McIntyre DD, Vogel HJ (1992) Pulse release of sugars and polyols from canopy bryophytes in tropical montane rain forest (Guadeloupe, French West Indies). Biotropica 24:121–133
Crespi BJ (2001) The evolution of social behavior in microorganisms. Trends Ecol Evol 16:178–183
Crowder LB, Cooper WE (1982) Habitat structural complexity and the interaction between bluegills and their prey. Ecology 63:1802–1813
Cuthill I, Guilford T (1990) Perceived risk and obstacle avoidance in flying birds. Anim Behav 40:188–190
Dagosto M, Yamashita N (1998) Effect of habitat structure on positional behavior and support use in three species of lemur. Primates 39:459–472
Dahl AL (1973) Surface area in ecological analysis: quantification of benthic coral–reef algae. Mar Biol 23:239–249

Danjon F, Barker DH, Drexhage M, Stokes A (2008) Using three–dimensional plant root architecture in models of shallow–slope stability. Ann Bot 101:1281–1293
Darwin C (1839) Journal of researches into the geology and natural history of the various countries visited by H.M.S. Beagle. Colburn, London
Davies DG, Parsek MR, Pearson JP, Iglewski BH, Costerton JW, Greenberg EP (1998) The involvement of cell–to–cell signals in the development of a bacterial biofilm. Science 280:295–298
Dawkins R (1982) The extended phenotype. Oxford University Press, Oxford, UK
Dawson EY (1966) Cacti in the Galapagos Islands with special reference to their relations with tortoises. In: Bowman RI (ed) The Galapagos, proceedings of the symposium of the California international scientific project, University of California Press, Los Angeles
Dayton PK (1971) Competition, disturbance, and community organization: the provision and subsequent utilization of space in a rocky intertidal community. Ecol Monogr 41:351–389
Dayton PK (1975a) Experimental studies of algal canopy interactions in a sea otter–dominated kelp community at Amchitka Island, Alaska. Fish Bull 73:230–237
Dayton PK (1975b) Experimental evaluation of ecological dominance in a rocky intertidal algal community. Ecol Monogr 45:137–159
Dayton PK (1985) Ecology of kelp communities. Ann Rev Ecol Syst 16:215–245
Dayton PK, Tegner MJ, Edwards PB, Riser KL (1999) Temporal and spatial scales of kelp demography: the role of oceanographic climate. Ecol Mongr 69:219–250
De Deyn GB, Van der Putten WH (2005) Linking aboveground and belowground diversity. Trends Ecol Evol 20:625–633
de Kroon H, Hendriks M, van Ruijven J, Ravenek J, Padilla FM, Jongejans E, Visser EJW, Mommer L (2012) Root responses to nutrients and soil biota: drivers of species coexistence and ecosystem productivity. J Ecol 100:6–15
DeVries PJ, Alexander LG, Chacon IA, Fordyce JA (2012) Similarity and difference among rainforest fruit–feeding butterfly communities in Central and South America. J Anim Ecol 81:472–482
Dean RL, Connell JH (1987) Marine invertebrates in an algal succession. III. Mechanisms linking habitat complexity with diversity. J Exp Mar Biol Ecol 109:249–273
Dean TA, Thies K, Lagos SL (1989) Survival of juvenile giant kelp: the effects of demographic factors, competitors, and grazers. Ecology 70:483–495
Decaëns T (2010) Macroecological patterns in soil communities. Global Ecol Beogeogr 19:287–302
Dejean A, Corbara B, Orival J (1999) The arboreal ant mosaic in two Atlantic rain forests. Selbyana 20:133–145
Demes B, Jungers WL, Gross TS, Fleagle JG (1995) Kinetics of leaping primates: influence of substrate orientation and compliance. Am J Phys Anthropol 96:419–429
Dial R, Bloodworth B, Lee A, Boyne P, Heys J (2004) The distribution of free space and its relation to canopy composition at six forest sites. Forest Sci 50:312–325
Dieckmann U, Law R, Metz JAJ (eds) (2000) The geometry of ecological interactions: simplifying spatial complexity. Cambridge University Press, Cambridge, UK
Dodd ME, Silvertown J, Chase MW (1999) Phylogenetic analysis of trait evolution and species diversity variation among angiosperm families. Evolution 53:732–744
Dodds WK, Biggs BJF, Lowe RL (1999) Photosynthesis–irradiance patterns in benthic microalgae: variations as a function of assemblage thickness and community structure. J Phycol 35:42–53
Doemel WN, Brock TD (1977) Structure, growth, and decomposition of laminated algal–bacterial mats in alkaline hot springs. Appl Environ Microbiol 34:433–452
Erwin TL (1982) Tropical forests: their richness in Coleoptera and other arthropod species. Coleopt Bull 36:74–75
Fang S, Clark R, Liao H (2012) 3D quantification of plant root architecture in situ. In: Mancuso S (ed) Measuring roots. Springer, New York
Fierer N, Nemergut D, Knight R, Craine JM (2010) Changes through time: integrating microorganisms into the study of succession. Res Microbiol 161:635–642
Flemming HC, Wingender J (2010) The biofilm matrix. Nat Rev Microbiol 8:623–633
Foottit RG, Adler PH (2009) Insect biodiversity: science and society. Wiley–Blackwell, Oxford, UK

Foster MS (1975) Regulation of algal community development in a *Macrocystis pyrifera* forest. Mar Biol 32:331–342
Foster K (2010) Social behaviour in microorganisms. In: Székely T, Komdeur J, Moore AJ (eds) Social behaviour: genes, ecology and evolution. Cambridge University Press, Cambridge, UK
Foster K, Bell T (2012) Competition, not cooperation, dominates interactions among culturable microbial species. Curr Biol 22:1845–1850
Fourcassie V, Beugnon G (1988) How do red wood ants orient when foraging in a three dimensional system? I. Laboratory experiments. Insect Soc 35:92–105
Franco M (1986) The influence of neighbours on the growth of modular organisms with an example from trees. Phil Trans R Soc Lond B 313:209–225
Frank DA, Pontes AW, Maine EM, Caruana J, Raina R, Raina S, Fridley JD (2010) Grassland root communities: species distributions and how they are linked to aboveground abundance. Ecology 91:3201–3209
Fraser SB, Sedberry GR (2008) Reef morphology and invertebrate distribution at continental shelf edge reefs in the south Atlantic bight. Southeast Nat 7:191–206
Freiberg M (1997) Spatial and temporal pattern of temperature and humidity of a tropical premontane rain forest tree in Costa Rica. Selbyana 18:77–84
Gaines SD, Roughgarden J (1987) Fish in offshore kelp forests affect recruitment to intertidal barnacle populations. Science 235:479–481
Garcia-Pichel F, Mechling M, Castenholz RW (1994) Diel migrations of microorganisms within a benthic, hypersaline mat community. Appl Environ Microbiol 60:1500–1511
Gauslaa Y (1995) The *Lobarion*, an epiphytic community of ancient forests threatened by acid rain. Lichenologist 27:59–76
Ge Z, Rubio G, Lynch JP (2000) The importance of root gravitropism for inter-root competition and phosphorus acquisition efficiency: results from a geometric simulation model. Plant Soil 218:159–171
Gebauer R, Martinková M (2005) Structure and functions of the types of Norway spruce (*Picea abies* [L.] Karst.) roots. J Forest Sci 51:305–311
Geiger R (1965) The climate near the ground. Harvard University Press, Cambridge, MA
Gerard VA (1984) The light environment in a giant kelp forest: influence of *Macrocystis pyrifera* on spatial and temporal variability. Mar Biol 84:189–194
Givnish TJ (1983) Convergent evolution of crown form in woody plants of southwestern Australia and New Caledonia. Am Phil Soc Yearb 1983:136
Givnish TJ (1984) Leaf and canopy adaptations in tropical forests. In: Medina E, Mooney HA, Vásquez-Yánes C (eds) Physiological ecology of plants of the wet tropics. Dr. Junk, The Hague
Givnish TJ (1995) Plant stems: biomechanical adaptation for energy capture and influence on species distributions. In: Gartner BL (ed) Plant stems: physiology and functional morphology. Chapman and Hall, New York
Graham NAJ, Nash KL (2012) The importance of structural complexity in coral reef ecosystems. Coral Reefs. doi:10.1007/s00338-012-0984-y
Graham NAJ, Wilson SK, Pratchett MS, Polunin NVC, Spalding MD (2009) Coral mortality versus structural collapse as drivers of corallivorous butterflyfish decline. Biodivers Conserv 18:3325–3336
Guerrero R, Mas J (1989) Multilayered microbial communities in aquatic ecosystems: growth and loss factors. In: Cohen Y, Rosenberg E (eds) Microbial mats: physiological ecology of benthic microbial communities. American Society for Microbiology, Washington, DC
Guerrero R, Piqueras M, Berlanga M (2002) Microbial mats and the search for minimal ecosystems. Int Microbiol 5:177–188
Haack TK, McFeters GA (1982) Nutritional relationships among microorganisms in an epilithic biofilm community. Microb Ecol 8:115–126
Hacker SD, Steneck RS (1990) Habitat architecture and the abundance and body-size-dependent habitat selection of a phytal amphipod. Ecology 71:2269–2285
Hackney JM, Carpenter RC, Adey WH (1989) Characteristic adaptations to grazing among algal turfs on a Caribbean coral reef. Phycologia 28:109–119

Haila Y (1990) Toward an ecological definition of an island: a northwest European perspective. J Biogeogr 17:561–568
Hallé F (1990) Tropical rain forests: structure and growth dynamics relative to utilization by birds. In: Keast A (ed) Biogeography and ecology of forest bird communities. SPB Academic, The Hague
Hallé F, Oldeman RAA, Tomlinson PB (1978) Tropical trees and forests: an architectural analysis. Springer, Berlin
Hallingbäck T (1990) Transplanting *Lobaria pulmonaria* to new localities and a review on the transplanting of lichens. Windahlia 18:57–64
Hamrick MW (1998) Functional and adaptive significance of primate pads and claws: evidence from New World anthropoids. Am J Phys Anthropol 106:113–127
Hansen SK, Rainey PB, Haagensen JA, Molin S (2007) Evolution of species interactions in a biofilm community. Nature 445:533–536
Hay ME (1986) Functional geometry of seaweeds: ecological consequences of thallus layering and shape in contrasting light environments. In: Givnish TJ (ed) On the economy of plant form and function. Cambridge University Press, Cambridge, UK
Heck KL Jr, Wetstone GS (1977) Habitat complexity and invertebrate species richness and abundance in tropical seagrass meadows. J Biogeogr 4:135–142
Herwitz SR, Slye RE (1992) Spatial variability in the interception of inclined rainfall by a tropical rainforest canopy. Selbyana 13:62–71
Hill WR (1996) Effects of light. In: Stevenson RJ, Bothwell ML, Lowe RL (eds) Algal ecology: freshwater benthic ecosystems. Academic, New York
Hill JL, Hill RA (2001) Why are tropical rain forests so species rich? Classifying, reviewing and evaluating theories. Prog Phys Geogr 25:326–354
Hillebrand H, Frost P, Liess A (2008) Ecological stoichiometry of indirect grazer effects on periphyton nutrient content. Oecologia 155:619–630
Hirose T, Werger MJA (1995) Canopy structure and photon flux partitioning among species in a herbaceous plant community. Ecology 76:466–474
Hoagland KD, Roemer SC, Rosowski JR (1982) Colonization and community structure of two periphyton assemblages, with emphasis on the diatoms (Bacillariophyceae). Am J Bot 69:188–213
Hodoki Y (2005) Bacteria biofilm encourages algal immigration onto substrata in lotic systems. Hydrobiologia 539:27–34
Holbrook NM (1995) Stem water storage. In: Gartner BL (ed) Plant stems: physiology and functional morphology. Chapman and Hall, New York
Holbrook SJ, Denny MW, Koehl MAR (1991) Intertidal "trees":consequences of aggregation on the mechanical and photosynthetic properties of sea-palms *Postelsia palmaeformis* Ruprecht. J Exp Mar Biol Ecol 146:39–67
Holbrook SJ, Brooks AJ, Schmitt RJ (2003) Variation in structural attributes of patch–forming corals and in patterns of abundance of associated fishes. Mar Freshw Res 53:1045–1053
Hölldobler B, Lumsden CJ (1980) Territorial strategies in ants. Science 210:732–739
Hölldobler B, Wilson EO (1977) Weaver ants: social establishment and maintenance of territory. Science 195:900–902
Hölldobler B, Wilson EO (1990) The ants. Harvard University Press, Cambridge, MA
Horn H (1971) The adaptive geometry of trees. Princeton University Press, Princeton
Hudon C, Bourget E (1983) The effect of light on the vertical structure of epibenthic diatom communities. Bot Mar 26:317–330
Hull DL (1988) Science as a process. University of Chicago Press, Chicago
Hurby T (1976) Observations of algal zonation resulting from competition. Estuar Coast Mar Sci 4:231–233
Hutchings MJ, De Kroon H (1994) Foraging in plants: the role of morphological plasticity in resource acquisition. Adv Ecol Res 25:159–238
Hyvärinen M, Halonen P, Kauppi M (1992) Influence of stand age and structure on the epiphytic lichen vegetation in the middle–boreal forests of Finland. Lichenologist 24:165–180

Ingram SW, Nadkarni NM (1993) Composition and distribution of epiphytic organic matter in a neotropical cloud forest, Costa Rica. Biotropica 25:370–383
Irie Y, Parsek MR (2008) Quorum sensing and microbial biofilms. Curr Top Microbiol Immunol 322:67–84
Irving AD, Connell SD (2006) Predicting understorey structure from the presence and composition of canopies: an assembly rule for marine algae. Oecologia 148:491–502
Jackson CR (2003) Changes in community properties during microbial succession. Oikos 101:444–448
Jackson JBC, Goreau TF, Hartman WD (1971) Recent brachiopod-coralline sponge communities and their paleoecological significance. Science 173:623–625
Jackson RB, Canadell J, Ehleringer JR, Mooney HA, Sala OE, Schulze ED (1996) A global analysis of root distributions for terrestrial biomes. Oecologia 108:389–411
Jander R (1990) Arboreal search in ants: search on branches (Hymenoptera: Formicidae). J Insect Behav 3:515–527
Jander R, Voss C (1963) Die Bedeutung von Streifenmustern fuer das Formensehen der Roten Waldameise (*Formica rufa* L.). Z Tierpsychol 20:1–9
Janzen DH (1968) Host plants as islands in evolutionary and contemporary time. Am Nat 102:592–595
Janzen DH (1973) Host plants as islands. II. Competition in evolutionary and contemporary time. Am Nat 107:786–790
Jeník J (1978) Roots and root systems in tropical trees. In: Tomlinson PB, Zimmerman MH (eds) Tropical trees as living systems. Cambridge University Press, Cambridge, UK
Jenkins B, Kitching RL (1990) The ecology of water-filled treeholes in Australian rainforests: food web reassembly as a measure of community recovery. Aust J Ecol 15:199–205
Johansen JL, Bellwood DR, Fulton CJ (2008) Coral reef fishes exploit flow refuges in high–flow habitats. Mar Ecol Prog Ser 360:219–226
Johnson RE, Tuchman NC, Peterson CG (1997) Changes in the vertical microdistribution of diatoms within a developing periphyton mat. J N Am Benthol Soc 16:503–519
Jones GP (1992) Interactions between herbivorous fishes and macro–algae on a temperate rocky reef. J Exp Mar Biol Ecol 159:217–235
Jones CG, Lawton JH, Shachak M (1997) Positive and negative effects of organisms as physical ecosystem engineers. Ecology 78:1946–1957
Jørgensen BB, Revsbech NP, Cohen Y (1983) Photosynthesis and structure of benthic microbial mats: microelectrode and SEM studies of four cyanobacterial communities. Limnol Oceanogr 28:1075–1093
Jørgensen BB, Revsbech NP, Blackburn TH, Cohen Y (1979) Diurnal cycle of oxygen and sulfide microgradients and microbial photosynthesis in a cyanobacterial mat sediment. Appl Environ Microbiol 38:46–58
Jørgensen BB, Cohen Y, Revsbech NP (1986) Transition from anoxygenic to oxygenic photosynthesis in a *Microcoleus chthonoplastes* cyanobacterial mat. Appl Environ Microbiol 51:408–417
Judd WS, Sanders RW, Donoghue MWJ (1994) Angiosperm family pairs: preliminary phylogenetic analyses. Harv Pap Bot 5:1–51
Karlson RH (1999) Dynamics of coral communities. Kluwer Academic, Dordrecht
Kastendiek J (1982) Competitor-mediated coexistence: interactions among three species of benthic macroalgae. J Exp Mar Biol Ecol 62:201–210
Kesanakurti PR, Fazekas AJ, Burgess KS, Percy DM, Newmaster SG, Graham SW, Barrett SCH, Hajibabaei M, Husband BC (2011) Spatial patterns of plant diversity below-ground as revealed by DNA barcoding. Mol Ecol 20:1289–1302
Kierek-Pearson K, Karatan E (2005) Biofilm development in bacteria. Adv Appl Microbiol 57:79–111
Kikuchi T, Peres JM (1977) Consumer ecology of seagrass beds. In: McRoy CP, Helfferich C (eds) Seagrass ecosystems. Marcel Dekker, New York

Kim S, Soltis DE, Soltis PS, Zanis MJ, Suh Y (2004) Phylogenetic relationships among early-diverging eudicots based on four genes: were the eudicots ancestrally woody? Mol Phylogenet Evol 31:16–30

Kitching JA, Macan TT, Gilson HC (1934) Studies in sublittoral ecology. I. A submarine gully in Wembury Bay, South Devon. J Mar Biol Assoc UK 19:677–705

Klausen M, Aaes-Jorgensen A, Molin S, Tolker-Nielsen T (2003) Involvement of bacterial migration in the development of complex multicellular structures in *Pseudomonas aeruginosa* biofilms. Mol Microbiol 50:61–68

Klotz JH, Reid BL (1992) The use of spatial cues for structural guideline orientation in *Tapinoma sessile* and *Camponotus pennsylvanicus*. J Insect Behav 5:71–82

Korber DR, Lawrence JR, Lappin-Scott HM, Costerton JW (1995) Growth of microorganisms on surfaces. In: Lappin-Scott HM, Costerton JW (eds) Microbial biofilms. Cambridge University Press, Cambridge, UK

Korte VL, Blinn DW (1983) Diatom colonization on artificial substrata in pool and riffle zones studied by light and scanning electron microscopy. J Phycol 19:332–341

Kricher J (1997) A neotropical companion. Princeton University Press, Princeton

Kuenen JG, Jørgensen BB, Revsbech NP (1986) Oxygen microprofiles of trickling filter biofilms. Water Res 20:1589–1598

Kühl M, Polerecky L (2008) Functional and structural imaging of phototrophic microbial communities and symbioses. Aquat Microb Ecol 53:99–118

Kühl M, Fenchel T (2000) Bio-optical characteristics and the vertical distribution of photosynthetic pigments and photosynthesis in an artificial cyanobacterial mat. Microb Ecol 40:94–103

Kühl M, Glud RN, Ploug H, Ramsing NB (1996) Microenvironmental control of photosynthesis and photosynthesis-coupled respiration in an epilithic cyanobacterial biofilm. J Phycol 32:799–812

Kühl M, Fenchel T, Kazmierczak J (2003) Growth, structure and calcification potential of an artificial cyanobacterial mat. In: Krumbein WE, Paterson D, Zavarzin G (eds) Fossil and recent biofilms, a natural history of life on Earth. Kluwer Academic, Dordrecht

Külla T, Lõhmus K (1999) Influence of cultivation method on root grafting in Norway spruce (*Picea abies* (L.) Karst). Plant Soil 217:91–100

Lakoff G (1987) Women, fire, and dangerous things: what categories reveal about the mind. University of Chicago Press, Chicago

Langdale JA (2008) Evolution of developmental mechanisms in plants. Cur Opin Genet Develop 18:368–373

Lassen C, Ploug H, Kühl M, Jørgensen BB (1994) Oxygenic photosynthesis and light distribution in marine microbial mats. In: Stal LJ, Caumette P (eds) Microbial mats: structure, development, and environmental significance. Springer, Berlin

Lawrence JR, Korber DR, Hoyle BD, Costerton JW, Caldwell DE (1991) Optical sectioning of microbial biofilms. J Bacteriol 173:6558–6567

Lawrence JR, Korber DR, Wolfaardt GM, Caldwell DE (1995) Behavioral strategies of surface–colonizing bacteria. Adv Microb Ecol 14:1–75

Lawrence JR, Swerhone GDW, Kuhlicke U, Neu TR (2007) In situ evidence for microdomains in the polymer matrix of bacterial microcolonies. Can J Microbiol 53:450–458

Lawton, JH (1983) Plant architecture and the diversity of phytophagous insects. Ann Rev Entomol 28:23–29

Leigh EG Jr (1999) Tropical forest ecology: a view from Barro Colorado Island. Oxford University Press, Oxford, UK

Levin PS, Hay ME (1996) Responses of temperate reef fishes to alterations in algal structure and species composition. Mar Ecol Prog Ser 134:37–47

Lieberman M, Lieberman D, Peralta R (1989) Forests are not just swiss cheese: canopy stereogeometry of non-gaps in tropical forests. Ecology 70:550–552

Liira J, Zobel K, Mägi R, Molenberghs G (2002) Vertical structure of herbaceous canopies: the importance of plant growth–form and species–specific traits. Plant Ecol 163:123–134

Lindo Z, Whiteley JA (2011) Old trees contribute bio–available nitrogen through canopy bryophytes. Plant Soil 342:141–148

Losos JB (2009) Lizards in an evolutionary tree: ecology and adaptive radiation of anoles. University of California Press, Berkeley

Lovett GM, Lindberg SE (1992) Concentration and deposition of particles and vapors in a vertical profile through a forest canopy. Atmos Environ 26A:1469–1476

Lowe RL, Guckert JB, Belanger SE, Davidson DH, Johnson DW (1996) An evaluation of periphyton community structure and function on tile and cobble substrata in experimental stream mesocosms. Hydrobiologia 328:135–146

Lowe RL, Shavit U, Falter JL, Koseff JR, Monismith SG (2008) Modeling flow in coral communities with and without waves: a synthesis of porous media and canopy flow approaches. Limnol Oceanogr 53:2668–2680

Lyford WH (1975) Rhizography of non-woody roots of trees in the forest floor. In: Torrey JG, Clarkson DT (eds) The development and function of roots. Academic, New York

Lynch JP, Nielsen KL, Davis RD, Jablokow AG (1997) SimRoot: modelling and visualization of root systems. Plant Soil 188:139–151

MacArthur R, MacArthur JW (1961) On bird species diversity. Ecology 42:594–598

MacArthur R, Wilson EO (1967) The theory of island biogeography. Princeton University Press, Princeton

MacIntyre HL, Geider RJ, Miller DC (1996) Microphytobenthos: the ecological role of the "secret garden" of unvegetated, shallow–water marine habitats. I. Distribution, abundance and primary production. Estuaries 19:186–201

Madsen JS, Burmølle M, Hansen LH, Sørensen SJ (2012) The interconnection between biofilm formation and horizontal gene transfer. FEMS Immu Med Microbiol 65:183–195

Mahall BE, Callaway RM (1992) Root communication mechanisms and intracommunity distributions of two mojave desert shrubs. Ecology 73:2145–2151

Maida M, Coll JC, Sammarco PW (1994) Shedding new light on scleractinian coral recruitment. J Exp Mar Biol Ecol 180:189–202

Maraun M, Alphei J, Bonkowski M, Buryn R, Migge S, Peter M, Schaefer M, Scheu S (1999) Middens of the earthworm *Lumbricus terrestris* (Lumbricidae): microhabitats for micro- and mesofauna in forest soil. Pedobiologia 43:276–286

Marks JC, Power ME (2001) Nutrient induced changes in the species composition of epiphytes on *Cladophora glomerata* Kütz. Hydrobiologia 450:187–196

Massol-Deyá AA, Whallon J, Hickey RF, Tiedje JM (1995) Channel structures in aerobic biofilms of fixed–film reactors treating contaminated groundwater. Appl Environ Microbiol 61:769–777

McCormick PV (1996) Resource competition and species coexistence in freshwater benthic algal assemblages. In: Stevenson RJ, Bothwell ML, Lowe RL (eds) Algal ecology: freshwater benthic ecosystems. Academic Press, New York

McCormick PV, Stevenson RJ (1991) Grazer control of nutrient availability in the periphyton. Oecologia 86:287–291

McCune B (1993) Gradients in epiphyte biomass in three *Pseudotsuga–Tsuga* forests of different ages in western Oregon and Washington. Bryologist 96:405–411

McCune B, Antos JA (1981) Diversity relationships of forest layers in the Swan Valley, Montana. Bull Torrey Bot Club 108:354–361

McElhinny C, Gibbons P, Brack C, Bauhus J (2005) Forest and woodland stand structural complexity: its definition and measurement. Forest Ecol Manag 218:1–24

McGraw WS (1998) Posture and support use of Old World monkeys (Cercopithecidae):the influence of foraging strategies, activity patterns, and the spatial distribution of preferred food items. Am J Primatol 46:229–250

Meulemans JT, Roos PJ (1985) Structure and architecture of the periphytic community on dead reed stems in Lake Maarsseveen. Arch Hydrobiol 102:487–502

Meyers TP, Huebert BJ, Hicks BB (1989) HNO_3 deposition to a deciduous forest. Boundary-Layer Meteorol 49:395–410

Miller AR, Lowe RL, Rotenberry JT (1987) Succession of diatom communities on sand grains. J Ecol 75:693–709
Miller RJ, Reed DC, Brzezinski MA (2011) Partitioning of primary production among giant kelp (*Macrocystis pyrifera*), understory macroalgae, and phytoplankton on a temperate reef. Limnol Oceanogr 56:119–132
Mitri S, Xavier JB, Foster KR (2011) Social evolution in multispecies biofilms. Proc Nat Acad Sci 108:10839–10846
Moe B, Botnen A (1997) A quantitative study of the epiphytic vegetation on pollarded trunks of *Fraxinus excelsior* at Havrå, Osterøy, western Norway. Plant Ecol 129:157–177
Moffett MW (1994) The high frontier: exploring the tropical rainforest canopy. Harvard University Press, Cambridge, MA
Moffett MW (1999) Life on vegetation: a framework for canopy biology. In: Hallé F (ed) Biologie d'une canopée de forêt équatoriale — IV. Pro–Natura and Opération Canopée, Montpellier, France
Moffett MW (2000) What's "up?" A critical look at the basic terms of canopy biology. Biotropica 32:569–596
Moffett MW (2001) The nature and limits of canopy biology. Selbyana 22:155–179
Moffett MW (2002) The highs and lows of tropical forest canopies. J Biogeogr 29:1264–1265
Moffett MW (2010) Adventures among ants. University of California Press, Berkeley
Moffett MW, Lowman MD (1995) Canopy access techniques. In: Lowman MD, Nadkarni NM (eds) Forest canopies. Academic, San Diego
Molin S, Tolker-Nielsen T (2003) Gene transfer occurs with enhanced efficiency in biofilms and induces enhanced stabilisation of the biofilm structure. Curr Opin Biotechnol 14:255–261
Molin S, Nielsen AT, Christensen BB, Andersen JB, Licht TR, Tolker-Nielsen T, Sternberg C, Hansen MC, Ramos C, Givskov M (2000) Molecular ecology of biofilms. In: Bryers JD (ed) Biofilms II: process analysis and applications. Wiley-Liss, New York
Mommer L, van Ruijven J, de Caluwe H, Smit-Tiekstra AE, Wagemaker CAM, Ouborg NJ, Bogemann GM, van der Weerden GM, Berendse F, de Kroon H (2010) Unveiling below–ground species abundance in a biodiversity experiment: a test of vertical niche differentiation among grassland species. J Ecol 98:1117–1127
Mommer L, Dumbrell AJ, Wagemaker CAM, Ouborg NJ (2011) Belowground DNA–based techniques: untangling the network of plant root interactions. Plant Soil 348:115–121
Monteith JL (1975/1976) Vegetation and the atmosphere, vol 1 and 2. Academic, London
Montgomery RA, Chazdon RL (2001) Forest structure, canopy architecture, and light transmittance in tropical wet forests. Ecology 82:2707–2718
Moore JC, McCann K, Setälä H, de Ruiter PC (2003) Top-down is bottom-up: does predation in the rhizosphere regulate aboveground dynamics? Ecology 84:846–857
Morin PJ (2000) Biodiversity's ups and downs. Nature 406:463–464
Morris CE, Monier JM, Jacques MA (1997) Methods for observing microbial biofilms directly on leaf surfaces and recovering them for isolation of culturable microorganisms. Appl Environ Microbiol 63:1570–1576
Mulholland PJ (1996) Role in nutrient cycling in streams. In: Stevenson RJ, Bothwell ML, Lowe RL (eds) Algal ecology: freshwater benthic ecosystems. Academic, New York
Mundy CN, Babcock RC (1998) Role of light intensity and spectral quality in coral settlement: implications for depth-dependent settlement? J Exp Mar Biol Ecol 223:235–255
Nadell CD, Xavier JB, Levin SA, Foster KR (2008) The evolution of quorum sensing in bacterial biofilms. PLoS Biol 6:e14
Neushul M (1971) Submarine illumination in *Macrocystis* beds. Beihefte zur Nova Hedwigia 32:241–254
Neushul M (1972) Functional interpretation of benthic marine algal morphology. In: Abbott IA, Kurogi M (eds) Contributions to the systematics of benthic marine algae of the North Pacific. Japanese Society of Phycology, Kobe
Niklas KJ (1997) Evolutionary biology of plants. University of Chicago Press, Chicago
Niklas KJ (2000) The mechanical stability of vertical stems. In: Kurmann MH, Hemsley AR (eds) The evolution of plant architecture. Royal Botanical Gardens, Kew

Niklas KJ, Kerchner V (1984) Mechanical and photosynthetic constraints on the evolution of plant shape. Paleobiology 10:79–101

O' Connell T, Bolger T (1997) Fungal fruiting bodies and the structure of fungus-micro-arthropod assemblages. Proc R Irish Acad Sect B 97:249–262

Paerl HW, Pinckney JL, Steppe TF (2000) Cyanobacterial–bacterial mat consortia: examining the functional unit of microbial survival and growth in extreme environments. Environ Microbiol 2:11–26

Pagès L (2000) How to include organ interactions in models of the root system architecture? The concept of endogenous environment. Ann For Sci 57:535–541

Pagès L, Doussan C, Vercambre G (2000) An introduction on below–ground environment and resource acquisition, with special reference on trees: simulation models should include plant structure and function. Ann For Sci 57:513–520

Paine RT, Suchanek TH (1983) Convergence of ecological processes between independently evolved competitive dominants: a tunicate–mussel comparison. Evolution 37:821–831

Parker GG (1983) Throughfall and stemflow in the forest nutrient cycle. Adv Ecol Res 13:57–133

Parker GG, Brown MJ (2000) Forest canopy stratification — is it useful? Am Nat 155:473–484

Passy SI, Larson CA (2011) Succession in stream biofilms is an environmentally driven gradient of stress tolerance. Microb Ecol 62:414–424

Pearce D, Bazin MJ, Lynch JM (1995) The rhizosphere as a biofilm. In: Lappin-Scott HM, Costerton JW (eds) Microbial biofilms. Cambridge University Press, Cambridge, UK

Pearse JS, Hines AH (1979) Expansion of a central California kelp forest following the mass mortality of sea urchins. Mar Biol 51:83–91

Perry DR (1978) Factors influencing arboreal epiphytic phytosociology in Central America. Biotropica 10:235–237

Peterson CG (1996) Response of benthic algal communities to natural physical disturbance. In: Stevenson RJ, Bothwell ML, Lowe RL (eds) Algal ecology: freshwater benthic ecosystems. Academic, San Diego

Peterson CG, Grimm NB (1992) Temporal variation in enrichment effects during periphyton succession in a nitrogen–limited desert stream ecosystem. J N Am Benthol Soc 11:20–36

Pike LH (1978) The importance of epiphytic lichens in mineral cycling. Bryologist 81:247–257

Pogoreutz C, Kneer D, Litaay M, Asmus H, Ahnelt H (2012) The influence of canopy structure and tidal level on fish assemblages in tropical Southeast Asian seagrass meadows. Estuar Coast Shelf Sci 107:58–68

Poorter L, Hawthorne W, Bongers F, Sheil D (2008) Maximum size distributions in tropical forest communities: relationships with rainfall and disturbance. J Ecol 96:495–504

Poorter L, Niklas KJ, Reich PB, Oleksyn J, Poot P, Mommer L (2012) Biomass allocation to leaves, stems and roots: meta-analyses of interspecific variation and environmental control. New Phytol 193:30–50

Porter JW (1974) Zooplankton feeding by the Caribbean reef-building coral *Montastrea cavernosa*. Proc Second Int Coral Reef Symp 1:111–125

Porter JW (1976) Autotrophy, heterotrophy, and resource partitioning in Caribbean reef-building corals. Am Nat 110:731–742

Postma JA, Lynch JP (2012) Complementarity in root architecture for nutrient uptake in ancient maize/bean and maize/bean/squash polycultures. Ann Bot 110:521–534

Povinelli DJ, Cant JGH (1995) Arboreal clambering and the evolution of self–conception. Quart Rev Biol 70:393–421

Pringle CM (1985) Effects of chironomid (Insecta: Diptera) tube-dwelling activities on stream diatom communities. J Phycol 21:185–194

Putz FE (1995) Vines in treetops: consequences of mechanical dependence. In: Lowman MD, Nadkarni NM (eds) Forest canopies. Academic, San Diego

Ramage G, Mowat E, Jones B, Williams C, Lopez-Ribot J (2009) Our current understanding of fungal biofilms. Crit Rev Microbiol 35:340–355

Ramsing NB, Kühl M, Jørgensen BB (1993) Distribution of sulfate–reducing bacteria, O_2, and H_2S in photosynthetic biofilms determined by oligonucleotide probes and microelectrodes. Appl Environ Microbiol 59:3840–3849

Ramsing NB, Ferris MJ, Ward DM (2000) Highly ordered vertical structure of *Synechococcus* populations within the one-millimeter-thick photic zone of a hot spring cyanobacterial mat. Appl Environ Microbiol 66:1038–1049
Randlkofer B, Obermaier E, Hilker M, Meiners T (2010) Vegetation complexity: the influence of plant species diversity and plant structures on plant chemical complexity and arthropods. Basic Appl Ecol 11:383–395
Rasmussen L (1975) The bryophytic epiphyte vegetation in the forest, Slotved Skov, Northern Jutland. Lindbergia 3:15–38
Reed DC, Foster MS (1984) The effects of canopy shadings on algal recruitment and growth in a giant kelp forest. Ecology 65:937–948
Reichenbach H (1951) The rise of scientific philosophy. University of California Press, Berkeley
Reiners WA, Olson RK (1984) Effects of canopy components on throughfall chemistry: an experimental analysis. Oecologia 63:320–330
Reynolds TB, Fink GR (2001) Baker's yeast, a model for fungal biofilm formation. Science 291:878–881
Richards AJ (1986) Plant breeding systems. George Allen and Unwin, London
Richards PW (1996) The tropical rain forest, an ecological study, 2nd edn. Cambridge University Press, Cambridge, UK
Richardson BA (1999) The bromeliad microcosm and the assessment of faunal diversity in a neotropical forest. Biotropica 31:321–336
Robinson CT, Minshall GW (1986) Effects of disturbance frequency on stream benthic community structure in relation to canopy cover and season. J N Am Benthol Soc 5:237–248
Roemer SC, Hoagland KD, Rosowski JR (1984) Development of a freshwater periphyton community as influenced by diatom mucilages. Can J Bot 62:1799–1813
Roeselers G, van Loosdrecht MCM, Muyzer G (2007) Heterotrophic pioneers facilitate phototrophic biofilm development. Microbial Ecol 54:578–585
Roos PJ (1979) Architecture and development of periphyton on reed-stems in Lake Maarsseveen. Hydrobiol Bull 13:117
Roxburgh SH, Watkins AJ, Wilson JB (1993) Lawns have vertical stratification. J Veg Sci 4:699–704
Ruinen J (1961) The phyllosphere. I. An ecologically neglected milieu. Plant Soil 15:81–109
Ruinen J (1975) Nitrogen fixation in the phyllosphere. In: Stewart WDP (ed) Nitrogen fixation by free–living micro–organisms. Cambridge University Press, Cambridge, UK
Rupp CJ, Fux CA, Stoodley P (2005) Viscoelasticity of *Staphylococcus aureus* biofilms in response to fluid shear allows resistance to detachment and facilitates rolling migration. Appl Environ Microbiol 71:2175–2178
Russell BC (1977) Population and standing crop estimates for rocky reef fishes of North–Eastern New Zealand. N Z J Mar Freshw Res 11:23–36
Russell BD (2007) Effects of canopy-mediated abrasion and water flow on the early colonisation of turf-forming algae. Mar Freshw Res 58:657–665
Russell G, Marshall B, Jarvis PG (eds) (1989) Plant canopies: their growth, form and function. Cambridge University Press, Cambridge, UK
Sagan D, Margulis L (1988) Garden of microbial delights. Harcourt Brace Jovanovich, Boston
Santelices B, Ojeda FP (1984) Effects of canopy removal on the understory algal community structure of coastal forests of *Macrocystis pyrifera* from southern South America. Mar Ecol Prog Ser 14:165–173
Schaefer DA, Reiners WA (1989) Throughfall chemistry and canopy processing mechanisms. In: Lindberg SE, Page AL, Norton SA (eds) Acidic precipitation. Vol. 3: sources, deposition, and canopy interactions. Springer, New York
Schaudinn C, Stoodley P, Kainovic A, O'Keeffe T, Costerton B, Robinson D, Baum M, Ehrlich G, Webster P (2007) Bacterial biofilms, other structures seen as mainstream concepts. Microbe 2:231–237
Schenk HJ (2006) Root competition: beyond resource depletion. J Ecol 94:725–739
Schenk HJ, Jackson RB (2002) The global biogeography of roots. Ecol Monog 72:311–328

Schmitt RJ, Holbrook SJ (1990) Contrasting effects of giant kelp on dynamics of surfperch populations. Oecologia 84:419–429
Schowalter TD (2011) Insect ecology: an ecosystem approach, 3rd edn. Academic, New York
Scott JT, Back JA, Taylor JM, King RS (2008) Does nutrient enrichment decouple algal–bacterial production in periphyton? J N Am Benthol Soc 27:332–344
Seifert R (1975) Clumps of *Heliconia* inflorescences as ecological islands. Ecology 56:1416–1422
Sekar R, Venugopalan VP, Nandakumar K, Nair KVK, Rao VNR (2004) Early stages of biofilm succession in a lentic freshwater environment. Hydrobiologia 512:97–108
Shaw DC (1996) Northwest forest canopies: preface. NW Sci 70:i–ii
Shugart HH, Saatchi S, Hall FG (2010) Importance of structure and its measurement in quantifying function of forest ecosystems. J Geophys Res 115:G00E13
Sillett SC, Gradstein SR, Griffin D (1995) Bryophyte diversity of *Ficus* tree crowns from cloud forest and pasture in Costa Rica. Bryologist 98:251–260
Sinoquet H, Pincebourde S, Adam B, Donès N, Phattaralerphong J, Combes D, Ploquin S, Sangsing K, Kasemsap P, Thanisawanyangkura S, Groussier-Bout G, Casas J (2009) 3-D maps of tree canopy geometries at leaf scale. Ecology 90:283
Sládecková A (1962) Limnological investigation methods for the periphyton ("aufwuchs") community. Bot Rev 28:286–350
Smit AL, Bengough AG, Engels C, van Noordwijk M, Pellerin S, van de Geijn SC (eds) (2000) Root methods: a handbook. Springer, Berlin
Sobczak WV (1996) Epilithic bacterial responses to variations in algal biomass and labile dissolved organic carbon during biofilm colonization. J N Am Benthol Soc 15:143–154
Sousa WP (1979) Experimental investigations of disturbance and ecological succession in a rocky intertidal algal community. Ecol Monogr 49:227–254
Spies TA (1998) Forest structure: a key to the ecosystem. N W Sci 72:34–39
Staal M, Borisov S, Rickelt LF, Klimant I, Kühl M (2011) Ultrabright planar optodes for luminescence life-time based microscopic imaging of O_2 dynamics in biofilms. J Microbiol Methods 85:67–74
Stal LJ (2000) Cyanobacterial mats and stromatolites. In: Whitton BA, Potts M (eds) Ecology of cyanobacteria. Kluwer Academic, Dordrecht
Steele MA (1999) Effects of shelter and predators on reef fishes. J Exp Mar Biol Ecol 233:65–79
Steneck RS (1997) Crustose corallines, other algal functional groups, herbivores and sediments: complex interactions along reef productivity gradients. In: Proceedings of 8th international coral reef symposium, Panama City 1:695–700
Steneck RS, Dethier MN (1994) A functional group approach to the structure of algal–dominated communities. Oikos 69:476–498
Sterck FJ, Schieving F (2007) 3-D growth patterns of trees: effects of carbon economy, meristem activity, and selection. Ecol Monog 77:405–420
Sterck FJ, Bongers F, During HJ, Martínez-Ramos M, de Kroon H (2003) Module responses in a tropical forest tree analyzed with a matrix model. Ecology 84:2751–2761
Stevenson RJ (1983) Effects of current and conditions simulating autogenically changing microhabitats on benthic diatom immigration. Ecology 64:1514–1524
Stevenson RJ (1996) An introduction to algal ecology in freshwater benthic habitats. In: Stevenson RJ, Bothwell ML, Lowe RL (eds) Algal ecology: freshwater benthic ecosystems. Academic, New York
Stewart PS, Franklin MJ (2008) Physiological heterogeneity in biofilms. Nat Rev Microbiol 6:199–2010
Stimson J (1985) The effect of shading by the table coral *Acropora hyacinthus* on understory corals. Ecology 66:40–53
Stock MS, Ward AK (1991) Blue–green algal mats in a small stream. J Phycol 27:692–698
Stolzenbach KD (1989) Particle transport and attachment. In: Characklis WG, Wilderer PA (eds) Structure and function of biofilms. Wiley, New York
Stoodley P, Boyle JD, DeBeer D, Lappin-Scott HM (1999a) Evolving perspectives of biofilm structure. Biofouling 14:75–90

Stoodley P, Lewandowski Z, Boyle JD, Lappin-Scott HM (1999b) The formation of migratory ripples in a mixed species bacterial biofilm growing in turbulent flow. Environ Microb 1:447–455
Stork NE, Adis J, Didham RK (1997) Canopy arthropods. Chapman and Hall, London
Svenning J-C (2000) Small canopy gaps influence plant distributions in the rain forest understory. Biotropica 32:252–261
Teske A, Stahl DA (2001) Microbial mats and biofilms: evolution, structure and function of fixed microbial communities. In: Staley JT, Reysenbach A-L (eds) Biodiversity of microbial life: foundation of earth's biosphere. Wiley, New York
Tilman D (1982) Resource competition and community structure. Princeton University Press, Princeton
Tomlinson PB (1983) Tree architecture. Am Sci 71:141–149
Tsegaye T, Mullins CE, Diggle AJ (1995a) An experimental procedure for obtaining input parameters for the rootmap root simulation program for peas (*Pisum sativum* L). Plant Soil 172:1–16
Tsegaye T, Mullins CE, Diggle AJ (1995b) Modeling pea (*Pisum sativum*) root growth in drying soil: a comparison between observations and model predictions. New Phytol 131:179–189
Tuchman NC (1996) The role of heterotrophy in algae. In: Stevenson RJ, Bothwell ML, Lowe RL (eds) Algal ecology: freshwater benthic ecosystems. Academic, New York
Tuchman NC, Stevenson RJ (1991) Effects of selective grazing by snails on benthic algal succession. J N Am Benthol Soc 10:430–443
Turner JS (2000) The extended organism: the physiology of animal-built structures. Harvard University Press, Cambridge, MA
Vadas RL, Steneck RS (1988) Zonation of deep water benthic algae in the Gulf of Maine. J Phycol 24:338–346
Van der Putten WH, Vet LEM, Harvey JA, Wackers FL (2001) Linking above- and belowground multitrophic interactions of plants, herbivores, pathogens, and their antagonists. Trends Ecol Evol 16:547–554
Van Gemerden H (1993) Microbial mats, a joint venture. Mar Geol 113:3–25
Venugopalan VP, Kuehn M, Hausner M, Springael D, Wilderer PA, Wuertz S (2005) Architecture of a nascent *Sphingomonas* sp. biofilm under varied hydrodynamic conditions. Appl Environ Microbiol 71:2677–2686
Vogel S (1996) Blowing in the wind: storm-resisting features of the design of trees. J Arboriculture 22:92–98
Vytopil E, Willis B (2001) Epifaunal community structure in Acropora spp. (Scleractinia) on the Great Barrier Reef: implications of coral morphology and habitat complexity. Coral Reefs 20:281–288
Wahl M (1997) Living attached: aufwuchs, fouling, epibiosis. In: Nagabhushanam R, Thompson M-F (eds) Fouling organisms of the Indian Ocean. A.A. Balkema, Rotterdam
Walla TR, Engen S, DeVries PJ, Lande R (2004) Modeling vertical beta–diversity in tropical butterfly communities. Oikos 107:610–618
Walter H (1973) Vegetation on earth. Springer, New York
Ward DM, Bateson MM, Ferris MJ, Kühl M, Wieland A, Koeppel A, Cohan FM (2006) Cyanobacterial ecotypes in the microbial mat community of Mushroom Spring (Yellowstone National Park, Wyoming) as species–like units linking microbial community composition, structure and function. Philos Trans R Soc B 361:1997–2008
Wardle DA, Bardgett RD, Klironomos JN, Setälä H, van der Putten WH, Wall DH (2004) Ecological linkages between aboveground and belowground biota. Science 304:1629–1633
Watnick P, Kolter R (2000) Biofilm, city of microbes. J Bacteriol 182:2675–2679
Weaver JE, Clements FE (1929) Plant ecology. McGraw-Hill, New York
Wernberg T, Kendrick GA, Toohey BD (2005) Modification of the physical environment by an *Ecklonia radiata* (Laminariales) canopy and implications for associated foliose algae. Aquat Ecol 39:419–430
Weitzel RG (1979) Periphyton measurements and applications. In: Weitzel RG (ed) Methods and measurements of periphyton communities: a review. American Society for Testing and Materials, Philadelphia

Wetzel RG (1983) Opening remarks. In: Wetzel RG (ed) Periphyton of freshwater ecosystems. Dr. W. Junk, The Hague

Wetzel RG (1993) Microcommunities and microgradients: linking nutrient regeneration, microbial mutualism, and high sustained aquatic primary production. Neth J Aquat Ecol 27:3–9

Whipps JM, Hand P, Pink D, Bendin GD (2008) Phyllosphere microbiology with special reference to diversity and plant genotype. J Appl Microbiol 105:1744–1755

Williams SL, Carpenter RC (1990) Photosynthesis/photon flux density relationships among components of coral reef algal turfs. J Phycol 26:36–40

Williams SL, Heck KL Jr (2001) Seagrass Community ecology. In: Bertness M, Gaines S, Hay M (eds) Marine community ecology. Sinauer Press, New York

Williams GA, Seed R (1992) Interactions between macrofaunal epiphytes and their host algae. In: John DM, Hawkins SJ, Price JH (eds) Plant–animal interactions in the marine benthos. Clarendon, Oxford, UK

Wilson JB, Steel JB, Steel SLK (2007a) Do plants ever compete for space? Folia Geobotanica 42:431–436

Wilson SK, Graham NAJ, Polunin NVC (2007b) Appraisal of visual assessments of habitat complexity and benthic composition on coral reefs. Mar Biol 151:1069–1076

Wiman BLB, Agren GI, Lannefors HO (1985) Aerosol concentration profiles within a mature coniferous forest–model versus field results. Atmos Environ 19:363–367

Wimpenny JWT, Colasanti R (1997) A unifying hypothesis for the structure of microbial biofilms based on cellular automaton models. FEMS Microbiol Ecol 22:1–16

Wimpenny JWT, Kinniment SL (1995) Biochemical reactions and the establishment of gradients within biofilms. In: Lappin-Scott HM, Costerton JW (eds) Microbial biofilms. Cambridge University Press, Cambridge, UK

Wing SR, Leichter JJ, Denny MW (1993) A dynamic model for wave–induced light fluctuations in a kelp forest. Limnol Oceanogr 38:396–407

Wohlgemuth S, Ronacher B, Wehner R (2001) Ant odometry in the third dimension. Nature 411:795–798

Yeh P-Z, Gibor A (1970) Growth patterns and motility of *Spirogyra* sp. and *Closterium acerosum*. J Phycol 6:44–48

Young JW (2009) Substrate determines asymmetrical gait dynamics in marmosets (*Callithrix jacchus*) and squirrel monkeys (*Saimiri boliviensis*). Am J Phys Anthropol 138:403–420

Young TP, Isbell LA (1991) Sex differences in giraffe feeding ecology: energetic and social constraints. Ethology 87:79–89

Zotz G (2006) Substrate use of three epiphytic bromeliads. Ecography 20:264–270

Chapter 4
Forest Canopies as Earth's Support Systems: Priorities for Research and Conservation

Reinmar Seidler, Kamaljit S. Bawa, Margaret Lowman, and Nalini M. Nadkarni

Keywords Forest conservation • Stratification • Ecosystem services • Disturbance • Capacity building

Background

Forest canopies, particularly those in the tropics, are the main repositories of earth's biodiversity, providing ecosystem services used by billions of people throughout the world. Services include climate regulation, thousands of renewable biological products, and entities of spiritual and aesthetic value (Ozanne et al. 2003; Devy and Ganesh 2003; Lowman 2009). Despite decades of effort to reduce deforestation, degradation of tropical forests—particularly the loss of intact canopy surface—continues unabated. Although accurate estimates of both forest area loss and forest carbon stock volumes are difficult, Harris et al. (2012) compared results of two studies which, despite divergent methodological approaches and independent data sets, agreed that gross carbon emissions from pantropical deforestation over the period 2000–2005 amounted to 3.0 ± 1.1 Gt CO_2 year^{-1} (0.8 ± 0.3 Pg C year^{-1}). Note

R. Seidler (✉)
University of Massachusetts, Boston, MA 02125, USA
e-mail: Reinmar.Seidler@gmail.com

K.S. Bawa
Ashoka Trust for Research in Ecology and the Environment (ATREE), Bangalore, KA, India
e-mail: kamal.bawa@umb.edu

M. Lowman
North Carolina Museum of Natural Sciences, North Carolina State University,
121 W. Jones Street, Raleigh, NC 27603, USA
e-mail: canopymeg@gmail.com

N.M. Nadkarni
Center for Science and Mathematics Education, University of Utah,
Salt Lake City, UT 84103, USA
e-mail: Nalini.nadkarni@utah.edu

M. Lowman et al. (eds.), *Treetops at Risk: Challenges of Global Canopy Ecology and Conservation*, DOI 10.1007/978-1-4614-7161-5_4,

that this figure includes only estimates of emissions from areas of *complete* deforestation and excludes estimates of tropical forest degradation short of forest conversion. The most recent FAO figures for net *global* forest loss (UN-FAO 2010) are a mean 5.2 million ha $year^{-1}$ over 2000–2010. That total incorporates an estimated increase in the annual rate of gross deforestation from 3.7 million ha $year^{-1}$ in the 1990s to 4.9 million ha $year^{-1}$ in the 2000s.

Almost all of this loss took place in the tropical forests of South America and Africa and was only partly offset by large tree-planting programs in a few countries. Even in a country such as India, where the government has made strong efforts to maintain and enhance forest cover, actual rates of natural forest loss remain significantly high (Davidar et al. 2010; Ravindranath et al. 2012). Existing global forest threats associated with population growth, trade, governance, and land use policies are augmented by conversion of forest to plantations of soya and palm oil (Scharlemann and Laurance 2008; Laurance et al. 2010; Laurance, this volume). Climate change will further intensify these pressures throughout the world.

Tropical forest canopy researchers are now addressing these urgent challenges. In the late 1980s, canopy science emerged as a distinct subdiscipline of biology (Lowman 2009). At the time of the first international canopy conference in 1994, canopy science was at an early stage of development, characterized by isolated researchers, lack of harmonized methods, and the perception that questions addressed were predominantly "observational" (Nadkarni 2001; Lowman et al. 1995). In the last decade, however, canopy studies have coalesced through formal and informal networks (Nadkarni and Parker 1994; Mitchell et al. 2002).

With the emerging threats to whole forests, canopy research is at a crossroads. On the one hand, there are pressing scientific challenges to increase our basic understanding of structure and compositions of forest canopies and their functional roles in ecosystem processes. On the other hand, an urgent and growing need exists for conserving, restoring, and sustainably managing whole-forest systems (Nadkarni et al. 2011). Although these two challenges need not conflict, economic and human resources are limited. Therefore, canopy scientists, in partnership with policy-makers, educators, and economists, must prioritize the issues they study in the next decade.

One question is to what extent canopy research can or should remain distinct from whole-forest biology and conservation. The initial launch of canopy research as a distinct discipline was necessary to highlight the unique and important roles of forest canopies. As canopy science has matured, however, there has been a growing need to integrate canopy studies into our understanding of forest ecosystems as wholes—thereby embracing the social, political, and economic dimensions of land cover change.

Here, we address three sets of questions. First, we review the recent research that documents how physical, physiological, or ecological parameters might either promote or constrain the separation of canopy science from the larger field of forest science. For instance, does canopy science naturally constitute a separate field of study because the canopy is to some significant extent a distinct ecosystem? If so, what are the priorities for canopy research as distinct from priorities for tropical forest ecosystems as a whole?

Second, we examine what is known about how canopies are being altered by human use of tropical forests and consider priorities for research in human interactions with forest canopies.

Third, to integrate forest canopy research programs more effectively with work on whole-forest ecosystems, we ask how canopy science can (1) directly address the emerging, urgent need to understand patterns of human impacts on and sustainability of tropical forest ecosystems; (2) help society interpret, mitigate, and adapt to global environmental changes, particularly land use and climate change; and (3) contribute to building human capacity to meet contemporary challenges in the tropics. It is evident from this volume that canopy researchers are now shifting their priorities toward forest canopy conservation by embracing whole-forest approaches with reference to ecosystem services, forest health, climate change, sustainability science, economics, education, and the social sciences. Many of the issues raised here are in line with earlier discussions of research priorities for tropical ecosystems (Bawa et al. 2004). Thus, this work is an extension of previous recent exercises on priorities in tropical biology and conservation.

1 Forest Canopies: Real or Imagined?

This question evokes long-standing debates in ecology about what defines a community, where to locate community boundaries, and how to balance the strength of local interactions versus the influence of broader regional processes. For nearly a century, there has been a trend in ecology *away* from the focus on dynamics within local communities and *toward* incorporating the complexities of wider forces acting on assemblages of organisms. The merging of canopy ecology into forest system science may thus be an element in what has (approvingly) been termed the "disintegration of the (local) ecological community" (Ricklefs 2008). The growth and increasing profile of conservation biology as a distinct discipline, with its need to acknowledge and account for human influences, has contributed to this trend. Nevertheless, many ecologists have hesitated to abandon the local concept of community, partly because the tractability of alternative models remains uncertain (Cuddington and Beisner 2005).

1.1 Trees and Forest Structure

A. R. Wallace, in his 1878 collection of essays on tropical nature, described tree stands in the virgin forests of the "equatorial zone" as being so clearly triple-stratified that the upper branches of the lower strata "do not touch the lowermost branches of those above them." This may have been literally true of the ancient forests of preindustrial Sarawak, but stratification is much less obvious today. The extent to which it can be discerned and quantified in different tropical forests has been a recurrent theme for forest ecologists (e.g., Smith 1973; Whitmore 1984; Popma et al. 1988; Ashton and Hall 1992; Richards 1996; Baker and Wilson 2000; Parker and Brown 2000). In the context of modern tropical forests—none is untouched by disturbances of various kinds—there is consensus on some objectively demonstrable structural stratification, mostly into either two or three strata, but this varies considerably among forests and is rarely consistent over large areas. Peter Ashton (in press) suggests four possible nonexclusive mechanisms for tree stratification: aggregation of maximum tree species heights, branch architecture, crown architecture, and leaf layering. Each of these emphasizes structural and growth characteristics of the trees themselves. Other hypotheses for nonrandom aggregation of forest trees into strata (e.g., Smith 1973) include possible selective advantages to trees in vertically stratified communities, such as enhanced availability of light, CO_2, and pollinators.

1.2 Forest Canopy as a Physical System

Clark et al. (2008) showed there was ubiquitous stratification among trees in the wet tropical rain forest of La Selva (Costa Rica) but that it varied among sites. The horizontal spatial extent of each stratification pattern could not be determined from point samples, and there were indications that individual stands tended to fall into individual stratification patterns, probably for reasons of common stand history and ecological legacies. This old-growth forest was probably affected, as are most forests, by the history of neighboring forest patches. In some cases, neighboring patches were disturbed enough to allow significant amounts of lateral light—and doubtless other influences—to enter the vertical transect sites, perhaps obscuring or altering stratification patterns. The apparent implication of such work is that forest strata are based on physical parameters such as light availability or understory wind patterns.

However, not all significant structural characteristics of the trees themselves respond to vertical variation in such parameters. Cavaleri et al. (2010) showed that leaf dry mass per unit area (LMA—a measure of the thickness and relative surface area of leaves) varies *less* with the light microenvironment than simply with height above the ground. In addition, LMA responds *linearly* to height. This implies that water relations within the tree must be having a more direct influence on leaf development and physiology than do other variables; but leaf water potential changes in a smooth gradient from ground to outer canopy surface and therefore cannot contribute to a stepped stratification pattern.

Direct micrometeorological tests—for instance, research in an old-growth neotropical rain forest (Szarzynski and Anhuf 2001) and in an old-growth deciduous temperate forest (Parker and Brown 2000)—show that the upper two-thirds of the within-canopy spatial volume are closely coupled to the diurnal cycle of atmospheric changes above, due primarily to wind gusts and light entering the canopy. The lowest third, in contrast, remains relatively isolated from the daily cycle, staying cooler during the day and warmer at night. In contrast, a vertical transect in a steeply sloping neotropical montane forest showed no such decoupling of the understory stratum (Motzer 2005). In such a sloping forest, light and wind tend to reach ground level at a different angle and the treetops remain at a smaller perpendicular distance from the ground surface. In all cases, a complete vertical section of forest shows an extremely complex, convoluted, and variable profile of microclimatic zonation (e.g., Parker and Brown 2000).

1.3 Other Organismal Groups

If physical parameters serve to structure tree assemblages into quasi-regular patterns, they should also influence other organisms. Aside from trees, are forest communities vertically stratified? There are many accounts of some kind of vertical differentiation for taxa as diverse as arthropods (Basset et al. 2003; Ellwood et al. 2009), lepidoptera (Schulze et al. 2001), butterflies (DeVries et al. 1997), beetles (Grimbacher and Stork 2007), cicadas (Sanbord et al. 2011), ants (Longino and Nadkarni 1990), Collembola (Rodgers and Kitching 1998), spiders in cocoa agroforestry plots (Stenchly et al. 2012), bees and wasps (Sobek et al. 2009), dung beetles (Tregidgo et al. 2010), bats (Bernard 2001; Hodgkison 2004), small mammals (Grelle 2003), birds (Pearson 1971; Walther 2002), nematodes (Powers et al. 2009), lichens (Komposch et al. 2000), and fungi (Lodge and Cantrell 1995), though no support was found for stratified habitat division among crickets (Jain and Balakrishnan 2012) or geometroid moths (Intachat and Holloway 2000). However, these taxa clearly respond to a variety of resource parameters, including temperature, humidity, insolation, food resources, structural characteristics of foliage, branch architecture, and perch sites. Some of these parameters covary with each other, with height above the ground or with depth beneath the canopy surface; others vary independently.

1.4 Forest Ecological Interactions: Are They Vertically Segregated?

Of course, ecological communities should be defined not only by spatial structuring of habitats and resources but also by some segregation of local ecological *interactions*. Apparent stratification has been observed in a variety of ecological interactions, such as herbivory, predation, and plant–pollinator interactions (Bawa 1990; Kang and Bawa 2003). Herbivory, however, appears to be significantly less host specific in tropical forests than once thought (Novotny et al. 2002; Novotny et al. 2007). Similarly, butterflies, bark beetles, and treehoppers may not be more host specific in tropical than in temperate forests (Mawdsley and Stork 1997; Novotny et al. 2006). It would be interesting to know how this revised picture of somewhat greater functional homogeneity in tropical forests than previously thought might be reflected in the spatial distribution of herbivore habitats.

1.5 How Meaningful Is the Concept of Stratification?

In summary, evidence has accreted in favor of some degree of vertical habitat segregation within many taxa and functional groups, dividing the forest spatial volume into habitat zones along the vertical as well as the horizontal dimension. However, stratification appears to ensue from multiple causes, to be present to different degrees in different taxa within different forest types, and to be an overwhelmingly local phenomenon—even varying among individual stands within a single forest. Thus it remains unclear whether such stratification should be considered a characteristic of individual trees, of species, or of higher taxa, or whether it is an emergent structural property of stands.

In fact, it appears that the terminology of stratification and layering is being invoked with several different meanings in this literature (cf. Parker and Brown 2000). Perhaps most commonly it means simply that in respect of many measurable parameters, there are gradients from the forest floor to the canopy surface. When the vertical space is sampled along a transect, different heights produce unlike samples. This is not the same as stating the vertical space is divided into relatively homogeneous units or layers delimited by detectable border zones—which is the only ecologically meaningful definition of a habitat patch, unit, or system (Cadenasso et al. 2003; Strayer et al. 2003).

Parker and Brown (2000) pointed out a basic weakness of some forest stratification work: height above the ground is really only a proxy for a complex gradient extending from the outer atmosphere–vegetation interface into the most sheltered forest interior. As a proxy, height may or may not be ecologically meaningful, i.e., there is little reason to assume that ecological strata will consistently run parallel to the ground. Indeed, many of the microclimatic conditions characteristic of the "upper" canopy layer—relatively intense insolation, gusting wind turbulence, desiccation—are those also found near forest edges at ground level and throughout small forest fragments (Murcia 1995).

In summary, where vertical structuring of forest habitats does exist, it is highly variable—and increasingly so in disturbed forests. Disturbance has a variety of direct and indirect effects, tending to increase the homogeneity of forest communities along both vertical and horizontal dimensions and generally reducing the distinctness of microclimates and of vertical strata. Selective logging and other disturbances that leave sharply defined forest edges tend to lower the "canopy" level, even as far as to the ground, making canopy resources available to other groups of resource users and blurring community boundaries.

Canopy science has made strong contributions to our understanding of forest community function at the local level. It may make greater contributions by reducing its emphasis on the parameter of distance from the ground—which is perhaps most directly relevant for the choice of tools and techniques needed to study it—and increasingly relating ecological assemblages or communities to the availability of the biophysical conditions necessary for their persistence. Forest "stratification" can be treated as a more or less continuous gradient in respect of functional and environmental characteristics, by simply grouping sample points into segments within a given range of conditions (Parker and Brown 2000). This retains the focus on the environmental parameters themselves and allows for more pertinent comparisons among sites—especially among sites subject to varying degrees of disturbance. Such an approach should encourage a more flexible redefining of concepts such as "canopy," "mid-canopy," and "understory" to match the (micro) habitat requirements of the taxon, interaction, or community of interest. It draws attention to functional similarities and parallel or divergent pressure-response patterns within forest systems under related threats. A goal is to improve our ability to model and predict outcomes of novel successional pathways, including retrograde pathways, under anthropogenic pressures or climate change.

2 Human Disturbance and Forest Canopies

Space does not allow a full recounting of the multiple ways in which human activities regularly disturb tropical forest canopies. Selective logging creates large and small artificial gaps, clear-cuts create artificial unbuffered edges, roads provide avenues for invasive species, intense fires burn drying forest canopies, hunting alters food webs and seed dispersal dynamics, and cattle reduce regeneration and alter successional pathways within gaps. Depending on the details of harvest techniques, NTFP harvest and fuelwood collection may or may not reduce canopy foliage density, alter canopy structure, or degrade gene pools. Selective logging may be both more pervasive and more broadly destructive in continental tropical forests than has previously been understood (Foley et al. 2005; Koltunov et al. 2009).

2.1 Canopy Ecosystem Services

Canopy research has shown that intact canopies provide a variety of ecosystem services to human economies but that many of these are indirect and not easily perceptible or quantifiable. Hence, local communities may not always be aware of them or of their value—especially those services, such as local and regional weather amelioration (Costa and Foley 2000; Ellison et al. 2011), that aggregate widely over space and time. Disturbed canopies produce a separate suite of services in many tropical forests, including attracting a variety of folivores and frugivores—some of which may become "ecosystem services" to the human dinner plate. Disturbed canopies also generate a set of significant *disservices,* such as human–wildlife conflict or increasing insect pest and disease vector populations (Vittor et al. 2006). Similarly, the role of forest systems in regulating the supply of water downstream is not straightforward. Trees and vegetation intercept water and return it to the atmosphere via evapotranspiration. Clear-cuts upstream release a pulse of runoff, which may be welcomed downstream. From this perspective, forests may be seen as "competitors" with humans for freshwater. Considered at larger temporal and geographical scales, however, they may be perceived as "biotic pumps" driving terrestrial water cycles (e.g., Makarieva et al. 2009).

Forest science has not yet been successful in evaluating such trade-offs, partly because they require integration of social, economic, and ecological data and partly because the trade-offs are set in motion by multiple interacting influences. Such questions will need to be approached from the standpoint of the whole forest rather than the canopy subsystem.

2.2 *Climate Change in the Canopy*

Under climate change, many tropical forests are likely to undergo heightened incidence of strong storms and droughts, both of which may already be contributing to hot fires in rain forests that historically have burned rarely or only mildly (Stork et al. 2009). In many places, destructive wildfire is now a perennial problem associated with high levels of human use and insecure tenure. Yet low levels of fire have been a stabilizing presence in some of these same forests for millennia (Cochrane 2003). Acting synergistically with the expansion of invasive species populations, climate changes can be expected to produce novel species assemblages (Hobbs et al. 2006). These need not impair canopy biophysical functioning in every case over the long run, but there will doubtless be an extended transition period during which many canopies will be broken and gapped, vastly altering microclimate regimes within (e.g., Ray et al. 2005). Added to these changes in physical parameters will be biotic changes as species respond differentially to climate change. A particularly high area of priority is possible decoupling of biotic interactions due to changes in phenology (Bawa and Dayanandan 1998; Pau et al. 2011; Diez et al. 2012).

2.3 *Canopies and Carbon Budgets*

Considering the importance of carbon to our global future, surprisingly little is known about how tropical forests interact with the atmosphere and soils in driving carbon cycles or about how these patterns might change in the future. One of the crucial "services" performed by forest canopies is regulating and minimizing soil organic carbon loss from oxidative mineralization and erosion. Although the canopy is known as the "interface of vegetation with the atmosphere," above- and belowground processes are tightly linked and the structure of root systems may have more influence on carbon uptake than canopy traits (Wood et al. 2012). Progress in nondestructive microscopy has made possible increasingly detailed analyses of time series of tropical forest soils before and up to a century after conversion to traditional agricultural use (e.g., Solomon et al. 2007 in Kenya: cropping of maize and millet without inputs of inorganic fertilizers). The soil organic matter that confers fertility is labile, and its loss after disturbance of the forest cover may be surprisingly rapid. But this only adds further urgency to the need to curtail forest cover degradation.

One area of considerable disagreement is the potential of forest biomes to sequester carbon and under what circumstances this potential might be approached. Optimistic estimates of global forest C-sequestration potential such as those of Lal (2005) suggest 1–3 Pg C $year^{-1}$, but they are heavily qualified by the intensive management procedures they prescribe in order to reach that goal. More recently, Malhi (2010) calculated that over the period 2000–2005, tropical forests absorbed 1.1 ± 0.3 Pg C $year^{-1}$, accounting for 9–15 % of total global anthropogenic emissions or nearly half of the total terrestrial carbon sink. During the same period, however, conversion of tropical biomes was a source of 1.3 ± 0.2 Pg C $year^{-1}$ into the atmosphere, resulting in a rough source-sink balance. This implies that future forest policy and implementation will be critical in tipping the balance.

What would it take to move carbon accounts toward the positive side of the ledger? We lack a complete picture of the processes governing C fluxes (Wood et al. 2012). Perhaps more importantly, the track record of governments in appreciating the need for forest sector support is weak. Even in a country such as India, where good soil management has been projected as a national priority and where ambitious reforestation targets have been in place for years, national investments in forest management have been hovering at or below 1 % of GDP (Balooni and Singh 2007). It is not clear that major improvements in forest management will come without a significant reordering of national- and state-level priorities throughout the tropics. Solving that dilemma also calls for a much closer integration of the political, economic, and forest sciences.

3 Toward an Integrated Science of Canopy and Tropical Forest Ecosystems

Human interactions with tropical forests constitute key elements of sustainability science and are recognized as future drivers of science and policy agendas. Key issues include the identification of local and regional drivers of change that threaten the integrity of forests and their canopies; analyses of human impacts on forest structure and ecosystem services in which canopies perform key functions (particularly the dynamics of carbon stocks, nutrients, water, and climate regulation); and integration of stand-level processes with landscape-level processes. In addition, there is a need for innovative new approaches to the inventory of canopy biotas. The human and financial resources currently being devoted to systematic biology are grossly inadequate, especially in tropical forests. We distill four key research and action issues from discussions at international canopy conferences: restoration, monitoring, policies and institutions, and governance.

Restoration offers great potential to enhance forest canopies and ultimately whole-forest cover, but the ability of that cover to sustain biodiversity and ecosystem services of canopy biota may not be equivalent to the primary forest. Species composition does not directly determine productivity levels for most ecosystem services, but early successional assemblages do not provide the same suite of benefits as closed-canopy forests. Although forests can regenerate on former agricultural land, and forest plantations are being established for commercial and restoration purposes, this vegetation does not match the composition and structure of the original forest cover (Chazdon 2008). In countries such as India and China, the massive reforestation efforts underway require immediate research inputs to ensure that restored forest canopies mimic the original diverse canopy structure (Guardiola-Claramonte et al. 2008).

In Australia, long-term studies of seedling regeneration indicate that hundreds, if not thousands, of years may be required for seedlings to fully restore the forest canopy (Lowman 1999). On the other hand, there are indications that for at least some tropical forests, the dispersal limitations of native late-successional trees may be a greater constraint on reestablishment than microclimatic conditions (e.g., for a neotropical moist forest, Martinez-Garza and Howe 2003). If dispersal limitations can be circumvented by planting or through artificial dispersal, it may be possible to reduce the time needed to reestablish closed-canopy conditions even on open pasture or large gaps (i.e., in 30–100 years less than would be the case under conditions of strictly natural regeneration). Reforestation on degraded lands—of the types common in South Asia, for instance (NRSC 2010)—can often benefit from the establishment of a protective layer or canopy of fast-growing species (Chazdon 2008). Even nonnative species can have positive effects on soil quality (e.g., Lugo 2009). In many restoration situations, the greatest challenge is to effectively yet equitably control resource use of the stand for at least several years or a decade.

Monitoring is one of the keys to finding appropriate responses to loss of forest canopies. Current estimates of deforestation and changes in canopy cover remain uncertain at every scale (Hansen et al. 2008; Wright 2010). Moreover, although monitoring helps identify symptoms and proximate drivers of change, it rarely succeeds in highlighting the social and economic parameters responsible for that change. Several Millennium Development Goals (MDGs) can be realized by incorporating the Convention on Biological Diversity (CBD) targets for curtailing biodiversity loss and encouraging sustainable use of resources. However, this requires comprehensive monitoring systems at the landscape level. Yet most countries (including India, despite its huge scientific and technical capacity) lack programs for systematic monitoring of human interactions with natural ecosystems. The 2009 International Canopy Conference organizers in Bangalore held a workshop to launch a forest monitoring program in India, with a very successful post-conference graduate training course. But in general, despite isolated efforts to monitor canopies with existing tools, the canopy science community has not yet embraced this approach despite the success of ground-based forest monitoring.

Free access to global datasets derived from remotely sensed imagery and advances in applications of remotely sensed data offer unprecedented opportunities to monitor the impact of global change on forest canopies. Thus, for example, Koltunov et al. (2009) have used MODIS data to examine the impact that altered forest structure due to selective logging may have on phenology. Asner et al. (2011, and this volume) show how remotely sensed imagery can pick up chemical differences among constituent tree species of humid tropical forest, contributing to an understanding of the functional and biological diversity of tropical forests. Remote sensing can also help in estimating carbon sequestration and changes thereto from land use and climate change (Gonzalez et al. 2010).

Policies and *institutions* ultimately determine the success of mitigation outcomes. Both formal and informal institutions play vital roles in the success of mitigation measures and policies for arresting ecosystem degradation (Ostrom 2000), but the success of forest conservation is hindered by inadequate attention to grassroots institutions. Indigenous knowledge systems should play a more critical role in designing and implementing the management of forest canopies—the hotspot for biodiversity and ecosystem services. Payment for Ecosystem Services (PES) and Reducing Emissions from Deforestation and Degradation (REDD) are being widely presented as principal mechanisms for curtailing deforestation and mitigating climate change at the international level (Sukhdev 2011). However, it remains unclear how they will have an impact on ongoing efforts to decentralize conservation and management of tropical forests in many countries (Agarwal et al. 2008; Lele et al. 2010) and to support rural livelihoods (Agarwal et al. 2012).

Governance often determines the effectiveness of institutions as well as policies for conservation of biodiversity, much of which inhabits the canopy (Barrett et al. 2006). Key governance debates to which canopy science can contribute are the seeming incompatibility between global agendas associated with PES/REDD and the local needs of decentralized biodiversity management (Phelps et al. 2010; Lele et al. 2010); critical evaluation of the effects of decentralized governance on forest resource management; and incorporation of MDGs and CBD targets into local governance regimes. As stakeholders who produce knowledge on a critical component of whole forests, canopy scientists must expand their role in mainstreaming societal debates on forest governance options, rather than remaining isolated in "scientific silos" (Porter and Rafols 2009).

4 Need for Building Human Capacity

Limited human resources constitute a major constraint of tropical forest canopy conservation. Bibliometric analyses indicate that the vast majority of published papers on forest canopies are from scientists based in North America, Europe and Japan (Nadkarni et al. 2011). Similarly, institutions funding tropical research tend to be based in the temperate zones rather than in situ within tropical countries (the phenomenon of "temperate bias," Lowman 1999, 2009). Much research in tropical forest canopies remains ecological in nature due to funding constraints, but many researchers recognize that the social and economic issues underlying management and conservation require urgent attention.

We make three recommendations to strengthen human capacity in tropical countries. First, increase investment in forest canopy research by governments and funding agencies within the context of comprehensive programs to monitor forest ecosystems. In most tropical countries, the need to establish such monitoring systems is paramount, because we do not know what is happening to tropical forests and uncertainty will only grow with advancing climate change. Second, support institutions in tropical countries to impart interdisciplinary training in forest research, including South–South interactions. We have described in other forums the kinds of institutions that are needed (Bawa et al. 2004, 2008). Third, create and nurture more collaborations among groups involved in canopy research through formal and informal networks to enhance training and capacity building in countries with critical and emerging forest conservation priorities.

Significant degradation of global forests over the decades has inspired the will to change research trajectories toward a fuller embrace of resource conservation approaches, toward more focused education of stakeholders in the importance of forest resources, and toward the search for new collaborations. Our objective is to integrate canopy science fully into a whole-forest approach and not to allow it to languish as an isolated field of study. We require a new determination to integrate knowledge across scales, disciplines, and knowledge systems including the areas of research, policy, and education. We must build a better understanding of the dynamics of forest canopies as critical components of whole forests to mitigate deleterious human impacts. Finally, heightened involvement of the public—through fostering "citizen science," engaging local stakeholders, and informing policy-makers—is a critical emerging responsibility of canopy scientists.

Acknowledgments The Bangalore workshop and the work on this manuscript were supported by grants from the U.S. National Science Foundation (DEB 0542130 and DEB 0956301).

References

Agarwal A, Chhatre A, Hardin R (2008) Changing governance of the world's forests. Science 320:1460–1462

Agarwal A et al (2012) Cool heads for a hot world–Social sciences under a changing sky. Global Environmental Change 22:329

Ashton PS, Hall P (1992) Comparisons of structure among mixed dipterocarp forests of north–western Borneo. J Ecol 80:459–481

Ashton PS, Seidler RG (in press) Reflections on the forests of tropical asia: lest the memory fade. Arnold Arboretum at Harvard University/Royal Botanic Gardens at Kew/Natural History Publications Sdn. Bhd., Cambridge, MA/London/Borneo

Asner G, Martin RE, Knapp DE et al (2011) Spectroscopy of canopy chemicals in humid tropical forests. Remote Sens Environ 115:3587–3598

Baker PJ, Wilson JS (2000) A quantitative technique for the identification of canopy stratification in tropical and temperate forests. For Ecol Manage 127:77–86

Balooni K, Singh K (2007) Prospects and problems of afforestation of wastelands in India: A synthesis of macro- and micro-perspectives. Geoforum. doi:10.1016/j.geoforum.2007.02.007

Barrett CB, Gibson CC, Hoffman G, Mc Cubbins MD (2006) The complex links between governance and biodiversity. Conserv Biol 20:1358–1366

Basset Y et al (2003) Vertical stratification of arthropod assemblages. In: Basset Y, Novotny V, Miller S, Kitching RL (eds) Arthropods of tropical forests: spatio-temporal dynamics and resource use in the canopy. Cambridge University Press, Cambridge

Bawa KS (1990) Plant pollinator interactions in tropical rain forests. Ann Rev Ecol Syst 21:399–422

Bawa KS, Balachander G, Raven P (2008) A case for new institutions. Science 319:136

Bawa KS, Dayanandan S (1998) Climate change and tropical forest genetic resources. Clim Change 23:449–466

Bawa KS, Kress WJ, Nadkarni NM, Lele S, Raven PH, Janzen H, Lugo AE, Ashton PS, Lovejoy TE (2004) Tropical ecosystems into the 21st. Science 306:227–228

Bernard E (2001) Vertical stratification of bat communities in primary forests of Central Amazon, Brazil. J Trop Ecol 17:115–126

Cadenasso ML et al (2003) A framework for a theory of ecological boundaries. Bioscience 53:750–758

Cavaleri MA et al (2010) Height is more important than light in determining leaf morphology in a tropical forest. Ecology 91:1730–1739

Chazdon RL (2008) Beyond deforestation: restoring forests and ecosystem services on degraded lands. Science 320:1458–1460

Clark DB et al (2008) First direct landscape scale measurement of tropical rain forest leaf area index, a key driver of global primary productivity. Ecol Lett 11:163–172

Cochrane MA (2003) Fire science for rainforests. Nature 421(6926):913–919

Costa MH, Foley JA (2000) Combined effects of deforestation and doubled atmospheric CO_2 concentrations on the climate of Amazonia. J Clim 13:18–34

Cuddington K, Beisner BE (2005) Ecological paradigms lost: roots of theory change. Elsevier Academic, Burlington/London

Davidar P et al (2010) Assessing the extent and causes of forest degradation in India: Where do we stand? Biol Cons 143:2937–2944

DeVries PJ et al (1997) Species diversity in vertical, horizontal, and temporal dimensions of a fruit–feeding butterfly community in an Ecuadorian rainforest. Biol J Linn Soc 62:343–364

Devy MS, Ganesh T (2003) Canopy science and its relevance to India. Curr Sci 85:581–584

Diez JM et al (2012) Forecasting phenology: from species variability to community patterns. Ecol Lett 15:545–553

Ellison D et al (2011) On the forest cover–water yield debate: from demand– to supply–side thinking. Glob Chang Biol. doi:10.1111/j.1365-2486.2011.02589.x

Ellwood F, Manica A, Foster WA (2009) Stochastic and deterministic processes jointly structure tropical arthropod communities. Ecol Lett 12:277–284

Foley JA, DeFries R, Asner GP et al (2005) Global consequences of land use. Science 309:570–574

Gonzalez P et al (2010) Forest carbon densities and uncertainties from Lidar, QuickBird, and field measurements in California. Rem Sens Envi 114:1561–1575

Grelle CEV (2003) Forest structure and vertical stratification of small mammals in a secondary Atlantic forest, Southeastern Brazil. Stud Neotrop Fauna Environ 38:81–85

Grimbacher PS, Stork NE (2007) Vertical stratification of feeding guilds and body size in beetle assemblages from an Australian tropical rainforest. Austral Ecol 32:77–85

Guardiola-Claramonte M et al (2008) Local hydrologic effects of introducing non–native vegetation in a tropical catchment. Ecohydrology 1:13–22

Hansen MC et al (2008) Humid tropical forest clearing from 2000 to 2005 quantified by using multitemporal and multiresolution remotely sensed data. Proc Nat Acad Sci 105:9439–9444

Harris N et al (2012) Progress toward a consensus on carbon emissions from tropical deforestation. Policy Brief. Winrock International, WHOI & Meridian Institute

Hobbs RJ et al (2006) Novel ecosystems: theoretical and management aspects of the new ecological world order. Glob Ecol Biogeogr 15:1–7

Hodgkison R (2004) Habitat structure, wing morphology, and the vertical stratification of Malaysian fruit bats (Megachiroptera: Pteropodidae). J Trop Ecol 20:667–673
Intachat J, Holloway JD (2000) Is there stratification in diversity or preferred flight height of geometroid moths in Malaysian lowland tropical forest? Biodivers Conserv 9:1417–1439
Jain M, Balakrishnan R (2012) Does acoustic adaptation drive vertical stratification? A test in a tropical cricket assemblage. Behav Ecol 23:343–354
Kang SH, Bawa KS (2003) Effects of successional status, habit, sexual systems, and pollinators on flowering patterns in tropical rain forest trees. Am J Bot 90:865–876
Koltunov A et al (2009) Selective logging changes forest phenology in the Brazilian Amazon: evidence from MODIS image time series analysis. Remote Sens Environ 113:2431–2440
Komposch H et al (2000) Diversity and vertical distribution of lichens in a Venezuelan tropical lowland rain forest. Selbyana 21:11–24
Lal R (2005) Forest soils and carbon sequestration. For Ecol Manage 220:242–258
Laurance WF, Kakul T, Keenan RJ, Sayer J, Passingan S, Clements GR, Villegas F, Sodhi NS (2010) Predatory corporations, failing governance, and the fate of forests in Papua New Guinea. Conservation Letters, Published online:16 Dec 2010, dOI:10.1111/j.1755–263X.2010.00156.x
Lele S, Wilshusen P, Brockington R, Seidler R, Bawa KS (2010) Beyond exclusion: alternative approaches to biodiversity conservation in the developing tropics. Curr Opin Environ Sustain 2:1–7
Lodge DJ, Cantrell S (1995) Fungal communities in wet tropical variation in time and space. Can J Bot 73(supp. 1):S1391–S1398
Longino JT, Nadkarni NM (1990) A comparison of ground and canopy leaf litter ants (Hymenoptera: Formicidae) in a neotropical montane forest. Psyche 97:81–93
Lowman MD (1999) Life in the treetops. Yale University Press, New Haven
Lowman MD (2009) Canopy research in the twenty-first century: a review of arboreal research. J Trop Ecol 50:125–136
Lowman MD, Bouricius B, Coley P, Halle F, Nadkarni NM, Parker G, Saterson K, Wright J (1995) What's up? Perspectives from the 1st international forest canopy conference at Sarasota FL. Selbyana 26:1–11
Lugo AE (2009) The emerging era of novel tropical forests. Biotropica 41(5):589–591
Makarieva AM, Gorshkov VG, Li BL (2009) Precipitation on land versus distance from the ocean: evidence for a forest pump of atmospheric moisture. Ecol Complex 6:302–307
Malhi Y (2010) The carbon balance of tropical forest regions, 1990–2005. Curr Opin Environ Sustain 2:237–244
Martinez-Garza C, Howe HF (2003) Restoring tropical diversity: beating the time tax on species loss. J Appl Ecol 40(3):423–429
Mawdsley NA, Stork NE (1997) Host–specificity and the effective specialization of tropical canopy beetles. In: Stork NE, Adis J, Didham RK (eds) Canopy arthropods. Chapman and Hall, London
Mitchell A, Secoy WK, Jackson T (eds) (2002) Global canopy handbook. Global Canopy Programme, Oxford, UK
Motzer T (2005) Micrometeorological aspects of a tropical mountain forest. Agr Forest Meteorol 135(1):230–240
Murcia C (1995) Edge effects in fragmented forests: implications for conservation. Trends Ecol Evol 10:58–62
Nadkarni NM (2001) Enhancement of forest canopy research, education, and conservation in the new millennium. Plant Ecol 153:361–367
Nadkarni NM, Parker GG (1994) A profile of forest canopy science and scientists–who we are, what we want to know, and obstacles we face: results of an international survey. Selbyana 15:38–50
Nadkarni M, Parker GG, Lowman MD (2011) Forest canopy studies as an emerging field of science. Ann Forest Sci 68:217–224
National Remote Sensing Centre (NRSC), India (2010) Wastelands atlas of India. Ministry of Rural Development, Government of India, New Delhi
Novotny V et al (2007) Low beta diversity of herbivorous insects in tropical forests. Nature 448

Novotny V et al (2002) Low host specificity of herbivorous insects in a tropical forest. Nature 416:841–844
Novotny V et al (2006) Why are there so many species of herbivorous insects in tropical rainforests? Science 313:1115
Ostrom E (2000) People and forests: communities, institutions, and governance. MIT Press, Cambridge, MA
Ozanne CMP et al (2003) Biodiversity meets the atmosphere: a global view of forest canopies. Science 301:183–186
Parker GG, Brown MJ (2000) Forest canopy stratification—is it useful? Am Nat 155(4)
Pau S et al (2011) Predicting phenology by integrating ecology, evolution and climate science. Glob Chang Biol 17:3633–3643
Pearson DL (1971) Vertical stratification of birds in a tropical dry forest. Condor 73:46–55
Phelps J, Webb E, Agarwal A (2010) Does REDD+threaten to recentralize forest governance? Science 328:312–313
Popma J et al (1988) Patterns in the vertical structure of the tropical lowland rain forest of Los Tuxtlas, Mexico. Vegetatio 74:81–91
Porter AL, Rafols I (2009) Is science becoming more interdisciplinary? Measuring and mapping six research fields over time. Scientometrics 81(3):719–745
Powers TO et al (2009) Tropical nematode diversity: vertical stratification of nematode communities in a Costa Rican humid lowland rainforest. Mol Ecol. doi:10.1111/j.1365-294X.2008.04075.x
Ravindranath NH et al (2012) Deforestation and forest degradation in India – implications for REDD+. Current Sci 102(8):1117–1125
Ray D et al (2005) Micrometeorological and canopy controls of fire susceptibility in a forested Amazon landscape. Ecol App. 15(5):1664–1678
Richards PW (1996) The tropical rain forest: an ecological study, 2nd edn. Cambridge University Press, Cambridge
Ricklefs RE (2008) Disintegration of the ecological community. Am Nat 172(6):741–750
Rodgers DJ, Kitching RL (1998) Vertical stratification of rainforest collembolan (Collembola: Insecta) assemblages: description of ecological patterns and hypotheses concerning their generation. Ecography 21:392–400
Sanbord AF et al (2011) Thermal adaptation and diversity in tropical ecosystems: evidence from cicadas (Hemiptera, Cicadidae). PLoS One 6(12):e29368. doi:10.1371/journal.pone.0029368
Scharlemann J, Laurance WF (2008) How green are biofuels? Science 319:43–44
Schulze CH et al (2001) Understorey versus canopy: patterns of vertical stratification and diversity among Lepidoptera in a Bornean rain forest. Plant Ecol 153:133–152
Smith AP (1973) Stratification of temperature and tropical forests. Am Nat 107(957):671–683
Sobek S et al (2009) Canopy vs. understory: does tree diversity affect bee and wasp communities and their natural enemies across forest strata? For Ecol Manag 258:609–615
Solomon D et al (2007) Long–term impacts of anthropogenic perturbations on dynamics and speciation of organic carbon in tropical forest and subtropical grassland ecosystems. Glob Chang Biol 13:1–20
Stenchly K et al (2012) Spider species richness in cocoa agroforestry systems, comparing vertical strata, local management and distance to forest. Agri Ecosys Environ 149:189–194
Stork NE et al (2009) Vulnerability and resilience of tropical forest species to land–use change. Conserv Biol 23(6):1438–1447
Strayer DL et al (2003) A classification of ecological boundaries. Bioscience 53(8):723–729
Sukhdev P (2011) Putting a price on nature: the economics of ecosystems and biodiversity. Solutions 1:34–43
Szarzynski J, Anhuf D (2001) Micrometeorological conditions and canopy energy exchanges of a neotropical rain forest (Surumoni–Crane Project, Venezuela). Plant Ecology 153:231–239
Tregidgo DJ et al (2010) Vertical stratification responses of an arboreal dung beetle species to tropical forest fragmentation in Malaysia. Biotropica 425:521–525
UN-FAO (2010) Global Forest Resources Assessment 2010: Main Report. FAO Forestry Paper 163. Food and Agriculture Organization of the United Nations, Rome

Vittor AY et al (2006) The effect of deforestation on the human–biting rate of *Anopheles darlingi*, the primary vector of *falciparum* malaria in the Peruvian Amazon. Am J Trop Med Hyg 74(1):3–11

Wallace AR (1878) Tropical nature and other essays. Macmillan, New York/London

Walther BA (2002) Vertical stratification and use of vegetation and light habitats by neotropical forest birds. J Ornithol 143:64–81

Whitmore TC (1984) A vegetation map of Malesia at scale 1:5 million. J Biogeog 11(6):461–471

Wood TE, Cavaleri MA, Reed C (2012) Tropical forest carbon balance in a warmer world: a critical review spanning microbial– to ecosystem–scale processes. Biol Rev 87(4):912–927

Wright SJ (2010) The future of tropical forests. Ann N Y Acad Sci 1195:1–27

Chapter 5
Emerging Threats to Tropical Forests

William F. Laurance

Keywords Biodiversity • Biofuels • Carbon emissions • China • Climate change • Deforestation • Emerging pathogens • Globalization • Industrialization • Tropical forests

Bullet Points

1. Emerging threats to tropical forests are changing, and their impacts are increasing.
2. Deforestation is increasingly being driven by industrial forces such as large-scale agriculture, logging, mining, and infrastructure expansion, and these impacts are being aggravated by climate change, emerging pathogens, and other environmental perils.
3. There are glimmers of hope for the conservation of global forests, including an increase in protected areas, increased public awareness of forest protection, and implementation of conservation policies such as carbon credits.

1 Introduction

Tropical forests are important for many reasons—for sustaining biodiversity and indigenous cultures, buffering climatic change, and providing other valuable ecosystem services (Laurance 1999; Lowman et al. 2006). Nearly half of the world's tropical forests have been lost in the last few centuries and, at present, another 10 million hectares or so of native forest are being felled annually—the equivalent of

W.F. Laurance (✉)
Centre for Tropical Environmental and Sustainability Science (TESS) and School of Marine and Tropical Biology, James Cook University, Cairns, QLD 4870, Australia
e-mail: bill.laurance@jcu.edu.au

M. Lowman et al. (eds.), *Treetops at Risk: Challenges of Global Canopy Ecology and Conservation*, DOI 10.1007/978-1-4614-7161-5_5,

40 football fields a minute (Laurance 2010). Beyond this, many native forests are being altered by threats such as habitat fragmentation, selective logging, overhunting, surface fires, and harvests of fuelwood and other forest products (Cochrane and Laurance 2002; Peres et al. 2006).

The forces that drive land-use change in the tropics are complex and continually evolving. Here I identify some emerging threats to tropical forests and their biota and briefly highlight some of their implications for nature conservation. My focus is on threats or trends that have increased markedly in the past decade or so or are looming just on the horizon.

2 Increasing Globalization and Industrialization

The ultimate and proximate drivers of deforestation are changing. In the 1980s and 1990s, an expanding human populace, as manifested by hundreds of millions of small-scale farmers and rural residents living in tropical frontier regions, was often seen as the primary driver of forest loss (e.g., Myers 1993). More recently, industrial drivers of deforestation—such as large-scale agriculture, plantations, and ranching—have risen sharply in importance (Butler and Laurance 2008; Rudel et al. 2009). Industrial activities such as selective logging, infrastructure expansion, and oil, gas, and mineral projects are also fueling a proliferation of roads in frontier regions, which also promote forest loss (see below).

The rising importance of industrial drivers is effecting several important changes in the tropics. First, it is accelerating the per capita rate of forest loss in some regions (Wright and Muller-Landau 2006) because forests are increasingly being felled by bulldozers and other heavy equipment, rather than by small-scale landowners armed only with machetes and chainsaws. Second, it may weaken the historically strong relationship between a country's population density and its remaining forest cover (Laurance 2007a). A country like Gabon or Cameroon, for instance, does not need a high population density to effect considerable deforestation when virtually every cut log the country produces is being exported to China. Industrialization will surely continue to grow globally as scores of developing nations—such as China, India, Brazil, South Africa, and many others—continue to expand economically. The Millennium Ecosystem Assessment projects a 300–500 % increase in global industrial activity by the year 2050 (MEA 2005). In the future, industrial drivers will continue to be a major driver of forest loss.

3 Road and Highway Expansion

We live in an era of unprecedented road and highway expansion. Many tropical regions that were remote and largely inaccessible just a decade ago have now been penetrated by networks of roads (Laurance et al. 2009). The problem with such roads is that their impacts often extend far beyond the road surface itself, unleashing a Pandora's Box of environmental problems that are difficult or impossible for governments to control. These include illegal land colonization, rampant land speculation, and influxes of illegal loggers, hunters, and gold miners. Fires and deforestation tend to increase dramatically near roads (Laurance et al. 2002; Kirby et al. 2006; Adeney et al. 2009). Such effects are especially severe and extensive around paved highways, which provide year-round access to forest regions and tend to spawn large networks of secondary and tertiary roads (Laurance et al. 2002).

Road expansion is one of the most important factors determining the magnitude and rate of forest destruction. For instance, new roads are projected to have far-reaching effects on future forest disruption in the Amazon (Fig. 5.1) (Laurance et al. 2001). Fortunately, road expansion is highly amenable to policy modification. Carbon-offset funds, for instance, might be used to promote the rerouting or closure of frontier roads in vulnerable regions, in order to reduce deforestation and atmospheric carbon emissions (Laurance et al. 2009).

4 China's Role in Illegal Logging

In the tropics, one of the most profound changes in the last decade is the stunning rise of timber consumption by China, which has transformed tropical timber markets. China is now overwhelmingly the biggest global consumer of tropical timber, importing around 40–45 million m^3 of timber annually (Laurance 2008a). More than half of all timber being shipped anywhere in the world today is destined for China (Globaltimber 2011).

China is being criticized because it has been remarkably aggressive in pursuing timber supplies while being little concerned with social or environmental equity. It has also focused mostly on buying raw timber (round logs) from timber-producing countries, which provides little employment or value adding for timber-producing nations, with most of the profits being realized by foreign loggers, log shippers, and wood-products manufacturers. Finally, it has been a poor global citizen in attempting to combat the scourge of illegal logging (Stark and Cheung 2006; Laurance 2008a), which is an enormous problem in many tropical nations. Illegal logging is slowly diminishing globally but this is despite, rather than because of, China's influence (Lawson and MacFaul 2010). For such reasons, China is vulnerable to a major boycott of its wood products, two-thirds of which are exported internationally (Laurance 2007b).

5 Expansion of Biofuels

As petroleum costs rise, biofuels are emerging as the most likely near-term alternative (Roberts 2005). Most expansion of biofuel production will occur in the tropics. It is where plants, which are the main feedstocks for biofuels, grow the fastest and where

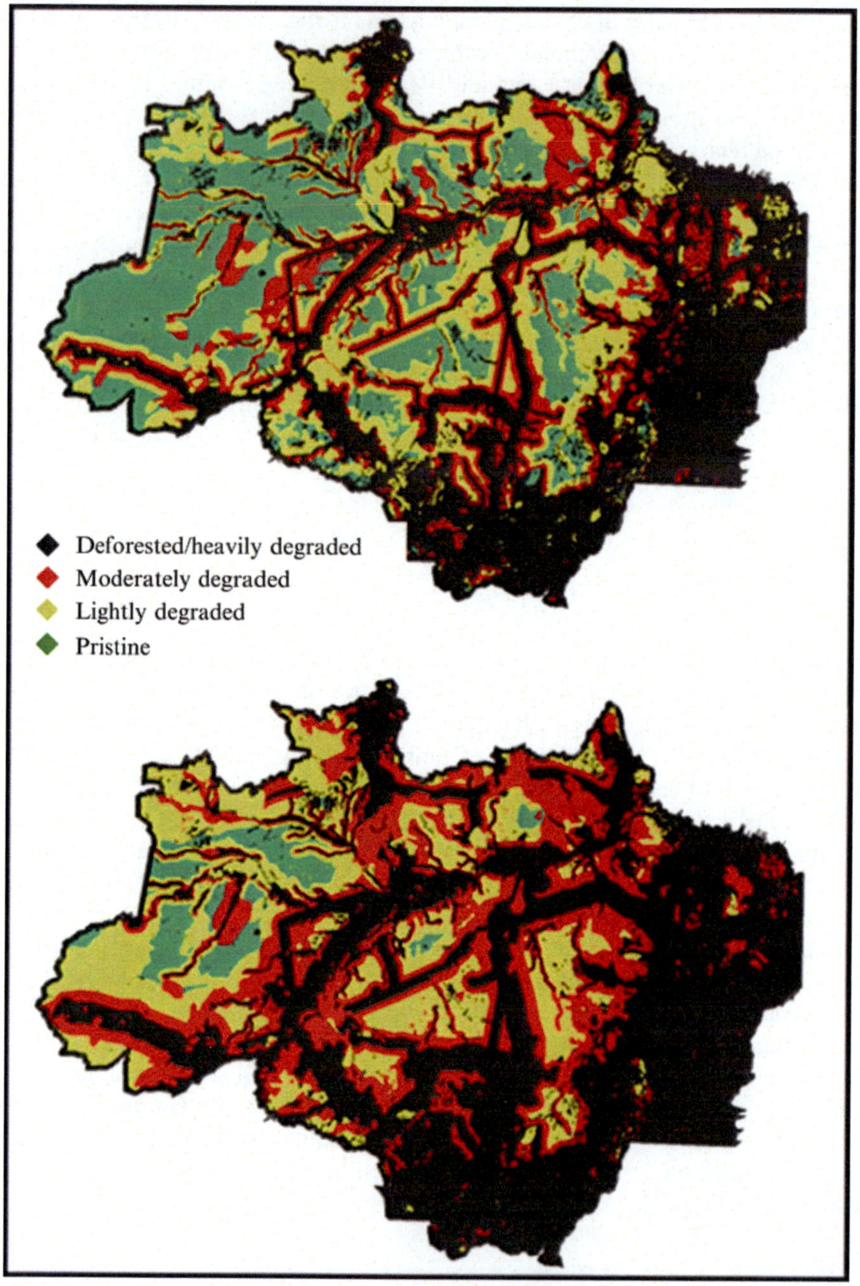

Fig. 5.1 Roads are projected to have a major impact on future patterns of forest loss, fragmentation, and degradation in the Brazilian Amazon. Shown are "optimistic" (*above*) and "nonoptimistic" (*below*) scenarios for the region in the year 2020 (From Laurance et al. (2001))

land is typically the cheapest. At least 35 million hectares of land—an area the size of Germany—is likely to be devoted to biofuel production by the year 2030, mostly in tropical countries (Sunderlin et al. 2008).

The proliferation of tropical biofuels will promote the conversion of forests and other native ecosystems for biofuel or food production. It will also drive up competition for land and elevating land prices, thereby increasing opportunity costs for conservation (Laurance 2008b). Under such circumstances, carbon trading will become less viable as a means to reduce deforestation and is likely to be competitive only for remote areas or those with poor, unproductive soils (Scharlemann and Laurance 2008). Unfortunately, such areas are under the least under threat of forest conversion.

6 Human Population Growth

The United Nations (2011) projects that the global population will peak at 10.1 billion people near the end of this century, from its current total of seven billion. Most of the additional three billion people on Earth will be added to the rosters of tropical nations (PAI 2006). Population growth is one of the most pervasive, ultimate drivers of deforestation and environmental degradation. Population pressures manifest themselves both locally and globally. For instance, growing international demand for wood products, paper pulp, and foodstuffs is helping to spur deforestation across much of the tropics. Population densities and population growth rates both tend to be very high in biodiversity hotspots (Cincotta et al. 2000), which contain large concentrations of species that are both locally endemic and imperiled by extinction. Among hotspot nations, population growth rates are a strong and positive predictor of deforestation rates (Jha and Bawa 2006), whereas population density is a strong predictor of endangered bird and mammal species among continental nations (McKee et al. 2003). Growing human numbers will worsen pressures on tropical forests in many different ways.

7 Emerging Pathogens

We live in an era of astonishing international mobility—one in which people and their associated goods and species are in a constant state of flux. As a result, natural ecosystems are being bombarded by potential new pathogens. The resulting "pathogen pollution" could be a far more serious threat for biodiversity than many appreciate (Cunningham et al. 2003).

Some of the most dramatic wildlife mortality events ever witnessed were caused by exotic pathogens or new disease vectors (Daszak et al. 2000). Examples include the mass die-off of African ungulates from rinderpest, the global extinctions or widespread extirpations of endemic Hawaiian birds from avian malaria, and the near extirpation of Diadema sea urchins in the Caribbean Sea. Today, ebola hemorrhagic fever is burning its way across the Congo Basin, causing mass mortality of gorillas and chimpanzees. Perhaps the most devastating of all contemporary wildlife pathogens is the chytrid fungus currently driving global amphibian declines (Laurance et al. 1996; Berger et al. 1998). Around 200 frog species, mostly from tropical and subtropical regions, have been driven to global extinction or been severely reduced in numbers by the fungus (Skerratt et al. 2007).

Such wildlife pathogens may be merely the tip of the iceberg. As molecular genetics and other techniques to detect foreign pathogens improve, we may discover that emerging diseases are among the most serious threats to biodiversity.

8 Climatic and Atmospheric Changes

Climate change could affect tropical forests in a variety of ways. The biotas of tropical mountains, which contain many local endemics, seem especially vulnerable to global warming. Warming could shrink and fragment the geographic ranges of montane species adapted to cool, wet conditions, increasing the likelihood of extinction (Williams et al. 2003; Thomas et al. 2004). Warming will also tend to elevate the cloud base on tropical mountains, reducing moisture inputs and increasing thermal radiation and desiccation stress (Pounds et al. 1999). Montane species may also have to contend with the arrival of lower-elevation competitors, predators, and pathogens as climates warm (Epstein 2001; Harvell et al. 2002).

Another key concern is that, in many parts of the tropics, even small reductions in dry-season rainfall could greatly increase the vulnerability of forests to droughts and fires (Cochrane and Laurance 2008). Unfortunately, our capacity to project future precipitation and temperature—and particularly to downscale projections for specific locations—is generally poor (Vera et al. 2006).

9 Environmental Synergisms

Combinations of environmental insults could be the downfall for many imperiled species. The tiger, for example, is not merely being endangered by destruction of its habitats or by widespread persecution and hunting, but by the combination of these environmental insults operating in concert. An analysis of endangered and extinct species from the IUCN Red Data Book (Fig. 5.2) provides strong support for the idea that synergisms are a key driver of species decline (Laurance and Useche 2009). In the tropics, many synergisms—between logging and hunting (Wilkie et al. 1992), habitat fragmentation and fire (Cochrane and Laurance 2002), and habitat disruption and climatic change (Thomas et al. 2004), among others—are likely to have serious impacts on biodiversity.

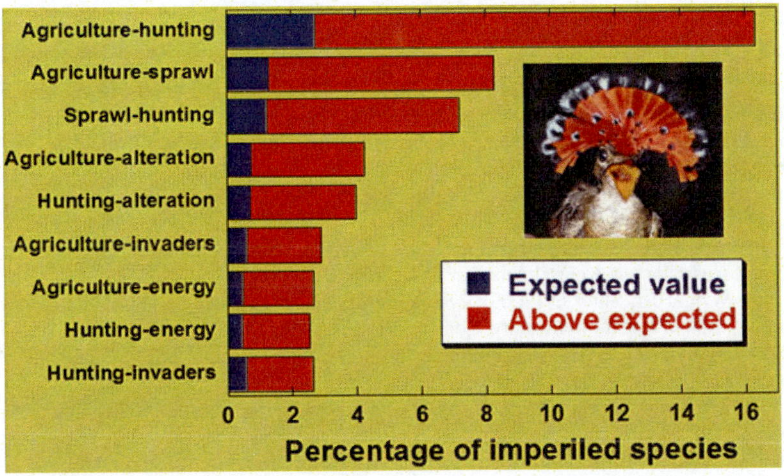

Fig. 5.2 Synergisms are a major driver of species declines. Shown are the expected vs. observed frequencies of mammal species driven to endangerment or extinction by key combinations of environmental threats, using data from the IUCN Red Data Book (Adapted from Laurance and Useche (2009))

10 Conclusion

While this review has identified a number of emerging threats to tropical forests, it is important to emphasize that "it is not all bad news." In fact, one can highlight many positive developments in tropical nature conservation. Fully protected areas now encompass nearly 6 % of the global land area, with nearly 15 % having some level of protection (Jenkins and Joppa 2009). Public awareness of the importance of tropical forests is clearly increasing (Baranzini et al. 2009). Carbon trading is becoming an important mechanism to slow tropical deforestation while reducing greenhouse gas emissions (Laurance 2007c; Venter et al. 2009). Conservation actions are having a positive impact on biodiversity protection at many levels (Brooks et al. 2009; Sodhi et al. 2011).

Rather than viewing the tropics as a lost cause, one should view this as a time of great opportunity for tropical conservationists. Having a better understanding of the emerging threats to tropical ecosystems should help to focus our conservation efforts and increase their effectiveness.

Acknowledgments I thank Meg Lowman, Mark W. Moffett, and T. Ganesh for useful comments on this chapter.

References

Adeney JM, Christensen NL, Pimm SL (2009) Reserves protect against deforestation fires in the Amazon. Plos One 4:5014

Baranzini A, Faust AK, Huberman D (2009) Tropical forest conservation: attitudes and preferences. Centre de Recherche Appliquée en Gestion, Geneva, http://ssrn.com/abstract=1431951

Berger L, Speare R, Daszak P, Greene DE, Cunningham AA, Gogging CL, Slocombeh R, Ragani MA, Hyatt AD, McDonald KR, Hines HB, Lips KP, Marantellim G, Parkes H (1998) Chytridiomycosis causes amphibian mortality associated with population declines in the rain forests of Australia and Central America. Proc Natl Acad Sci USA 95:9031–9036

Brooks TM, Wright SJ, Sheil D (2009) Evaluating the success of conservation actions in safeguarding tropical forest biodiversity. Conserv Biol 23:1448–1457

Butler RA, Laurance WF (2008) New strategies for conserving tropical forests. Trends Ecol Evol 23:469–472

Cincotta RP, Wisnewski J, Engelman R (2000) Human population in the biodiversity hotspots. Nature 404:990–992

Cochrane MA, Laurance WF (2002) Fire as a large-scale edge effect in Amazonian forests. J Trop Ecol 18:311–325

Cochrane MA, Laurance WF (2008) Synergisms among fire, land use, and climate change in the Amazon. Ambio 37:522–527

Cunningham AA, Daszak P, Rodriguez JP (2003) Pathogen pollution: defining a parasitological threat to biodiversity conservation. J Parasitol 89:878–883

Daszak P, Cunningham AA, Hyatt AD (2000) Emerging infectious diseases of wildlife—threats to biodiversity and human health. Science 287:443–449

Epstein PR (2001) Climate change and emerging infectious diseases. Microbes Infect 3:747–754

Globaltimber (2011) China: illegal imports and exports. www.globaltimber.org.uk/Chinaillegalimpexp.htm. Accessed 22 Sep. 2011

Harvell CD, Mitchell CE, Ward J, Altizer S, Dobson A, Ostfeld R, Samuel M (2002) Climate warming and disease risk for terrestrial and marine biota. Science 296:2158–2162

Jenkins CN, Joppa L (2009) Expansion of the global terrestrial protected area system. Biol Conserv 142:2166–2174

Jha S, Bawa K (2006) Population growth, human development, and deforestation in Biodiversity hotspots. Conserv Biol 20:906–912

Kirby KR, Laurance WF, Albernaz AK, Schroth G, Fearnside PM, Bergen S, Venticinque VM, da Costa C (2006) The future of deforestation in the Brazilian Amazon. Futures 38:432–453

Laurance WF (1999) Reflections on the tropical deforestation crisis. Biol Conserv 91:109–117

Laurance WF (2007a) Have we overstated the tropical biodiversity crisis? Trends Ecol Evol 22:65–70

Laurance WF (2007b) The dragon and the rainforest. Tropinet 18(1):1–2

Laurance WF (2007c) A new initiative to use carbon trading for tropical forest conservation. Biotropica 39:20–24

Laurance WF (2008a) The need to cut China's illegal timber imports. Science 319:1184

Laurance WF (2008b) How green are biofuels? A tropical perspective, p. 19–28. In: Biofuels and neotropical forests: trends, implications, and emerging alternatives. Smithsonian Tropical Research Institute, Panama

Laurance WF (2010) Habitat destruction: death by a thousand cuts. In: Sodhi N, Ehrich PR (eds) Conservation biology for all. Oxford University Press, Oxford, UK

Laurance WF, Useche DC (2009) Environmental synergisms and extinctions of tropical species. Conserv Biol 23:1427–1437

Laurance WF, McDonald KR, Speare R (1996) Epidemic disease and the catastrophic decline of Australian rain forest frogs. Conserv Biol 10:406–413

Laurance WF, Cochrane MA, Bergen S, Fearnside PM, Delamonica P, Barber C, D'Angelo S, Fernandes T (2001) The future of the Brazilian Amazon. Science 291:438–439

Laurance WF, Albernaz AKM, Schroth G, Fearnside GP, Venticinque EP, Da Costa C (2002) Predictors of deforestation in the Brazilian Amazon. J Biogeogr 29:737–748
Laurance WF, Goosem M, Laurance SG (2009) Impacts of roads and linear clearings on tropical forests. Trends Ecol Evol 24:659–669
Lawson S, MacFaul L (2010) Illegal logging and related trade: indicators of the global response. Chatham House, London
Lowman MD, Burgess J, Burgess E (2006) It's a jungle out there—more tales from the treetops. Yale University Press, New Haven
McKee JK, Sciulli PW, Fooce CD, Waite TA (2003) Forecasting global biodiversity threats associated with human population growth. Biol Conserv 115:161–164
MEA (Millennium Ecosystem Assessment) (2005) Ecosystems and human well-being: synthesis. Island Press, Washington, DC
Myers N (1993) Tropical forests: the main deforestation fronts. Environ Conserv 20:9–16
PAI (2006) Mapping the future of world population. Population Action International, Washington, DC
Peres CA, Barlow J, Laurance WF (2006) Detecting anthropogenic disturbance in tropical forests. Trends Ecol Evol 21:227–229
Pounds JA, Fogden M, Campbell J (1999) Biological response to climate change on a tropical mountain. Nature 398:611–615
Roberts P (2005) The end of oil: on the edge of a perilous new world. Houghton-Mifflin, New York
Rudel TK, DeFries R, Asner GP, Laurance WF (2009) Changing drivers of tropical deforestation create new challenges and opportunities for conservation. Conserv Biol 23:1396–1405
Scharlemann J, Laurance WF (2008) How green are biofuels? Science 319:52–53
Skerratt LF, Berger L, Speare R, Cashins S, McDonald KR, Phillott AD, Hines HB, Kenyon N (2007) Spread of chytridiomycosis has caused the rapid global decline and extinction of frogs. EcoHealth 4:125–134
Sodhi NS, Butler S, Laurance WF, Gibson L (2011) Conservation successes at micro, meso and macro scales. Trends Ecol Evol 26:585–594
Stark T, Cheung SP (2006) Sharing the blame. Greenpeace International and Greenpeace China (www.illegal-logging.info/uploads/sharing_the_blame.pdf)
Sunderlin WD, Hatcher J, Liddle M (2008) From exclusion to ownership? Challenges and opportunities in advancing forest tenure reform. Rights and Resources Initiative, Washington, DC
Thomas CD, Cameron A, Green R, Bakkenes M, Beaumont L, Collingham Y, Erasmus B, De Siqueira M, Grainger A, Hannah L, Hughes L, Huntley B, Van Jaarsveld A, Midgley G, Miles L, Ortega-Huerta M, Peterson A, Phillips O, Williams SE (2004) Extinction risk from climate change. Nature 427:145–148
United Nations (2011) World population prospects: the 2010 revision. U.N. Population Division, New York
Venter O, Laurance WF, Iwamura T, Wilson K, Fuller R, Possingham H (2009) Harnessing carbon trading to protect biodiversity. Science 326:1368
Vera C, Silvestri G, Liebmann B, González P (2006) Climate change scenarios for seasonal precipitation in South America from IPCC-AR4 models. Geophys Res Lett 33:L13707. doi:10.1029/2006GL025759
Wilkie DS, Sidle JG, Boundzanga GC (1992) Mechanized logging, market hunting, and a bank loan in Congo. Conserv Biol 6:570–580
Williams SE, Bolitho E, Fox S (2003) Climate change in Australian tropical rainforests: an impending environmental catastrophe. Proc R Soc Lond B 270:1887–1892
Wright SJ, Muller-Landau HC (2006) The future of tropical forest species. Biotropica 38:287–301

Chapter 6
Rethinking the Role of Tropical Forest Science in Forest Conservation and Management

Alex Racelis and James Barsimantov

Keywords Bibliometry • Deforestation

1 Introduction

The rapid deterioration of the global environment is no longer a surprise to anyone (Dirzo and Raven 2003; Bawa et al. 2004; Wright 2005; Barnosky et al. 2012; Cardinale et al. 2012), but not much tangible progress has occurred to curb the deterioration and unsustainable exploitation of our natural resources (Tollefson and Gilbert 2012). There is no better example of this than tropical forests (FAO 2005), which continue to disappear despite the serious implications of this loss on important ecosystems services such as carbon cycling, climate regulation, and biodiversity (Costanza et al. 1997; Bonan 2008; Chazdon 2008; Sugden et al. 2008).

But does a better scientific understanding of tropical forests contribute to better conservation and management? There are countless local or regional case studies that exemplify research-driven conservation. However, the difficulty in scaling up this work has become increasingly apparent (Farrington and Boyd 1997; Berkes 2006), challenging the assumption that the environmental sciences can effectively

A. Racelis (✉)
Department of Environmental Studies, University of California Santa Cruz, 1156 High Street, Santa Cruz, CA 95064, USA

Department of Biology, University of Texas Pan American, 1201 West University Drive, Edinburg, TX 78539, USA
e-mail: racelisae@utpa.edu

J. Barsimantov
Department of Environmental Studies, University of California Santa Cruz, 1156 High Street, Santa Cruz, CA 95064, USA
e-mail: jbarsima@soe.ucsc.edu

M. Lowman et al. (eds.), *Treetops at Risk: Challenges of Global Canopy Ecology and Conservation*, DOI 10.1007/978-1-4614-7161-5_6,

drive better management (Hillborn and Ludwig 1993; Shellenberger and Nordhaus 2004). In this chapter, we analyze the relationship between scientific publications on tropical forests and the effectiveness of regional conservation and management, as a didactic tool to illustrate the limitations of the current research paradigm in influencing environmental outcomes on a global scale. Given the reality that tropical resources are imperiled (Wright 2005; Gibbs et al. 2010) despite increasing research, we quantified the role of forest science in forest management using publications as a metric. We discuss the notion that a shift may be required in the application of research outcomes because research productivity does not necessarily correlate with effective conservation.

2 Methods

We conduct a simplified regression analysis of publication records and country-level deforestation rates in countries with tropical forests. A regression analysis, while unable to demonstrate causality, may uncover relationships between publication records and changes in country-level deforestation. To test the hypothesis that research on tropical forests has had an impact on deforestation at a global scale, we examined whether country-level publication records are related to changes in country-level deforestation rates. A bibliometric approach of publication record was recorded using the Internet-based search engine Web of Science, the world's leading citation database which provides multidisciplinary coverage to subscribers of over 10,000 international journals in the sciences, social sciences, and arts and humanities (including research articles, notes, conference proceedings, and reviews), hereafter referred to as publications. We examined 84 countries with >10,000 ha of tropical forest area in 1990 (Food and Agriculture Organization 1995).[1] For each country, we tabulated the number of publications about forests from 1980 to 2005 by restricting our search to each country name (and common country name alternatives[2]) and the search parameter "forest*".[3] Web of Science uses search parameters to scan subject matter, keywords, and abstract, eliminating the concern for unsubstantiated hits from authors' country affiliation. With a subset of the most-studied case-study countries, we experimented with other search parameters such as "deforest*,"

[1]A total of 90 countries are considered tropical countries by FAO. Excluded from the model were countries with under 10, 000 ha of forest: Singapore, Grenada, St. Kitts and Nevin, Djibuti, Antigua and Barbuda, and St. Vincent.

[2]Alternative country names were explored using the Wikipedia site "alternative country names." Examples of countries with publications under alternative country names included Brazil (Brasil), Madagascar (Malagasy Republic), and Congo (Zaire), although the percent of total publications under alternative names was negligible.

[3]The asterisk signifies a truncated word and includes all variants of the truncation in the search. For example, "forest*" includes forests, forestry, and forester. We also examined both Spanish and French variants of the search parameters and found that this did not significantly alter search results.

"manag*," and "conserv*" to examine how well our original search parameter "forest*" captures the forest change/conservation/management-related literature and we report these results below.

From this search, we constructed two main variables for publications: total number of publications and total publications per hectare of forest (publications/ha) for each country. These were included in our regression analysis as main predictor variables. Theoretically, using the total publications per country variable might imply that academic research influences policy at a national level, including improvements in legal systems, enforcement, and sustainable development programs. On the other hand, the "publications per hectare" variable would imply a direct effect on forest management. To explore both possible relationships, we ran separate regression models for each predictor variable. Additionally, to explore the expectation of a lagged effect of science on forest conservation and management, we summed publications per country for each 10-year interval beginning in 1980 and include separate runs of the model to account for the potential of a time-lagged influence.

We compiled deforestation rates from the FAO (2005) Global Forest Resources Assessment (Food and Agriculture Organization 2006). The FAO defines countries with tropical forests as those containing the following ecosystems: tropical rain, tropical dry, tropical moist deciduous, tropical montane, and subtropical forests. Information on total forest area, total forest loss, and deforestation rate was compiled for the years 1990, 2000, and 2005, corresponding to the three most recent Global Forest Assessments. Thus, the time periods of our change in deforestation rate variable are 1990–2000 and 2000–2005. Since we are interested in the impact of scientific literature on changes in deforestation rates, our main response variable is the difference in deforestation rates between the two time periods.

We also included demographic and socioeconomic independent variables of population and income to explore how our publications variables compare to variables that influence deforestation (Kaimowitz and Angelsen 1998). We used FAO statistics on population density, population growth, rural population, GDP per capita, and annual economic growth (Food and Agriculture Organization 1995, 2006).

In all, we included the following variables in our analysis:

- *Change* in deforestation rate ($\text{deforest}_{1990\text{–}2000}$—$\text{deforest}_{2000\text{–}2005}$)
- Total $\text{publications}_{1980\text{–}2005}$
- Publications per million hectares of $\text{forest}_{1980\text{–}2005}$
- *Change* in population growth rate ($\text{popgrowth}_{2000\text{–}2005}$—$\text{popgrowth}_{1990\text{–}2000}$)
- *Change* in percent rural population (ruralpop_{2005}—ruralpop_{1995})[4]
- *Percent change* in GDP (GDP_{2004}—GDP_{1990})
- *Change* in illiteracy rate ($_{2000-1990}$)

We estimated a multivariate regression model using both of the publications variables described above in order to explore our hypothesis that a high rate of scientific research will correlate to decreasing rates of deforestation. We also explored the

[4]Data not available for 1990.

potential for any delayed influence scientific research may have on deforestation rates by substituting our two publications variables with publications from 1980 to 1989 in the first time-lag model and from 1990 to 1999 in the second. Although some studies have modeled nonlinear relationships for economic variables (Cropper and Griffiths 1994; Rock 1996), we chose a linear relationship in accordance with the majority of studies (Palo et al. 1987; Burgess 1993; Gullison and Losos 1993; Inman 1993) and because nonlinear relationships reported mixed results. We estimated this linear multivariate regression model in a stepwise analysis using the JMP statistical software package.

3 Results: Publication Trends

Our bibliometric analysis yielded more than 23,390 records from 1980 to 2005 across all countries. The publication record for all countries has grown geometrically over the past 25 years: From 1980 to 1989 an average of 72 publications were produced a year about tropical forests, and in the following decade (1990–1999), that average rose to 891 publications per year. In the last decade (2009–2009), we produced an average of 2,995 publications a year about tropical forests and this number undoubtedly will continue to rise.

To examine how well our main search parameter "forest*" captures forest ecology, conservation, and management-related literature, we conducted an exploratory search using other parameters. For example, the correlation coefficient between forest* and deforest* was 0.928 for the 15 most-studied countries, which accounted for two-thirds of all forest-related publications in our analysis (see Fig. 6.3). We found that between (forest*) and (forest* AND (manag* OR conserv*)),[5] the correlation coefficient was 0.929, suggesting that using only "forest*" as the search parameter adequately reflected publication trends for articles that specifically deal with forest management or forest conservation (Fig. 6.1).

Publications pertaining to ecology (31 %) represented the largest portion of knowledge produced about forests in the most-studied tropical countries, although studies in environmental sciences, plant sciences, zoology, forestry, and biodiversity conservation were also consistently represented in our search.[6] This rapid growth and diversity in research on tropical forests suggests increased scientific knowledge to inform conservation and sustainable management of tropical forests. Admittedly, our data indicated that increased research has not led to increased conservation.

[5]This search parameter would yield any publications that had the word "forest" (or its variants) as well as either the words that contain "manag" (such as management) or "conserv" (conservation). These added parameters reduced total bibliometric records by an average of 75 %.

[6]Search was conducted among the top 15 most-studied countries, which represents 66 % of total publications record across all countries. The aforementioned subject areas, when used as search parameters in Web of Science, were all highly correlated with the search parameter "forest*" (0.88, 0.97, 0.95, 0.81, 0.91, and 0.92, respectively).

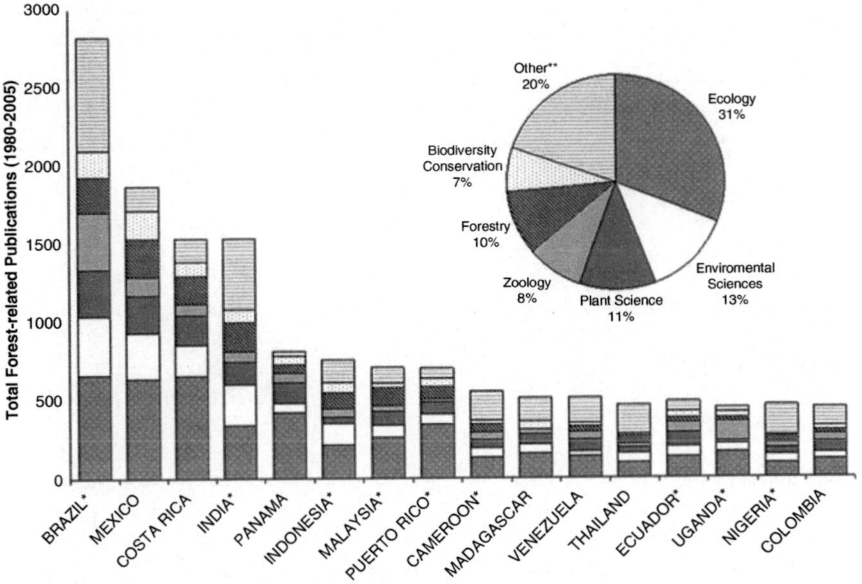

*Asterisk denotes country with worsening deforestation rate over the same time period
**'Other' subject areas include tropical medicine, biology, entomology, geography, ornithology, soil science, agronomy, among others

Fig. 6.1 Fifteen most-studied tropical countries given search parameters by subject area

3.1 Scientific Publications and Deforestation

None of our predictor variables were found to significantly predict deforestation rates in a linear regression analysis (see Table 6.1). Time-lag versions of the model with either publications variable did not produce results different from the original model. Tests for colinearity among independent variables showed no correlations between independent variables higher than 0.22.

These regression results have important implications for scientific research on tropical forests. In effect, neither total publications nor publications per hectare were found to significantly predict deforestation rates, nor did we find a science-policy time-lag effect. These results bring into question the impact and relevance that scientific publications have on deforestation and suggest that we should begin to ask ourselves whether a continued exponential increase in publications is acceptable if it is unrelated to more effective forest management.

To further understand the relationship between publications, deforestation, and other factors, we take a closer look at a few cases. Costa Rica had 1,679 total publications between 1980 and 2005, i.e., 655 publications per million ha of forest. Over the same time span, Costa Rica's deforestation rate improved by 0.9 %. Brazil, with a change in deforestation rate of –0.1, had 2,593 publications in the same time period (but that amounts to only five publications per million ha of forest). In order to match Costa Rica's rate of publications/ha, over 340,000 publications would need to have

Table 6.1 Regression coefficients of three models on change in deforestation rates from 1990–2000 to 2000–2005 using predictor variables from a bibliometric analysis and other socioeconomic and demographic variables

	No time lag	Time lag (90–99)	Time lag (80–89)
Bibliometric variables			
Publications per hectare of forest	0.00028		
	(0.00050)		
Total publications	0.00001		
	(0.00023)		
Publications per ha (90–99)		0.0007	
		(0.0011)	
Total publications (90–99)		0.0000	
		(0.0006)	
Publications per ha (80–89)			0.0023
			(0.011)
Total publications (80–89)			−0.0042
			(0.0075)
Socioeconomic/demographic variables			
Change in population growth rate (90–04)	0.026	0.027	0.014
	(0.068)	(0.067)	(0.062)
Change in percent rural population (95–05)	−0.017	−0.016	−0.018
	(0.015)	(0.015)	(0.015)
Change in GDP (90–04)	−0.045	−0.045	−0.049
	(0.075)	(0.075)	(0.075)
Change in illiteracy rates (90–00)	0.012	0.011	0.015
	(0.028)	(0.028)	(0.028)
Change in percent of GDP from trade (90–95)	−0.005	−0.005	−0.003
	(0.012)	(0.012)	(0.013)
Constant	0.017	0.008	0.091
	(0.275)	(0.276)	(0.278)
Observations	84	84	84
F-ratio	0.319	0.333	0.323
R square	0.0285	0.0297	0.0289

[a]Significant at $p<.05$, robust t statistics in parentheses

been written on Brazilian forests between 1980 and 2005. Second, a high level of publications was clearly not enough to stem deforestation. As a comparison to Costa Rica, neither Puerto Rico (#8 at 705 total publications and #1 at 1,745 publications/ha) nor Ecuador (#13 at 539 total publications and #26 at 39 publications/ha)—two other Latin American countries with relatively high level of publications—has managed to reduce their deforestation rates. It is likely that the decrease in Costa Rica's deforestation rates may be linked to the economics of a burgeoning tourism-based economy. Despite its similar publication trends and budding tourism sector, Puerto Rico has not been able to similarly reverse its rate of forest loss. Cattle and sugar production still remain as an important source of income in the agriculture sector and the biggest anthropogenic threat to remaining forests (Central Intelligence Agency 2008).

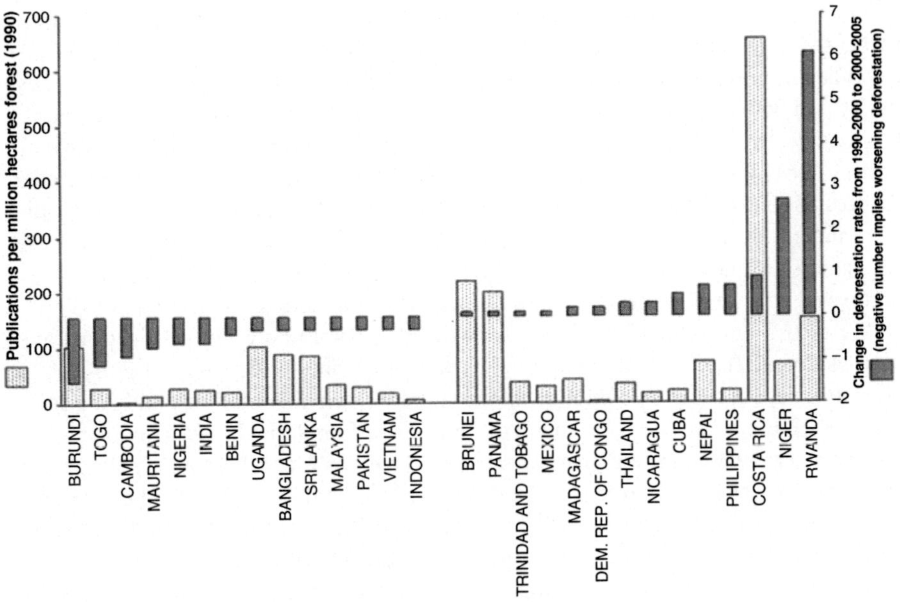

Fig. 6.2 Countries with highest positive and negative change in deforestation rates and related publications per million hectares (including all countries with lower deforestation rates)

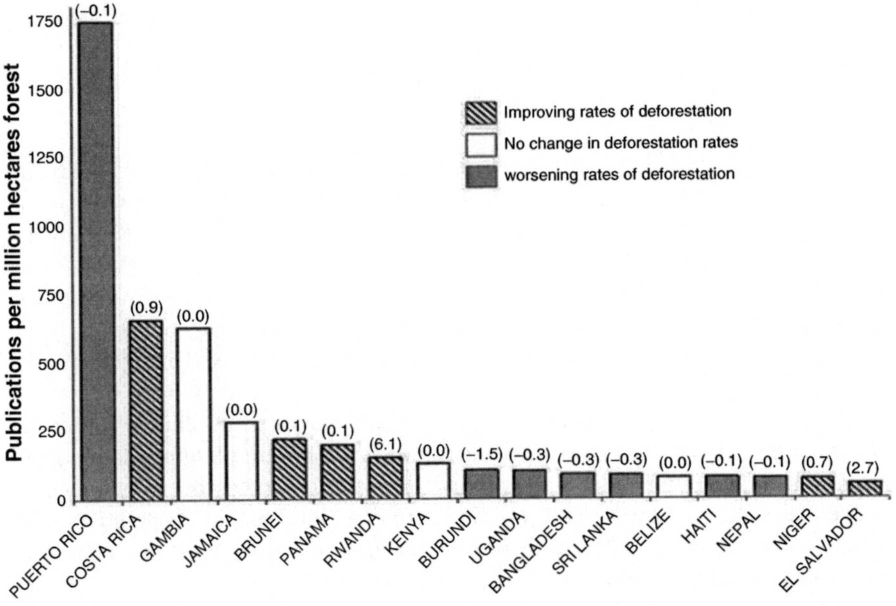

Fig. 6.3 Top 15 most intensively researched countries, with the highest rates of publications per million hectares. Deforestation trends (change in deforestation rates between 1990–2000 and 2000–2005) reported in parentheses

For another comparison, we examined two African countries, Uganda and Rwanda, both with high rates of publications per hectare (103 and 150, respectively). While Rwanda's deforestation rate has improved by 0.6 %, Uganda lost forest cover by –0.3 %. In both countries, deforestation is linked to fuelwood collection and charcoal production driven by population growth (Kayanja and Byarugaba 2001; Masozeraa et al. 2006). However, in Uganda, ineffective government action has facilitated deforestation (Tenywa 2008), while in Rwanda, a massive reforestation effort by the government has contributed to a dramatic rise in forest cover (Butler 2006). These case studies suggest that research itself does lead to effective conservation but likely involves a multitude of social, economic, and political factors (Figs. 6.2 and 6.3).

We also found evidence that the lack of relationship between publications and deforestation exists on a regional level as well. In a finer-scale study of publication rates, Pitman et al. (2007) compiled all literature written about biology and conservation in the Madre de Dios department of Peru. They found roughly 2048 publications written since 1970, with the yearly rate of publications more than quadrupling from 1980 to the present. This gives Madre de Dios a total ranking of 240 publications per million hectares rate, extremely high on a global scale. However, Madre de Dios lost about 0.16 % of its forest per year between 1991 and 2000 and about 0.20 % per year between 2000 and 2005 (Vuohelainen et al. 2012). Despite its relatively high rate of publications, deforestation increased in recent years. This is probably due primarily to expansion of an international highway into the region, which suggests the importance of social and political decisions influencing scientific findings, and the inability of scientific research on its own to drive policy outcomes in the context of international development (Vuohelainen et al. 2012).

4 Discussion

Numerous regression models of deforestation at the global level have been conducted using different datasets, variables, methods, and definitions of forest cover and deforestation (see Kaimowitz and Angelsen 1998 for a review). Kaimowitz and Angelsen sharply criticize global deforestation models because country-level forest cover data available from the United Nations' Food and Agriculture Organization (FAO) are unreliable, having been estimated for many countries using population models rather than calculated using remotely sensed images. While more recent FAO assessments conducted in 2000 and 2005 may have improved slightly, we cannot claim that country-level forest cover data are reliable and accurate due to incomplete data. Furthermore, the complex nature of deforestation may mask the accuracy of global deforestation assessment. This study does not intend to resolve these inconsistencies, but our aim is to question the current practices and applications of research on tropical forests and how to inspire best practices that will insure conservation actions.

Our bibliometric evaluation indicates the lack of a discernible relationship between published research and the trajectory of tropical forest conservation. Based on these findings, we propose changing the way researchers allocate funds, time, and energy to global conservation. We challenge researchers not only to consider significant changes in the approach, design, execution, and dissemination of research but to also question the traditional hands-off approach of academics in catalyzing positive environmental outcomes. Certain studies and publications, especially those in high-profile journals that capture media attention, can focus attention on specific regions or subjects, but these are the exception rather than the rule. Rather, most peer-reviewed research is inaccessible to local forest managers. One key to increasing the impact of forest-related research is to make it relevant and appropriate to local land managers and regional, national, and international policy makers (Salwasser 1993; Ludwig et al. 2001; Bawa et al. 2004; Pitman et al. 2007).

Recommendations to insure that scientific research leads to effective conservation outcomes include:

1. *Changing incentives in academia.* Publication record in scientific literature has long been a principal priority for scientists (Bazzaz et al. 1998) and is a primary measure for professional advancement. This trend promotes emphasis on the intellectual merit and prolific production of research, instead of on its broader impacts and social relevance. To note, large granting institutions such as the National Science Foundation (NSF) are now requiring that the broader impacts of the research be more pragmatic instead of conceptual (e.g., National Science Foundation 2002).
2. *Applied interdisciplinary research.* There is increased awareness that social, economic, and cultural contexts ultimately lead to better more sustainable forest management solutions. Graduate programs that emphasize training in interdisciplinary environmental research (Martin and Umberger 2003) should be encouraged.
3. *Collaborative research with host tropical forest countries.* Studies of tropical forests are conducted disproportionately by US, Australian, and European research intuitions (Stocks et al. 2008). Researchers from these countries working in tropical countries should prioritize collaboration with local researchers and managers, increasing the likelihood of successful forest management.
4. *Open access to information.* The "business" of publishing has led to increasingly expensive journal subscription rates, making scientific works prohibitively inaccessible to smaller institutions, developing country researchers, land managers, and the public. Using open access publications and publishing in appropriate languages is critical.
5. *Innovative teaching, training, and outreach.* Training scientists to become better educators and communicators is critical to foster awareness of ecological issues in current and future generations and develop critically minded citizens. Scientists should be professionally trained to collaborate with science educators such as K-12 teachers, as to develop their own skills in effective education outreach while sharing their wealth of knowledge of environmental sustainability.
6. *Empowering local stakeholders.* Local participation is key for sustainable resource management (Rodriguez et al. 2007). Research investment should be

set aside for strengthening local capacity through (a) participatory approaches, including local involvement in question defining, experimental design, and training; (b) involvement and ownership as collaborators in knowledge production; and (c) dissemination of research through local media and outreach to local and global communities through social media.

References

Barnosky AD, Hadly EA, Bascompte J, Berlow EL, Brown JH, Fortelius M, Getz WM, Harte J, Hastings A, Marquet PA, Martinez ND, Mooers A, Roopnarine P, Vermeij G, Williams JW, Gillespie R, Kitzes J, Marshall C, Matzke N, Mindell DP, Revilla E, Smith AB (2012) Approaching a state shift in Earth's biosphere. Nature 486:52–58

Bawa KS, Kress WJ, Nadkarni NM (2004) Beyond paradise—meeting the challenges in tropical biology in the 21st century. Biotropica 36:276–284

Bazzaz F, Ceballos G, Davis M et al (1998) Ecological science and the human predicament. Science 282:879

Berkes F (2006) From community-based resource management to complex systems. Ecol Soc 11:45

Bonan GB (2008) Forests and climate change: forcings, feedbacks, and the climate benefits of forests. Science 320:1444–1449

Burgess JC (1993) Timber production, timber trade and tropical deforestation. Ambio 22:136–143

Butler RA (2006) Deforestation rates jump in Uganda and Burundi, fall in Rwanda. www.mongabay.com. Accessed March 08, 2008

Cardinale BJ, Duffy JE, Gonzalez A, Hooper DU, Perrings C, Venail P, Narwani A, Mace GM, Tilman D, Wardle DA, Kinzig AP, Daily GC, Loreau M, Grace JB, Larigauderie A, Srivastava DS, Naeem S (2012) Biodiversity loss and its impact on humanity. Nature 486:59–67

Central Intelligence Agency (2008) The World Factbook – Puerto Rico. ISSN 1553–8133

Chazdon RL (2008) Beyond deforestation: restoring forests and ecosystem services on degraded lands. Science 320:1458–1460

Costanza R, d'Arge R, Groot RD, Farberk S, Grasso M, Hannon B, Limburg K, Naeem S, O'Neill RV, Paruelo J, Raskin RG, Sutton P, Belt MVD (1997) The value of the world's ecosystem services and natural capital. Nature 387:253–260

Cropper M, Griffiths C (1994) The interaction of population growth and environmental quality. Am Econ Rev 84:250–254

Dirzo R, Raven PH (2003) Global state of biodiversity and loss. Annu Rev Environ Res 28:137–167

FAO (2005) State of the World's forests. Food and Agriculture Organization of the United Nations, Rome

Farrington J, Boyd C (1997) Scaling up the participatory management of common pool resources. Dev Policy Rev 15:371–391

Food and Agriculture Organization (1995) Forest resources assessment 1990: global synthesis. FAO forestry paper no. 124, Rome

Food and Agriculture Organization (2006) Global forest resources assessment 2005. FAO forestry paper no. 147, Rome

Gibbs HK, Ruesch AS, Achard F, Clayton MK, Holmgren P, Ramankutty N, Foley JA (2010) Tropical forests were the primary sources of new agricultural land in the 1980s and 1990s. Proc Natl Acad Sci 107:16732–16737

Gullison RE, Losos EC (1993) The role of foreign debt in deforestation in Latin-America. Conserv Biol 7:140–147

Hillborn R, Ludwig D (1993) The limits of ecological research. Ecol App. 3:550–552

Inman K (1993) Fueling expansion in the third-world—population, development, debt, and the global decline of forests. Soc Nat Res Intl J 6:17–39
Kaimowitz D, Angelsen A (1998) Economic models of tropical deforestation: a review. CIFOR, Bogor
Kayanja F, Byarugaba D (2001) Disappearing forests of Uganda: the way forward. Curr Sci 81:936–947
Ludwig D, Mangel M, Haddad B (2001) Ecology, conservation, and public policy. Ann Rev Ecol Syst 32:481–517
Martin PE, Umberger BR (2003) Trends in interdisciplinary and integrative graduate training: an NSFIGERT example. Quest 55:86–94
Masozeraa MK, Alavalapatib JR, Jacobson SK, Shresthab RK (2006) Assessing the suitability of community-based management for the Nyungwe Forest Reserve, Rwanda. For Policy Econ 8:206–216
National Science Foundation (2002) Merit review broader impacts criterion: representative activities. Available from http:// www.nsf.gov/pubs/2002/nsf022/bicexamples.pdf
Palo M, Salmi J, Mery G (1987) Deforestation in the tropics: pilot scenarios based on qualitative analysis. In: Palo M, Salmi J (eds) Deforestation or development in the third World. Finnish Forest Research Institute, Helsinki
Pitman NC, Azaldegui ML, Salas K, Vigo GT, Lutz DA (2007) Written accounts of an Amazonian landscape over the last 450 years. Conserv Biol 21:253–262
Rock MT (1996) The stork, the plow, rural social structure and tropical deforestation in poor countries? Ecol Econ 18:113–131
Rodriguez JP, Taber AB, Daszak P, Sukumar R, Valladares-Padua C, Padua S, Aguirre LF, Medellin RA, Acosta M, Aguirre AA, Bonacic C, Bordino P, Bruschini J, Buchori D, Gonzalez S, Mathew T, Mendez M, Mugica L, Pacheco LF, Dobson AP, Pearl M (2007) Environment—globalization of conservation: a view from the south. Science 317:755–756
Salwasser H (1993) Sustainability needs more than better science. Ecol App. 3:587–589
Shellenberger M, Nordhaus T (2004) The death of environmentalism. Available from http://www.thebreakthrough.org/images/Death_of_Environmentalism.pdf. Accessed March 2007
Stocks G, Seales L, Paniagua F, Maehr E, Bruna EM (2008) The geographical and institutional distribution of ecological research in the tropics. Biotropica 40:397–404
Sugden A, Smith J, Pennisi E (2008) The future of forests. Science 320:1435
Tenywa G (2008) Ugandan president continues to undermine national forest reserves. www.mongabay.com. Accessed March 18, 2008
Tollefson J, Gilbert N (2012) Earth summit: rio report card. Nature 486:20–23
Vuohelainen AJ, Coad L, Marthews TR, Malhi Y, Killeen TJ (2012) The effectiveness of contrasting protected areas in preventing deforestation in Madre de Dios, Peru. Environ Manag 50:645–663
Wright SJ (2005) Tropical forests in a changing environment. Trends Ecol Evol 20:553–560

Chapter 7
REDD: How Can Scientists Change the Political Jungle?

Andrew W. Mitchell

Keywords REDD • Carbon capture • Forest conservation • Climate change • Forest policy

Many years ago, I dreamed of researching the tropical forest canopy. Later, I built walkways there and found myself 70 m up in an enchanted world—where life meets the atmosphere. Little did I think that this might one day hold a key to the survival of the rain forest.

In global markets today, rain forests are worth more dead than alive. That simple fact accounts for most of their destruction, 80 % of which is caused by large-scale agribusiness growing beef, soy, palm oil, and biofuels or extracting timber. The global diaspora of corporate supply chains feeding European pigs and chickens with soy from the Amazon or providing beef for Americans and increasingly the rising Indian and Chinese middle class, or palm oil in anything from cookies, cosmetics, and vehicle fuel, creates a heavy investment transforming forest worldwide. Essentially, it is a Newtonian problem, with nothing economically recognizable pushing outwards from inside the forest. So what could possibly have the economic prowess to compete on equal terms to give a real value to forests standing up, rather than cut down? Many mechanisms such as the payment for ecosystem services and carbon credits have emerged to offset forest reduction. More recently, the REDD (reducing emissions from forest degradation and deforestation), a United Nations mechanism to reduce emissions from deforestation and degradation in developing countries, is emerging as an important economic mechanism to conserve forests.

The true economic value of REDD originates in the forest canopy. It is here that tropical forest leaves sequester about one ton of carbon per hectare per year, collectively about a billion tons annually. They store it in branches, trunks, and roots—anything from 100 to 400 t per hectare. The forest canopy is the world's

A.W. Mitchell (✉)
Global Canopy Programme (GCP), Oxford, UK
e-mail: a.mitchell@globalcanopy.org

M. Lowman et al. (eds.), *Treetops at Risk: Challenges of Global Canopy Ecology and Conservation*, DOI 10.1007/978-1-4614-7161-5_7,

most threatened natural carbon-capture-and-storage machine on earth. Today emissions from deforestation equal those from the entire global transport sector. Governments now agree that curbing this source of emissions, by paying developing countries and poorer communities not to deforest, is one of the most effective ways to combat climate change between now and 2020. But it was not always that way.

Back in 2007, 190 countries were preparing to meet in Bali at the UN Climate Conference of the Parties (COP) to decide on a road map to reduce the human emissions driving climate change. Deforestation was barely on the agenda. Most delegates were from energy sectors and did not understand the importance of forests. Some big NGOs (nongovernmental organizations) were also lining up to exclude forests as a carbon sink mechanism, believing that conventional donor-funded conservation could still win the day (even though 13 million square kilometers of destroyed forest annually over the previous decade, attested otherwise). If forests could be included in the emerging global UN carbon trading mechanism, the Global Canopy Program (GCP), a conservation organization, felt it offered a chance to raise US$ billions to help save forests from a new scalable source. Canopy researchers felt we had to act now, to readdress the balance and explain why science indicates that forests should be part of the solution, not excluded.

Just 3 months ahead of the crucial world meeting, GCP contacted researchers all over the world and invited them to sign the Forests Now Declaration, rapidly drafted with like-minded organizations. It explained how vital forests were in regulating climate at local-to-global scales and how all other efforts to mitigate climate change would be inadequate, if emissions from forests were not curbed as well. It called on governments to include the forest sector in the Bali policy action plan as a means to combat emissions through what was then called avoided deforestation, a concept put forward by Brazil and Papua New Guinea. It also made clear that forests were like the earth's beating heart, pumping water and heat worldwide, housing 60–70 % of all life on earth, and supporting the livelihoods of 1.4 billion of the world's poor.

As a prelude to the COP meeting, GCP launched the Declaration to media in Brazil situated high on a forest tower in the heart of the Amazon, flanked by representatives of the Amazonas state government, Amerindian communities, and Brazilian scientists (Fig. 7.1). The first signatures were appended 50 m up in the canopy to highlight its importance. Three months later, after taking the Declaration to New York, London, Paris, and worldwide through the Internet gathering signatures, GCP presented the Declaration in Bali at the UN Conference, by then endorsed by two heads of State, Ministers, Nobel Prize winners, many NGOs, and some of the world's most influential tropical scientists. There, along with the support of many other organizations and some bold governments, the detractors were outvoted and the forests were included under emissions and this led to the emergence of REDD.

So, what has happened since 2007? The idea of REDD to pay rain forest nations to reduce deforestation and their emissions is complicated and expensive. Sir Nicholas Stern, the Eliasch Review, and the European Union came out with reports estimating a cost ranging from US$17 to 33 billion a year to halve deforestation by 2020. So the first problem is how to raise the money; the second is how to spend it. Many issues were raised. A great proportion of rain forest nations have weak governance; how could donors be sure the money was being well used? Who should

Fig. 7.1 Launch of Forests Now Declaration on canopy tower in the Amazon. Pedro Garcia signs the Declaration with the author (far right) on behalf of the Confederation of Indigenous Organizations of the Brazilian Amazon (COIAB) which represents some 180 tribes in the region. 10 September 2007

receive it? In many cases, the tenures of the forests are often disputed: forest communities, ranchers, and the government who owns it. Would REDD lead to a cash-driven landgrab by elites, forcing poor communities out? What happens if one country pays another for a REDD credit, and then, the forest burns down or does not actually exist at all? How would one monitor, report, and verify accurately that the emissions reductions were real? Government aid alone would likely fall short of the financing required, so an additional notion is for polluting companies to pay fines, thereby offsetting their emissions, through buying forest carbon "credits." But was this not letting them off the hook? And why would they pay unless government regulation forced them to do so?

REDD rapidly became bogged down in its own mechanics and an endless global merry-go-round of international conversations. It needed a major boost to gain political momentum. Unlike a factory, a rain forest cannot be rebuilt inside 100 years—the situation of tropical deforestation needed a sense of emergency. In 2008, in response to the call of the Prince of Wales, GCP outlined a plan of action and later he launched The Prince's Rainforests Project and installed a team of 25 experts within his own home of Clarence House in London. The mission of the expert team was to write an emergency plan for forests and present this to governments to raise the substantial funds needed to halt deforestation. The Prince was extremely active, calling influential leaders worldwide to galvanize action. The resulting document "An Emergency Plan for Tropical Forests" was presented to 18 Heads of State and political leaders in St James Palace alongside the London meeting of the G20 in 2009. It provided the basis of a joint government proposal, agreed upon at the UN climate change meeting in Copenhagen, which pledged US$4.5 billion of "fast start" funds for forests. This was less than needed, but at least a beginning.

Today REDD readiness plans are being mapped out by many forested nations with the help of the UN-REDD program and the World Bank's Forest Carbon Partnership Facility. The Norwegian government pledged US$2.5 billion towards the process. REDD is not out of the woods yet, but in just 3 years, it has gone from being the outsider to as the favorite front runner in the UN climate negotiations. The coming years will tell, though, if REDD really can be a game-changing intervention, capable of turning the tide of increasing tropical forest destruction worldwide. Its evolution owes a lot to rainforest researchers, especially in Brazil, who were prepared to get out of the proverbial trees and make a "noise," in the unfamiliar jungle of global politics (www.theREDDdesk.org).

1 Explanatory Box

1.1 Community MRV

How would forest communities benefit from REDD payments? For REDD to work, indigenous and rural farming families will need to manage their forests differently to keep them standing, in return for payments for ecosystem services (PES). Funds could come from government emissions reduction schemes or private sector carbon trading. Families would be paid to "stop-move-intensify," i.e., stop deforestation at the frontier, move food production to restored, degraded land, and intensify production on existing land.

Communities will need to monitor, report, and verify (MRV) how they are doing, in order to qualify for payments. Actions will need to be accounted for within a national "low carbon development plan" whereby governments will have to report their verifiable emissions reductions to the United Nations, as well as other metrics. GCP is working with 16 Makushi indigenous communities in Guyana as a pilot region. Monitors are being trained to use rugged handheld mobile phones with android platforms to run data collection software under development by Google. Using both traditional and scientific knowledge, the communities are mapping their lands, developing indicators of ecosystem services that their forests deliver, determining local drivers of forest loss, and monitoring community well-being. This service could form the basis of receiving payments under the future REDD scheme and lead to increased employment in remote forested areas (www.globalcanopy.org).

Chapter 8
Narrowing Global Species Estimates

Nigel E. Stork and Andrew J. Hamilton

Keywords Global species estimates • Arthropods • How many species

Bullet Points

1. Recent estimates of global species richness are converging on 5 ± 3 million species, suggesting that higher estimates of 30 million species or more are extremely unlikely.
2. One reason that global species estimates are lower than many previously considered is because scaling up from local species richness to regional species richness rarely more than a factor of 10 or 20.
3. Further, at least one study indicates that both canopy and ground strata are of tropical rainforests equally likely to produce new undescribed species.

Summary

In the early 1980s, new discoveries suggested the tropical rainforest canopy was the 'last biotic frontier' and was believed to be home for many millions of undescribed species. It was thought that 30 million species or more tropical forest arthropods, mostly residing in the canopy, might exist. New estimates for all species on Earth, including insects and other arthropods, are converging on lower and a recent review suggests that 5 ± 3 million species might be a useful summary estimate. For the first

N.E. Stork (✉)
Environmental Futures Centre, Griffith School of Environment, Griffith University, Nathan 170 Kessels Road, Nathan, QLD 4111, Australia
e-mail: nigel.stork@griffith.edu.au

A.J. Hamilton
Department of Agriculture and Food Systems, The University of Melbourne, Dookie Campus, 940 Dookie-Nalinga Road, Dookie College, VIC 3647, Australia
e-mail: andrewjh@unimelb.edu.au

M. Lowman et al. (eds.), *Treetops at Risk: Challenges of Global Canopy Ecology and Conservation*, DOI 10.1007/978-1-4614-7161-5_8,

time, these estimates all have statistically robust confidence intervals which suggest that estimates of 30 million species or more are exceedingly unlikely. Other studies of the relationship of local species richness to regional species diversity show regional numbers of species are rarely more than a factor of 10–20 higher than local measures, demonstrating why many are misled into believing high local species richness might mean high global richness.

1 Introduction

In the latter half of the twentieth century, biologists using knockdown insecticides discovered the previously unrecognised and enormous diversity of life, particularly of insects and arthropods, in the canopy of tropical and temperate forests (Erwin 1982; Stork 1988). Such discoveries led to the suggestion that the canopy is 'the last biotic (or biological) frontier' (Stork 1995). Some speculated that the Earth held 30 million or more arthropod species and that most of these were undescribed species in the canopy of tropical forests (Erwin 1982, 1988, 1991). These discoveries coincided with revelations that tropical forests were being harvested or cleared at alarming rates around the world and led to many leading biologists suggesting that this might lead to the extinction of a large proportion of tropical species. These issues were highlighted by Wilson and Peter (1988) and were also of critical importance in driving the signing of the Convention on Biological Diversity by 193 countries (subsequently ratified by 168 countries) in 1992.

Past estimates of global species richness are derived principally from (1) relatively simple extrapolations from the number of insect species specific to different species of trees, (2) extrapolations of the relative proportions of known against unknown faunas, (3) simple models of body sizes of species, and (4) best guesses from specialist taxonomists for their groups (Stork 1988; May 1988, 1990, 1997; Stork 1993, 1997, 1999). All of these methods have major weaknesses, mostly the level of uncertainty in the validity of the assumptions made and also the need for extreme extrapolations due to the lack of initial observations. What all of these lacked were measures of confidence in the assumptions.

Determining how many species there are on Earth including those have yet to be discovered and described is a complex task since taxonomists are not certain how many have been described already. Estimates suggest that 1.5–1.8 million species have been named and described so far and about two-thirds of these have been catalogued in various catalogues. Taxonomists sometimes disagree on what constitutes a species because individuals can vary so much and because some species are accidentally described several times. The proportions of such synonyms range from over 80 % to 90 % in some algal genera, 7–80 % (32 % overall) for insects, 33–88 % for groups of seed plants, 38 % for world molluscs, 50 % for marine fish, and 40 % for all marine species (Costello et al. 2013). Considering past rates of synonymy, it is probable that overall at least 20 % of the currently estimated described species are yet to be discovered synonyms (May 1997; Costello et al. 2013). There are only a few groups of organisms where it is likely all or virtually all species on Earth have been described, such as the birds and large mammals. For others, such as the insects, nematodes, and fungi, the task is still far from complete.

2 Recent New Estimates of Global Species Richness

In a recent review of how many species there are and current extinction rates, it was argued that overall estimates of global species numbers are converging on around 5 ± 3 million species (Costello et al. 2013). This is largely based on four new estimates (Table 8.1). Two of these use the new and extensive (but as yet incomplete) catalogues of described species to provide estimates for all species. The first used extrapolation of the rate of discovery of selected higher taxa to predict 8.7 million (±1.3 million SE) eukaryotic species globally, of which 25 % (2.2 million ± 0.18 million SE) were marine (Mora et al. 2011). The second modelled rates of species description and predicted that 23.7–30.9 % (CI_{95}) of marine species and 21.3–29.1 % of terrestrial species await description. From this and using estimates of species described to date, they suggest that there are about 1.8–2.0 million species on Earth, 0.3 million of which are marine (Costello et al. 2012). A third approach used uncertainty analysis of the specificity of arthropods on tropical trees. Two related models were presented, with one predicting 6.1 [CI_{90} 3.6, 11.4] million species and other 7.8 [CI_{90} 3.9, 13.7] (Hamilton et al. 2010, 2011). Sensitivity analyses on these models revealed that the proportion of arthropods found in the canopy had the largest influence on uncertainty in the prediction. The first of these two models was subsequently subject to further analysis using probability bounds analysis, so as to remove assumptions about distributional form for parameters and dependence between parameters (Hamilton et al. 2013). This arguably overconservative approach predicted bounds around the median estimate of 2.4–20.0 million species.

Table 8.1 Four recent models pertinent to the understanding of global species richness

Study	Quantity directly estimated	Method	Estimation (millions of species)
Hamilton et al. (2010, 2011)	Terrestrial tropical arthropod spp. richness (millions)	Extrapolation from samples, latin hypercube sampling	Model A: 6.1 [CI_{90} 3.6, 11.4] Model B: 7.8 [CI_{90} 3.9, 13.7]
Hamilton et al. (2013)	Terrestrial tropical arthropod spp. richness (millions)	Extrapolation, probability bounds analysis	2.4–20.0 at the median prediction of 6.1
Mora et al. (2011)	Global spp. richness (millions)	Extrapolation from higher taxa	8.7 (SE ±0.18)
Costello et al. (2012)	% yet to be discovered (percentage)	Modelling rates of description	Marine [CI_{95} 23.7, 30.9] Terrestrial [CI_{95} 21.3, 29.1]

Two of the recent estimates suggest that marine organisms comprise only about 16 % of all species on Earth. This is surprising given that, first, oceans cover 70 % of the surface and, second, that at the higher taxonomic level of phylum, most phyla are found in the marine environment with the terrestrial fauna being less rich. The largest group in terms of species, after the arthropods, is the plants with an estimated 400,000 species, and most of these are terrestrial. The explosion of diversity that arose through the evolution of the flowering plants and the associated insects that live and feed on them may account for the apparent disproportionate proportions of these two groups from the total.

3 Extrapolating from Local to Global Species Richness

One of the interesting problems in estimates of species richness is the relationship between local species estimates or counts and regional or global estimates and counts. What many studies show is that where local sites are well studied or sampled, the local species counts are surprisingly high and that regional estimates are usually less than an order of magnitude greater, demonstrating that local species pools can be very large. Here, we discuss a few examples that demonstrate these findings.

Most flowering plants have been described and the number of angiosperms is usually estimated to be around 400,000 species. The Smithsonian's Centre for Forest Science (CFS) coordinates plant species data for some 40–50 forest plots each of about 20–50 ha in size from around the world. A few of the 25–50 ha tropical rainforest plots have very high numbers of species. The Yasuni 25 ha plot in Ecuador, the richest studied plot in the world, has 1,104 tree morphospecies species (Valencia et al. 1994). Hubbell et al. (2008) extrapolated that there might be 11,210 tree species with trunks reaching above 10 cm in diameter in the entire Amazon basin, an area 7 million km^2 and 25 million times as large as the Yasuni plot. Collectively, all of the CFS plots measure just a few square kilometres in size but contain almost 20,000 species of plants. In other words, about 5 % of the world's plant species have been counted from less than 0.0001 % of the Earth' terrestrial surface (148.94 million km^2). Clearly, as additional areas are added, particularly locally, the accumulative number of plant species slows down very rapidly. Of course, just how many additional local endemic species there are may change these figures.

For insects and other arthropods, there are few equivalent studies, but again, regional species richness may not be much less than local species richness. For example, Robins and colleagues (Robbins et al. 1996) collected 1,300 species of butterfly from one small area of lowland tropical forest in south-east Peru through a few weeks sampling over 2–3 years, and yet, there are probably only 5,000 species of butterfly in South America. Peter Hammond (pers. comm.) collected over 1,000 species of beetles over more than 20 years in an urban but ancient oak woodland in the outskirts of London (Richmond Park), and yet, the total beetle fauna for the British Isles is little more than 4,000 species. Elsewhere, Hammond (Hammond 1990) led a team of entomologists from the Natural History Museum to extensively sample a small area of forest in north Sulawesi, Indonesia, for 1 year. The beetles were the most comprehensively sampled taxon with over 6,000 species. What Hammond and others demonstrated is that a wide range of sampling methods was required to adequately sample the insect fauna of an area. Collecting methods such as Malaise traps, light traps, Berlese funnels, or canopy fogging sampled the local fauna in their own peculiar ways with some species being collected by several or all of these methods and many other species only being collected by one method. Furthermore, the scale of the taxonomic challenge in sorting the very large samples to species required a team of specialists and was beyond the ability of single individuals.

Basset and colleagues (Basset et al. 2012) recently undertook the most comprehensive sampling program of insects at any single site in the tropical world. They collected 6,144 arthropod species, sampling from the ground up into the canopy, in just 0.48 ha and estimated that the surrounding 6,000 ha forest reserve most likely contained around 25,000 species of arthropods. Further, they found that models of insect-plant diversity fitted the species accumulation curve well giving support to estimates of global species richness of arthropods cited above (Hamilton et al. 2010, 2011, 2013).

4 Is the Forest Canopy the Last Biotic Frontier?

Earlier, we discussed the finding that more than 80 % of species are found in terrestrial environments but the question still remains: Is the rainforest canopy still the last biotic frontier? One way to answer this question would be to sample equally from the canopy and the ground and to determine which had the greater species richness and which held a greater proportion of undescribed species. This is not a simple task as the sampling method needs to be equally efficient in both strata. In addition, determining whether or not a morphospecies in a sample is one that has been described and named is not a simple task. One such test (Stork et al. 2008) used Malaise traps modified to also be flight interception traps to sample the canopy and the ground at the canopy crane site in the Daintree in Australia. Over 4 years of sampling, 1,473 beetle species were collected. A subsample of 17 families was selected for analysis. These included 156 species, 96 of which were found to be previously described and 60 which were considered to be undescribed. Stork et al. (2008) showed that described species were significantly more likely to be in the canopy and were more likely to be larger. Undescribed species were found to be just as likely to be found near the ground as in the canopy. So, from this study at least, it appears that both the ground and the canopy are important frontiers for discovery of species. The Daintree forests are relatively well studied compared to other tropical rainforests, and hence, the proportion of known to unknown species in the canopy compared to the ground may differ elsewhere.

Acknowledgements We thank Bill Laurance for his comments on an early draft of this chapter.

References

Basset Y et al (2012) Arthropod diversity in a tropical forest. Science 338:1481

Costello MJ, Wilson S, Houlding B (2012) Predicting total global species richness using rates of species description and estimates of taxonomic effort. Syst Biol 61:871

Costello MJ, May RM, Stork NE (2013) Can we name Earth's species before they go extinct? Science 339:413

Erwin TL (1982) Tropical forests: their richness in Coleoptera and other arthropod species. Coleopts Bull 36:74

Erwin TL (1988) The tropical forest canopy: The heart of biotic diversity. In: Wilson EO, Peter FM (eds) Biodiversity. National Academy Press, Washington, DC, pp. 123–129

Erwin TL (1991) How many species are there?: Revisited. Conserv Biol 5:1

Hamilton AJ, Basset Y, Benke KK, Grimbacher PS, Miller SE, Novotny V et al (2010) Quantifying uncertainty in estimation of global arthropod species richness. Am Nat 176:90–95

Hamilton AJ, Basset Y, Benke KK, Grimbacher PS, Miller SE, Novotny V et al (2011) Correction. Am Nat 177:544–545

Hamilton AJ, Basset Y, Benke KK, Grimbacher PS, Miller SE, Novotny V, et al (2013) Estimating global species richness: refining probabilistic models using probability bounds analysis. Oecologia 171:591–600

Hammond PM (1990) Insect abundance and diversity in the Dumoga-Bone National Park, N. Sulawesi, with special reference to the beetle fauna of lowland rainforest in the Toraut region. In: Knight WJ, Holloway JD (eds) Insects and the rain forests of South East Asia (Wallacea). Royal Entomological Society, London, pp. 197–254

Hubbell SP et al (2008) How many tree species are there in the Amazon and how many of them will go extinct? Proc Natl Acad Sci 105:11498

May RM (1988) How many species are there on Earth? Science 241:1441

May RM (1990) How many species? Philos Trans R Soc Lond B 330:293

May RM (1997) The dimensions of life on earth. In: Raven PH, Williams T (eds) Nature and human society: the quest for a sustainable world. National Academy Press, Washington, DC, pp. 30–45

Mora C, Tittensor DP, Adl S, Simpson AGB, Worm B (2011) How many species are there on Earth and in the ocean? Plos Biol 9

Robbins RK, Lamas G, Mielke OHH, Harvey DJ, Casagrande MM (1996) Taxonomic composition and ecological structure of the species-rich butterfly community at Pakitza. Parque Nacional del Manu, Peru

Stork NE (1988) Insect diversity: facts, fiction and speculation. Biol J Linn Soc 35:321

Stork NE (1993) How many species are there? Biodivers Conserv 2:215

Stork NE (1995) Last biological frontier. Sci NY 35:3

Stork NE (1997) Measuring global biodiversity and its decline. In: Reaka-Kudla ML, Wilson DE, Wilson EO (eds) Biodiversity II: understanding and protecting our biological resources. Joseph Henry, Washington, DC, pp. 41–68

Stork NE (1999) The magnitude of biodiversity and its decline. In: Cracraft J, Grifo FT (eds) The living planet in crisis: biodiversity science and policy. Columbia University Press, New York, pp. 3–32

Stork NE et al (2008) What determines whether a species of insect is described? Evidence from a study of tropical forest beetles. Insect Conserv Divers 1:114

Valencia R, Balslev A, Paz YMC (1994) High tree alpha diversity in Amazonian Ecuador. Biodivers Conserv 3:21–28

Wilson EO, Peter FM (eds) (1988) Biodiversity. National Academy Press, Washington, DC

Part II
Climate Change

Introduction

Climate change is upon us, from the hot and humid tropics to the frozen poles. As global leaders debate on how to mitigate the effects of climate change, there is a growing body of knowledge on how climate change interferes with ecosystem functioning, especially in the forest canopies. Situated at the interface between the biosphere and the atmosphere, forest canopies not only influence microclimatic changes in a forest but also have implications at regional levels. Though a higher frequency of abnormal climatic conditions have been recorded in recent years, climate change is still a relatively slow process. Modelling responses of the biota to climate change has become an increasingly critical (and challenging) process as witnessed in the chapters of this section. However, it is critical that the models are validated with long-term monitoring; such efforts remain almost non-existent for forest canopies.

Areas of knowledge that have been overlooked but innovative insights to forest health are summarised in this section. These include the frequency and impact of lightning and cyclones, as well as reviews of some of the innovative studies in canopy construction cranes and with collaborative teams. All of these approaches are critical and may serve to inspire more active climate mitigation responses from policy makers, thereby minimising the risk to our canopies and planet earth in general.

Chapter 9
Tropical Cyclones and Forest Dynamics Under a Changing Climate: What Are the Long-Term Implications for Tropical Forest Canopies in the Cyclone Belt?

Stephen M. Turton

Keywords Tropical cyclones • Forests • Disturbance • Climate change • Dynamics • Structure • Fragmentation

Summary

Tropical cyclones (also known as hurricanes and typhoons) affect moist tropical forests in eight ocean basins around the world: northwest Pacific, north Indian, southwest Indian, southwest Pacific, southeast Pacific, northeast Pacific, north Atlantic/Caribbean and south Atlantic. Frequency and intensity of tropical cyclones varies greatly across these regions, as well as being strongly affected by the El Niño-Southern Oscillation cycle.

Outside equatorial forest regions, tropical cyclones are the major disturbance phenomena for moist tropical forests in exposed coastal areas where they play a significant role in shaping forest structure and species composition. Extensive forest fragmentation and habitat loss in many forest regions in the global cyclone belt have increased the vulnerability of remaining forest remnants to tropical cyclones due to their high forest edge to area ratios.

Tropical cyclone intensity is predicted to increase significantly over this century, while cyclone frequency is expected to remain stable or decline, depending on the oceanic basin under consideration. Changes in cyclone intensity have profound implications for tropical forest structure and composition in forested landscapes located in the global cyclone belt. Moreover, many of these landscapes contain highly fragmented forests (e.g. eastern Madagascar, northeast Australia, northeast Philippines), thereby increasing their vulnerability to cyclonic winds due to enhanced edge effects and risks of post-disturbance fires and weed

S.M. Turton (✉)
Centre for Tropical Environmental & Sustainability Sciences, School of Earth and Environmental Sciences, James Cook University, PO Box 6811, Cairns, QLD 4870, Australia
e-mail: steve.turton@jcu.edu.au

M. Lowman et al. (eds.), *Treetops at Risk: Challenges of Global Canopy Ecology and Conservation*, DOI 10.1007/978-1-4614-7161-5_9,

invasions. There is a real risk of a phase shift to vegetation types dominated by disturbance species, at the expense of cyclone intolerant species, which may lead to declines in biodiversity. This chapter evaluates the latest climate change science in regard to tropical cyclones and discusses the implications of increased cyclone intensity for forest structure and composition, particularly for what are now dominant fragmented forest landscapes.

1 Introduction

Natural and anthropogenic disturbances shape forest ecosystems by modifying their structure and species composition and functional processes, such as nutrient cycling (Dale et al. 2001). Tropical moist forests of the world are regularly affected by a plethora of natural disturbances, including infrequent but damaging wild fires, droughts, floods, occasional landslides, severe thunderstorms and tropical cyclones (Turton and Stork 2008). All these natural disturbances interact in complex ways with anthropogenic disturbances across the landscape, such as forest clearing and habitat fragmentation and degradation (Turton 2012). Outside equatorial forests, tropical cyclones are among the most significant disturbance agents regularly upsetting equilibrium of tropical moist forests in exposed coastal and near-coastal regions.

Tropical cyclone intensity is predicted to increase significantly over the twenty-first century under anthropogenic climate change, while cyclone frequency is expected to remain stable or decline, depending on the oceanic basin under consideration (Knutson et al. 2010). Changes in cyclone intensity have profound implications for tropical forest structure and composition in forested landscapes located in the world's tropical cyclone belt. Moreover, many of these landscapes contain highly fragmented forests, thereby increasing their vulnerability to cyclonic winds due to enhanced edge effects and risks of post-disturbance fires and weed invasions. There is a real risk of a phase shift to vegetation types dominated by disturbance species (including weeds), at the expense of cyclone intolerant species, which may lead to declines in biodiversity and profound changes in forest structure (Turton 2012). This chapter will evaluate the state of knowledge of interactions between tropical cyclones and moist tropical forests and will then discuss the implications of increased cyclone intensity under projected anthropogenic climate change for forest structure and floristic composition, particularly for the remaining moist forest landscapes in the cyclone belt.

2 Tropical Cyclones

Tropical cyclones (also known as hurricanes and typhoons) form in response to an initial atmospheric disturbance over warm tropical oceans with sea surface temperatures over 27 °C and range in intensity from comparatively weak systems, where maximum wind speeds do not generally exceed 160 km h^{-1}, to extremely destructive, where maximum wind gusts have been recorded in excess of 350 km h^{-1} (Turton 2008). In addition to differences in cyclone intensity, the size of tropical

Table 9.1 Regions of the world containing remaining tropical moist forest that is subject to disturbance from tropical cyclones at a range of temporal scales

Ocean (season)	Moist tropical forest regions affected by cyclones	Frequency
Northwest Pacific (all months)	Northern and central Philippines, southern China, Taiwan, Vietnam	High–very high
North Indian (all months)	Western and eastern India, Bangladesh, Myanmar, Sri Lanka	Low
Southwest Indian (November–May)	Eastern Madagascar, Mozambique, southern Tanzania, Mauritius, La Reunion	Medium–high
Southwest Pacific (November–April)	Northeast Australia, southern Papua New Guinea, Solomon Islands, Vanuatu, New Caledonia, Fiji, Samoa	Medium
Southeast Pacific (November–April)	Cook Islands, French Polynesia	Very low (strong El Niño years only)
Northeast Pacific (May–November)	Western Central America, Hawaiian Islands (rarely)	High
North Atlantic/Caribbean (June–November)	Outer and lesser Antilles, Nicaragua, Honduras, Belize, Guatemala, western Mexico, south and southeast USA	Medium–high
South Atlantic (November–April)	Southeast Brazil	Extremely low

cyclones (as determined by the radius of gale force winds extending around the center) varies greatly, from one system to another. For example, Super Typhoon Tip had a massive radius of 1,110 km compared with Severe Cyclone Tracy's very small 48 km radius. Hence, large-diameter systems will affect a much greater area of the Earth's surface compared with small-diameter systems. However, wind speed near the center of the cyclone is a function of the intensity of the cyclone and is independent of the spatial size of the system.

Tropical cyclones affect all tropical ocean regions of the world but very rarely the southeast Pacific and South Atlantic where sea surface temperatures are generally too low throughout the summer months (McGregor and Nieuwolt 1998). Due to the lack of Coriolis force, tropical cyclones do not occur within about 5–7° of the equator. The northwest Pacific has the highest frequency of tropical cyclones of all the ocean basins in the world followed by the northeast Pacific, while the north Indian has the lowest frequency (McGregor and Nieuwolt 1998).

2.1 *Tropical Cyclones and Tropical Moist Forests*

Table 9.1 summarises regions of the world where tropical moist forests are affected by tropical cyclones at varying frequencies. With the exception of the northwest Pacific and north Indian, where tropical cyclones have been recorded in every month of the year, the remaining regions experience cyclones in the summer months with a peak in the late summer to early autumn (McGregor and Nieuwolt 1998).

2.1.1 Landscape and Local-Scale Patterns and Processes

Tropical cyclones affect forest landscapes at both the landscape and local scales (Boose et al. 1994; Grove et al. 2000; Turton 2008). At the landscape scale (>10 km), effects are the result of complex interactions of anthropogenic (e.g. forest fragmentation), meteorological (e.g. storm intensity and size), topographical (e.g. windward versus leeward slopes) and biotic factors (e.g. forest type). By comparison, at the more local scale (<1 km), effects are largely influenced by forest structure and composition, as well as smaller-scale topographical features.

2.1.2 Ecological Effects

Tropical cyclones have both visible and invisible effects on forests (Everham and Brokaw 1996; Lugo 2008). Visible effects are pervasive and include widespread defoliation of canopy trees and removal of vines and epiphytes, along with breakage of crown stems and significant tree falls (Turton 2012). Significant changes in canopy cover and density result in profound changes in under- and mid-storey microclimates and complex responses to newly created light, temperature and moisture regimes (Everham and Brokaw 1996). Cyclonic disturbance also accelerates the invasion of exotic tree and understorey weed species leading to a decline in native species (Bellingham et al. 2005).

Table 9.2 summaries the main ecological effects of Tropical Cyclone Larry on multi-use landscapes in northeast Australia. This body of research is restricted to evaluating the immediate visible effects of the storm on vegetation and selected fauna and tracks their state and condition 6–12 months after the cyclonic event. Interestingly, the size of forest remnants did not appear to correlate with vegetation damage after Cyclone Larry; rather proximity to the cyclone's track was a better indicator of forest damage. A number of endemic and rare rainforest mammal species (several ringtail possum species, tree kangaroos and flying foxes) showed remarkable resilience to the impacts of the cyclone, while the flightless southern cassowary was found to be highly vulnerable, particularly in areas of highly fragmented forest in a mixed agricultural/urban matrix. After the cyclone, 82 % of cassowary deaths occurred on the roads as birds wandered out in search of food. Riparian vegetation along the region's waterways was particularly susceptible to wind damage with subsequent invasions by weeds.

Compared to the visible effects of tropical cyclones on forests, invisible effects are less well documented in the literature (Lugo 2008). Examination of invisible or hidden effects requires well-focused research both in the short- and long-term time scales, which is challenging. Lugo (2008) proposes that the ecological role of tropical cyclones in forests involves six principal effects:

- They change the ecological space available to organisms.
- They set organisms in motion.

Table 9.2 How Cyclone Larry affected fragmented and intact rainforests, plantations and restoration plantings in the Wet Tropics region of northeast Australia (Turton 2012)

Main ecological effects of Cyclone Larry on rainforest ecosystems and their biota
• Forest structure, measured up to 6 months after Cyclone Larry, did not differ between small (<40 ha) fragments and larger (intact) forested areas; the severity of effects in both was dictated by proximity to the cyclone's eye and track.
• Forest structure was not different between small linear fragments and edges and interiors of larger forest tracts—a year after the cyclone, its main effect was increased local spatial variability in all forests.
• A surprising finding was that secondary vegetation that developed along two artificial edges created during forest fragmentation (a road and a power line lane) may have buffered forest interiors against changes in understorey moisture and light more effectively than a long-standing natural edge along a stream course.
• Thirty-five percent of the regional endangered cassowary population was killed directly during Cyclone Larry but those birds that survived and ventured beyond the fragments suffered even higher mortality—struck by motor vehicles or attacked by dogs.
• In coastal areas, planted trees in riparian sites were more severely damaged by the cyclone than trees in nearby forest fragments. However, on the Atherton Tableland where wind speeds were lower, planted rainforest trees were not severely damaged.
• There were no reductions in populations of five species of folivorous marsupials, measured 6–8 months after the cyclone, in fragmented Mabi forests within pastoral landscapes compared with pre-cyclone estimates.
• Numbers of frugivorous birds were much reduced 2 weeks after Cyclone Larry in severely affected forest fragments but were at pre-cyclone levels by seven months after the cyclone.
• Before Cyclone Larry, the regional population of the flying fox *Pteropus conspicillatus* roosted in large camps but afterwards altered its roosting behaviour so that there were many small camps, presumably in response to food limitation.
• Non-native plant species (weeds) germinated in the rainforest in the first 6–9 months after cyclone disturbance, but little since. Most weed species were ephemeral and unlikely to persist during forest recovery.
• Trees with high woody density in rainforests affected by Cyclone Larry showed greater resistance.
• The abrupt decrease in wind speeds away from the eye of Cyclone Larry meant that damage to widespread tree species could be assessed across a gradient of wind severity, and most showed consistent responses.

- They increase the heterogeneity of the landscape and the variability in ecosystem processes.
- They rejuvenate the landscape and its ecosystems and redirect succession.
- They shape forest structure, influence their species composition and diversity and regulate their function.
- They induce evolutionary change through natural selection and ecological creativity through self-organisation.

When considering these ecological effects, Lugo (2008) distinguishes between immediate effects (0–3 years), immediate responses (0–20 years), trajectories of responses (0–100 years) and long-term legacies (>100 years).

3 Likely Effects of Climate Change on Tropical Cyclones and Tropical Forests

Lugo's (2008) heuristic model provides an excellent framework for studying interactions between tropical cyclones and forest ecosystems across a range of spatial and temporal scales. One of the research priorities listed in his global review urges more research on the anticipated effects of anthropogenic climate change on tropical cyclone frequency and intensity and the likely consequences for forest ecosystems in the global tropical cyclone belt.

Globally, climate change and associated warming of tropical oceans and the atmosphere is predicted to increase the intensities of tropical cyclones this century while having largely neutral effects on cyclone frequency (see review by Turton 2012). Recent research by Knutson et al. (2010), based on theory and high-resolution dynamical models, consistently suggests that greenhouse warming will cause the globally averaged intensity of tropical cyclones to shift towards stronger storms, with intensity increases of 2–11 % by 2100. However, existing modelling studies also project decreases in the globally average frequency of tropical cyclones, by 6–34 %, depending on the oceanic basin under consideration (Knutson et al. 2010). Changes in cyclone intensity and frequency have profound implications for tropical forests in the cyclone belt and raise many questions about the long-term sustainability of tropical forests:

- Will tropical cyclones become much more common in forest regions, such as southeast Brazil and the central Pacific, as a consequence of global warming (see Table 9.1)?
- Will there be a poleward shift in the range of tropical cyclones bringing them into contact with forest ecosystems that currently do not experience severe cyclones?
- What are the implications of more intense tropical cyclones for our tropical forests and adjacent agricultural and plantation landscapes?
- Will forest biodiversity decline in these landscapes?
- Will forest structure change over time towards a generally lower stature?
- Will a shift in the severity spectrum for tropical cyclones tend to favour species more resistant to strong winds, such as palms and some tree species, and species more tolerant of post-disturbance stresses, such as pioneers, vines and weeds?
- How will contiguous forests fare compared with forest remnants largely surviving in agricultural matrices?
- How might changes in cyclone intensity act synergistically with other ongoing changes in forest ecosystems, such as those driving habitat loss and fragmentation and associated desiccation and increased fire risk?

Finally, perhaps we should adopt Lugo's (2008) stance that given predicted changes in cyclone intensity, coupled with forest fragmentation and habitat loss, we need to rethink terminology applied to ecological effects of tropical cyclones on forest ecosystems and their biota (Fig. 9.1). Many of the terms used in the past and current ecological disturbance literature in relation to tropical cyclones and forests would seem to be highly inappropriate and perhaps better suited to describing

Fig. 9.1 Rethinking ecological terminology applied to forest ecosystems in the aftermath of disturbance from a tropical cyclone (After Lugo 2008)

'impacts'	→	'affects'
'destroyed'	→	'modified'
'catastrophic'	→	'changed'
'damaged'	→	'effects'
'recovery'	→	'trajectories of response'

interactions between tropical cyclones and human communities, infrastructure and primary production systems.

Moreover, synergies between anthropogenic climate change, more extreme weather events, habitat loss and fragmentation mean we may longer expect most forest ecosystems to recover to their original state in the aftermath of a severe tropical cyclone. Instead, we should prepare ourselves for witnessing trajectories of responses that may culminate, over time, in a forest ecosystem that is structurally and floristically very different to its original pre-disturbance state.

References

Bellingham PJ, Tanner EVJ, Healey JR (2005) Hurricane disturbance accelerates invasion by the alien tree *Pittosporum undulatum* in Jamaican montane rain forests. J Veg Sci 16:675–684

Boose ER, Foster DR, Fluet M (1994) Hurricane impacts to tropical and temperate forest landscapes. Ecol Monogr 64:369–432

Dale VH, Joyce LA, McNulty S, Neilson RP, Ayres MP, Flannigan BJ, Watton BM (2001) Climate change and forest disturbances. Bioscience 51:723–34

Everham EM, Brokaw NVL (1996) Forest damage and recovery from catastrophic wind. Bot Rev 62:113–73

Grove SJ, Turton SM, Siegenthaler DT (2000) Mosaics of canopy openness induced by tropical cyclones in lowland rain forests with contrasting management histories in northeastern Australia. J Trop Ecol 15:883–894

Knutson TR, John L, McBride JL, Chan J, Emanuel K, Holland G, Landsea C, Held I, Kossin JP, Srivastava AK, Sugi M (2010) Tropical cyclones and climate change. Nat Geosci 3:157–163

Lugo AE (2008) Visible and invisible effects of hurricanes on forest ecosystems: an international review. Austral Ecol 33:368–98

McGregor GR, Nieuwolt S (1998) Tropical climatology: an introduction to the climates of the low latitudes. Wiley, Chichester

Turton SM (2008) Landscape-scale impacts of Cyclone Larry on the forests of northeast Australia including comparisons with previous cyclones impacting the region between 1858 and 2006. Austral Ecol 33:409–16

Turton SM (2012) Securing landscape resilience to tropical cyclones in Australia's wet tropics under a changing climate: lessons from Cyclones Larry (and Yasi). Geogr Res 50:15–30

Turton SM, Stork NE (2008) Impacts of tropical cyclones on forests in the wet Tropics of Australia. In: Stork NE, Turton SM (eds) Living in a Dynamic Tropical Forest Landscape, Blckwell Publishing, Oxford, UK pp 47–58

Chapter 10
Canopies and Climate Change

Claire M.P. Ozanne

Keywords Climate change • Canopy • Biodiversity • Invertebrates • Elevated CO_2

Bullet Points

1. Forest canopies are a critical interface between terrestrial Earth and the atmosphere.
2. Trees and their inhabitants are invariably subject to changes in climate, and documentation of their responses is an emerging priority for forest research.

1 Introduction

At the interface between the biosphere and the atmosphere, the forest canopy both experiences and influences changes in temperature and humidity, gas fluxes and air chemistry and the physical effects of wind and precipitation. The canopy is exposed to conditions not found elsewhere in the forest structure; thus trees, epiphytes, vertebrates, invertebrates and microbes inhabiting the edges of the canopy must be adapted physiologically and behaviourally to either tolerate or avoid these extremes. Here we focus on the impact of the key factors in the climate change models. From our understanding of forest ecosystem dynamics and recent experimental and observation studies, what can be said about responses within the canopy to the changes in climatic parameters predicted by the models?

C.M.P. Ozanne (✉)
Department of Life Sciences, Centre for Research in Ecology, University of Roehampton, Holybourne Avenue, London SW15 4JD, UK
e-mail: c.ozanne@roehampton.ac.uk

M. Lowman et al. (eds.), *Treetops at Risk: Challenges of Global Canopy Ecology and Conservation*, DOI 10.1007/978-1-4614-7161-5_10,

1.1 Tree Responses to Elevated Temperature and CO_2

Climate change models estimate a range of temperature changes from 1.8° to 4.0° over the next 100 years (IPCC 2007) and forests are already responding to recent changes in average global temperature. Upward movement of range margins have been recorded for a number of trees in Sweden (e.g. *Picea abies*—240 m over 50 years (Kullman 2002)). However, recent work on large-scale data sets in the USA predicts range contractions of many tree species (Zhu et al. 2011), which would result in significant changes in the species complement of forests with marked implications for canopy composition and structure.

In the canopy itself, phenological response to temperature change is particularly significant, especially the events associated with bud burst, flowering and leaf fall. In temperate regions, first leaf dates advanced by 1.2 days per decade between 1955 and 2002 (Scheifinger et al. 2003), and in Europe, autumn has shown a delay of 1.3 days per decade (Menzel et al. 2006). In European forests, a longer growing season could be advantageous for some species across the elevation gradient (e.g. *Quercus petraea* over *Fagus sylvatica)* (Vitasse et al. 2011). In the tropics, phenology is more sensitive to precipitation than temperature changes per se; however, flowering on a mass scale, particularly amongst the Dipterocarpaceae, may be triggered by a drop in nighttime temperatures of about 2 °C (Appanah 1993). Nighttime temperatures in the tropics have risen more than daytime temperatures, partly as a result of increased cloud cover, signalling potential problems for the persistence of these biologically and commercially important species.

At the extreme ends of climate change predictions where drought occurs, tree mortality will increase. There is the potential for marked spatial variation in dieback, with trees on the forest edge suffering greater drought-induced mortality. The complex interactions between climate change, tree mortality and the feedback that would result if dieback occurred on a large scale (carbon balance and loss of canopy cover) have been cited as a potential non-linear tipping element in the climate system (Allen et al. 2010).

The initial response of forest trees to elevated CO_2 is predicted to be an average increase in net primary productivity of about 23 %—the so-called CO_2 fertilisation effect (Norby et al. 2005). However, whether this increase is sustained will depend on interactions with other factors including soil nutrient availability. In particular, the uptake and availability of nitrogen could become limiting, but the effects are likely to remain highly ecosystem specific (Lukac et al. 2010) and perhaps tree species specific where there are variations in physiology and mycorrhizal relationships. Lukac et al. (2010) suggest that there will be complex responses to the interactions between elevated CO_2, temperature and water availability. The broader reach of the response of forests to climate change are hard to track in detail, but the outcomes will impact on physical structure of the canopy and on the vertebrates, epiphytes, invertebrates and microbes associated with trees that become more or less prevalent.

2 Canopy Invertebrates and Climate Change

Invertebrates are major contributors to the diversity of organisms in forest canopies. Twenty-five percent of species in the canopy are predicted to be invertebrates and 10 % of canopy invertebrates are predicted to be specialists (Ozanne et al. 2003). These animals play essential roles in fundamental ecosystem processes in forests such as pollination, herbivory, decomposition and parasitism/predation. As with forest trees, forest insects will respond at the individual and population level to changes in the key environmental parameters, and this will impact on their contribution to the forest ecosystem. Range shifts are already documented for some insects such as dragonflies (Paulson 2001), but not yet for those associated with forest canopies. Here, I focus on pivotal relationships between forests and insects.

2.1 Pollination

Because of their high reproductive rates and potentially short generation times, most insects are able to respond to aspects of climate change more rapidly than their host plants, and this can result in life cycle asynchrony developing between the two. For forest trees, the canopy is a key space for pollination and climate-induced changes in relationships between host plants and insect pollinators have already been recorded. For example, in the Iberian Peninsula, the appearance of the pollinating bee *Apis mellifera* has moved by up to 35 days over a 30-year period, and it now frequently appears 25 days earlier than host plant flowers (Gordo and Sanz 2005). Schweiger et al. (2010) have suggested that a more subtle effect of climate change may result from a warming-induced downward shift in body size of pollinating insect communities. This change could disrupt specialised plant-pollinator relationships where pollinator morphology is key to success (e.g. for access to nectar or matched in energy requirements) and plants respond more slowly than their pollinators. However, as Hegland et al. (2009) state, we are far from being able to predict with any certainty whether there will be serious disruption in pollinator-plant relationships as a result of climate change.

2.1.1 Herbivory

The percentage of forest primary production consumed by insect herbivores is highly variable with extremes of near 0–100 % leaf area consumption (Lowman 1995). The development of asynchrony between insects and host plants has significant impact in these relationships too, with largely negative effects for the herbivores. For example, insect emergence that tracks temperature change may occur earlier in the season when the necessary plant resources (buds, leaves, nectar sources) are not available;

alternatively, drought may result in early senescence of host plants. There are documented examples of asynchrony resulting from the impact of temperature and moisture effects on butterflies such as *Euphydryas editha* (Parmesan 2006), resulting in dramatic population crashes and local extinction of species populations.

Another key factor for herbivores is the quality of available food resources. Experimental work using free-air CO_2 enrichment, or FACE, indicates a change in the C:N balance of leaves under elevated CO_2. Since for insect herbivores the conversion of plant to animal material is relatively inefficient, any shift towards lower nitrogen values reduces food quality. Low food quality may be compensated for by increased consumption, longer generation times and/or lower reproductive capacity. Early work on leaf feeders in woody plants suggested that increased levels of herbivory were likely under elevated CO_2 with resulting higher levels of leaf tissue loss (Whittaker 2001). However, recent reviews of the effects of elevated CO_2 on herbivory in field-based experiments with trees have demonstrated a consistent decrease in the number of leaves damaged and the overall level of damage by insect herbivores. In these more complex systems, the declining trend may be the result of a number of interacting factors including a reduction in the fitness and fecundity of leaf-chewing herbivores as food quality declines, together with the impact of natural enemies on population size (Knepp et al. 2005; Wang et al. 2008). Indeed there are potentially complex interactions across trophic levels that suggest we need to know more about the response of top-down control mechanisms in future climate scenarios (Barton et al. 2009).

However, insect populations do not always follow a pattern of decline, and a number of phloem-feeding aphid species show increases in population density and faster reproductive rates or development in elevated CO_2 environments (Sun and Ge 2011). In tree-feeding aphids such as the green spruce aphid (*Elatobium abietinum*) which is limited by low winter temperatures, studies suggest that temperature rise would enable populations to expand. However, if generalist predators also expand their ranges, these will limit the extent to which the aphid populations can respond (Straw et al. 2009). Sap feeders are not adversely affected by changes in leaf C:N and it is speculated that increased rates of reproduction and development may be explained by a combination of increases in ambient temperature and leaf temperature (caused by higher levels of leaf respiration in elevated CO_2) (Smith and Jones 1998).

Where herbivores are considered pests, there are concerns about the potential for increasing impact with climate change. We have already noted the potential for sap-feeding insects to increase density with temperature and CO_2 rise. The mountain pine beetle (*Dendroctonus ponderosae*) is another pest that has already reduced the time taken to complete its life cycle from 2 to 1 year in the Rocky Mountains, thereby increasing its potential direct impact and indirect impact via the rust fungus it transmits (Logan et al. 2003). Kurz et al. (2008) suggest that the negative effect of forest insect outbreaks on uptake and storage of atmospheric carbon is so significant that it should be factored into climate change models.

3 Complexity and Future Questions

In all of the above scenarios, the biodiversity resilience argument suggests that primary forest will be more resilient to climate change than secondary forest and plantation, although boreal forests are likely to be resilient even though they are not biodiversity rich because they are already highly disturbance adapted (Thompson et al. 2009). However, it is also clear that whilst we have some clues to individual species responses, the influence of climate change on the interactions between them is far from understood and this is a crucial area for future research. Gilman et al. (2010) propose a useful framework from which this complexity could be tackled in the laboratory and forest. We cannot assume either that forests are passive recipients of change; they are dynamic, evolving and contributing to atmospheric processes. For example, biogenic VOCs (volatile organic chemicals) are highly reactive in the troposphere even at the levels released from leaves by feeding herbivores (e.g. the weevil *Strophosoma melanogrammum* feeding on young Norway spruce trees (Prieme et al. 2000)). VOCs react with atmospheric oxidants to form aerosols that act either directly by reflecting or absorbing solar radiation or indirectly by acting as cloud condensation nuclei. The variation in VOC composition caused by herbivore activity may well affect the ways in which VOCs regulate oxidative capacity of the troposphere. This dynamic interaction at the interface of the biosphere and the atmosphere is a rich area for future research.

References

Allen CD et al (2010) A global overview of drought and heat-induced tree mortality reveals emerging climate change risks for forests. For Ecol Manag 259:660–684

Appanah S (1993) Mass flowering of dipterocarp forests in the aseasonal tropics. J Biosci 18:457–474

Barton BT, Beckerman AP, Schmitz OJ (2009) Climate warming strengthens indirect interactions in an old-field food web. Ecology 90:2346–2351

Gilman SE, Urban MC, Tewksbury J, Gilchrist GW, Holt RD (2010) A framework for community interactions under climate change. Trends Ecol Evol 25:325–331

Gordo O, Sanz JJ (2005) Phenology and climate change: a long term study in a Mediterranean locality. Oecologia 146:484–495

Hegland SJ, Nielsen A, Lázaro A, Bjerknes AL, Totland Ø (2009) How does climate warming affect plant–pollinator interactions? Ecol Lett 12:184–195

IPCC (2007) Climate Change 2007: the physical science basis. In: S Solomon, D Qin, M Manning, Z Chen, M Marquis, KB Averyt, M Tignor, HL Miller (eds) Contribution of working group I to the fourth assessment report of the intergovernmental panel on climate change. Cambridge University Press, Cambridge, United Kingdom/New York, 996 pp

Knepp RG, Hamilton JG, Mohan JE, Zanger AR, Berenbaum MR, DeLucia EH (2005) Elevated CO_2 reduces leaf damage by insect herbivores in a forest community. New Phytol 167:207–218

Kullman L (2002) Rapid recent range-margin rise of tree and shrub species in the Swedish Scandes. J Ecol 90:68–77

Kurz WA, Dymond CC, Stinson G, Rampley GJ, Neilson ET, Carroll AL, Ebata T, Safranyik L (2008) Mountain pine beetle and forest carbon feedback to climate change. Nature 452:987–990

Logan JA, Régnière J, Powell JA (2003) Assessing the impacts of global warming on forest pest dynamics. Front Ecol Environ 1:130–137

Lowman MD (1995) Herbivory as a canopy process in rain forest trees. In: Lowman MD, Nadkarni NM (eds) Forest canopies. Academic, San Diego

Lukac M, Calfapietra C, Lagomarsino L, Loreto F (2010) Global climate change and tree nutrition: effects of elevated CO_2 and temperature. Tree Physiol 30:1209–1220

Menzel A et al (2006) European phenological response to climate change matches the warming pattern. Global Change Biol 12:1969–1976

Norby RJ et al (2005) Forest response to elevated CO_2 is conserved across a broad range of productivity. Proc Natl Acad Sci 102:18052–18056

Ozanne CMP, Anhuf D, Boulter SL, Keller M, Kitching RL, Körner C, Meinzer FC, Mitchell AW, Nakashizuka T, Silva Dias PL, Stork NE, Wright SJ, Yoshimura M (2003) Biodiversity meets the atmosphere: a global view of forest canopies. Science 301:183–186

Parmesan C (2006) Ecological and evolutionary responses to recent climate change. Annu Rev Ecol Evol Syst 37:637–69

Paulson DR (2001) Recent odonata records from southern Florida: effects of global warming? Int J Odonatol 4:57–69

Prieme A, Knudsen TB, Glasius M, Christensen S (2000) Herbivory by the weevil, *Strophosoma melanogrammum*, causes severalfold increase in emission of monoterpenes from young Norway spruce (*Picea abies*). Atmos Environ 34:711–718

Scheifinger H, Menzel A, Koch E, Peter CH (2003) Trends of spring time frost events and phenological dates in central Europe. Theor Appl Climatol 74:41–51

Schweiger O et al (2010) Multiple stressors on biotic interactions: how climate change and alien species interact to affect pollination. Biol Rev 85:777–795

Smith PHD, Jones TH (1998) Effects of elevated CO_2 on the chrysanthemum leaf miner, *Chromatomyia syngenesiae*: a greenhouse study. Global Change Biol 4:287–291

Straw NA, Timms JEL, Leather SR (2009) Variation in the abundance of invertebrate predators of the green spruce aphid *Elatobium abietinum* (Walker) (Homoptera: Aphididae) along an altitudinal transect. For Ecol Manag 258:1–10

Sun Y, Ge F (2011) How do aphids respond to elevated CO_2? J Asia-Pac Entomol 14:217–220

Thompson I et al (2009) Forest resilience, biodiversity, and climate change. A synthesis of the biodiversity/resilience/stability relationship in forest ecosystems. Convention on biological diversity technical series no. 43. Secretariat of the Convention on Biological Diversity, Montreal

Vitasse Y, Francois C, Delpierre N, Dufrêne E, Kremer A, Chinee I, Delzona S (2011) Assessing the effects of climate change on the phenology of European temperate trees. Agric For Meteorol 151:969–980

Wang X, Ji L, Wang G, Liu Y (2008) Potential effects of elevated carbon dioxide on leaf-feeding forest insects. Front Biol China 3:68–77

Whittaker JB (2001) Insects and plants in a changing atmosphere. J Ecol 89:507–518

Zhu K, Woodall CW, Clark JS (2011) Failure to migrate: lack of tree range expansion in response to climate change. Global Change Biol. doi:10.1111/j.1365-2486.2011.02571

Chapter 11
Church Forest Status and Carbon Sequestration in Northern Ethiopia

Catherine L. Cardelús, M. Baimas-George, Margaret Lowman, and Alemayu Wassie Eshete

Keywords Sacred grove • Forest fragment • Soil status • Ethiopia • Dry montane forest

Bullets

1. Approximately 5 % of original forests in northern Ethiopia remain, mostly as a result of subsistence agriculture.
2. The remaining forest fragments exist under the stewardship of the Christian Orthodox Church, called "church forests."
3. These fragments, ranging from 3 to 300 ha in size, represent important carbon sequestration sites, as well as important conservation sites for biodiversity and fresh water.

1 Introduction

Ethiopia boasts one of tropical Africa's richest biodiversity, predominantly in her forest fragments (Tolera et al. 2008). However, forests in northern Ethiopia have undergone severe deforestation, with an estimated 4 % remaining (Gatzweiler 2007;

C.L. Cardelús (✉) • M. Baimas-George
Department of Biology, Colgate University, Hamilton 13346, NY, USA
e-mail: ccardelus@colgate.edu; mbaimas@colgate.edu

M. Lowman
North Carolina Museum of Natural Sciences, North Carolina State University, 121 W. Jones Street, Raleigh, NC 27603, USA
e-mail: canopymeg@gmail.com

A.W. Eshete
Department of Forestry, Bahir Dar University, Bahir Dar, Ethiopia
e-mail: Alewas2008@yahoo.com

M. Lowman et al. (eds.), *Treetops at Risk: Challenges of Global Canopy Ecology and Conservation*, DOI 10.1007/978-1-4614-7161-5_11,

Wassie et al. 2009). Human activities ranging from subsistence agriculture to collecting firewood are causal factors in a country where population has nearly doubled in 20 years from 43 million in 1984 to almost 80 million by 2000 (Feoli et al. 2002). The last remaining forest fragments in northern highlands of Ethiopian (called the Amhara Region) are housed in some 35,000 forest fragments called "church forest" 'and "sacred groves" that range from 3 to 300 ha in size and date back 1500 years. Protected by the Orthodox Tewahido Church (EOTC, Wassie 2007), these fragments represent spiritual as well as biodiversity sanctuaries of both flora (Bongers et al. 2006; Wassie and Teketay 2006) and fauna (Lowman 2010a). For example, some plants species are found only within a few fragments (Wassie 2002). The church leadership view biodiversity conservation as one of its primary stewardships, but the lack of perimeter delineation of these forest fragments threatens their future.

Biodiversity surveys of Ethiopian church forests are underway, with particular focus on pollinators (Lowman 2010a, b). The role of church forests in providing additional ecosystem services—soil conservation, fresh water protection, and carbon sequestration—is unstudied. Of further concern is the impact of this magnitude of deforestation on climate regulation (West et al. 2010). Conversion of forest to pasture increases surface temperatures, decreasing precipitation and evapotranspiration in surrounding areas and can even increase the duration of the dry season (Foley et al. 2005; Shukla et al. 1990; Walton 1980). Whether forests that have undergone such severe damage can be maintained (and/or ultimately restored) is in doubt (Laurance and Williamson 2001; Shukla et al. 1990), and rough estimates using Google Earth indicate that many of these forests will disappear completely in the next 10 years without immediate conservation action (Lowman 2010b). The Amhara State contains ~ 84,466 ha of forest, approximately 0.53 % of the total area of the region; deforestation rates were estimated at 9.8 % loss (8,278 ha) in the last 5 years (FAO 2010).

2 Results

We compared carbon stocks in the interior, edge, and clearing surrounding two church forests in the south Gondor Administrative Zone in the Amhara National Regional State of Ethiopia. Significant carbon losses were measured between interior forest floor soils and surrounding pasture of up to 72 % of the carbon from the top 10 cm of the forest floor (Cardelús et al. 2013), and these losses were mirrored in nitrogen losses. Estimates indicate that the soil carbon stock losses from deforestation in this region in the last 5 years alone could be as high as 0.211 teragrams.

The increased pressures on these forests from farmers who both reduce the forest size and also exacerbate the edge effects are compromising the ecosystem integrity of these remnants. Species richness is relatively low (Wassie, pers. comm.) and thus resilience. Aerial images of northeastern Ethiopia illustrate the dominance of arid agricultural lands, with tiny forest fragments dotting the landscape (Fig. 11.1). The conversion of forest to pasture can affect climate on both the local and the global scale (Laurance 2004; Shukla et al. 1990).

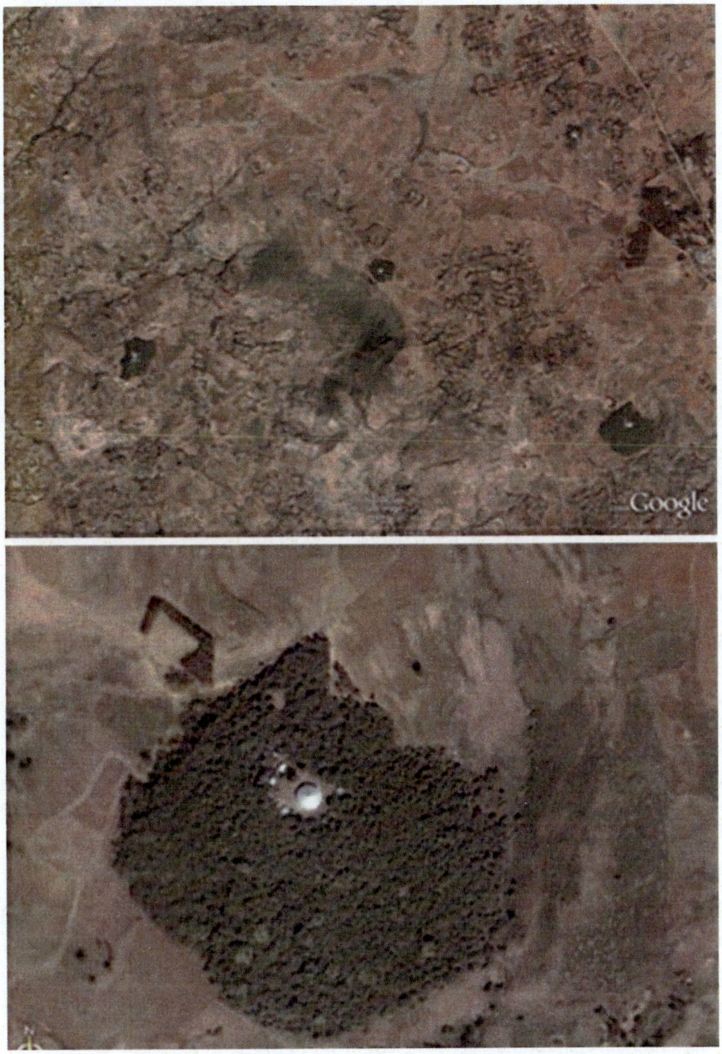

Fig. 11.1 Google Earth image of church forests in northern Ethiopia (**a**) and a close-up view of the church forest on *lower right* of (**a**), (**b**)

Given the severe extent of deforestation in Ethiopia, it is essential to conserve these last forest fragments (see www.treefoundation.org). They provide the important function of carbon sequestration, as well as the conservation of fresh water, biodiversity, pollinators, soil, and a spiritual heritage.

References

Bongers F, Wassie A, Sterck FJ, Bekele T, Teketay D (2006) Ecological restoration and church forests in northern Ethiopia. J Drylands 1:35–44

Cardelús CL, Scull P, Hair J, Baimas-George M, Lowman MD, Eshete AW (2013) A preliminary assessment of Ethiopian sacred grove status at the landscape and ecosystem scales. Diversity 5:320–334

FAO (2010) Global forest resources assessment 2010. Country report: Ethiopia; Forest Department: Food and Agriculture Organization of the United Nations. Rome, Italy, p 43

Feoli E, Vuerich LG, Zerihun W (2002) Evaluation of environmental degradation in northern Ethiopia using GIS to integrate vegetation, geomorphological, erosion and socio-economic factors. Agric Ecosys Environ 91:313–325

Foley JA, DeFries R, Asner GP, Barford C, Bonan G, Carpenter SR, Chapin FS, Coe MT, Daily GC, Gibbs HK (2005) Global consequences of land use. Science 309:570

Gatzweiler FW (2007) Deforestation of Ethiopia's Afromontane rainforests. Reasons for concern. Bonn, ZEF policy brief no. 7. Center for Development Research

Laurance WF (2004) Forest-climate interactions in fragmented tropical landscapes. Philos Trans Biol Sci 359:345–352

Laurance WF, Williamson GB (2001) Positive feedbacks among forest fragmentation, drought, and climate change in the Amazon. Conserv Biol 15:1529–1535

Lowman MD (2010a) CSI in Ethiopia: children survey insects. North C Nat 18:8–9

Lowman MD (2010b) Finding sanctuary – saving the biodiversity of Ethiopia one church forest at a time. Explor J 88:26–35

Shukla J, Nobre C, Sellers P (1990) Amazon deforestation and climate change. Science 247:1322

Tolera M, Asfaw Z, Lemenih M, Karltun E (2008) Woody species diversity in a changing landscape in the south-central highlands of Ethiopia. Agric Ecosys Environ 128:52–58

Walton S (1980) Replacing trees with wasteland: the demise of tropical moist forests. Bioscience 30:377–380

Wassie A (2002) Opportunities, constraints, and prospects of the Ethiopian Orthodox Tewahido churches in conserving forest resources: the case of churches in South Gonder, Northern Ethiopia. Swedish University of Agricultural Sciences, Uppsala

Wassie A (2007) Ethiopian church forests: opportunities and challenges for restoration. Wageningen University, Wageningen

Wassie A, Teketay D (2006) Soil seed banks in church forests of northern Ethiopia: implications for the conservation of woody plants. Flora Morphol Distrib Funct Ecol Plants 201:32–43

Wassie A, Sterck F, Teketay D, Bongers F (2009) Tree regeneration in church forests of Ethiopia: effects of microsites and management. Biotropica 41:110–119

West PC, Narisma GT, Barford CC, Kucharik CJ, Foley JA (2010) An alternative approach for quantifying climate regulation by ecosystems. Front Ecol Environ 9:126–133

Chapter 12
A Novel Approach to Simulate Climate Change Impacts on Vascular Epiphytes: Case Study in Taiwan

Rebecca C.-C. Hsu and Jan H.D. Wolf

Keywords Dispersal limitation • Global climate change • Maximum entropy method (MaxEnt) • Subtropical island • Tree persistence

Bullet Points

1. In comparison to terrestrial rooted plants, canopy epiphytes represent a promising model system to study the impact of climate change on living organisms because they are sensitive to the atmospheric climate, have relatively short life cycles, and show great potential for dispersal and colonization.
2. Epiphytes are adapted to highly dynamic forest canopies by producing many, mostly wind-dispersed diaspores, whereas long-lived dominant canopy trees have relatively limited dispersal ability and are more resistant to unfavorable climatic conditions. Accordingly, the response of tree-dependent epiphytes to climate change is likely slowed down by persistent forest trees and the slow immigration rate of trees into suitable new habitat.
3. The incorporation of forest trees into epiphyte species distribution models under climatic change scenarios in Taiwan confirmed that many known sensitive epiphytes (specialists) had restricted distributions (i.e., they were confined to mid-elevation/montane cloud forests), whereas insensitive species (generalists) were widespread and included several pantropical species.

R.C.-C. Hsu (✉)
Taiwan Forestry Research Institute, No. 67, Sanyuan Street, Taipei 10079, Taiwan

Universiteit van Amsterdam, Institute for Biodiversity and Ecosystem Dynamics (IBED), PO Box 94248, Amsterdam, GE 1090, The Netherlands
e-mail: ecogarden@tfri.gov.tw

J.H.D. Wolf
Universiteit van Amsterdam, Institute for Biodiversity and Ecosystem Dynamics (IBED), PO Box 94248, Amsterdam, GE 1090, The Netherlands
e-mail: J.H.D.Wolf@uva.nl

M. Lowman et al. (eds.), *Treetops at Risk: Challenges of Global Canopy Ecology and Conservation*, DOI 10.1007/978-1-4614-7161-5_12,

Summary

Forest vascular epiphytes have a critical influence on forest hydrology and nutrient cycling, so it is vital to assess how climate change affects these inhabitants. This study describes a novel hierarchical modeling approach, incorporating forest migration velocity and forest type–epiphyte interactions into classical species distribution models (SDMs). In Taiwan, eight major forest-type distributions were modeled under climate change scenarios and tailored to include dominant tree species' dispersal limitations and hypothesized persistence under unfavorable climate conditions. The forest projections, together with 16 environmental variables, were used to predict the likely distribution in the year 2100 of 237 vascular epiphyte species. The hierarchical modeling approach produced ecologically interpretable results. Biotic interactions must be considered when developing SDMs for dependent species, such as epiphytes.

1 Epiphytes in a Changing World

In the wet tropics, epiphytes form a conspicuous layer in the forest canopy, support abundant coexisting biota, and are known to have a critical influence on forest hydrology and nutrient cycling. Since canopy-dwelling plants have no vascular connection to the ground or their host plants, they are likely more sensitive to environmental changes than their soil-rooted counterparts (Benzing 1998), subsequently regarded as one of the groups most vulnerable to global climate change. Epiphytes have adapted to life in highly dynamic forest canopies by producing many, mostly wind-dispersed, seeds or spores. Consequently, epiphytes should colonize trees rapidly, which, in addition to atmospheric sensitivity and short life cycles, make epiphytes suitable climate change indicators. In this study, we assess the impact of climate change on Taiwanese epiphytes using a modeling approach.

2 Epiphyte Distribution Modeling

In the field of climate change impact research, species distribution models (SDMs) have been used increasingly to estimate potential species range shifts under paleontological and/or future climate change conditions. SDMs attempt to recognize species' realized niches, which are used to construct potential geographic distributions by relating species occurrences with values of predictor variables across a series of observation sites. Although classical SDMs may yield satisfactory results, modeling dependent epiphyte distributions is conducive to including additional factors. Since epiphyte performance also depends on the presence and characteristics of host trees, host tree composition probably has a significant influence on epiphyte assemblages. Other relevant factors include (1) information on regional topography and climate, (2) the microclimate associated with forest types and the specific epiphyte–tree biotic interactions, (3) the persistence of host trees (to sustain epiphyte populations under adverse climate change conditions), and (4) the dispersal ability of host trees compared to epiphytes as linked to deceleration of epiphyte colonization of suitable new habitat. We propose a novel modeling approach tailored for epiphytes, demonstrated on subtropical island of Taiwan.

3 Methods

3.1 Data Preparation

We obtained the localities of 237 strictly arboreal epiphyte species (Hsu and Wolf 2009) from herbarium records, published plant inventories, and our own botanical observations and entered these in a database (MsAccess). The final database comprised 18,239 epiphyte records with occurrences ranging from 5 to 1,083 grid cells (1 km^2). In addition, we obtained the localities of dominant canopy trees (11,700 records in total) from the third national forest resource inventory, conducted by the Taiwan Forest Bureau in 1993. Taiwanese major forest types were grouped as (1) lowland broad-leaved forest (BLL), (2) midland broad-leaved forest (BLM), (3) highland broad-leaved forest (BLH), (4) cypress forest, (5) *Pinus* forest, (6) *Tsuga* forest, (7) *Picea* forest, and (8) *Abies* forest.

We derived present climate data from an array of weather stations (recorded from 1900 to 1990) and future projected climate data (A2 and B2 scenarios, decadal average) from the Intergovernmental Panel on Climate Change (IPCC) Third Assessment Report (http://www.grida.no/publications/other/ipcc_tar/). We statistically downscaled the latter to a resolution of 1 km^2 to match the present-day data. We used monthly temperature and rainfall data to calculate ecologically relevant climate variables and applied a correlation test to exclude highly correlated (Pearson's $r>0.75$) factors. We selected one edaphic, four topographic factors, and 16 environmental variables exhibiting low correlation to build our model (Table 12.1).

3.2 Model Building

We built the SDMs of epiphytes with the maximum entropy method (MaxEnt, version 3.3.3). This program was developed to model species' geographic distributions with presence-only data and has been shown to outperform the majority of other modeling applications, especially when sample sizes are small. MaxEnt is particularly suited for epiphytes, since most epiphyte species (especially orchids) are notoriously rare and the program places no weight on the absence of an epiphyte in a forest, which is difficult to ensure, especially for high-canopy species. MaxEnt calculates a probability distribution over the grid, which may be interpreted as an index of habitat suitability for a species (Elith et al. 2011). The program also provides an estimate of the relative contribution of each environmental variable to the model and the relative magnitudes of environmental variables, which are derived from one training set of data that can be

Table 12.1 Environmental variables used in model building

No.	Environmental variable	Unit	Calculation
1	Annual mean temperature	°C	Average monthly mean temperature
2	Annual precipitation	Millimeter	Average monthly precipitation
3	Temperature seasonality	Decimal fraction	The standard deviation of the monthly mean temperatures
4	Precipitation seasonality	Decimal fraction	The coefficient of variation of the monthly mean precipitation
5	Total water deficiency	Millimeter minus °C	Monthly precipitation minus twice the monthly mean temperature
6–11	Monthly rainfall (P01, P04, P05, P06, P07, P10)	Millimeter	Monthly rainfall in January, April, May, June, July, and October
12	Inclination	Degree	Average terrain slopes of 1 km^2 land area
13–14	Aspect (eastness, northness)	Ordinal numbers: 0 ~ 8	Aspect measurement transformed by sin(aspect rad), cos(aspect rad), and assigned ordinals: 0, flat; 1, (−1)–(−0.75); 2, (−0.75)–(−0.5); 3, (−0.5)–(−0.25); 4, (−0.25)–0; 5, 0–0.25; 6, 0.25–0.5; 7, 0.5–0.75; 8, 0.75–1
15	Distance to 3,000 m	Meter	The distance to the nearest location above 3,000 m asl
16	Soil category	Cardinal numbers: 0 ~ 9	No soil (0), inceptisols (1), oxisols (2), alfisols (3), spodosols (4), mollisols (5), entisols (6), ultisols (7), andisols (8), vertisols (9)

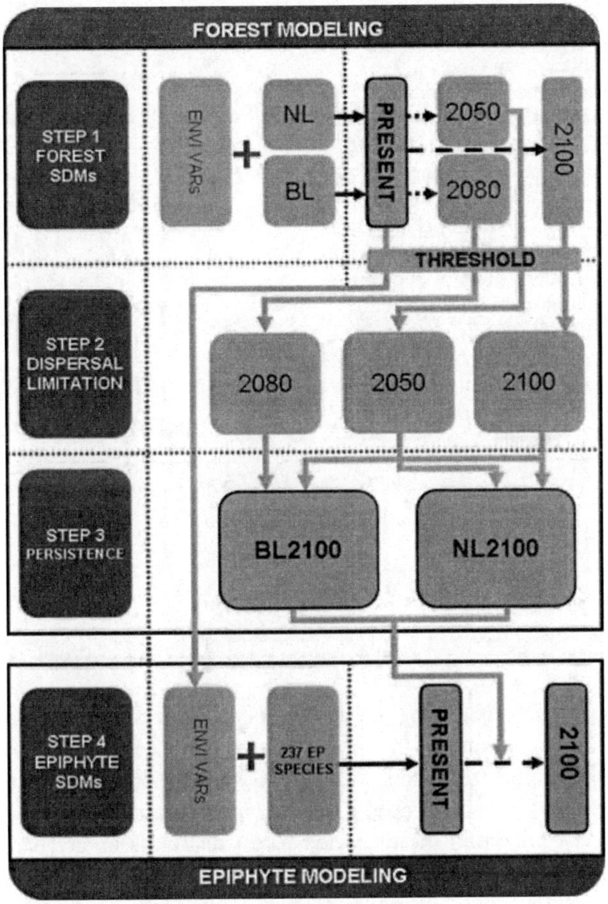

Fig. 12.1 The stepwise hierarchical modeling approach used in this study. The procedure THRESHOLD removed species distributions below thresholds. The years 2050, 2080, and 2100 are target years for our models. Solid line arrows indicate SDM modeling; broken line arrows indicate SDM projection. Framed squares indicate our final SDMs. *NL*, needle forests: *Abies*, *Picea*, *Tsuga*, cypress, and *Pinus*; *BL*, broad-leaved forests: highland (BLH), midland (BLM), and lowland (BLL); *ENVI VARs*, environmental variables; *EP* epiphyte

"projected" on another set of environmental data, thus enabling MaxEnt to model species distribution under different climate conditions such as future climate simulations. We refined the classical SDM by considering dispersal limitation, tree persistence, and biotic interactions between epiphytes and host trees using a stepwise hierarchical modeling approach (Fig. 12.1).

- *Step 1, forest species distribution modeling*: We modeled forest distributions under present climatic conditions and subsequently made projections for future scenarios (specifically A2 and B2 scenarios, for the years 2050, 2080, and 2100). We used the modeled forest projections at year 2050 and 2080 as intermediate steps, incorporating divergent persistence abilities for needle- (NL) and broad-leaved trees (BL) (see step 3).
- *Step 2, forest species dispersal limitation*: We calculated the maximum expanded range of each forest type with the age of the tree at maturity in target years (1 km y^{-1}; Engler and Guisan 2009). The projected forest distributions were then bounded by species dispersal limitations (Fig. 12.2).

Fig. 12.2 An example of a model incorporating dispersal limitation and tree persistence. The modeled distributions (year 2050 = *blue*; year 2100 = *red*) outside dispersal ranges (*grey bubbles*) were removed from the result. *Black dots* indicate present-day plant occurrence. Considering tree persistence, the tree distribution in year 2050 (*blue grids*) was assigned a threshold value (lowest suitability) and added to the 2100 distribution (*red grids*)

- *Step 3, forest species persistence*: We included the persistence ability of forests, as a measure of the time that trees can tolerate unfavorable climate conditions. We hypothesized a persistence of 20 years for broad-leaved trees (BLL, BLM, and BLH) and 50 years for needle trees (*Abies*, cypress, *Picea*, *Pinus*, and *Tsuga*). Accordingly, we modified the projected forest distributions at year 2100 by incorporating BL distributions at year 2080 and NL distributions at year 2050 (Fig. 12.2). The persistent/extended distributions were assigned threshold values of distribution probability (i.e., minimum habitat suitability).
- *Step 4, epiphyte species distribution modeling*: We used the resulting eight forest distributions (eight variables) together with the 16 abiotic variables (Table 12.1) to model the distribution of 237 epiphyte species. For each species, we simulated present-day conditions and then modeled projections for the year 2100 under both A2 and B2 climate change scenarios.

3.3 Species Richness Map

To create a species richness map, we first applied a threshold of sensitivity-specificity sum maximization to convert the MaxEnt probability distribution to a predicted presence map for each species. Next, every single species map was overlaid to produce a species richness map for epiphytes. The richness map was corrected for land-use change to eliminate species distributions in urbanized regions (assuming this remains unchanged in 2100).

4 Results, Applications, and Prospects

The SDM-generated forest distribution patterns agreed strongly with observed data (area under curve [AUC] values ranging from 0.809 to 0.967). All forest types were highly sensitive to mean annual temperature: this is expected because mean annual temperature is a driving force for altitudinal vegetation stratification in Taiwan. Consistent with observations, the models indicated that epiphyte distribution was strongly correlated with forest type; forest type was one of the three most significant factors contributing to the modeled distribution of each epiphyte species. The projected epiphyte richness map suggested a notable future shift in the altitudinal distribution of epiphytes. Epiphyte diversity is currently greatest at 1,000–1,500 m above sea level (asl) (nearly 100 species per 1 km^2), but, under simulated climate change conditions, our model indicated that this belt of maximum species richness would shift to elevations of 1,500–2,000 m asl (Fig. 12.3).

We also identified the epiphytes that were most and least sensitive to the climate change scenarios by ranking their range turnover rates. Generally, relatively insensitive species were correlated with lowland forest (BLL), while species that were more sensitive to climate change were associated with mid-elevation forests (cypress and BLM). Moreover, our projection also inferred a distinct decline in cypress forest, a major component of montane cloud forest at mid-altitudes (1,800–2,500 m asl). Tropical montane cloud forests are unique among terrestrial ecosystems for their particular hydrological regime and typically occur in narrow altitude belts characterized by high endemism and abundant epiphytes (Bruijnzeel et al. 2010). Our model results showed that many

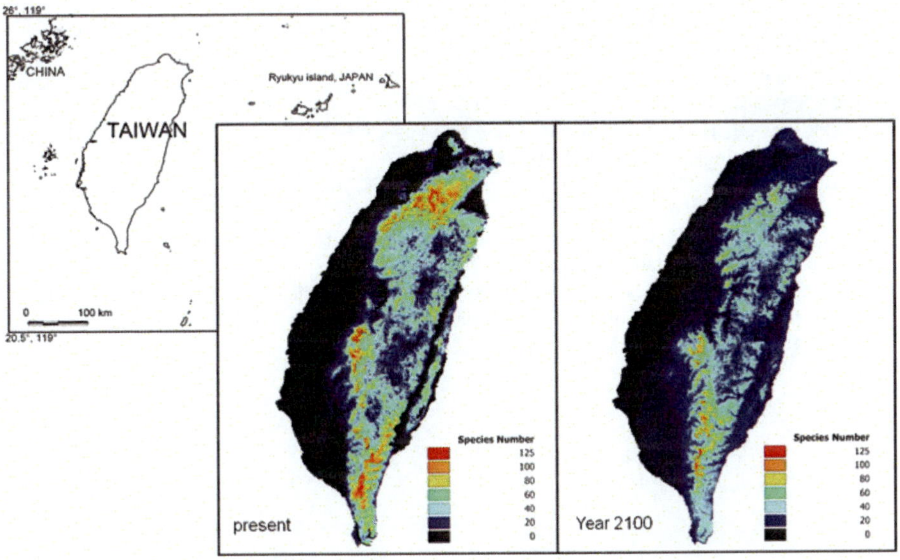

Fig. 12.3 The species richness maps of epiphytes in Taiwan under present and climate change conditions (values are the average of scenarios A2 and B2, IPCC)

epiphytes of the mid-elevation cloud forests (e.g., *Bulbophyllum chitouense* and *Mecodium badium*) are relatively sensitive to climate change conditions and may thus be suitable indicators for climate change. Conversely, our models suggested that species that were insensitive were generally widespread; these included several pantropical species (e.g., *Psilotum nudum* and *Hoya carnosa*) that are likely to demonstrate broader tolerances to climate change than species with a more restricted distribution (for a detailed species list and model results, see Hsu et al. 2012).

Finally, our model indicated a considerable decline in the surface area covered by mid-elevation forests and their associated epiphytic species under both climate change scenarios. Despite this perceived threat, most mid-altitude forests currently fall outside established protective reserves. We recommend establishing long-term plots for epiphyte monitoring outside and inside conservation areas, especially in those areas where climate change is projected to result in dramatic species composition changes.

The prediction of future species distributions is challenging because the species range shift is an ongoing process, reflecting dynamic relationships with environmental parameters, both biotic and abiotic (Elith et al. 2010). Our approach integrates knowledge on forest dynamics and host tree–epiphyte relationships and yielded easily interpretable results that will be crucial to building reliable and realistic species distribution models for dependent species such as epiphytes.

References

Benzing DH (1998) Vulnerabilities of tropical forests to climate change: the significance of resident epiphytes. Clim Change 39:519–540

Bruijnzeel LA, Scatena FN, Hamilton LS (2010) Tropical montane cloud forests: science for conservation and management. Cambridge University Press, Cambridge, UK, 740 Pp

Elith J, Kearney M, Phillips S (2010) The art of modelling range-shifting species. Methods Ecol Evol 1:330–342

Elith J, Phillips SJ, Hastie T, Dudík M, Chee YE, Yates CJ (2011) A statistical explanation of MaxEnt for ecologists. Divers Distributions 17:43–57

Engler R, Guisan A (2009) MigClim: predicting plant distribution and dispersal in a changing climate. Divers Distributions 15:590–601

Hsu RCC, Wolf JHD (2009) Diversity and phytogeography of vascular epiphytes in a tropical-subtropical transition island, Taiwan. Flora 204:612–627

Hsu RCC, Tamis WLM, Raes N, de Snoo GR, Wolf JHD, Oostermeijer G, Lin SH (2012) Simulating climate change impacts on forests and associated vascular epiphytes in a subtropical island of East Asia. Divers Distributions 18:334–347

Chapter 13
Sensitivity and Threat in High-Elevation Rainforests: Outcomes and Consequences of the IBISCA-Queensland Project

R.L. Kitching, L.A. Ashton, C.J. Burwell, S.L. Boulter, Penelope Greenslade, M.J. Laidlaw, C.L. Lambkin, S.C. Maunsell, A. Nakamura, and F. Ødegaard

Keywords Insects • *Nothofagus* forest • IBISCA • Climate change • Australia

Summary

Multiple taxon surveys of plant and animal diversity along a subtropical altitudinal gradient in rainforest in South-east Queensland identify a distinct assemblage for every focal taxon investigated in the *Nothofagus*-dominated cloud forest above

---

R.L. Kitching (✉) • L.A. Ashton • S.L. Boulter • S.C. Maunsell • A. Nakamura
Environmental Futures Centre, Griffith School of the Environment, Griffith University, Nathan, QLD 4111, Australia
e-mail: r.kitching@griffith.edu.au; l.ashton@griffith.edu.au; s.boulter@griffith.edu.au; s.maunsell@griffith.edu.au; a.nakamura@griffith.edu.au

C.J. Burwell
Environmental Futures Centre, Griffith School of the Environment, Griffith University, Nathan, QLD 4111, Australia

Biodiversity Program, Queensland Museum, PO Box 3300, South Brisbane, QLD 4101, Australia
e-mail: Chris.burwell@qm.qld.gov.au

P. Greenslade
Centre for Environmental Management, School of Science and Engineering, Mt Helen, University of Ballarat, Ballarat, VA 3350, Australia
e-mail: Pgreenslade@staff.ballarat.edu.au

M.J. Laidlaw
Queensland Department of Science, Information Technology, Innovation and the Arts, Queensland Herbarium, Toowong, QLD 4066, Australia
e-mail: Melinda.Laidlaw@science.dsitia.qld.gov.au

C.L. Lambkin
Biodiversity Program, Queensland Museum, PO Box 3300, South Brisbane, QLD 4101, Australia
e-mail: christine.lambkin@qm.qld.gov.au

F. Ødegaard
Norwegian Institute for Nature Research, PB 5685 Sluppen NO-7485, Trondheim, Norway
e-mail: frode.odegaard@nina.no

M. Lowman et al. (eds.), *Treetops at Risk: Challenges of Global Canopy Ecology and Conservation*, DOI 10.1007/978-1-4614-7161-5_13,

1,000 m altitude. These high-elevation forests are not only biotically unique but they represent an endangered community under even moderate levels of predicted global warming. Although assemblages associated with lower elevations may adapt to a warmer world by moving upwards, these high-elevation sets of species have nowhere to go. At the continental scale, these forest types are rare. Some species or their near-analogues may persist in more southerly locations, but those at the northern ecosystemic limits appear doomed.

1 Introduction

The IBISCA[1] approach to biodiversity assessment in forests was, initially, the brainchild of Yves Basset, Bruno Corbara and Hector Barrios (Basset et al. 2007). The four IBISCA projects carried out to date have examined selected aspects of beta-diversity in tropical, subtropical and temperate forests. In each case a set of research questions were defined and a sampling design executed. Researchers with interests in particular taxa or ecological processes were invited to join one or more of the proposed field expeditions to carry out sub-projects of their choice within the general experimental design. When successful, this approach not only provides individual researchers or groups of researchers with analyzable and publishable data sets in their specific areas of interest but it also facilitates comparative and other meta-analyses with homogeneous criteria.

The first IBISCA project examined canopy–ground comparisons in the lowland tropical rainforest of Panama (Basset et al. 2007). This was followed by two projects which examined altitudinal changes in arthropod and plant diversity: IBISCA-Queensland, in the subtropical rainforest of South-east Queensland (Kitching et al. 2011), and IBISCA Santo, in tropical insular forest on Santo Island, Vanuatu (Bouchet et al. 2011). The fourth project, IBISCA Auvergne, examined the interaction between arthropod biodiversity and woodland structure in a managed forest in Central France.

The IBISCA-Queensland Project was carried out between 2006 and 2008 in Lamington National Park in South-east Queensland. Within a single wholly forested catchment, 20 sampling sites were established, four at each of five altitudes (300, 500, 700, 900, and 1,100 m asl). The lowest elevations were characterised by warm subtropical rainforest (notophyll vine forest sensu Webb 1959) with higher elevations grading into 'cool subtropical rainforest' through to 'cool temperate rainforest' at the highest elevations (i.e. grading from 'notophyll vine forest' to 'microphyll fern forest' in the terminology of Webb 1959). McDonald and Hunter (2010) discuss the detailed composition of the rainforest vegetation of the region, and Laidlaw et al. (2011a, b) describe and analyse in detail the vegetation of the IBISCA sites.

[1]The IBISCA designation was originally the acronym for 'Investigating the Biodiversity of Soil and Canopy Arthropods' but has evolved into a general description of multidimensional, multi-researcher, multinational projects examining forest diversity.

The rationale for this project is based on the logic that studying adjacent altitudes in a continuously forested catchment (with sampling sites standardised for substrate and aspect) will provide insights as to how biodiversity responds under various scenarios of future climate change. Strong et al. (2011) discuss the physics and meteorology behind this logic as well as presenting results on the actual altitudinally related lapse rates encountered.

Results are now catalogued for a range of taxa including plants, springtails, moths, ants, flies, beetles and bugs. In this chapter we present a comparative summary of some of these results and discuss emergent conservation issues. Detailed methodologies and analyses have already been documented (see Ashton et al. 2011; Burwell and Nakamura 2011; Boulter et al. 2011; Greenslade and Kitching 2011; Lambkin et al. 2011; Laidlaw et al. 2011a, b; Kitching et al. 2011; Ødegaard and Diserud 2011). In some instances, data have been reanalyzed and re-presented in slightly different form for visual comparability.

2 The Uniqueness of the High-Elevation Forests: Results from Disparate Taxa

Figure 13.1 compares overall patterns of assemblage composition across six taxa, using nonmetric multidimensional scaling ordination based on presence/absence data. Samples were collected from four independent locations at each of five elevations along our Lamington altitudinal gradient. Figure 13.1a is derived from surveys of woody plants with stem diameters at 1.3 m height (dbh) of 5 cm or greater (see Laidlaw et al. 2011a). Figure 13.1b summarises Collembola sampled using pitfall traps (Greenslade and Kitching 2011). Figure 13.1c shows the distribution of moth species trapped using light traps (Ashton et al. 2011). Figure 13.1d reflects a similar analysis of beetles collected by sweeping (Ødegaard and Diserud 2011). Figure 13.1e is an ordination of the ant species sampled by a variety of methods (primarily hand collecting, litter extraction and bark spraying) (Burwell and Nakamura 2011). Finally, Fig. 13.1f presents an ordination of samples of higher flies (i.e. Diptera minus the Nematocera) collected in Malaise traps, sorted and analyzed at the family level (Lambkin et al. 2011).

In each of these six examples, we performed permutational multivariate ANOVA (PERMANOVA, Anderson et al. 2008) to test for differences in assemblage composition among elevations (using 19999 unrestricted permutations of the raw data, with five elevations incorporated as a fixed factor). There is statistically significant evidence that the assemblages of organisms under study are far from random and that altitude is strongly associated with the patterns generated by the ordinations (see individual papers referenced above).

The flora and fauna associated with the samples from 1,100 m asl (called microphyll fern forest by Webb (1959) or 'cool temperate rainforest' as used by Floyd (1990)) were the best defined assemblages and more isolated from the lower forest sites. The results of post hoc pairwise comparisons were consistent with observed patterns,

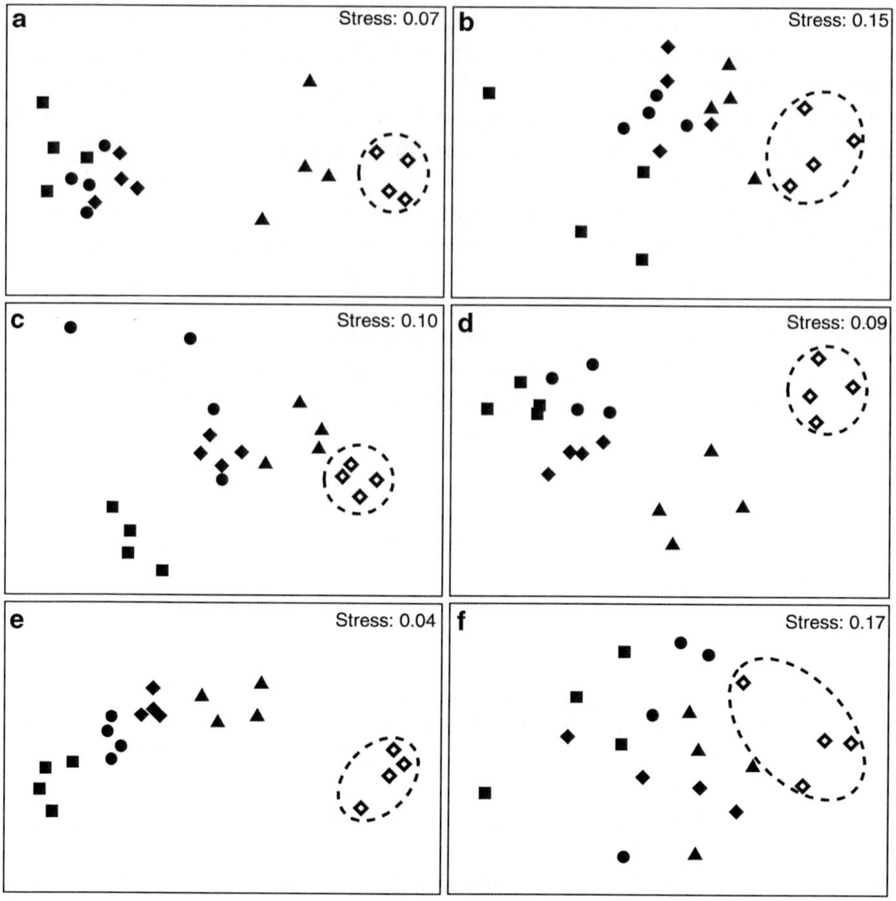

Fig. 13.1 Results of nonmetric multidimensional scaling ordinations on presence/absence data for taxa along an altitudinal gradient in subtropical rainforest in Lamington National Park, South-east Queensland, Australia: (**a**) trees with dbh greater than 5 cm; (**b**) Collembola from pitfall traps (October 2006); (**c**) night-flying Lepidoptera with forewing length >1cm, from light traps (October 2006 and March 2007); (**d**) Coleoptera collected by beating to head height (October 2006); (**e**) Formicidae based on a protocol including pitfall traps, bark spraying, litter extraction and hand collecting (October 2006, January 2007, March 2007, July 2007 and January 2008); (**d**) families of Diptera (excluding Nematocera) from ground zone Malaise traps (October 2006, January 2007 and July 2007) (For further details of methods used, see references in the chapter text)

with assemblages at 1,100 m significantly different from those from lower elevations. These unique higher-elevation assemblages have, embedded within them, species characteristic of, and restricted to, these cooler forests. Table 13.1 lists the characteristic taxa for each of the six examples illustrated in Fig. 13.1.

Although we focused on the 1,100 m sites because of their special conservation significance, we noted that the ordinations (with the exception of fly families) showed clear, progressive changes from 300 to 900 m. For trees and beetles, 900 m also

Table 13.1 Species or morpho-species unique to 1,100 m sites surveyed during IBISCA-Queensland with their wider distribution where known (abbr.: *NSW*, New South Wales; *Qld*, Queensland)

Species or morpho-species	Known distribution
Angiospermae[a]	
Angiospermae[a]	
Acronychia octandra	Regional endemic: South-east Qld, North-eastern NSW only
Arthropteris beckleri	Widespread: wet tropics of Qld to Southern NSW
Blechnum wattsii	Widespread: Southern Qld to Tasmania
Blechnum patersonii	Widespread: Northern Qld to Southern NSW
Berberidopsis beckleri	At northern limit, spreads south to central, coastal NSW
Callicoma serratifolia	Widespread: Central Qld to Southern NSW
Dendrobium falcorostrum	Local endemic: border ranges only, *Nothofagus moorei* associated
Cyperus disjunctus	At northern limit, spreads south to central, coastal NSW
Cyathea australis	Widespread: Northern Qld to Tasmania
Cryptocarya foveolata	At northern limit, spreads south to central, coastal NSW
Dockrillia pugioniformis	Widespread: South-east Qld to South-eastern NSW only
Drymophila moorei	Regional endemic: South-east Qld, North-eastern NSW only
Helmholtzia glaberrima	Regional endemic: South-east Qld, North-eastern NSW only
Marsdenia rostrata	Widespread: Northern Qld to Victoria
Melicope hayesii	Regional endemic: South-east Qld, North-eastern NSW only
Parsonsia induplicata	Regional endemic: South-east Qld, North-eastern NSW only
Nothofagus moorei	Regional endemic: South-east Qld, North-eastern NSW only
Pennantia cunninghamii	Widespread: Northern Qld to Southern NSW
Parsonsia tenuis	Local endemic: Border ranges only
Polyosma cunninghamii	Widespread: South-east Qld to South-eastern NSW
Ripogonum fawcettianum	Regional endemic: South-east Qld, North-eastern NSW only
Ripogonum discolor	Widespread: Northern Qld to central NSW
Quintinia sieberi	At northern limit, spreads south to South-eastern NSW
Collembola	
Acanthanura sp. nov.	Local endemic to this site, under process of description
Pseudachorutinae gen.nov., sp. nov.1	Probably local endemic to this site—known only from this site
Pseudachorutinae gen.nov., sp. nov. 2	Probably local endemic to this site—known only from this site
Lobellini sp.	Probably local endemic to this site—known only from this site
Folsomina sp.	Probably or likely on other moist summits
Proisotoma sp.	Probably or likely on other moist summits
Calvatomina pagoda	Probably or likely on other moist summits, including Melanesia
Acanthomurus sp. 2	Unknown, insufficient information
Rastriopes sp.	Unknown, insufficient information
Cryptopygus sp.	Unknown, insufficient information
Coleoptera	
21 species of which >3 specimens were found (out of a grand total of 1,219 spp.) were restricted to the 1,100 m sites. These belonged to the families Carabidae (3 spp.), Ptiliidae (1 sp.), Scydmaenidae (2 spp.), Staphylinidae (8 spp.), Scaphiidae (1 sp.), Pselaphidae (1 sp.), Scirtidae (2 spp.), Buprestidae (1 sp.) and Byrrhidae (1 sp.). These have not been identified further at this stage	

(continued)

Table 13.1 (continued)

Species or morpho-species	Known distribution
Lepidoptera	
Geometridae, Ennominae	
Dyscheralcis crimnodes	A high-elevation species in tropical and subtropical Australia, close to its southern range limit here. Southern limit in Northern NSW
Lychnographa heroica	Restricted to South-east Queensland and northern half of NSW. Southern limit is near Barrington Tops National Park, NSW
Middletonia hemichroma	Restricted to South-east Queensland and northern half of NSW. Southern limit is near Barrington Tops National Park, NSW
Geometridae, Larentiinae	
Heterochasta conglobata	A high-elevation species in tropical and subtropical Australia, close to its southern range limit here. Southern limit in Northern NSW
Noctuidae, Amphipyrinae	
Thalatha trichroma	Restricted to South-east Queensland and northern half of NSW. Southern limit is in Blue Mountains National Park, NSW
Hymenoptera, Formicidae	
Pachycondyla IBISCA1	Uncertain—to date known only from this site
Diptera families	
Helosciomyzidae (124 of 130 specimens at 1,100 m)[b]	In Australia restricted to higher elevation rainforest from the tropics southward. Also in New Zealand and Chile

[a]Based on an assessment of the entire vascular flora, not just the canopy trees used in Fig. 13.1a. List includes only taxa which could confidently be identified to species and excludes one widespread herbaceous species

[b]These were all *Helosciomyza ferruginea*—the same species identified by Wilson et al. (2007) as a potential indicator of future climate change within the wet tropics of far North Queensland

exhibited distinctively different assemblages. The lower degree of separation found in the fly assemblages is attributable to the family-level resolution of the data, yet assemblages at 1,100 m were different from those of lower elevations. In some instances better separation was achieved when relative abundances were included in the analyses. The results of such analyses are discussed elsewhere (Burwell et al. 2011).

3 Endemism and Options: Conservation Implications for a High-Elevation Specialist

In undisturbed tracts of natural vegetation such as the rainforests of Lamington National Park, fauna and flora characteristics of lower elevations have the option to move to higher elevations as the climate warms. Allowing for the 1.5 °C lapse rate per 200 m elevation and assuming nothing but temperature is involved (a big assumption), this means that the biota currently encountered below the 900 m elevation mark has at least the potential to adapt to the range of changes predicted over the coming century. In contrast, 1,100 m biotas are different because they have nowhere to go.

Cool temperate rainforest dominated by *Nothofagus* or *Eucryphia* spp. occurs in Australia from extreme South-east Queensland stretching south through New South Wales to Victoria and is the dominant wet forest type in Tasmania. In the north the canopy of these forests is generally dominated by *Nothofagus moorei,* in the central part of their distribution by *Eucryphia moorei* and in Victoria and Tasmania by *Nothofagus cunninghamii.* In all, Australia has about 820,000 ha of this forest type but only about 200, 000 ha on continental Australia and 624,100 ha in Tasmania (Williams 1974). The 200,000 ha of cool temperate rainforest on mainland Australia is made up of isolated patches, few of which exceed more than a 1,000 ha in extent. These patches are separated from each other by warmer rainforest types, by wet and dry scleromorphic woodlands and by large areas of lowland developed for agriculture and settlement. Taxa that are strong fliers or sufficiently small to be wind assisted in their movements could, in principle, move south in the face of a warming climate. Many of the local endemics though are far from highly vagile and simply do not have that option. Bass Strait, separating Tasmania from mainland Australia, represents a significant barrier for species seeking cooler *Nothofagus* forests.

4 Future Work and Conservation Options

To assess the species-level conservation threat to high-elevation forests of Lamington National Park, we need to quantify the wider distributions of the species that make up these unique assemblages (Table 13.1). For the arthropods, many gaps exist in the database. Museum-based surveys of species of Lepidoptera and ants are underway. Other groups must wait.

Policymakers are faced with questions about management actions to insure stewardship of these high-elevation rainforests. Clearly, species of invertebrates cannot be systematically moved to more southerly habitat (although this is an option actively discussed for vertebrates). Even if this were possible, the unpredictable consequences concerning their impacts on the more southerly sites make it a risky option. For invertebrates, ex situ conservation is not practical. Steffen et al. (2009) have made it quite clear that reducing the impacts of 'other stressors' in environmental terms (such as clearing, inappropriate fire regimes and invasive weed impacts) will likely maximise the intrinsic resilience of natural ecosystems to change driven by global warming. We are not optimistic about the long-term survival of these unique mountain communities. In contrast, species in lower-elevation rainforests (see Burwell and Nakamura 2011; Ødegaard and Diserud 2011) have greater plasticity to shift up elevational gradients, even if their levels of local endemism are less than the cool temperate ecosystems.

Acknowledgements The IBISCA-Queensland Project was funded by a Smart State grant from the State of Queensland with other funding from Griffith University; the Queensland Museum; the Queensland Herbarium; the Global Canopy Programme, NRM Queensland; O'Reillys' Rainforest Resort; and the Australian Department of Environment, Heritage and the Arts. We are grateful for this support. David Putland and Heather Christensen provided invaluable and impeccable research and administrative assistance. Other individuals and organisations have been acknowledged elsewhere.

References

Anderson MJ, Gorley RN, Clarke KR (2008) PERMANOVA+ for PRIMER: guide to software and statistical methods, Primer-E Ltd. Plymouth

Ashton LA, Kitching RL, Maunsell SC, Bito D, Putland DA (2011) Macrolepidopteran assemblages along an altitudinal gradient in subtropical rainforest – exploring indicators of climate change. Memoirs Qld Museum Nat 55:375–389

Basset Y, Corbara B, Barrios H, Cuenoud P, Leponce M, Aberlenc HP, Bail J, Bito D, Bridle JR, Castano-Meneses G, Cizek L, Cornejo A, Curletti G, Belabie JHC, Dejean A, Didham RK, Dufrene M, Fagan LL, Floren A, Frame DM, Halle F, Hardy OJ, Hernandez A, Kitching RL, Lewinsohn TM, Lewis OT, Manumbor E, Medianero E, Missa O, Mitchell AW, Mogia M, Novotny V, Odegaard F, Oliviera EG, Orivel J, Ozanne C, Pascal O, Pinzon S, Rapp M, Ribeiro S, Roisin Y, Roslin T, Roubik DW, Samaniego M, Schmidl J, Sorensen LL, Tischeshkin A, Van Rosselaer C, Winchester NN (2007) IBISCA-Panama, a large-scale study of arthropod beta diversity and vertical stratification in a lowland rainforest: rationale, study sites and field protocols. Bulletin de l'Institut Royal des Sciences Naturelles de Belgique-Entomologie 77:39–69

Bouchet P, Le Guyader H, Pascal O (eds) (2011) The natural history of Santo Island. Museum National D'Histoire Naturelle, Paris

Boulter SL, McDonald WJF, Kitching RL, Zalucki JM, Jessop LW (2011) Plant reproductive phenology and floral resources of an Australian subtropical rainforest. Memoirs Qld Museum Nat 55:463–479

Burwell CJ, Nakamura A (2011) Distribution of ants species along an altitudinal transect in continuous rainforest in subtropical Queensland, Australia. Memoirs Qld Museum Nat 55:391–411

Burwell CJ, Nakamura A, Kitching RL (2011) Biodiversity, altitude and climate change in an Australian subtropical rainforest. Results from the IBISCA-Queensland project. Memoirs Qld Museum Nat 55:235–493

Floyd A (1990) Australian rainforests in New South Wales, 2 vol. Surrey Beatty, Chipping Norton

Greenslade P, Kitching RL (2011) Potential effects of climatic warming on the distribution of Collembola along an altitudinal transect in Lamington National Park, Queensland, Australia. Memoirs Qld Museum Nat 55:333–347

Kitching RL, Putland D, Ashton LA, Laidlaw MJ, Boulter SL, Christensen H, Lambkin CL (2011) Detecting biodiversity changes along climatic gradients: the IBISCA-Queensland project. Memoirs Qld Museum Nat 55:235–250

Laidlaw MJ, McDonald WJF, Hunter RJ, Kitching RL (2011a) Subtropical rainforest turnover along an altitudinal gradient. Memoirs Qld Museum Nat 55:271–290

Laidlaw MJ, McDonald WJF, Hunter RJ, Putland DA, Kitching RL (2011b) The potential impacts of climate change on Australian subtropical rainforest. Aust J Bot 59:440–449

Lambkin CL, Boulter SL, Starick NT, Cantrell BK, Bickel DJ, Wright SG, Power N, Schutze MK, Turco F, Nakamura A, Burwell CJ (2011) Altitudinal and seasonal variation in the family-level assemblages of flies (Diptera) in an Australian subtropical rainforest: one hundred thousand and counting! Memoirs Qld Museum Nat 55:315–331

McDonald WJF, Hunter RJ (2010) The Vegetation. In: Kitching RL, Braithwaite R, Cavanaugh J (eds) Remnants of Gondwana: a natural and social history of the Gondwana rainforests of Australia. Surrey Beatty, Baulkham Hills

Ødegaard F, Diserud OH (2011) Taxonomic composition of Coleoptera, Hemiptera (Heteroptera and Coleorrhyncha) and Mutillidae (Hymenoptera) at five different altitudes in Lamington National Park (Queensland, Australia). Memoirs Qld Museum Nat 55:359–374

Steffen W, Burbidge AA, Hughes L, Kitching RL, Lindenmayer D, Musgrave W, Stafford Smith M, Werner P (2009) Australia's biodiversity and climate change: a strategic assessment of the vulnerability of Australia's biodiversity to climate change. CSIRO, Melbourne

Strong CL, Boulter SL, Laidlaw MJ, Maunsell SC, Putland D, Kitching RL (2011) The physical environment of an altitudinal gradient in the rainforest of Lamington National Park, southeast Queensland. Memoirs Qld Museum Nat 55:251–270

Webb LJ (1959) A physiognomic classification of Australian rainforests. J Ecol 47:551–570
Williams WD (1974) Biogeography and ecology in Tasmania. Junk, The Hague
Wilson R, Williams SE, Trueman JWH, Yeates DK (2007) Communities of Diptera in the Wet Tropics of North Queensland are vulnerable to climate change. Biodivers Conserv 16:3163–3177

Chapter 14
A Mature Forest Canopy in a CO_2-Rich Future: An Experiment at the Swiss Canopy Crane Research Site

Ch. Körner

Keywords Biodiversity • Canopy CO_2 enrichment • Carbon • Deciduous forest • Soils • Stable isotopes • Water

Bullet Points

Eight years of CO_2 enrichment of temperate-zone deciduous tree crowns caused:

1. No stimulation of tree growth, suggesting carbon saturation of forest trees at current atmospheric CO_2 concentrations of c. 390 ppm
2. Indirect effects of canopy water savings (less transpiration) and root carbon release on soil processes

With a 50 m tall construction crane (Fig. 14.1), tree crowns of over 100-year-old temperate deciduous forest trees were exposed to a future scenario of 540 ppm CO_2 atmosphere for 8 years. Porous tubing was woven into the canopies of 30–35 m tall dominant trees and CO_2 released via computer control. This first and still only mature forest canopy CO_2 enrichment experiment addressed whether trees in a natural setting are (1) taking up more CO_2 when more is offered, (2) investing the extra photosynthates into more biomass, (3) undergoing qualitative changes, and/or (4) saving water.

This experiment proved the feasibility of canopy-scale CO_2 enrichment without intrusive infrastructure (Pepin and Körner 2002). The CO_2 consumption of 1.5–2 t food quality CO_2 (cleaned industrial waste gas) on a 16 h bright midsummer day by c. 500 m^2 enriched forest canopy of 12 tall trees sets financial and practical constraints. One major limitation is that plot-scale replication would require additional cranes and would multiply CO_2 costs. One great benefit of working with tall trees is that their complete life history is engraved in their annual growth rings, including the period immediately before and during the treatment. Hence, responses of trees

Ch. Körner (✉)
Institute of Botany, University of Basel, Basel, Switzerland
e-mail: ch.koerner@unibas.ch

M. Lowman et al. (eds.), *Treetops at Risk: Challenges of Global Canopy Ecology and Conservation*, DOI 10.1007/978-1-4614-7161-5_14,

Fig. 14.1 Swiss Canopy Crane in an over 100-year-old mixed deciduous forest near Basel

can be standardized by their individualistic growth in the recent past. Because it is known that tree species differ in their responses to any environmental change, studying a single species is at risk to arrive at inconclusive answers. Therefore, this experiment aimed at including as many different temperate tree species as possible, inevitably losing statistical power at the individual species level but arriving at an overall "tree response" that can be expected to be closer to average responses likely to occur at a landscape scale. (During the course of the experiment, we lost two slender *Acer campestre* individuals due to wind throw, and *Prunus avium* was too close to the periphery so that CO_2 enrichment was unreliable.) Eleven trees of oak (*Quercus petraea*), beech (*Fagus sylvatica*), hornbeam (*Carpinus betulus*), and lime (*Tilia platyphyllos*), each with a canopy of c. 50 m^2 and a stem diameter of 30–50 cm, were used for the final analysis. All root and soil data were collected at defined distances from the stems around each individual. Since the fossil fuel-derived CO_2 gas used to enrich the forest canopy contained reliably and consistently less of the rare 13C isotope (compared to the natural 12C isotope) than is contained in ambient air, all new carbon incorporated into trees by photosynthesis carried a label that permitted tracking the path of "new" CO_2 in the ecosystem.

An impressive example of fast and intense canopy–soil coupling was the observation that fruit bodies of mycorrhizal fungi were composed of 70 % new carbon only a few weeks after trees were first exposed to the 540 ppm atmosphere. These fungi are living in symbiosis with trees and have direct access to sugar transported to the roots. Leaves and tree rings took 2–4 years to reach a steady-state new carbon signal (Keel et al. 2006). Tree canopies kept absorbing more CO_2 over the entire experimental period (Bader et al. 2013 unpubl.), but as in other experiments, it was hard to identify where the extra carbon went. It did not lead to faster tree growth, greater leaf production, more roots, or more fruits (Körner et al. 2005; Asshoff and Hättenschwiler 2006). Early in the experiment, we observed enhanced belowground activity under trees receiving more CO_2 (Keel et al. 2006). For instance, the air in the soil pores was enriched in CO_2, and the respiratory CO_2 emitted from the ground carried a clear isotope signal. We could map, on the ground, where canopies received extra CO_2 (Steinmann et al. 2004). As time progressed, these signals vanished. After 7 years neither soil CO_2 concentration nor soil CO_2 evolution were enhanced, but the isotope signal was sustained, so new carbon was flowing into the soil (Bader and Körner 2010).

The actual responses to increased CO_2 were more delicate in nature. For instance, oaks accumulated more starch and sugar in their leaves, whereas beech and hornbeam did not. Canopy insects changed their food preferences in response to elevated CO_2 (Hättenschwiler and Schafellner 2004; Bignucolo and Körner 2010). Soil fauna lived on much older organic material than expected and did not incorporate the new carbon isotope signal after 6 years (Pollierer et al. 2007), similar to saprobic fungi that retained the "old" carbon signal throughout (Keel et al. 2006). Trees as a whole consumed less water when exposed to elevated CO_2, but these expected responses of leaf stomata were only significant in hornbeam, with no response in oak. The net outcome was that soil moisture stayed slightly higher during rainless periods in the CO_2-enriched area (Leuzinger and Körner 2007). This may explain why trees exposed to elevated CO_2 produced less fine roots after 6 years (Bader et al. 2009). The hydrological consequences of such canopy water savings in a CO_2-rich future could be substantial (e.g., more runoff), but a model using these data revealed that the signal almost disappears when scaling from trees to the landscape and from some warm summer days to a full-year water budget (Leuzinger and Körner 2010).

Soil scientists found evidence for carbon-priming effects at our site, which means that microbes fed with extra sugar released from tree roots decomposed organic pools in the soil that in turn released extra nitrate traceable in the soil solution (Schleppi et al. 2012). And more dissolved organic carbon was seeping into the subsoil. Why did the extra carbon supplied and absorbed by leaves not stimulate tree growth? Presumably this has to do with soil nutrients other than soluble nitrogen, such as phosphate. Mineral nutrients represent finite resources in any natural ecosystem and set the ultimate limit to any carbon "fertilization." Hence, plant growth cannot depart from what soils permit, irrespective of atmospheric CO_2 concentration.

One forest component did show a dramatic response to elevated CO_2: lianas. Tested in situ as well as under controlled conditions, *Hedera helix* (evergreen ivy) as well as *Clematis vitalba* (deciduous vine) showed a clear stimulation of growth as long as they grew in deep forest shade (Zotz et al. 2006; Grob, unpubl.; Fig. 14.2). Similar to tropical forests (Granados and Körner 2002), there is increasing evidence that liana growth into the canopy will also increase. In the case of *Hedera* and *Clematis*, the effects on forest dynamics will be very small, but in tropical forests, lianas play a key role in tree turnover, with the possibility of a long-term reduction in forest carbon stocks in response to atmospheric CO_2 enrichment (Körner 2004, 2009).

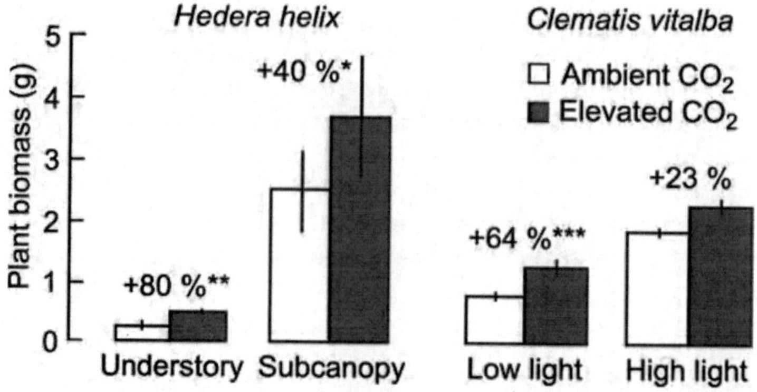

Fig. 14.2 The response of evergreen *Hedera helix* and deciduous *Clematis vitalba* to elevated CO_2. *Hedera* was studied in situ in the forest floor and the sub-canopy, and *Clematis* was studied in a phytotron experiment. Note the slower absolute but much greater relative growth stimulation in deep shade (in situ understorey 0.4 % and sub-canopy 12 % of full sunlight; low and high light in the growth chamber refer to ca. 5 % and 20 % of daily mean natural daylight)

References

Asshoff R, Hättenschwiler S (2006) Changes in needle quality and larch bud moth performance in response to CO_2 enrichment and defoliation of treeline larches. Ecol Entomol 31:84–90

Bader MKF, Körner C (2010) No overall stimulation of soil respiration under mature deciduous forest trees after 7 years of CO_2 enrichment. Glob Change Biol 16:2830–2843

Bader M, Hiltbrunner E, Körner C (2009) Fine root responses of mature deciduous forest trees to free air carbon dioxide enrichment (FACE). Funct Ecol 23(5):913–921

Bader MKF, Siegwolf R, Körner C (2013) Sustained enhancement of photosynthesis in mature deciduous forest trees after 8 years of free air CO_2 enrichment. Planta 232:1115–1125

Bignucolo O, Körner C (2010) Leaf miner activity and its effects on leaf chemistry in adult beech under elevated CO_2. Basic Appl Ecol 11:251–256

Granados J, Körner C (2002) In deep shade, elevated CO_2 increases the vigor of tropical climbing plants. Glob Change Biol 8:1109–1117

Hättenschwiler S, Schafellner C (2004) Gypsy moth feeding in the canopy of a CO_2-enriched mature forest. Glob Change Biol 10:1899–1908

Keel SG, Siegwolf TW, Körner C (2006) Canopy CO_2 enrichment permits tracing the fate of recently assimilated carbon in a mature deciduous forest. New Phytol 172:319–329

Körner C (2004) Through enhanced tree dynamics carbon dioxide enrichment may cause tropical forests to lose carbon. Phil Trans R Soc Lond 359:493–498

Körner C (2009) Responses of humid tropical trees to rising CO_2. Ann Rev Ecol Evol Syst 40:61–79

Körner C, Asshoff R, Bignucolo O, Hättenschwiler S, Keel SG, Pelaez-Riedl S, Pepin S, Siegwolf RTW, Zotz G (2005) Carbon flux and growth in mature deciduous forest trees exposed to elevated CO_2. Science 309:1360–1362

Leuzinger S, Körner C (2007) Water savings in mature deciduous forest trees under elevated CO_2. Glob Change Biol 13:2498–2508

Leuzinger S, Körner C (2010) Rainfall distribution is the main driver of runoff under future CO_2-concentration in a temperate deciduous forest. Glob Change Biol 16:246–254

Pepin S, Körner C (2002) web-FACE: a new canopy free-air CO_2 enrichment system for tall trees in mature forests. Oecologia 133:1–9

Pollierer MM, Langel R, Körner C, Maraun M, Scheu S (2007) The underestimated importance of belowground carbon input for forest soil animal food webs. Ecol Lett 10:729–736

Schleppi P, Bucher-Wallin I, Hagedorn F, Körner C (2012) Increased nitrate availability in the soil of a mixed mature temperate forest subjected to elevated CO_2 concentration (canopy FACE). Glob Change Biol 18:757–768

Steinmann K, Siegwolf RTW, Saurer M, Körner C (2004) Carbon fluxes to the soil in a mature temperate forest assessed by 13C isotope tracing. Oecologia 141:489–501

Zotz G, Cueni N, Körner C (2006) In situ growth stimulation of a temperate zone liana *(Hedera helix)* in elevated CO_2. Funct Ecol 20:763–769

Chapter 15
Shock Value: Are Lianas Natural Lightning Rods?

Stephen P. Yanoviak

Keywords Canopy • Electrical resistivity • Mortality • Trees • Tropical forest • Vines

Bullet Points

1. Lightning is commonly listed as a source of tropical tree mortality, but lightning effects on trees are often inconspicuous and are rarely quantified.
2. Lianas presumably are struck by lightning at least as often as trees but are more likely to be killed due to their small stem diameters.
3. The relatively low electrical resistivity of liana stems suggests that they are a source of lightning protection for trees.
4. Quantifying lightning-caused tree and liana mortality is critical to understanding current and future forest dynamics.

Summary

Forest canopies are principal contact points for lightning, which may be increasing in frequency due to climatic change. Although lightning is commonly listed as a source of tree mortality, many strikes are nonlethal, and lightning damage to tropical trees often is inconspicuous. Lianas are a major component of lowland tropical forest canopies but have been overlooked in the context of lightning. This chapter summarizes two hypotheses regarding liana–lightning interactions. First, lightning is an important agent of liana mortality. Lianas comprise a large fraction of upper canopy area but have relatively small stem diameters. Consequently, the probability that a liana will be struck is high relative to the likelihood it will survive the damage.

S.P. Yanoviak (✉)
Department of Biology, University of Louisville, 139 Life Sciences Building, Louisville, KY 40292, USA
e-mail: steve.yanoviak@louisville.edu

M. Lowman et al. (eds.), *Treetops at Risk: Challenges of Global Canopy Ecology and Conservation*, DOI 10.1007/978-1-4614-7161-5_15,

Second, lianas provide natural lightning protection for trees. The anatomy of liana stems should impart relatively low electrical resistivity. Preliminary data show that this is true for temperate woody vines compared to live tissues of similar-sized tree branches. If this is a widespread phenomenon, lianas may protect trees by being more attractive to incoming lightning or by bearing the bulk of the electrical current when lightning strikes a host tree. Support for these hypotheses could explain relatively high liana turnover rates. Resolving the importance of climate-driven sources of mortality like lightning is essential for understanding future tropical forest dynamics.

1 Introduction

1.1 Lightning Basics

Lightning is among the most powerful and awe-inspiring environmental phenomena on earth. It is prominent in human cultural history and relatively well understood scientifically (Rakov and Uman 2007). Individual lightning strokes vary in intensity and may occur from cloud to ground (CG), ground to cloud (GC), or within and between clouds (intra/inter-cloud; IC). The visible electrical discharge (the "return stroke" formed when ascending and descending leaders meet) is both hot (ca. 30,000 °C) and powerful (ca. 30,000A) and is the component of lightning that causes significant structural and biological damage (Fig. 15.1). Whereas the basic physics of lightning is relatively well understood (Rakov and Uman 2007), the ecology of lightning remains poorly studied.

Lightning frequency varies seasonally and geographically, but flash density generally is highest in the tropics (e.g., Williams 2005; Price 2009). For example, central Panama receives ca. 40 lightning flashes km^{-2} $year^{-1}$, with peak flash rates occurring between mid-July and mid-August (1995–2012 satellite data courtesy of Phillip Bitzer, University of Alabama in Huntsville). Under current climatic conditions, ca. 25 % of those flashes are potentially damaging to trees (as CG or GC lightning; Boccippio et al. 2001). Consequently, the forest on a 15 km^2 site in Panama (e.g., Barro Colorado Island) receives ca. 150 strikes per year. This number is expected to increase over the next few decades due to climate change. Specifically, for each 1 °C increase in average surface temperature (or [CO_2] doubling), lightning frequency may increase by at least 10 %, with some estimates exceeding 50 % (Williams 2005; Price 2009).

1.2 Lightning (Sometimes) Kills Trees

Widespread tree mortality shapes forest structure at the landscape scale via climate-driven catastrophic disturbance (e.g., fire, hurricanes; Lugo and Scatena 1996). Localized agents of tree mortality also affect forest structure (Lugo and Scatena

Fig. 15.1 Lightning damage to a mature *Tabebuia guayacan* in Panama. The upper trunk 10–15 m above the ground was deeply cracked and partially stripped of bark (*left image*), whereas the lower trunk within 3 m of the ground showed only exit wounds (*right image*). This intense strike caused group mortality (Magnusson et al. 1996)—almost all trees and shrubs within ca. 5 m of the focal tree also were killed. The strike occurred in late May 2012, ca. 2 weeks before the photo was taken. Coincidentally, this tree was one of the tallest within an 80×80 m research plot from which all lianas had been removed (Photo credits: S. P. Yanoviak)

1996), but most have only indirect links to climate (e.g., pollution, pathogens). Lightning is an exception. Lightning damages millions of trees worldwide each year (Taylor 1974). It is consistently listed as a source of tree mortality in the ecological literature and is hypothesized to be the main cause of death for the largest and oldest rainforest trees (Anderson 1964; Fig. 15.1). However, evidence for the latter is mostly circumstantial (Magnusson et al. 1996), and observations in temperate forests and tropical plantations show that trees often are not killed by lightning strikes (Taylor 1974).

Some lightning-struck trees exhibit no obvious physical damage (e.g., Furtado 1935), although they may be physiologically compromised or rendered more attractive to pathogens and herbivores (Taylor 1974). Thus, it is likely that lightning damage is frequently undocumented or misclassified by tropical ecologists; by default, the bulk of lightning-caused tree mortality is included in the ecologically ambiguous "standing dead" category. Even nonlethal lightning strikes ultimately affect tree fitness and forest turnover rates; thus, accurate quantification of lightning-induced mortality will improve forest dynamics models and facilitate predictions of future forest structure under conditions of increased lightning frequency.

1.3 *Lightning (Always) Kills Lianas*

Discussions of the ecological effects of lightning invariably focus on trees. However, lianas (woody vines) are major components of tropical forests (Schnitzer and Bongers 2002), and their abundance in the canopy makes them vulnerable to lightning damage. Liana–lightning interactions have been completely overlooked in the context of forest dynamics, yet both lianas and lightning will shape the ecology of tropical forests over the next century (Williams 2005; Schnitzer and Bongers 2011). Two hypotheses regarding liana–lightning interactions in tropical forests provide a point of entry for field research on this topic: (1) lightning is a significant source of mortality for lianas relative to trees and (2) liana stems act as natural lightning rods that inadvertently protect their host trees.

At least three aspects of the liana growth form suggest that lightning-caused mortality is relatively greater for lianas than for trees. First, individual lianas have disproportionately high lightning exposure relative to their basal area; their leaves are distributed like carpets in the uppermost forest canopy (Putz 1984). Collectively, lianas constitute a large proportion of highly exposed foliage in the canopy, and this fraction peaks (to > 30 %) during the wet season (Avalos and Mulkey 1999) when lightning frequency is also at a maximum. Moreover, liana tendrils commonly emerge above other canopy foliage (SY, pers. obs.), providing numerous origination points for ascending electrical leaders. Second, the small diameters and soft tissues of liana stems (Carlquist 1991) are unlikely to withstand the damage caused by lightning. The radius of a lightning return stroke ranges ca. 1–12 cm (Rakov and Uman 2007), which encompasses the diameters of most liana stems (Hegarty and Caballé 1991). Finally, the internal architecture and relatively high water-holding capacity of liana stems (Carlquist 1991) likely result in higher electrical conductivity than tree tissues. Although the process of lightning formation from specific ascending leaders is not completely resolved (Rakov and Uman 2007), more conductive substrates (e.g., liana stems) presumably generate stronger (more highly charged) or longer leaders, thereby increasing the local probability of a strike. The general characteristics of lianas suggest that they are especially susceptible to lightning-caused mortality, and lightning is likely to be a major contributor to the higher turnover rates observed for lianas relative to trees (Phillips et al. 2005).

1.4 *Lianas (Maybe) Protect Trees*

Lianas depend on trees for support. This close physical association is generally viewed as mechanical parasitism (Schnitzer and Bongers 2002), but it may also protect trees against lightning damage. A key assumption of this hypothesis is that liana stems have lower electrical resistivity than tree branches of similar diameter. No comparative resistivity data are currently available for tropical plants, but preliminary surveys in temperate forests support this assumption (Fig. 15.2). Such a difference in resistivity could protect trees via two mechanisms. First, as mentioned

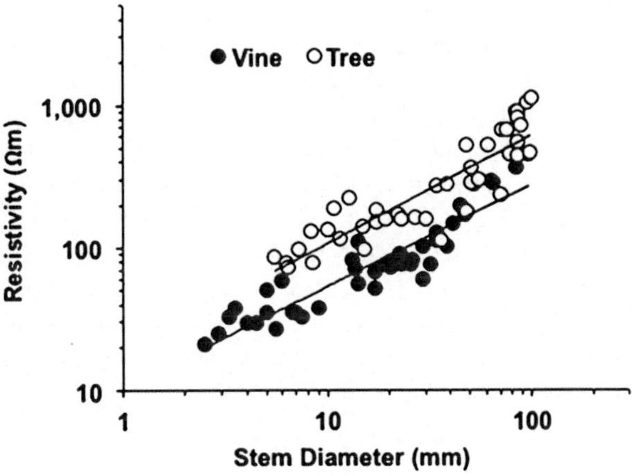

Fig. 15.2 Electrical resistivity of woody vines and tree branches up to 100 mm in diameter in temperate forests around Louisville, Kentucky. Data were recorded in October 2012

above, differential propagation of strike leaders could reduce the probability that a tree will be struck when lianas are present. Second, the bulk of the current in a lightning return stroke is distributed to the most conductive substrates (Rakov and Uman 2007). Thus, even if lightning initially strikes a tree crown, most of the charge (and damage) is likely to be borne by attached lianas. Although purely speculative, these mechanisms are not outside the realm of possibility.

Differential lightning conduction by lianas potentially explains why clear evidence of non-catastrophic (CG) strikes is uncommon among trees in tropical forests. Lianas provide a distributed network of low-resistance stems that effectively conduct the electrical current away from trees and to the ground. This phenomenon could be very common but completely unnoticed. Damage to an individual liana (including death) is likely to be inconspicuous to the ground-based observer in dense forest for various reasons. Liana stems are often very abundant in the understory, and changes in individual stems are easily overlooked. Liana foliage tends to be broadly and thinly distributed in the canopy; thus, leaf litter from a dead individual is unlikely to be conspicuously clumped when it lands on the ground. Moreover, leaves of many liana species do not fall until months after the stem is dead (SY, pers. obs.). Finally, the leaves and understory stems of lianas are often widely separated, such that the leaf litter from a dead individual may be tens of meters from its base. Whereas even low-intensity lightning damage is often conspicuous in temperate forests (e.g., Taylor 1974), the characteristic growth form of lianas may mask similar damage in tropical forests.

1.5 Electrical Properties of Lianas Versus Trees

To explore how electrical resistivity varies among different species and growth forms, my students and I used a megohmmeter to measure the electrical resistivity of living tissues of 43 vine stems (5–100 mm diameter) and 42 similarly sized branches of canopy trees in temperate oak–hickory forests around Louisville. Focal vine species were *Toxicodendron radicans*, *Parthenocissus quinquefolia*, and *Vitis aestivalis*, and focal trees were *Acer saccharum*, *Carya glabra*, *Juglans nigra*, *Quercus rubra*, and *Ulmus americana*. The average (± SE) resistivity of vine stems (167 ± 17.9 Ωm) was significantly lower than the resistivity of tree branches over the same size range (281 ± 18.1 Ωm; ANCOVA $F_{1,82} = 89.2$, $P < 0.0001$; Fig. 15.2), suggesting that vines would carry the bulk of the current from a lightning strike in the crown of their host trees. Resistivity differed among species of trees ($F_{4,36} = 2.84$, $P = 0.038$) and vines ($F_{2,39} = 5.43$, $P = 0.008$), and the range of resistivity values overlapped slightly (Fig. 15.2), suggesting that not all woody vine species are potentially protective of all tree species.

2 Research Needs

Research on the ecological effects of lightning poses many logistical challenges. Principal among these is accurate quantification of the role of lightning in forest dynamics. In particular, documenting nonlethal, inconspicuous strikes on individual trees requires continuous recording of strike activity in forests at smaller spatial scales than are currently possible via triangulation of electromagnetic signals. However, the rapid pace of technological advances in electronics and satellite monitoring capability may soon overcome this hurdle.

Apart from the hypotheses presented here, a number of basic questions remain to be answered regarding the ecology of lightning. First, are the tallest (i.e., emergent) trees in a forest most likely to be struck, as hypothesized by Anderson (1964)? As intuitive as the answer may seem, height differences among trees in continuous forest may be too small to be relevant at the scale of a typical lightning stroke, and evidence from temperate forests is inconclusive (Mäkelä et al. 2009). Second, are some tree species resistant to lightning strikes (Furtado 1935; Anderson 1964)? Lightning has been a selective force throughout the evolution of terrestrial life. Whereas no tree is expected to survive rare, intense (e.g., GC) lightning (Fig. 15.1), it is reasonable to suspect that some canopy tree species have anatomical or physiological traits that minimize damage from lower-intensity (e.g., CG) strikes, which are the most common. Finally, what is the effect of artificial lightning attractors on forest dynamics? Recent growth in wireless telecommunications and hydrocarbon drilling has led to a dramatic increase in the number of towers within forests, even in relatively remote locations. Such towers generally extend well above the surrounding forest canopy and function as lightning attractors (Rakov and Uman 2007), potentially changing local forest dynamics. Ironically, the canopy cranes that have been constructed around the world to study forest ecology should similarly disrupt natural strike frequencies.

References

Anderson JAR (1964) Observations on climatic damage in peat swamp forest in Sarawak. Commonw Forest Rev 43:145–158

Avalos G, Mulkey SS (1999) Seasonal changes in liana cover in the upper canopy of a Neotropical dry forest. Biotropica 31:186–192

Boccippio DJ, Cummins KL, Christian HJ, Goodman SJ (2001) Combined satellite-and surface-based estimation of the intracloud-cloud-to-ground lightning ratio over the continental United States. Mon Weather Rev 129:108–122

Carlquist S (1991) Anatomy of vine and liana stems: a review and synthesis. In: Putz FE, Mooney HA (eds) The biology of vines. Cambridge University Press, Cambridge

Furtado CX (1935) Lightning injuries to trees. J Malays Branch R Asiatic Soc 13:157–162

Hegarty EE, Caballé G (1991) Distribution and abundance of vines in forest communities. In: Putz FE, Mooney HA (eds) The biology of vines. Cambridge University Press, Cambridge

Lugo AE, Scatena FN (1996) Background and catastrophic tree mortality in tropical moist, wet, and rain forests. Biotropica 28:585–599

Magnusson WE, Lima AP, de Lima O (1996) Group lightning mortality of trees in a Neotropical forest. J Trop Ecol 12:899–903

Mäkelä J, Karvinen E, Porjo N, Mäkelä A, Tuomi T (2009) Attachment of natural lightning flashes to trees: preliminary statistical characteristics. J Lightning Res 1:9–21

Phillips OL, Vásquez Martínez R, Monteagudo Mendoza A, Baker TR, Núñez Vargas P (2005) Large lianas as hyperdynamic elements of the tropical forest canopy. Ecology 86:1250–1258

Price C (2009) Will a drier climate result in more lightning? Atmos Res 91:479–484

Putz FE (1984) The natural history of lianas on Barro Colorado Island, Panama. Ecology 65:1713–1724

Rakov VA, Uman MA (2007) Lightning: physics and effects. Cambridge University Press, Cambridge, UK

Schnitzer SA, Bongers F (2002) The ecology of lianas and their role in forests. Trends Ecol Evol 17:223–230

Schnitzer SA, Bongers F (2011) Increasing liana abundance and biomass in tropical forests: emerging patterns and putative mechanisms. Ecol Lett 14:397–406

Taylor AR (1974) Ecological aspects of lightning in forests. Proc Tall Timbers Fire Ecol Conf 13:455–482

Williams ER (2005) Lightning and climate: a review. Atmos Res 76:272–287

Chapter 16
Potential Impacts of Global Changes on Epiphytic Bryophytes in Subtropical Montane Moist Evergreen Broad-Leaved Forests, SW China

Liang Song and Wen-Yao Liu

Keywords Climate change • Epiphyte • Forest degradation • Nitrogen deposition • Subtropical evergreen broad-leaved forest

Summary

Subtropical montane moist evergreen broad-leaved forests (MMEBF) harbor a myriad of epiphytic bryophytes and lichens, which play an important role in the hydrology and mineral cycling of the whole forest ecosystems. Due to their sensitivity to atmospheric conditions, epiphytic bryophytes are suggested to be sensitive to global environmental changes such as forest degradation, global warming, and enhanced N pollution, but empirical evidence remains scarce. To assess the fate of epiphytic bryophytes under global changes, we reviewed the three experiments that were conducted in the subtropical montane forest recently, plus a few related studies. In MMEBF, forest degradation resulted in major loss of species richness of epiphytic bryophytes. Even slight changes in climate resulted in remarkably reduced rates of growth and detrimental effects on the vitality of the two transplanted bryophyte species. With the prediction that climate conditions will become warmer and drier, epiphytic bryophytes may be reduced in MMEBF in the future. N addition had significantly negative effects on species richness and cover of the epiphytic bryophyte community; the decline of some epiphytic bryophytes under relatively high N input may result from detrimental effects on degradation to photosynthetic pigments.

L. Song (✉) • W.-Y. Liu
Key Laboratory of Tropical Forest Ecology, Xishuangbanna Tropical Botanical Garden, Chinese Academy of Sciences, Kunming, Yunnan 650223, People's Republic of China
e-mail: songliang@xtbg.ac.cn

M. Lowman et al. (eds.), *Treetops at Risk: Challenges of Global Canopy Ecology and Conservation*, DOI 10.1007/978-1-4614-7161-5_16,

1 Introduction

Evergreen broad-leaved forests contribute to global biodiversity and are important to the sustainable development of subtropical regions of China (Wu 1980; Wang et al. 2005). Subtropical montane moist evergreen broad-leaved forests (MMEBF) are a subgroup of evergreen broad-leaved forests and occur at high altitude in Yunnan Province, southwestern China (Wu 1987). The total area of Ailao Mountains National Nature Reserve (NNR, Fig. 16.1) is 50,400 ha, considered the largest tract of natural evergreen broad-leaved forest in China. A total area of 5,100 ha of MMEBF is still present in the core area of Ailao Mountains NNR thanks to rugged topography, relatively difficult access, and effective protection (Young and Herwitz 1995; Qiu and Xie 1998).

Due to high water availability, the presence of large trees, and absence of widespread human disturbance, the MMEBF located in the Ailao Mountains NNR in southwestern China harbored abundant orchids, ferns, bryophytes, lichens, and suspended soils (Xu and Liu 2005) (Fig. 16.2). Nearly 11 t of epiphytic biomass has been measured in 1 ha of MMEBF in Ailao Mountains NNR (Xu and Liu 2005). Species richness of epiphytes in moist subtropical montane forest is also very high.

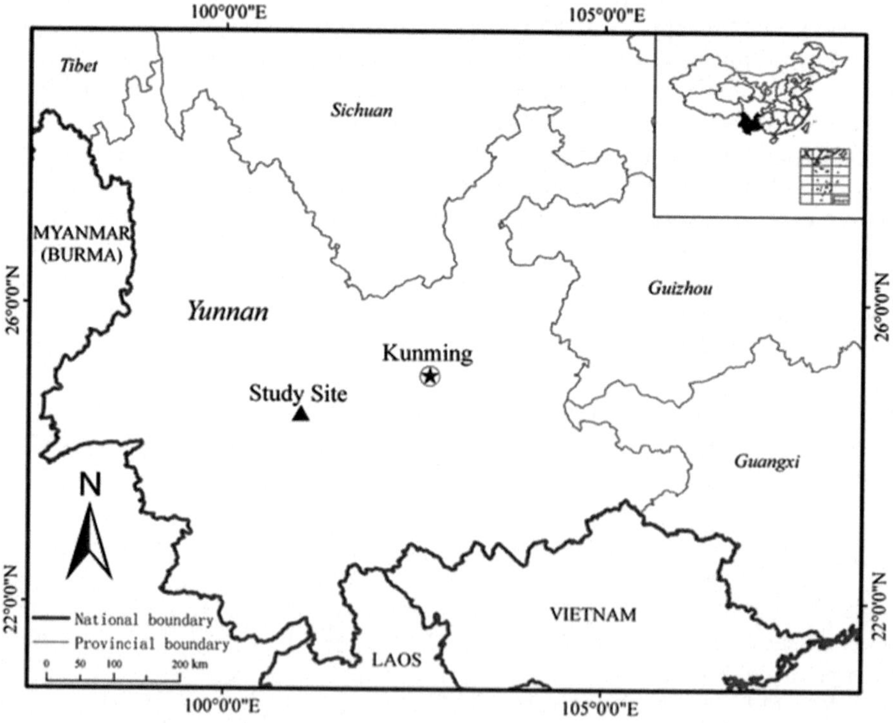

Fig. 16.1 Location of the study site in Ailao Mountain NNR, China

Fig. 16.2 Abundant epiphytic plants in MMEBF (Photo by Liang Song and Yuan-Lin Yao)

In the Ailao Mountains NNR, about 500 epiphytic species have been recorded (Xu and Liu 2005; Ma et al. 2009; Li et al. 2013). The forest is especially rich in nonvascular epiphytes, including 176 epiphytic bryophytes (Ma et al. 2009) and 217 epiphytic lichens (Li et al. 2013). These plants not only represent some of the most striking life forms of the forest but also play an important role in the hydrology and mineral cycling of the whole forest ecosystem in the Ailao Mountains (Liu et al. 2002; Chen et al. 2010; Han et al. 2010).

The rich diversity of subtropical epiphytes is now becoming increasingly endangered by human activities such as forest harvesting and fossil fuel combustion. Forest harvesting or other human disturbance such as road and dam building usually caused destruction of MMEBF. It is estimated that about 90 % of the montane forests of the Andes have already disappeared, even though great efforts have been made to conserve these forests and slow down the pace of deforestation (Gradstein 2008). The rapid conversion of these forests poses the following questions: what is the fate of the rich epiphytic communities? Will they survive in degraded forests? Further, can they reestablish in regrowing secondary forests? How fast is their recovery (Gradstein 2008)?

Another question to be addressed is the possible impacts of changes in atmospheric conditions due to human activities on epiphytes. On the one hand, growth in human activities such as the burning of fossil fuels has increased greenhouse gases emissions worldwide (e.g. CO_2), which has been suggested to cause global climate change (IPCC 2007). On the other hand, rapid economic development has also increased usage and emissions of nitrogen (N) in Asia, especially in China, and is predicted to continue to increase in the future (Zheng et al. 2002; Richter et al. 2005). These human-driven processes may have serious indirect impacts on arboreal plants because their reliance on atmospheric moisture and nutrients (Zechmeister et al. 2007). Epiphytic bryophytes are suggested to be particularly sensitive to atmospheric environmental changes, given their typically high leaf area indices and high surface area to volume ratios, the fact that they do not have a rooting system, and lack a cuticular barrier (Liu et al. 2007).

In this chapter we briefly review some of our epiphytic bryophytes work in subtropical MMEBFs in southwestern China.

2 Potential Impacts of Forest Degradation on Epiphytic Bryophytes

Many studies have dealt with the impacts of forest degradation on epiphyte diversity, and the results are controversial (Gradstein 2008). While some reported a marked decrease of species richness and diversity following forest disturbance (Turner et al. 1994; da Costa 1999; Vellak and Paal 1999), others found little or no change (Nkongmeneck et al. 2002; Holz and Gradstein 2005). Reasons for these different results remain unclear and prompted us to undertake the following investigation.

In order to assess the impact of forest degradation on species richness of epiphytes in MMEBF, we surveyed and analyzed the epiphytic bryophytes growing on *Lithocarpus xylocarpus* (Kurz) Markgr trunks with different diameter classes in typical primary and secondary *Lithocarpus* forests (110 years old) in Ailao Mountains NNR, SW China. Forest degradation resulted in marked loss of species richness of epiphytic bryophytes. If the comparison were made on common hosts with similar diameters (excluded large trees in the primary forest), ten more bryophyte species were recorded in the primary forest, which indicated that the primary forest had greatly favored certain epiphytes (Song et al. 2011). Further, species number and total coverage of epiphytic bryophytes per plot in the primary forest were remarkably higher than secondary forest in different diameter classes except for class two, in which bryophyte species numbers per plot in two forest types were nearly the same (Song et al. 2011). Typically, in secondary forest, phorophytes only provide limited complexity in structure (Barthlott et al. 2001), while microclimate is drier caused by the open canopy and the stronger radiation (Gradstein 2008). As a result, many epiphytic bryophytes could not survive in secondary forest. The long period of succession in primary forest, on the other hand, stimulated the hosts in primary forest to develop a diversified canopy structure and thus wetter microclimate. This habitat had in turn made it possible that epiphytic bryophytes with different niches could coexist. In addition, if larger host individuals in the primary forest were considered in the comparison, the advantage of primary forest was more obvious, which increased from 10 to 19. Lack of large diameter phorophytes seemed to be one of the main reasons that caused low richness in secondary forests.

Epiphytic bryoflora would be rather difficult to recover once destroyed. For example, few bryophytes returned after 25 years in subtropical rain forests of Australia (Chapman and King 1983). The 10–15 years old secondary montane forests in Costa Rica harbored similar diversity of cryptogamic epiphytes as primary forest in the same region, but the species composition was markedly different even after 40 years of succession (Holz and Gradstein 2005). In the MMEBF, we found that species richness and cover were remarkably lower in the secondary forest compared with the primary forest (Song et al. 2011). After nearly 110 years of succession, half (28 species) of the primary forest species still had not reestablished in the secondary forest (Song et al. 2011). All these results indicate that the epiphytic bryoflora has not completely recovered in the secondary forest. Slow recovery rates in epiphytic flora were also recorded in tropical forests (Acebey et al. 2003), as well as in temperate (Norris 1987) and boreal forests (Kuusinen and Siitonen 1998). The majority of these studies argued that at least 100 years were needed for the regeneration of most nonvascular epiphytic flora in the regrowing forest (Zotz and Bader 2009). We suggest that even a longer period, perhaps much more than 110 years, is needed for the reestablishment of microhabitats and epiphytic bryophyte communities in subtropical montane forests of SW China (Song et al. 2011).

3 Potential Impacts of Climate Change on Epiphytic Bryophytes

The world is warming (IPCC 2007), and most tropical land surfaces are predicted to be drier (Foster 2001). Over the past 40 years, the annual temperature in Yunnan Province (southwestern China) has increased at a rate of 0.3 °C decade^{-1} (Fan et al. 2011), while the air relative humidity has shown a decreasing trend in tropical regions of Yunnan (He et al. 2007).

The first empirical data on the impact of global warming on epiphytes arose from work on bryophytes and lichens in Europe, where the range extension of Atlantic and Mediterranean bryophyte species into the east and northeast in the last decades was attributed to global warming (Frahm and Klaus 1997, 2001). Although lack of long-term distribution records and monitoring limited an understanding of possible impacts of climate change on epiphytes, transplantation experiments provide a useful approach (Gradstein 2008). We conducted a transplantation experiment to assess the potential impacts of predicted climate change on the growth and vitality of two common epiphytic bryophytes (*Bazzania ovistipula* (Steph.) Mizut. and *Trismegistia undulate* Broth. et Yas) at three elevations in MMEBF in southwest China. Results indicated that the simulated climate change significantly negatively affected the growth and vitality of the two epiphytic bryophytes within 2 years (Figs. 16.3 and 16.4). Similarly, epiphyte mats collected from an upper site were transplanted to lower sites (from 1,480 to 1,410 and 1,330 m) in Costa Rica, where amounts of cloud water were reported to be smaller (Nadkarni and Solano 2002). In Bolivia, branches covered with dense bryophyte mats were transplanted to lower altitudes

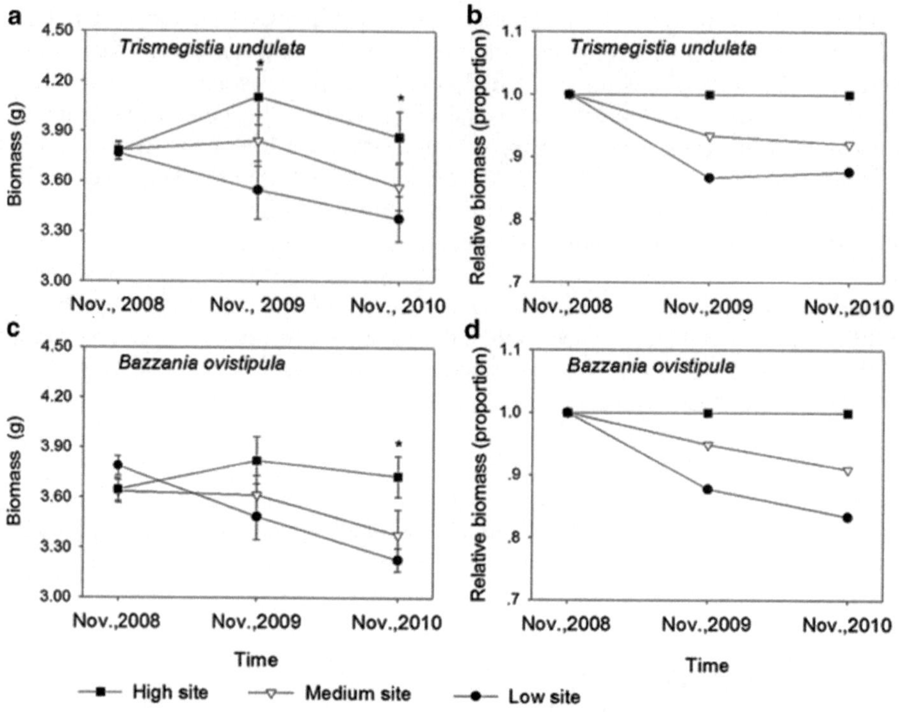

Fig. 16.3 Treatment effects of simulated climate change on growth of **a** (biomass)/**b** (relative biomass) *Trismegistia undulate* Broth. et Yas and **c** (biomass)/**d** (relative biomass) *Bazzania ovistipula* (Steph.) Mizut. (represented by biomass) (Note: *bars* indicate ± SE. * represent differences among transplantation sites within the observation period are significant at the 0.05 level using one-way ANOVA)

(from 3,000 to 2,700 and 2,500 m), which were naturally exposed to warmer and drier habitat. Two years later, the relative abundance of species in the community had been altered significantly (Jácome et al. 2011). Using a space-for-time gradient, the above translocation experiments confirmed the sensitivity of epiphytic bryophytes to climate change.

Increased drought is likely to become the most detrimental aspect of climate change for epiphytes that prefer cool and wet forests (Zotz and Bader 2009). Gradstein (2008) proposed that drought had directly deleterious effects on nonvascular epiphytes in cloud forests, which was also true for epiphytic bryophytes in our MMEBF. Our measurements showed that *T. undulate* and *B. ovistipula* exhibited negative growth rates, with a quantitative decline in vitality between November 2009 and November 2010, in contrast to positive growth rates between November 2008 and November 2009 (Figs. 16.3 and 16.4), which correlated with severe dryness during this period (Song et al. 2012b). Water availability may be the most important environmental factor affecting poikilohydric epiphytes (Bates et al.

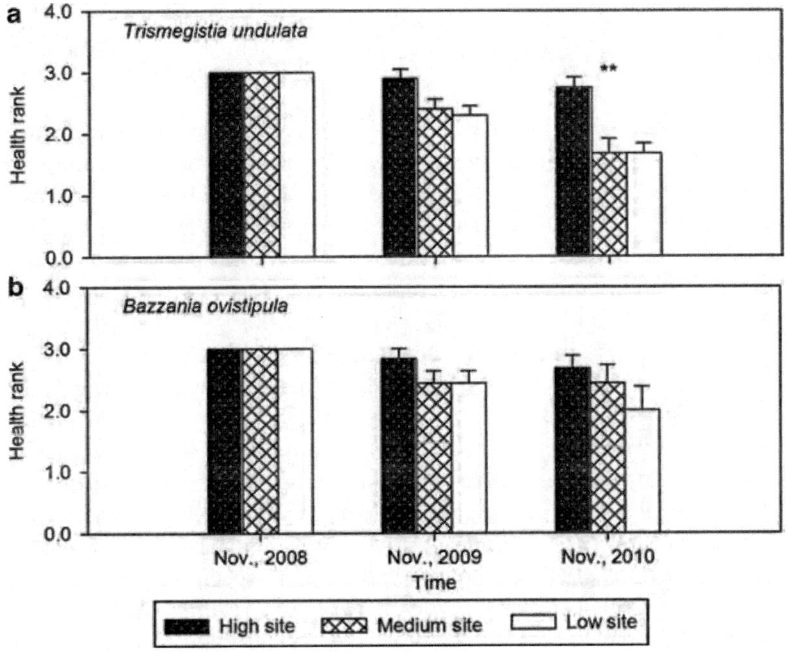

Fig. 16.4 Treatment effects of simulated climate change on health of (**a**) *Trismegistia undulate* Broth. et Yas and (**b**) *Bazzania ovistipula* (Steph.) Mizut. (Note: *bars* indicate ± SE. ** indicates that differences among transplantation sites within the observation period are significant at the 0.01 level using Kruskal-Wallis H's nonparametric test)

2004; Sillett and Antoine 2004). For example, in French Guiana, epiphytes were more abundant in a tropical lowland cloud forest compared with a nearby lowland rain forest, with higher RH and additional throughfall through fog in the cloud forest (Normann et al. 2010; Gehrig-Downie et al. 2011). Although epiphytic bryophytes can intercept cloud water droplets from moving air during low rainfall periods and respond to short-term drought by reducing physiological activity (Sillett and Antoine 2004; León-Vargas et al. 2006), they cannot tolerate drought long term.

As is mentioned above, epiphytic bryophytes in the MMEBF were under severe water shortage during the 2009–2010 drought. In contrast, Qi et al. (2012) reported that trees in the forest did not experience water deficits during that period. Epiphytic bryophytes have little or no buffered pools of moisture and nutrients, while trees can survive in dry seasons or droughts by absorbing stored underground water. Thus, epiphytic bryophytes were more responsive to subtle changes in climate than were terrestrially rooted plants (Nadkarni and Solano 2002). We infer that epiphytic bryophytes will respond much sooner to changes in the water cycle, especially dryness, than their terrestrial counterparts in the subtropical montane moist forests of southwestern China (Song et al. 2012b).

4 Potential Impacts of Atmospheric N Pollution on Epiphytic Bryophytes

Rapid economic development has increased usage and emissions of nitrogen (N) in Asia, especially in China, and this trend is predicted to continue in the future (Zheng et al. 2002; Richter et al. 2005). It is suggested that excessive N loads may affect the growth, structure, function, and dynamics of forests (Li et al. 2003). In recent years, there has been increasing concern about the effects of atmospheric N deposition on lower plants such as bryophytes, which represent the most sensitive types of vegetation to N pollution (Mitchell et al. 2005; Cape et al. 2009).

Mitchell et al. (2004) assessed the potential effects of increased or decreased N deposition on growth and tissue N of epiphytic bryophytes by reciprocal transplantation in Atlantic oak woods. Results showed that the growth rates of *Isothecium myosuroides*, *Dicranum scoparium*, and *Frullania tamarisci* were strongly reduced following an increase in atmospheric N deposition (Mitchell et al. 2004). In a further study, Mitchell et al. (2005) investigated the composition of epiphytic communities in seven Atlantic oak woods and found that bark pH and NH_4^+ concentration in the stemflow explained the greatest amount of variation in species composition among sites. They also identified *Isothecium myosuroides* and *Frullania tamarisci* as indicators of small N inputs and estimated that the critical load for N deposition for epiphytes in Atlantic oak woods reached 11–18 kg N ha^{-1} $year^{-1}$.

In the MMEBF in Ailao Mountain NNR in southwestern China, we conducted a field manipulation experiment to study the impact of N deposition on epiphytic bryophytes (Song et al. 2012a). Results showed that species richness, cover, and growth of epiphytic bryophytes were negatively related to the concentration of simulated N addition (Song et al. 2012a). Low N addition (15 kg N ha^{-1} $year^{-1}$) resulted in a remarkable decline in species richness and cover of the whole bryophyte communities (Table 16.1, Fig. 16.5), while medium N additions (30 kg N ha^{-1} $year^{-1}$) had a significant negative impact on the growth of *Plagiochila arbuscula* (Table 16.1) (Song et al. 2012a).

Table 16.1 Results of repeated measure ANOVAs for species richness (A) and cover (B) of epiphytic bryophyte communities and growth of *Plagiochila arbuscula* (C) under different N treatment levels from April 2009 to April 2010 (Based on Song et al. (2012a))

Source	F	P
(A) Species richness		
N treatment	16.69	<0.001
Time	53.15	<0.001
N treatment × time	36.84	<0.001
(B) Cover		
N treatment	53.52	<0.001
Time	302.46	<0.001
N treatment × time	45.96	<0.001
(C) Shoot length of *Plagiochila arbuscula*		
N treatment	16.54	<0.001
Time	592.35	<0.001
N treatment × time	20.88	<0.001

Control (0 kg N ha^{-1} year^{-1}) Low N treatment (15 kg N ha^{-1} year^{-1})

Medium N treatment (30 kg N ha^{-1} year^{-1}) High N treatment (60 kg N ha^{-1} year^{-1})

Fig. 16.5 Visual treatment effects of nitrogen solutions on epiphytic bryophytes

However, in a tropical forest located at a similar latitude in southern China, even medium levels of N addition (100 kg N ha^{-1} year^{-1}) did not alter the overall diversity of terrestrial plants over 5 years (Lu et al. 2010). Another study indicated that seedling growth of two tropical tree species exhibited positive effects under low N addition (50 kg N ha^{-1} year^{-1}) (Mo et al. 2008). The sharp contrast above demonstrates that epiphytic bryophytes are much more sensitive to N pollution than terrestrial trees.

Normally, when N input increases slightly, N may be taken up and then used to enhance the chlorophyll content of the cell, thus enlarging photosynthetic capacity (Baxter et al. 1992; Limpens and Berendse 2003). However, excessive N input is detrimental to chlorophyll, inducing an adverse effect on the photosynthetic system and thus negatively affects net photosynthesis (Van der Heijden et al. 2000). In our study, the PQ value of *P. assamica* was significantly decreased by N addition (Song et al. 2012a), an indication of chlorophyll degradation (Bignal et al. 2008). We conclude that direct deleterious effects of excess N on the physiology of some epiphytic bryophytes may be one of the mechanisms that explain negative impacts of increased atmospheric N deposition.

In addition, we monitored the recovery rate of bryophyte communities 12 months after cessation of N addition. Strengbom et al. (2001) found that N-induced decreases in the abundance of specific bryophytes may persist even after N input has been terminated. Our results supported their observation since we detected minimal changes in species richness, cover, and species composition of trunk-dwelling epiphytic bryophytes after 12 months of recovering from excess N input (Song et al. 2012a). This indicates that the epiphytic bryophytes studied here are vulnerable to high N input over a short period of time but that recovery over the same period is difficult.

5 Conclusions

In MMEBF, forest degradation led to major loss of species richness of epiphytic bryophytes. Changes of canopy structure, microclimate, and host characteristics may be the main causes of the decline of epiphytic bryophytes in the secondary forest.

In MMEBF, even slight changes in climate resulted in remarkably reduced rates of growth and detrimental effects on the vitality of the two transplanted bryophyte species over 2 years. Nonvascular epiphytes in this forest type responded much more rapidly to changes in water availability than terrestrial trees. Epiphytic bryophytes may be negatively affected or even severely damaged in the future in subtropical montane forests, as climate conditions are predicted to be warmer and drier.

In MMEBF, N addition had significantly negative effects on species richness and cover of the epiphytic bryophyte community. Harmful effects of high N loads were recorded for chlorophyll, growth, and vitality of the species tested. The decline of some epiphytic bryophytes may result from detrimental effects on degradation to photosynthetic pigments.

Based on current observations, it is likely that epiphytic bryophytes in MMEBF are currently at risk of decline under global changes such as forest degradation, global warming, and enhanced N pollution. Additionally, adverse effects of global changes on epiphytic bryophytes cannot be considered in isolation. Considering epiphytic bryophytes play important roles in hydrological and nutrient cycles of the forest ecosystem and facilitate other plants and animal lives in forest canopies, deterioration of epiphytes from global changes may have cascading negative effects on the whole forest ecosystem.

Acknowledgments These studies were supported by the National Natural Science Foundation of China (No. U1133605, 30870437), the QCAS Biotechnology Fund (No. GJHZ1130), Knowledge Innovation Program of Chinese Academy of Science (No. KSCX2-YW-Z-1019), and West Light Foundation of the Chinese Academy of Sciences. The Management Authority of Ailao Mts. National Nature Reserve and the Ecological Station of Ailao Mountains are thanked for granting permission and facilitating this research. We wish to thank Wen-Zhang Ma for help in identifying bryophyte species and Mr. Jia-Fu Wu for preparing Fig. 16.1. We thank Jin-Hua Qi, Yuan-Lin Yao, Gan-Can Song, Xi Chen, Su Li, Hua-Zheng Lu, and Ke Chen for assistance in the field surveying.

References

Acebey A, Gradstein SR, Krömer T (2003) Species richness and habitat diversification of bryophytes in submontane rain forest and fallows of Bolivia. J Trop Ecol 19:9–18

Barthlott W, Schmit–Neuerburg V, Nieder J, Engwald S (2001) Diversity and abundance of vascular epiphytes: a comparison of secondary vegetation and primary montane rain forest in the Venezuelan Andes. Plant Ecol 152:145–156

Bates JW, Roy DB, Preston CD (2004) Occurrence of epiphytic bryophytes in a tetrad transect across southern Britain. 2 analysis and modelling of epiphyte–environment relationships. J Bryol 26:181–197

Baxter R, Emes MJ, Lee JA (1992) Effects of an experimentally applied increase in ammonium on growth and amino–acid metabolism of *Sphagnum cuspidatum* Ehrh ex Hoffm from differently polluted areas. New Phytol 120:265–274

Bignal KL, Ashmore MR, Headley AD (2008) Effects of air pollution from road transport on growth and physiology of six transplanted bryophyte species. Environ Pollut 156:332–340

Cape JN, Van der Eerden LJ, Sheppard LJ, Leith ID, Sutton MA (2009) Evidence for changing the critical level for ammonia. Environ Pollut 157:1033–1037

Chapman WS, King GC (1983) Floristic composition and structure of a rainforest area 25 years after logging. Austr J Ecol 8:415–423

Chen L, Liu W-Y, Wang G-S (2010) Estimation of epiphytic biomass and nutrient pools in the subtropical montane cloud forest in the Ailao Mountains, south–western China. Ecol Res 25:315–325

da Costa DP (1999) Epiphytic bryophyte diversity in primary and secondary lowland rainforests in southeastern Brazil. Bryologist 102:320–326

Fan Z-X, Bräuning A, Thomas A, Li J-B, Cao K-F (2011) Spatial and temporal temperature trends on the Yunnan Plateau (Southwest China) during 1961–2004. Int J Climatol 31:2078–2090

Foster P (2001) The potential negative impacts of global climate change on tropical montane cloud forests Earth. Sci Rev 55:73–106

Frahm JP, Klaus D (1997) Moose als Indikatoren von Klimafluktuationen in Mitteleuropa. Erdkunde 51:181–190

Frahm JP, Klaus D (2001) Bryophytes as indicators of recent climate fluctuations in Central Europe. Lindbergia 26:97–104

Gehrig-Downie C, Obregón A, Bendix J, Gradstein SR (2011) Epiphyte biomass and canopy microclimate in the tropical lowland cloud forest of French Guiana. Biotropica 43:591–596

Gradstein SR (2008) Epiphytes of tropical montane forests—impact of deforestation and climate change. In: Gradstein SR, Homeier J (eds) The tropical montane forest—patterns and processes in a biodiversity hotspot biodiversity and ecology series. University of Göttingen Press, Göttingen

Han B, Zou X-M, Kong J-J, Sha L-Q, Gong H-D, Yu Z, Cao T (2010) Nitrogen fixation of epiphytic plants enwrapping trees in Ailao Mountain cloud forests, Yunnan, China. Protoplasma 247:1–8

He Y-L, Zhang Y-P, Yang X-B (2007) Climate change in tropical area of southwestern China since 1950s. Sci Geogr Sin (Chinese Version) 27:499–505

Holz I, Gradstein SR (2005) Cryptogamic epiphytes in primary and recovering upper montane oak forests of Costa Rica–species richness, community composition and ecology. Plant Ecol 178:89–109

IPCC (2007) Climate change 2007. Synthesis report, Geneva

Jácome J, Gradstein SR, Kessler M (2011) Responses of epiphytic bryophyte communities to simulated climate change. In: Tuba Z, Slack NG, Stark LR (eds) Bryophyte ecology and climate change. Cambridge University Press, Cambridge, NY

Kuusinen M, Siitonen J (1998) Epiphytic lichen diversity in old-growth and managed *Picea abies* stands in southern Finland. J Veg Sci 9:283–292

León-Vargas Y, Engwald S, Proctor MCF (2006) Microclimate, light adaptation and desiccation tolerance of epiphytic bryophytes in two Venezuelan cloud forests. J Biogeogr 33:901–913

Li D-J, Mo J-M, Fang Y-T, Pen S-L, Gundersen P (2003) Impact of nitrogen deposition on forest plants. Acta Ecol Sin (Chinese Version) 23:1891–1900

Li S, Liu W-Y, Li D-W (2013) Epiphytic lichens in subtropical forest ecosystems in southwest China: species diversity and implications for conservation. Biol Conserv 159: 88–95

Limpens J, Berendse F (2003) Growth reduction of *Sphagnum magellanicum* subjected to high nitrogen deposition: the role of amino acid nitrogen concentration. Oecologia 135:339–345

Liu W-Y, Fox JED, Xu Z-F (2002) Nutrient fluxes in bulk precipitation, throughfall and stemflow in montane subtropical moist forest on Ailao Mountains in Yunnan, south–west China. J Trop Ecol 18:527–548

Liu X-Y, Xiao H-Y, Liu C-Q, Li Y-Y (2007) [delta] ^{13}C and [delta] ^{15}N of moss *Haplocladium microphyllum* (Hedw) Broth for indicating growing environment variation and canopy retention on atmospheric nitrogen deposition. Atmos Environ 41:4897–4907

Lu X-K, Mo J-M, Gilliam F, Zhou G-Y, Fang Y-T (2010) Effects of experimental nitrogen additions on plant diversity in an old–growth tropical forest. Global Change Biol 16:2688–2700

Ma W-Z, Liu W-Y, Li X-J (2009) Species composition and life forms of epiphytic bryophytes in old–growth and secondary forests in Mt Ailao, SW China. Cryptogamie Bryol 30:477–500

Mitchell RJ, Sutton MA, Truscott AM, Leith ID, Cape JN, Pitcairn CER, Van D (2004) Growth and tissue nitrogen of epiphytic Atlantic bryophytes: effects of increased and decreased atmospheric N deposition. Funct Ecol 18:322–329

Mitchell RJ, Truscot AM, Leith ID, Cape JN, Van D (2005) A study of the epiphytic communities of Atlantic oak woods along an atmospheric nitrogen deposition gradient. J Ecol 93:482–492

Mo J-M, Li D-J, Gundersen P (2008) Seedling growth response of two tropical tree species to nitrogen deposition in southern China. Eur J Forest Res 127:275–283

Nadkarni NM, Solano R (2002) Potential effects of climate change on canopy communities in a tropical cloud forest: an experimental approach. Oecologia 131:580–586

Nkongmeneck BA, Lowman MD, Atwood JT (2002) Epiphyte diversity in primary and fragmented forests of Cameroon, Central Africa: a preliminary survey. Selbyana 23:121–130

Normann F, Weigelt P, Gehrig–Downie C, Gradstein SR, Sipman HJM, Obregon A, Bendix J (2010) Diversity and vertical distribution of epiphytic macrolichens in lowland rain forest and lowland cloud forest of French Guiana. Ecol Indic 10:1111–1118

Norris DH (1987) Long–term results of cutting on the bryophytes of the *Sequoia sempervirens* forest in northern California. Symp Biol Hung 35:467–473

Qi J-H, Zhang Y-J, Zhang Y-P, Liu H-Y, Yang Q-Y, Song L, Gong H-D, Lu Z-Y (2012) Water–holding capacity of an evergreen broadleaf forest in Ailao Mountain and its functions in mitigating effects of Southwest China drought. Acta Ecol Sin (Chinese Version) 32:1692–1702

Qiu X-Z, Xie C-X (1998) Studies on the forest ecosystem in Ailao Mountains, Yunnan. Yunnan Sciences and Technology Press, Kunming (in Chinese with English summary)

Richter A, Burrows JP, Nü H, Granier C, Niemeier U (2005) Increase in tropospheric nitrogen dioxide over China observed from space. Nature 437:129–132

Sillett SC, Antoine ME (2004) Lichens and bryophytes in forest canopies. In: Lowman MD, Nadkarni NM (eds) Forest canopies. Academic, Amsterdam

Song L, Liu W-Y, Ma W-Z, Tan Z-H (2011) Bole epiphytic bryophytes on *Lithocarpus xylocarpus* (Kurz) Markgr in the Ailao Mountains, SW China. Ecol Res 26:351–363

Song L, Liu W-Y, Ma W-Z, Qi J-H (2012a) Response of epiphytic bryophytes to simulated N deposition in a subtropical montane cloud forest in southwestern China. Oecologia 170:847–856

Song L, Liu W-Y, Nadkarni NM (2012b) Response of non–vascular epiphytes to simulated climate change in a montane moist evergreen broad–leaved forest in southwest China. Biol Conserv 152:127–135

Strengbom J, Nordin A, Sholm TN, Ericson L (2001) Slow recovery of boreal forest ecosystem following decreased nitrogen input. Funct Ecol 15:451–457

Turner IM, Tan HTW, Wee YC, Ibrahim AB, Chew PT, Corlett RT (1994) A study of plant species extinction in Singapore: lessons for the conservation of tropical biodiversity. Conserv Biol 8:705–712

Van der Heijden E, Verbeek SK, Kuiper PJC (2000) Elevated atmospheric CO_2 and increased nitrogen deposition: effects on C and N metabolism and growth of the peat moss *Sphagnum recurvum* P Beauv var *mucronatum* (Russ) Warnst. Global Change Biol 6:201–212

Vellak K, Paal J (1999) Diversity of bryophyte vegetation in some forest types in Estonia: a comparison of old unmanaged and managed forests. Biodivers Conserv 8:1595–1620

Wang X-H, Yan E-R, Yan X, Wang L-Y (2005) Analysis of degraded evergreen broad–leaved forest communities in Eastern China and issues in forest restoration. Acta Ecol Sin (Chinese Version) 25:1796–1803

Wu Z-Y (1980) Vegetation of China (In Chinese). Science Press, Beijing

Wu Z-Y (1987) Vegetation of Yunnan (In Chinese). Science Publishing Agent, Beijing

Xu H-Q, Liu W-Y (2005) Species diversity and distribution of epiphytes in the montane moist evergreen broad–leaved forest in Ailao Mountain, Yunnan Biodiversity. Science 13:137–147

Young SS, Herwitz SR (1995) Floristic diversity and co–occurrences in a subtropical broad–leaved forest and two contrasting regrowth stands in central–west Yunnan Province, China. Vegetation 119:1–13

Zechmeister HG, Dirnböck T, Hülber K, Mirtl M (2007) Assessing airborne pollution effects on bryophytes–lessons learned through long–term integrated monitoring in Austria. Environ Pollut 147:696–705

Zheng X-H, Fu C-B, Xu X-K, Yan X-D, Huang Y, Han S-H, Hu F, Chen G-X (2002) The Asian nitrogen cycle case study. Ambio 31:79–87

Zotz G, Bader MY (2009) Epiphytic plants in a changing world–global: change effects on vascular and non–vascular epiphytes. Prog Bot 70:147–170

Chapter 17
"Canopy-Less" Monitoring of Biodiversity and Climate Change: Signs of a Leaky Roof

Soubadra Devy, T. Ganesh, and Margaret Lowman

Keywords Ecosystem service • Landscape • Rainforest • Long-term

Bullet Points

1. Monitoring of biodiversity has gained importance in recent decades due to global climate change. Although we have a sizeable percentage of biota locked up in the forest canopies, most monitoring programs have neglected this realm.
2. Remote sensing and Lidar techniques allow for some landscape-level monitoring of canopies. For finer stand-level data, canopy biologists include passive ways of monitoring by collaborating with systems and instrumentation experts enabling dataflow to the ground.

Summary

Forest canopies are considered the last biotic frontier, and studies of canopy biota and related processes are just beginning to emerge in some parts of the world. Monitoring changes in biodiversity and related processes have gained much significance in the few last decades, particularly due to climate change. In addition, changes in biodiversity have been addressed by incorporating monitoring at various spatial and temporal scales that range from landscape-level changes, ecosystem dynamics, to population and species-level processes over extended time scales.

S. Devy (✉) • T. Ganesh
Ashoka Trust for Research in Ecology and the Environment (ATREE), Royal Enclave, Srirampura, Jakkur Post, Bangalore, KA 560064, India
e-mail: soubadra@atree.org; tganesh@atree.org

M. Lowman
North Carolina Museum of Natural Sciences, North Carolina State University, 121 W. Jones Street, Raleigh, NC 27603, USA
e-mail: canopymeg@gmail.com

M. Lowman et al. (eds.), *Treetops at Risk: Challenges of Global Canopy Ecology and Conservation*, DOI 10.1007/978-1-4614-7161-5_17,

Most efforts on forest monitoring remain "canopy less" and hence can give an eclipsed view. The main impediment is access for repeated sampling. It is now critical for canopy ecologists to set up a flagship monitoring effort in canopies to incorporate passive monitoring protocols through collaboration with system engineers and instrumentation experts and integrate into ground-level efforts.

1 Introduction

Several anthropogenic factors such as selective logging, agriculture, and fire have led to severe deforestation globally; these are considered as major forces of ecological change throughout tropical regions (Asne et al. 2006; Hansen et al. 2010). A number of studies show that such changes alter species distributions and abundance (Hansen and Defries 2004; Laurance and Peres 2006) and in the future may even lead to the extinction of a large fraction of the species on earth (e.g., Thomas et al. 2004; Raxworthy et al. 2008).

With increasing rise in global CO_2, large-scale changes in the functional aspects of the forests are also getting affected in terms of forest compositional change (Melillo et al. 1996; Houghton et al. 2001) and the dynamics of forest ecosystems (Phillips et al. 1994). Forest canopies are considered the last unexplored biotic frontier in many parts of the world but with potential of having unique biota. Studies of canopy biota and related processes are just beginning to emerge in some parts of the world (Devy and Ganesh 2003; Devy et al. 2012). Forest degradation and climate change are happening at a rapid rate that can affect species uniquely adapted to forest canopies (making them prone to local extinction). For instance, canopy discontinuity affects non-volant canopy mammals like primates, and the disappearance of fog and mist might affect the lichens, orchids, and other bryophytes in the canopy (Lowman and Rinker 2004).

Many of the canopy species provide important ecosystem services such as pollination and pest control and often travel long distances across the mountain landscapes tracking the food resources. For instance, bees in tropics forage on several canopy trees and help in pollination (Devy and Davidar 2003) but are rarely monitored; the emphasis has always been either trees (e.g., Center Tropical Forest Science, USA) or large charismatic mammals (Center for wildlife studies, India). Monitoring is also carried out at ground level, omitting the canopy biota (Devy et al. 2009).

2 "The Third Dimension"

Monitoring landscapes, ecosystems, populations, and species is a complex process (Singh et al. 2010; Müller et al. 2010). Despite these challenges, monitoring has gained much significance in the last decade, with recent emphasis by Hansen et al. (2010) using remotely sensed data. In addition, changes in biodiversity have been addressed by incorporating monitoring at various spatial and temporal scales that range from landscape-level changes, ecosystem dynamics, and population- and species-level processes over extended time scales (Lindenmayer and Likens 2010; Hobbie et al. 2003; Dearing et al. 2006). This "third dimension" comprising the vertical structure of a forest, especially in structurally complex rainforest ecosystems, is an important habitat where most species reside between ground and canopy (Ozanne et al. 2003; Lowman and Nadkarni 1996; Lowman and Rinker 2004; Lowman et al. 2012). The modification of this niche space can strongly affect species as seen in a study on avian assemblages in the rainforests of India where species loss was documented in insectivorous birds of mid-storey following forest disturbance (Vivek and Ganesh 2012). In most of the old world tropics, canopies are poorly explored even from a biodiversity point of view. Therefore, we must initiate short but focused collaborative studies (e.g., IBISCA, Investigating the Biodiversity of Soil and Canopy Arthropods) to collect base line inventory from selected sites before monitoring protocols can be set up.

3 Climate Change and Canopy Monitoring

Large-scale monitoring of tree canopies has been revolutionized by the use of high-resolution satellite imageries and LIDAR (Asner et al. 2007). Studies on carbon sinks in relation to forest canopies and climate change have determined how rapidly newly assimilated carbon(C) is invested into recalcitrant structures of the forests (Free-Air CO_2 Enrichment, FACE, Körner et al. 2005). Recently the Carnegie Institution and its collaborators have started developing new scientific approaches that integrate taxonomic, chemical, and spectral remote sensing perspectives—collectively called spectranomics—to map canopy function and diversity among species throughout tropical forests of the world (Asner and Martin 2009; see chapter by Asner in this volume). The necessity of ground truthing such expansive aerial monitoring is now almost entirely eliminated, given the extraordinary capability of airborne surveys.

Those long-term monitoring sites that currently exist or are planned in the future should involve monitoring components from "soil to canopy" as proposed by Global Canopy Programme (Fig. 17.1). Though some efforts have been done to establish these "Whole Forest Observatories" in parts of America, Australia, Europe, and Malaysia, many forest types in Asia, Africa, and South America have none (Mitchell et al. 2002; Ganesh et al. 2006). Such infrastructure could facilitate long-term data collection synchronously from the ground to canopy and would encourage a diverse range of researchers to participate and contribute to understand forest change.

4 Passive Techniques for Biodiversity Monitoring

Accessibility has been an obvious impediment for major global monitoring programs to have excluded the canopy in their sampling. But with the advent of many data loggers, camera traps, and sound recorders, passive canopy monitoring is

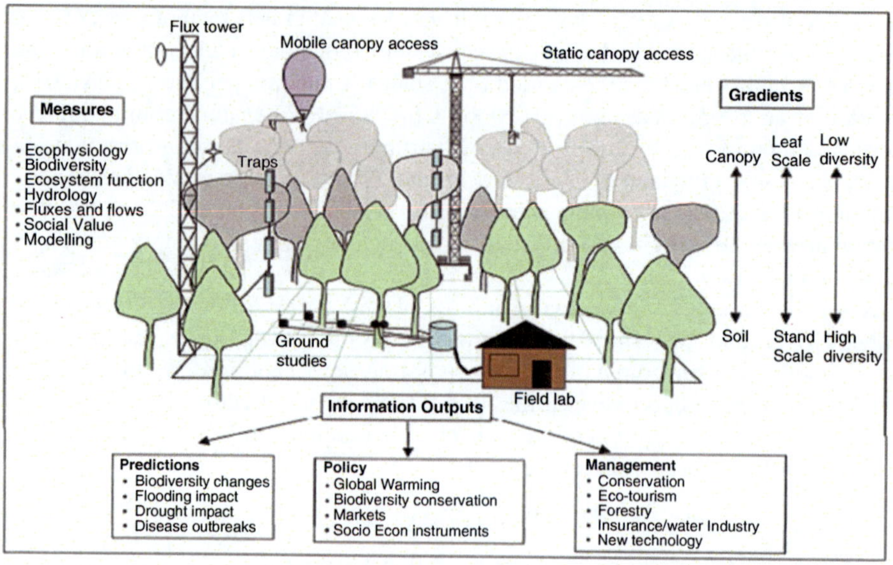

Fig. 17.1 Whole forest observatory for monitoring biodiversity and related processes from soil to canopy as visualized by Global Canopy Programme

simple and reliable (e.g., Lobo et al. 2005; Ganesh and Devy 2012). Even cryptic nocturnal taxa (e.g., frogs and crickets; see chapters in this volume by Seshadri & Ganesh and Jain & Balakrishnan) can be studied using these techniques. Our pilot study in collaboration with Sun Microsystems demonstrated that passive monitoring is feasible by downloading data from data loggers in the canopy from the ground. But there are still major drawbacks to accurately monitoring "the third dimension" with adequate statistics and models (see Nychka and Nadkarni 1990). Perhaps a flagship canopy project, which demonstrates passive monitoring with methodological rigor, could expand global monitoring efforts.

5 Conclusions

Although the above-mentioned efforts may seem an expensive venture, it is only a small percentage of what is invested on polar or space exploration. We strongly advocate for integrated forest canopy monitoring at multiple sites with diverse methods to address challenges including biodiversity loss, environmental change, and ecosystem services.

Acknowledgments We thank the Department of Science and Technology, New Delhi, Ministry of Environment and Forests, Govt. of India and Indian Academy of Sciences, Bangalore. National Science Foundation supported the Fifth International Canopy Conference in India. The article is a result of the proceedings of this conference and the subsequent workshops. Margaret Lowman acknowledges Fulbright funding that facilitated our efforts on canopy science in India. The GCP provided the figure of the Whole Forest Observatory, and M.B. Prashanth and K.S. Seshadri edited this chapter.

References

Asne GP, Broadbent EN et al (2006) Condition and fate of logged forests in the Brazilian Amazon. PNAS 103:12947–12950

Asner GP, Martin RE (2009) Airborne spectranomics: mapping canopy chemical and taxonomic diversity in tropical forests. Front Ecol Environ 7:269–276

Asner GP, Knapp DE et al (2007) Carnegie airborne observatory: in-flight fusion of hyperspectral imaging and waveform light detection and ranging for three-dimensional studies of ecosystems. J Appl Remote Sens 1:013536–013536-21

Dearing JA, Battarbee RW et al (2006) Human-environment interactions: learning from the past. Reg Environ Change 6:115–123

Devy MS, Davidar P (2003) Pollination systems of trees in Kakachi – a mid-elevation wet evergreen forest in Western Ghats, India. Am J Bot 90:650–657

Devy MS, Ganesh T (2003) Canopy science and its relevance to India. Curr Sci 85:101–104

Devy MS, Vivek R, Ganesh T (2009) Unexplored Indian forest canopies: are we dealing with an eclipsed view of our biodiversity? In http://www.ces.iisc.ernet.in/biodiversity/sahyadri_enews/newsletter/issue31. Accessed 5 Feb 2013

Devy MS, Ganesh T, Tripathy A (eds) (2012) Forest canopies of South Asia: a glimpse. Ashoka Trust for Research in Ecology and the Environment, Bengaluru

Ganesh T, Devy MS (2012) Sensing animals in the canopy. In: Lowman MD, Schowalter TD, Franklin JF (eds) Methods in forest canopy research. University of California Press, Berkeley

Ganesh T, Devy MS, Bawa KS (2006) Whole forest observatories: an international network for monitoring canopy biodiversity and global climate change. Information for the first planning workshop from country India, ATREE, Bengaluru, India

Hansen M, DeFries R (2004) Detecting long-term global forest change using continuous fields of tree-cover maps from 8-km advanced very high resolution radiometer (AVHRR) data for the years 1982–99. Ecosystems 7:695–716

Hansen MC, Stehman SV, Potapov PV (2010) Quantification of global gross forest cover loss. PNAS 107:8650–8655

Hobbie JE, Carpenter SR et al (2003) The US long term ecological research program. Bioscience 53:21–32

Houghton RA, Lawrence KT et al (2001) The spatial distribution of forest biomass in the Brazilian Amazon: a comparison of estimates. Global Change Biol 7:731–746

Körner C, Asshoff R et al (2005) Carbon flux and growth in mature deciduous forest trees exposed to elevated CO_2. Science 309:1360–1362

Laurance WF, Peres C (eds) (2006) Emerging threats to tropical forests. The University of Chicago Press, Chicago

Lindenmayer DB, Liken GE (2010) The science and application of ecological monitoring. Biol Conserv 143:1317–1328

Lobo JA, Quesada M, Stoner KE (2005) Effects of pollination by bats on the mating system of *Ceiba pentandra* (Bombacaceae) populations in two tropical life zones in Costa Rica. Am J Bot 92(2):370–376

Lowman MD, Nadkarni NM (eds) (1996) Forest canopies. Academic, San Diego

Lowman MD, Rinker HB (eds) (2004) Forest canopies. Elsevier Academic Press, San Diego

Lowman M, Schowalter T, Franklin J (eds) (2012) Methods in forest canopy research. University of California, Berkeley

Melillo JM, Houghton RA et al (1996) Tropical deforestation and the global carbon budget. Ann Rev Energy Environ 21:293–310

Mitchell AW, Secoy K, Jackson T (2002) The global canopy handbook: techniques of access and study in the forest roof. GCP, Oxford, UK

Müller F, Baessler C et al (2010) Long-term ecosystem research between theory and application–an introduction. In: Müller C, Baessler H et al (eds) Long-term ecological research. Springer, Dordrecht

Nychka D, Nadkarni NM (1990) Spatial analysis of points on tree structures: the distribution of epiphytes on tropical trees. Institute of Statistics Mimeograph Series No. 1971

Ozanne CMP, Anhuf D et al (2003) Biodiversity meets the atmosphere: a global view of forest canopies. Science 301:183–186

Phillips OL, Gentry AH et al (1994) Dynamics and species richness of tropical rain forests. PNAS 91:2805–2809

Raxworthy CJ, Pearson RG et al (2008) Extinction vulnerability of tropical montane endemism from warming and upslope displacement: a preliminary appraisal for the highest massif in Madagascar. Global Change Biol 14:1703–1720

Singh SJ, Haberl H et al (2010) Conceptualising long-term socio-ecological research (LTSER): integrating the social dimension. In: Müller C, Baessler H et al (eds) Long-term ecological research. Springer, Dordrecht

Thomas CD, Cameron A et al (2004) Extinction risk from climate change. Nature 427:145–148

Vivek R, Ganesh T (2012) Habitat structure and its effects on bird assemblages in the Kalakad-Mundanthurai Tiger Reserve, India. JBNHS 109(1&2):87–95

Part III
New Approaches

Introduction

Forest canopies are biodiversity hotspots but face severe challenges in the changing environment; therefore, innovative research is essential to the prosperity and future success of this precious ecosystem. While recent efforts have increased awareness and helped aid forest conservation, the canopy is still relatively unknown. Problems such as how to access the canopy and effective sampling techniques greatly affect research studies. Not only is obtaining samples difficult, but the forest is also structurally complex with each layer host to specialized species. This makes understanding the canopy even more challenging. Easier access into the canopy creates many new research studies including long-term monitoring, comparison of vertical and horizontal patterns, and noninvasive surveying and tracking techniques. This chapter presents a broad range of cutting-edge methodologies and preliminary findings that reflect new questions and priorities for canopy research – from sloths to crickets to pollinators to mistletoes to lianas and more. It is increasingly evident that new and improved access techniques facilitate asking questions about process and function. In addition, various remote-sensing techniques are increasingly used to monitor canopy changes at both the global and regional scale. Stay tuned, as forest canopies become a hotspot not just for biodiversity but as a scientific treasure trove where some of the important processes that regulate the health of our planet are unraveled by intrepid arbornauts.

Chapter 18
Mesoscale Exploration and Conservation of Tropical Canopies in a Changing Climate

Gregory P. Asner

Keywords Biodiversity • Biomass • Carbon • Deforestation • Carnegie airborne observatory • Forest degradation • LiDAR • Rainforest • Tropical forest

1 Introduction

Tropical forests harbor the majority of species found on land, yet we still know very little about the canopies that create habitat required for a vast array of flora and fauna. Tropical canopies remain a biological frontier, calling us to explore them in hopes of better understanding their composition, structure and function, and how they came to be. Yet we undertake this challenge at a time of very rapid change for tropical canopies, in terms of both their geographic extent and their condition. The global tropical forest landscape is changing at such a rate and with such complexity; the urgency of mapping and reporting on them has reached an all-time high.

We know that tropical forest canopies are in trouble. It does not take an expert to recognize the severity of at least part of the problem: deforestation. Particularly over the last 50 years, tropical forests have been cut down, burned, and even mulched in vast quantities that defy individual comprehension. The United Nations Food and Agriculture Organization estimates a global forest loss rate of about 13 million hectares *every year*, of which about 10 million hectares is lost from tropical forests (FAO 2007). These numbers are certainly very hard to put into perspective. Take, for example, a single soccer or football field, and imagine foresting it with a typical central Amazonian tree canopy. That field would cover just 0.4–0.7 ha (depending upon your sport of interest), but it would contain about 100 metric tons of tree

G.P. Asner (✉)
Department of Global Ecology, Carnegie Institution for Science, 260 Panama Street, Stanford, CA 94305, USA
e-mail: gpa@carnegiescience.edu

M. Lowman et al. (eds.), *Treetops at Risk: Challenges of Global Canopy Ecology and Conservation*, DOI 10.1007/978-1-4614-7161-5_18,

biomass above ground and another 50 t below the soil surface. Adding it up, we are losing many millions of tons of carbon (which is ~48 % of dry biomass) from tropical forests each year through deforestation. Now take about 100 of these football fields and put them together as one large "quadrat"—the word that ecologists usually give to a small (e.g., 1 m^2) sampling unit, and then take that 50-ha quadrat to the western Ecuadorian Amazon. You would be able to fit more than 1,000 different species of trees into this quadrat, not to mention myriad vines, epiphytes, and much more. Although many of these species occur over large areas, we think a very substantial proportion is extremely rare, and thus plowing over 10 million hectares of forest per year undoubtedly results in high rates of local and potentially global species extinction.

Deforestation, as it turns out, is actually trumped by other forms of land use. The persistence of forest disturbances caused by selective logging, fire, and even hunting has left a massive wake of degraded landscapes that harbor fewer (or different) species and less carbon than their predecessor primary forests (Peres et al. 2006). Today, quantitative estimates suggest that the global footprint of forest degradation is 20–30 times larger than that of wholesale deforestation (Asner et al. 2009b). In other words, 200–300 million hectares of forest may be degraded each year by the engine of human activity inside forests, but these changes go relatively unrecognized as the canopies continue to be called "forests."

Even given the devastating impacts of deforestation and forest degradation, it is the specter of climate change that brings the entire tropical forest commons to an all new level of concern. We know relatively little about how climate change has affected or will affect tropical forests, but some early consensus is just now being reached. First, it is clear that tropical forests are heating up: Widespread measurement-based evidence shows that nighttime temperatures are on the rise (Clark 2004) and that the trees are losing carbon due to increased temperature-driven respiration (Clark and Clark 1994). Geographic analyses also now suggest that just a 1–2 °C rise in temperature, which is now estimated to be exceeded by another 1–2 °C, will put many lowland tropical forests beyond their late-Holocene temperature range (Wright et al. 2009), with potential impacts on growth, mortality, carbon losses, and hydrological cycles. More recently, the effects of precipitation change have come into greater focus. In the Amazon basin which contains half of all tropical forest, the climate has been drying at a rate of about 0.3 % per year (Li et al. 2008). These findings are further bolstered by prognostic modeling studies that suggest the possibility of a major reconfiguration of the vegetation in the Amazon by 2080 (Asner et al. 2010a) (Fig. 18.1).

The climate change issue extends well beyond the average trends of increasing temperature and decreasing precipitation. Like hotspots of deforestation such as in Indonesia and Brazil (Hansen et al. 2008), we now recognize hotspots of extreme climate events. In 2005, a large portion of the Amazon basin underwent a massive drought associated with sea surface temperature anomalies in the North Atlantic Ocean (Phillips et al. 2009). Although 2005 was dubbed a once-in-a-century drought, another potentially even larger drought occurred in 2010 in Amazonia (Lewis et al. 2011). These mega-droughts were surprises, and they differed from that of the El Nino–Southern Oscillation (ENSO)-related droughts in the eastern

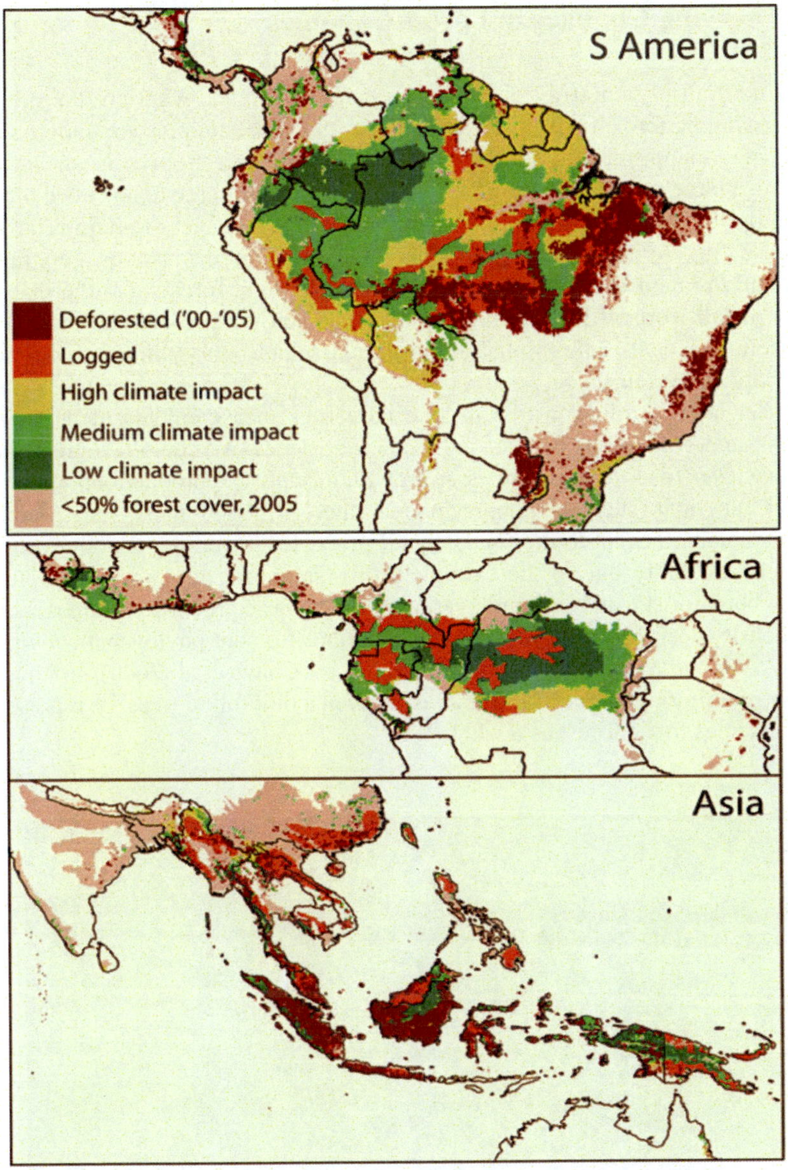

Fig. 18.1 The humid tropical forest biome is being subjected to a major reconfiguration resulting from a combination of land use and climate change. *Red* and *orange* areas show deforestation and forest degradation, respectively, mapped between 2000 and 2005. The *green* and *yellow* areas show forecasts for changes in precipitation and temperature sufficient to cause species migration or extinction (Data are taken from Asner et al. (2010a) and climate data from Loarie et al. (2009))

Brazilian Amazon (Asner et al. 2000). The 2005 and 2010 droughts reached further westward and deep into the Peruvian and Colombian Amazon. There seems to be relatively little safe harbor from climate change in the Amazon or anywhere else in the global humid tropics (Asner et al. 2010a).

2 Observing Changes in Forest Canopies

Given the multiplying array of threats to tropical forests, what are the options for assessing patterns, rates, and types of changes that canopies are undergoing? Measuring, mapping, and monitoring tropical forest canopies is an activity spanning a huge range of ecological scales, from a single tree to an entire biome. Yet the options for scientists, conservationists, and managers remain relatively narrow, owing to three particular challenges: (1) the enormous geographic extent and the natural ecological variability of tropical forests, (2) the inaccessibility of tall tropical forest canopies at any given point location, and (3) the overwhelming local-scale (alpha) and regional-scale (beta, gamma) diversity of many tropical systems.

We can labor to climb a tree to take numerous measurements of its foliage, branches, and possibly its flowers and fruits (Fig. 18.2). We could take those measurements over time, to assess rates of growth, mortality, reproduction, and physiological functions such as photosynthetic rate. Although these measurements provide valuable insight to the fundamental processes governing canopy function, they lack scalability due to the extremely diverse nature of tropical canopies in terms of both floristic and structural variation. Even basic properties of trees such as wood density, leaf thickness, and chemical composition are phylogenetically organized (Asner and Martin 2011; Asner et al. 2011c; Chave et al. 2006), so how can intensive monitoring of one or a few trees, or even a plot full of trees, be representative of changes to the forest as a whole?

Fig. 18.2 Climbing and measuring a single tree in a tropical forest is a major undertaking requiring experience. The enormous size and complex architecture of tropical canopies renders them difficult to study using ground-based approaches. Here Carnegie Institution botanist Nestor Jaramillo climbs a tree in the lowland Peruvian Amazon (Photo by Jake Bryant (www.envirofoto.com))

Fig. 18.3 The Carnegie Airborne Observatory (CAO) Alpha sensor package flew from 2006 to 2011 on a variety of aircraft, such as the Piper Navajo shown in the *upper left panel*. The Alpha system included a Visible-to-Near-Infrared (VNIR) imaging spectrometer and a waveform light detection and ranging (LiDAR) scanner (*right and bottom left panels*). The system was operated using computers and high-performance navigation equipment (*left center panel*)

At the other extreme, we can use satellites to observe and analyze the state of tropical forests. From the vantage point of space, many satellite sensors provide a way to assess the geographic extent of forests (Hansen et al. 2008; Skole and Tucker 1993). More recently, new instrumentation and analytical methods have allowed for monitoring the factors associated with forest degradation, such as selective logging and fire (Asner et al. 2005b; Morton et al. 2011; Souza et al. 2005) (Fig. 18.3). Maps derived from such approaches provide a critically important wall-to-wall perspective on the overall status of tropical forest cover and, to a lesser extent, forest condition. However, it is this aspect of the satellite-based approach—its broad brushing of the tropics in shades of green—that limits its use for exploring the forest, its interior composition, function, and structure. Is there a scale at which the forest *and* its trees can be imaged, mapped, and dissected with the detail one might expect after laboriously climbing an individual tree? It may seem impossible, but some radical new advances are making this a reality, and these advances are blurring the boundary between field-based and remotely sensed measurements.

3 Evolution of the Carnegie Airborne Observatory

Airborne observation is one of the oldest forms of remote sensing (Fensham and Fairfax 2002). The technology has advanced from simple monochromatic analog cameras to high-performance optical, RADAR, and LiDAR (light detection and ranging) instrumentation. Most studies using modern airborne sensors have typically focused on a specific observation goal, such as vegetation type from optical cameras, canopy height from LiDAR and RADAR, or canopy chemistry from imaging spectrometers (Lefsky et al. 2002; McGraw et al. 1998; Ustin et al. 2004). More recently, effort to combine data from different technologies has yielded multidimensional data that more closely serves emerging questions on the state of our biosphere (Asner and Vitousek 2005). Early multisensor studies provided strong justification for the use of these approaches as they greatly advanced our understanding of the structure and functioning of ecosystems.

In this context, we developed the Carnegie Airborne Observatory (CAO; http://cao.ciw.edu) to more fully probe the structural, functional, and compositional properties of ecosystems at the "mesoscale." Here I am referring to the ecological scale at which organisms assemble, compete, disperse, and migrate—a scale rarely probed by either field- or satellite-based methods. Following years of testing and development in the Hawaiian Islands, where topography, soils, and climate vary strongly and in predictable ways that provide a unique outdoor laboratory for ecological studies (Vitousek 1995), we were finally ready to build an operational system that would open scientific doors to understanding more fully the way that ecosystems are changing. In 2006, we integrated a Visible-to-Near-Infrared (VNIR; 365–1,050 nm) imaging spectrometer and a waveform LiDAR into a system called CAO Alpha (Asner et al. 2007) (Fig. 18.3). The premise behind the CAO, starting with the Alpha system and continuing to present, is that a fully integrated set of observational capabilities, incorporating measurements expressly selected for multidimensional analysis of ecosystems, would advance our understanding of the biosphere in two particular ways: (1) by resolving ecological properties at scales commensurate with the flows of energy, materials, and organisms (gene flow) within and among ecosystems and (2) by developing new understanding at the mesoscale, yet with fine biological resolution that is largely intractable from the orbital vantage point. The plan was to build a system that would facilitate fundamental discoveries that have eluded ecology in the past, and which would simultaneously advance conservation and management of ecosystems.

CAO Alpha flew from 2006 to 2011, mapping and analyzing ecosystems stretching from Hawaii to South Africa and from California to Peru (Asner et al. 2008b, 2009a, 2010b, 2011a, 2012a; Carlson et al. 2007; Colgan et al. 2012; Levick et al. 2010). The VNIR spectrometer and LiDAR combination proved to be a powerful combination that uncovered the way vegetation structure and some aspects of chemistry covary regionally and how the terrain beneath the canopy mediates the composition of species (Féret and Asner 2012). However, studies in the CAO Alpha era also highlighted a need to further increase the dimensionality of the airborne data, particularly in terms of vegetation function (chemistry) and biological diversity (Asner et al. 2011b). Following an activity to define technology and science–algorithm requirements, the next generation CAO-2 Airborne Taxonomic Mapping System (AToMS) was built and launched in June 2011 (Asner et al. 2012b) (Fig. 18.4a).

AToMS is comprised of three unique sensors and their associated onboard control and computing systems: (1) Visible-to-Shortwave-Infrared (VSWIR) imaging spectrometer, (2) waveform LiDAR, and (3) Visible-to-Near-Infrared (VNIR)

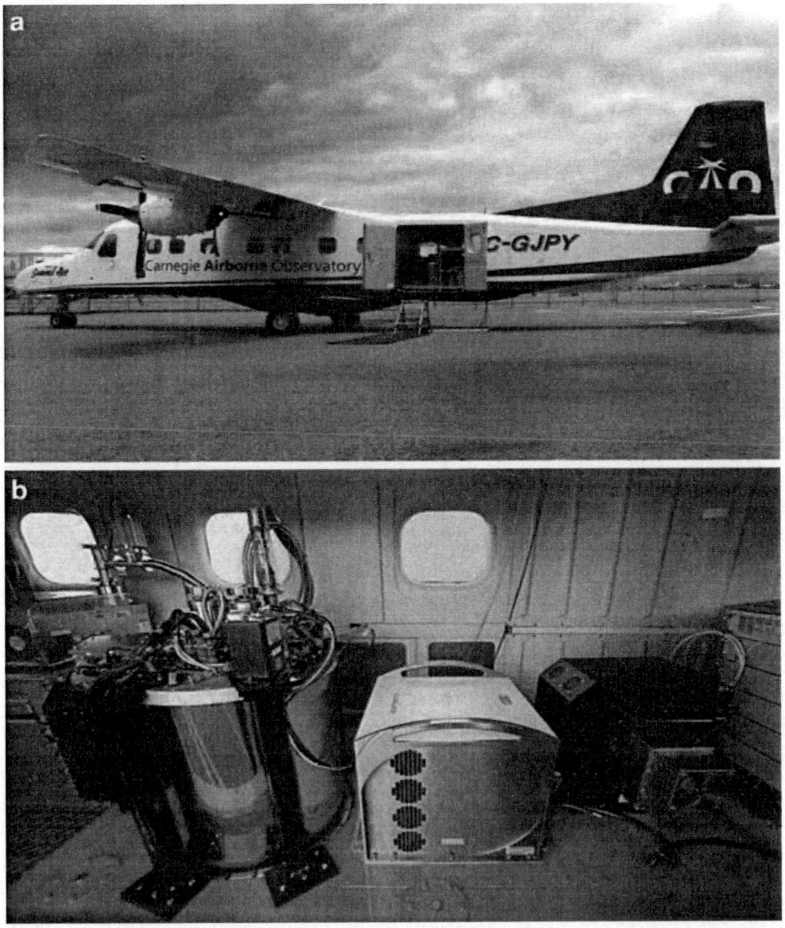

Fig. 18.4 CAO Airborne Taxonomic Mapping System (AToMS) is flown on a Dornier 228–200 twin turboprop aircraft providing global reach and an office-computing-instrument facility allowing for mapping up to 3,200 km^2 of tropical forest per day. The AToMS sensors include a very high-fidelity Visible-to-Shortwave-Infrared (VSWIR) imaging spectrometer (*left*), dual-laser LiDAR (*middle*), and zoom Visible-to-Near-Infrared (VNIR) imaging spectrometer (*right*), co-mounted on a stabilized plate straddling a large hole in the bottom of the aircraft (*bottom panel*)

imaging spectrometer (Fig. 18.4b). The VSWIR provides full-range (380–2,510 nm) spectroscopic radiance measurements with very high fidelity needed for canopy chemical and physiological applications, as well as for geological and atmospheric measurements. The LiDAR provides detailed three-dimensional (3-D) data on vegetation structure, sub-canopy ground surface elevation, and the 3-D structure of non-vegetated objects. The VNIR spectrometer provides increased spatial detail, with four pixels per VSWIR pixel, and covering the 365–1,052-nm wavelength range. This higher level of spatial resolution is often needed to discern individual life-forms and species.

4 Touring Canopies with the CAO

The range of topics covered by, and insights derived from, the CAO have been groundbreaking and perhaps even breathtaking at times. Recent studies have spanned the fields of ecology and evolutionary biology, geomorphology and biogeochemistry, atmospheric and aquatic sciences, and forest conservation and management (see http://cao.ciw.edu). Yet perhaps the most profound advances have been made by measuring, mapping, and simply appreciating forest canopies, particularly in remote tropical regions. The CAO data reveal the 3-D structure of many millions of trees, lianas, and other growth forms (Fig. 18.5a), allowing for viewing of and flying among the canopy species that inhabit forests such as the Amazon. When the data on the spectral properties of the canopies are integrated with the structural data from the LiDAR, the result is a new view of the treetops as never before experienced (Fig. 18.5b). Each tree in this image is a different species, owing to the high alpha-diversity of this lowland Amazonian forest on the Peru–Brazil border. As such, the canopy spectral properties differ by species, revealing unique chemical and physiological traits among the taxa that have assembled in this place.

Extensive field and laboratory work have determined that the spectral properties of canopies shown in Fig. 18.5b are driven by a combination of 21 chemical traits found in the foliage (Asner and Martin 2008; Asner et al. 2011b), and many of those traits are organized phylogenetically. The phylogenetic signal is sufficiently strong to indicate not only functional diversity in the forest but biological diversity as well, including floristic composition (Asner and Martin 2009, 2011). A typical example is shown in Fig. 18.6. The upper and lower panels show an oblique view and a plan view, respectively, of a lowland Amazon forest along the Tambopata River in the Madre de Dios watershed of Peru. Within the riverbed to the left in the image, there is low-statured vegetation shown mostly in green with some red near the center of the stands. These river islands are dominated by a common herbaceous species *Salix humboldtiana* (Salicaceae) that displays nearly constant canopy chemistry as shown in the green color. The red toward the island interiors is another species—*Tessaria integrifolia* (Asteraceae)—which replaces *Salix* successionally. Along the river's bank, the brilliant red colors indicate the presence of *Cecropia latiloba* and *C. membranacea* (Urticaceae), each of which has a chemical composition promoting rapid growth (high nitrogen, chlorophyll, and water content). Looking at the upper panel, it is clear that these successional stands are lower in height (and biomass) as compared to the larger canopies farther from the river's edge. As one moves away from the river into the inactive floodplain environment, the canopy quickly diversifies, becoming a mosaic of organisms with divergent (and sometimes convergent) canopy chemical traits, as indicated by many shades of red to indigo. To the far right in the image, an oxbow lake emerges with its unique vegetation as well as a different mixture of aquatic chemical traits. In all, this image reveals more about the functional diversity, and the spatial configuration of biological diversity, than any field or satellite image has ever produced. This image is the first, and it is one in thousands now compiled for Amazonia and other tropical regions covering millions of unexplored tropical landscape.

Fig. 18.5 The CAO can image a tropical forest in three dimensions and with detailed spectral information. The detailed 3-D imaging allows the viewer to view the canopies from any orientation, even from the *bottom up*, or in this case from an extreme oblique angle. (**a**) Natural color composite of a canopy located in a remote section of the Amazon basin along the Peru–Brazil border. Different colors indicate variation in leaf type and condition, as well as leafless tree crowns. Notice the branch structure and foliage throughout the canopy. (**b**) The canopy is now colored with spectral information on foliar pigment and nitrogen information. *Bright pink* canopies are those with the highest capacity to absorb and utilize sunlight for growth. *Gray–blue* canopies are those without foliage

5 CAO Applications for Science and Conservation

Given the unique suite of measurements provided by the CAO technology, what are some of the applications to scientific study and conservation of tropical forest canopies? While the full breadth of potential applications has hardly been realized, some early successes indicate that these new mapping approaches can greatly advance our knowledge while simultaneously supporting efforts to protect ecosystems.

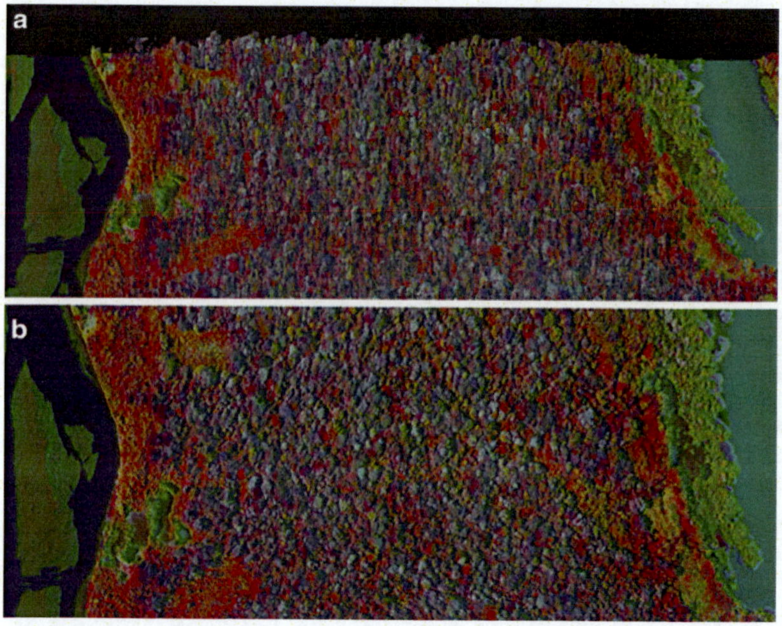

Fig. 18.6 The CAO imaged this lowland Amazonian landscape in the Madre de Dios watershed of Peru, with views shown obliquely in panel **a** to reveal variation in canopy height and architecture and in plan view in panel **b** to show crown shape. Image color is based on a combination of three groups of foliar chemicals regulating light capture and growth, longevity and defense, and metabolism (Asner and Martin 2011). To the *far left*, the Tambopata River is shown in *blue*. To the *far right*, a large oxbow lake is seen in *green blue*. These colors indicate differences in surface water chemistry. Notice the similar coloring (and thus chemical traits) of successional vegetation along the banks of the river and lake

One major science-to-conservation focus of the CAO has centered on quantifying spatial patterns of aboveground carbon stocks in tropical forests, savannas, and many other types of ecosystems. Carbon stock, or carbon density, is the amount of carbon stored in vegetation and soils on a per-area basis. Aboveground carbon density (or ACD) is the portion of the total carbon stock in live and standing dead plant tissues above the soil surface. In grasslands and open shrublands, imaging spectroscopy has proven highly accurate for mapping and monitoring ACD by measuring the size and extent of woody vegetation canopies and indices of their vertical depth (Asner et al. 2005a; Huang et al. 2009). Moreover, LiDAR has been key to estimating the size and density of the vegetation in woodlands, savannas, and forests, particularly as the canopies coalesce spatially (Colgan et al. 2012; Mascaro et al. 2011). From LiDAR-based measurements of canopy height and vertical profile, spatially explicit maps of forest carbon stocks have become commonplace (Fig. 18.7). The result is a much-enhanced accuracy of tropical forest carbon stocks, which in turn has facilitated more interest and financial investment in the protection of these forests via programs to reduce carbon emissions from tropical deforestation and

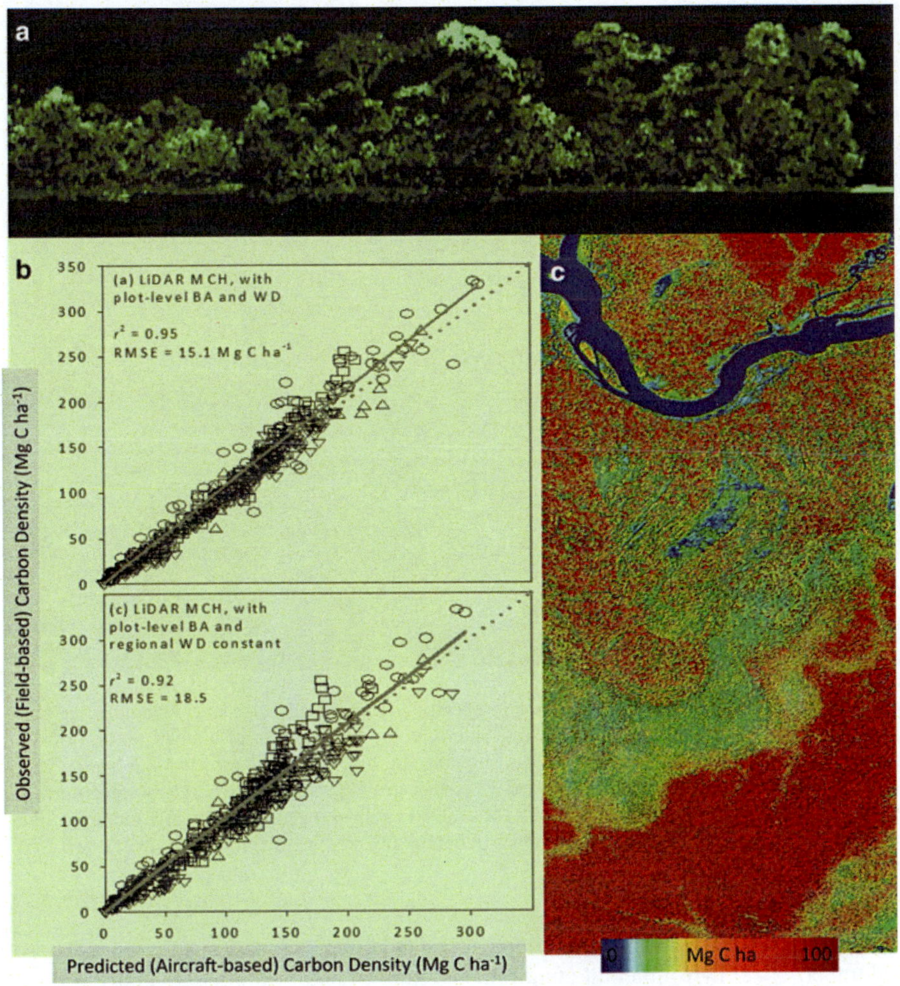

Fig. 18.7 (**a**) The CAO LiDAR measures the vertical distribution of plant tissues from top- of-canopy to ground level, revealing the canopy architecture and branch configuration. (**b**) Aboveground carbon density estimates are derived from the LiDAR Mean Canopy Profile Height (MCH), along with additional information on basal area (BA) and wood density (WD) derived at local (*top graph*) or regional (*bottom graph*) scales. Accuracies have proven to be very high for all forests tested in the tropics globally (Asner et al. 2011d). (**c**) The result of combination 3-D structure from LiDAR and calibrations to carbon density is a map of forest aboveground carbon stocks at high spatial resolution

forest degradation (REDD) (Asner 2009, 2011). Indeed, programs under the REDD policy instrument have already brought large amounts of conservation funding to tropical forests worldwide (Angelsen 2008; Herold and Skutsch 2009).

Another area of enormous growth with the CAO centers on biodiversity mapping, including species determination. One of the great challenges in ecology, as well as

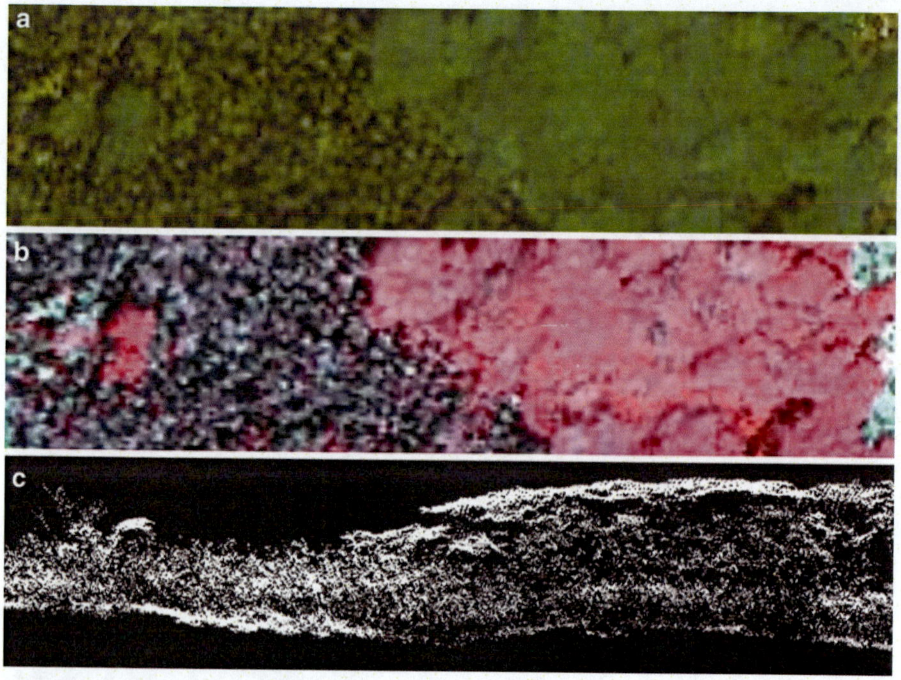

Fig. 18.8 This CAO image captured the presence of the highly invasive tree *Falcataria moluccana* (Fabaceae) in a Hawaii rainforest. (**a**) Natural color composition of a small portion of the landscape. (**b**) Spectrometer measurements showing the higher rate of growth of *F. moluccana* (*bright pink color*) against the background of slower-growing native *Metrosideros polymorpha* (Myrtaceae, *blue–purple colors*). (**c**) The CAO LiDAR then showed the vertical structure of the canopies, indicating the fundamental changes in habitat invoked by the biological invasion (Based on Asner et al. (2008a))

in conservation, rests in understanding causes and consequences of species diversity and under what environmental conditions maximum and/or unique diversity occurs (Tilman and Pacala 1993). Additionally, there is much interest in where species can be found on the landscape—species that may affect the success of higher trophic species. Recent work in Hawaii has shown that alien-invasive species fundamentally change the rules of existence for other species, and even for whole assemblages of species. For example, CAO spectrometer measurements afforded mapping of the top ten or so invasive canopy and sub-canopy species throughout Hawaii (Asner et al. 2008a). Following mapping of the nonnative plants, the CAO LiDAR was used to assess their ecological impact on the structure and composition of entire landscapes (Fig. 18.8). Results from such studies have worked their way into the management of Hawaii's alien-invasive species and into the restoration and protection of remaining Hawaiian flora and the ecosystems that harbor them (Kellner et al. 2011).

Of all the potential applications of the CAO, its role in understanding the effects of climate change on canopies may prove the most critical to science and conservation. Whereas field-plot studies cannot resolve the spatially heterogeneous impacts of climate variability and change over large geographic areas, satellite-based observations have been equally hindered by a lack of biological resolution. For example, the 2005 mega-drought in the Amazon led some to believe that decreased precipitation and cloud cover resulted in increased greenness—a remote-sensing term related to canopy leaf area and gross primary productivity (Saleska et al. 2007). These satellite measurements were later reanalyzed in such a way to show that the drought did not increase canopy greenness as much as it decreased cloud cover alone (Samanta et al. 2010). In short, the satellite measurements were, at best, inconclusive and more likely only sensitive to relative changes in cloud cover associated with drought (Asner and Alencar 2010). Yet the few sparsely distributed field plots in the region suggested that tree mortality increased during the 2005 drought (Phillips et al. 2009). By 2010, another mega-drought hit the Amazon basin, this time in a different regional pattern that struck deep into Peru (Lewis et al. 2011). The results of compounding droughts remain unknown, owing to the remoteness of the forests and the poor sensitivity of most satellite sensors to canopy biochemical and physical responses to drought.

With the CAO, we explored two regions thought to be negatively impacted by the 2005 and 2010 droughts. A region in northern Peru, stretching from the Cordillera Azul conservation area to the Brazil border, was explored. CAO maps revealed a vast area covering millions of hectares dotted with leafless trees in quantities that seem to far exceed those of the normal amount of dry-season deciduousness that we often experience in tropical forests (Fig. 18.9). Moreover, areas of rolling terrain as imaged by the CAO's LiDAR coincided with leafless and/or dead trees that covered up to 70 % of the landscape. The 3-D chemical imaging strongly suggested that, where terrain becomes variable, the imprint of drought is exacerbated. We think this may be caused by increased drainage and a lack of water retention capacity in rolling terrain as compared to the flatter surfaces found in other parts of the lowland Amazon. These measurements continue with repeated flights to assess the longer-term impact of these droughts and to remain prepared for additional droughts into the future. Without the unique 3-D chemical imaging of the CAO, it is likely that these droughts will go unanalyzed by any other means at the Amazon mesoscale.

6 The Future of Mesoscale Canopy Science

While doing any field work in the tropics, it does not take long to recognize the extreme challenges in carrying out ground-based carbon inventory, seeking new species, or understanding spatial and temporal impacts of climate variability and change. For botanists and ecologists fieldwork alone will likely not yield the data required to make the scientific breakthroughs or conservation advances needed now at the mesoscale. And remote sensing will not replace the critically important role that field botany and ecology play in identifying species or in studying species distributions in tropical forests. Alternatively, remote sensing serves a vital role in extending our view of tropical forest carbon stocks, floristic composition, and climate interactions to scales that are intractable on the ground and invisible from space.

Fig. 18.9 In 2005 and 2010, the western Amazon basin underwent severe drought, yet the short- and long-term impacts remain completely unknown. In these first-ever images of post-drought forest condition from the CAO, the mediating role of local topography is apparent: (**a**) An area of flatter terrain (*left*) is associated with mostly live tree canopy shown in *green* colors (*right*); however (**b**) in areas with rolling hills supported increased runoff and drainage (*left*), the canopy is mostly senescence or dead altogether (*pink colors, right*). The relief in the lower image is only 1–3 m, indicating that just slight increases in topographic variation can greatly exacerbate the impact of drought on tropical forest canopies

Perhaps the most profound impact of high resolution, mesoscale canopy exploration, and mapping rests in its role in facilitating human interaction with our environment. At a time when people are becoming less connected to nature, particularly with remote regions of the planet that most will never see, the CAO is providing the 3-D experience true to the composition and functioning of these ecosystems, while making the experience tractable for the nonexpert or non-adventurer. With this as a core component to the CAO's mission, I strongly feel that the perspectives of high-ranking decision makers can be positively influenced, while at the same time inspiring the youngest members of society to appreciate and conserve forest canopies in the tropics and worldwide.

Acknowledgments I thank the many colleagues who have influenced my thinking on the issues presented here, and I give a special thanks to Meg Lowman and Roberta Martin for their thoughtful advice and review. The Carnegie Airborne Observatory is made possible by the Gordon and Betty Moore Foundation, John D. and Catherine T. MacArthur Avatar Alliance Foundation, Grantham, Foundation for the Protection of the Environment, Avatar Alliance Foundation, Margaret A. Cargill Foundation, W. M. Keck Foundation, Mary Anne Nyburg Baker and G. Leonard Baker Jr., and William R. Hearst III.

References

Angelsen A (2008) Moving ahead with REDD: issues, options and implications. Center for International Forestry Research (CIFOR), Bogor

Asner GP (2009) Tropical forest carbon assessment: integrating satellite and airborne mapping approaches. Environ Res Lett 3:1748–9326

Asner GP (2011) Painting the world REDD: addressing scientific barriers to monitoring emissions from tropical forests. Environ Res Lett 6:021002

Asner GP, Alencar A (2010) Drought impacts on the Amazon forest: the remote sensing perspective. New Phytol 9999

Asner GP, Martin RE (2008) Spectral and chemical analysis of tropical forests: scaling from leaf to canopy levels. Remote Sens Environ 112:3958–3970

Asner GP, Martin RE (2009) Airborne spectranomics: mapping canopy chemical and taxonomic diversity in tropical forests. Front Ecol Environ 7:269–276

Asner GP, Martin RE (2011) Canopy phylogenetic, chemical and spectral assembly in a lowland Amazonian forest. New Phytol 189:999–1012

Asner GP, Vitousek PM (2005) Remote analysis of biological invasion and biogeochemical change. Proc Nat Acad Sci USA 102:4383–4386

Asner GP, Townsend AR, Braswell BH (2000) Satellite observation of El Nino effects on Amazon forest phenology and productivity. Geophys Res Lett 27:981–984

Asner GP, Elmore AJ, Hughes RF, Warner AS, Vitousek PM (2005a) Ecosystem structure along bioclimatic gradients in Hawai'i from imaging spectroscopy. Remote Sens Environ 96:497–508

Asner GP, Knapp DE, Broadbent EN, Oliveira PJC, Keller M, Silva JNM (2005b) Selective logging in the Brazilian Amazon. Science 310:480–482

Asner GP, Knapp DE, Kennedy-Bowdoin T, Jones MO, Martin RE, Boardman J, Field CB (2007) Carnegie Airborne Observatory: in flight fusion of hyperspectral imaging and waveform light detection and ranging (LiDAR) for three-dimensional studies of ecosystems. J Appl Remote Sens. doi:10.1117/1111.2794018

Asner GP, Hughes RF, Vitousek PM, Knapp DE, Kennedy-Bowdoin T, Boardman J, Martin RE, Eastwood M, Green RO (2008a) Invasive plants transform the 3-D structure of rainforests. Proc Nat Acad Sci 105:4519–4523

Asner GP, Jones MO, Martin RE, Knapp DE, Hughes RF (2008b) Remote sensing of native and invasive species in Hawaiian forests. Remote Sens Environ 112:1912–1926

Asner GP, Levick SR, Kennedy-Bowdoin T, Knapp DE, Emerson RE, Jacobson J, Colgan MS, Martin RE (2009a) Large-scale impacts of herbivores on the structural diversity of African savannas. Proc Nat Acad Sci 106:4947–4952

Asner GP, Rudel TK, Aide TM, Defries R, Emerson RE (2009b) A contemporary assessment of change in humid tropical forests. Conserv Biol 23:1386–1395

Asner GP, Loarie SR, Heyder U (2010a) Combined effects of climate and land use change on the future of humid tropical forests. Conserv Lett 3:395–403

Asner GP, Powell GVN, Mascaro J, Knapp DE, Clark JK, Jacobson J, Kennedy-Bowdoin T, Balaji A, Paez-Acosta G, Victoria E, Secada L, Valqui M, Hughes RF (2010b) High-resolution forest carbon stocks and emissions in the Amazon. Proc Nat Acad Sci 107:16738–16742

Asner GP, Hughes RF, Mascaro J, Uowolo AL, Knapp DE, Jacobson J, Kennedy-Bowdoin T, Clark JK (2011a) High resolution carbon mapping on the million hectare Island of Hawaii. Front Ecol Environ 9:434–439

Asner GP, Martin RE, Knapp DE, Tupayachi R, Anderson C, Carranza L, Martinez P, Houcheime M, Sinca F, Weiss P (2011b) Spectroscopy of canopy chemicals in humid tropical forests. Remote Sens Environ 115:3587–3598

Asner GP, Martin RE, Tupayachi R, Emerson R, Martinez P, Sinca F, Powell GVN, Wright SJ, Lugo AE (2011c) Taxonomy and remote sensing of leaf mass per area (LMA) in humid tropical forests. Ecol App. 21:85–98

Asner GP, Mascaro J, Muller-Landau HC, Vieilledent G, Vaudry R, Rasamoelina M, Hall J, van Breugal M (2011d) A universal airborne LiDAR approach for tropical forest carbon mapping. Oecologia. doi:10.1007/s00442—011-2165

Asner G, Clark J, Mascaro J, Vaudry R, Chadwick KD, Vieilledent G, Rasamoelina M, Balaji A, Kennedy-Bowdoin T, Maatoug L, Colgan M, Knapp D (2012a) Human and environmental controls over aboveground carbon storage in Madagascar. Carbon Balance Manag 7:2

Asner GP, Knapp DE, Boardman J, Green RO, Kennedy-Bowdoin T, Eastwood M, Martin RE, Anderson C, Field CB (2012b) Carnegie Airborne Observatory-2: increasing science data dimensionality via high-fidelity multi-sensor fusion. Remote Sens Environ 124:454–465

Carlson KM, Asner GP, Hughes FR, Ostertag R, Martin RE (2007) Hyperspectral remote sensing of canopy biodiversity in Hawaiian lowland rainforests. Ecosystems 10:536–549

Chave J, Muller-Landau HC, Baker TR, Easdale TA, ter Steege H, Webb CO (2006) Regional and phylogenetic variation of woody density across 2456 neotropical tree species. Ecol App. 16: 2356–2367

Clark DA (2004) Tropical forests and global warming: slowing it down or speeding it up? Front Ecol Environ 2:73–80

Clark DC, Clark DB (1994) Climate induced annual variation in canopy tree growth in a Costa Rican tropical rain forest. J Ecol 82:865–872

Colgan MS, Asner GP, Levick SR, Martin RE, Chadwick OA (2012) Topo edaphic controls over woody plant biomass in South African savannas. Biogeosciences 9:1809–1821

FAO (2007) State of the World's forests. FAO, Rome

Fensham RJ, Fairfax RJ (2002) Aerial photography for assessing vegetation change: a review of applications and the relevance of findings for Australian vegetation history. Aust JBot 50:415–429

Féret JB, Asner GP (2012) Tree species discrimination in tropical forests using airborne imaging spectroscopy. IEEE Transactions on Geoscience and Remote Sensing. doi:10.1109/TGRS. 2012.2199323

Hansen MC, Stehman SV, Potapov PV, Loveland TR, Townshend JRG, DeFries RS, Pittman KW, Arunarwati B, Stolle F, Steininger MK, Carroll M, DiMiceli C (2008) Humid tropical forest clearing from 2000 to 2005 quantified by using multitemporal and multiresolution remotely sensed data. Proc Natl Acad Sci 105:9439–9444

Herold M, Skutsch M (2009) Measurement, reporting and verification for REDD+: objectives, capacities and institutions. In: Angelsen A (ed) Realising REDD+: national strategy and policy options. CIFOR, Bogor Barat

Huang C, Asner GP, Martin RE, Barger NN, Neff JC (2009) Multiscale analysis of tree cover and aboveground carbon stocks in pinyon juniper woodlands. Ecol App. 19:668–681

Kellner JR, Asner GP, Kinney KM, Loarie SR, Knapp DE, Kennedy-Bowdoin T, Questad EJ, Cordell S, Thaxton JM (2011) Remote analysis of biological invasion and the impact of enemy release. Ecol App. 21:2094–2104

Lefsky MA, Cohen WB, Parker GG, Harding DJ (2002) Lidar remote sensing for ecosystem studies. BioScience 52:19–30

Levick SR, Asner GP, Chadwick OA, Khomo LM, Rogers KH, Hartshorn AS, Kennedy-Bowdoin T, Knapp DE (2010) Regional insight into savanna hydrogeomorphology from termite mounds. Nat Commun 1:65

Lewis SL, Brando PM, Phillips OL, van der Heijden GMF, Nepstad D (2011) The 2010 Amazon drought. Science 331:554

Li WH, Fu R, Juarez RIN, Fernandes K (2008) Observed change of the standardized precipitation index, its potential cause and implications to future climate change in the Amazon region. Philos Trans R Soc B Biol Sci 363:1767–1772

Loarie SR, Duffy PB, Hamilton H, Asner GP, Field CB, Ackerly DD (2009) The velocity of climate change. Nature 462:1052–1057

Mascaro J, Asner GP, Muller-Landau HC, van Breugal M, Hall J, Dahlin K (2011) Controls over aboveground forest carbon density on Barro Colorado Island, Panama. Biogeosciences 8:1615–1629

McGraw JB, Warner TA, Key TL, Lamar WR (1998) High spatial resolution remote sensing of forest trees. Tree 13:301–302

Morton DC, DeFries RS, Nagol J, Souza CM Jr, Kasischke ES, Hurtt GC, Dubayah R (2011) Mapping canopy damage from understory fires in Amazon forests using annual time series of Landsat and MODIS data. Remote Sens Environ 115:1706–1720

Peres CA, Barlow J, Laurance WF (2006) Detecting anthropogenic disturbance in tropical forest. Trends Ecol Evol 21:227–229

Phillips OL, Aragao LEOC, Lewis SL, Fisher JB, Lloyd J, Lopez-Gonzalez G, Malhi Y, Monteagudo A, Peacock J, Quesada CA, van der Heijden G, Almeida S, Amaral I, Arroyo L, Aymard G, Baker TR, Banki O, Blanc L, Bonal D, Brando P, Chave J, de Oliveira ACA, Cardozo ND, Czimczik CI, Feldpausch TR, Freitas MA, Gloor E, Higuchi N, Jimenez E, Lloyd G, Meir P, Mendoza C, Morel A, Neill DA, Nepstad D, Patino S, Penuela MC, Prieto A, Ramirez F, Schwarz M, Silva J, Silveira M, Thomas AS, ter Steege H, Stropp J, Vasquez R, Zelazowski P, Davila EA, Andelman S, Andrade A, Chao KJ, Erwin T, Di Fiore A, Honorio E, Keeling H, Killeen TJ, Laurance WF, Cruz AP, Pitman NCA, Vargas PN, Ramirez-Angulo H, Rudas A, Salamao R, Silva N, Terborgh J, Torres-Lezama A (2009) Drought sensitivity of the Amazon rainforest. Science 323:1344–1347

Saleska SR, Didan K, Huete AR, da Rocha HR (2007) Amazon forests green up during 2005 drought. Science 318:612

Samanta A, Ganguly S, Hashimoto H, Devadiga S, Vermote E, Knyazikhin Y, Nemami RR, Myneni RB (2010) Amazon forests did not green up during the 2005 drought. Geophys Res Lett 37:05401

Skole D, Tucker C (1993) Tropical deforestation and habitat fragmentation in the Amazon: satellite data from 1978 to 1988. Science 260:1905–1910

Souza C, Roberts DA, Cochrane MA (2005) Combining spectral and spatial information to map canopy damages from selective logging and forest fires. Remote Sens Environ 98:329–343

Tilman D, Pacala S (1993) The maintenance of species richness in plant communities. In: Ricklefs RE, Schluter D (eds) Species diversity in ecological communities: historical and geographical perspectives. University of Chicago Press, Chicago

Ustin SL, Roberts DA, Gamon JA, Asner GP, Green RO (2004) Using imaging spectroscopy to study ecosystem processes and properties. BioScience 54:523–534

Vitousek PM (1995) The Hawaiian Islands as a model system for ecosystem studies. Pac Sci 49:2–16

Wright SJ, Muller-Landau HC, Schipper J (2009) The future of tropical species on a warmer planet. Conserv Biol 23:1418–1426

Chapter 19
Why Do Sloths Poop on the Ground?

Bryson Voirin, Roland Kays, Martin Wikelski, and Margaret Lowman

Keywords Sloth • Defecation • Arboreal mammals

The bizarre ground-based defecation behavior of two- and three-toed sloths remains one of the most paradoxical and humorous mysteries in canopy biology. Both two-toed (*Choloepus* sp.) and three-toed (*Bradypus* sp.) sloths defecate and urinate only once every 4–8 days in the wild. Although all other arboreal mammals release their excrement from the forest canopy, all sloths climb down from the treetops and relieve themselves on the forest floor. In the canopy, sloths have few predators, but on the ground, sloths expose themselves to great variety of predators. There must be some defining reason behind this high-risk behavior.

For years after their discovery, sloths were thought to be strictly arboreal. Their body and limb design is specifically adapted towards a life in the canopy (Beebe 1926). Their

B. Voirin (✉)
Max Planck Institute for Ornithology, Radolfzell, Germany

Smithsonian Tropical Research Institute, Panama, República de Panamá
e-mail: brysonvoirin@gmail.com

R. Kays
North Carolina Museum of Natural Sciences, Raleigh, NC, USA

Smithsonian Tropical Research Institute, Panama, República de Panamá

North Carolina State University, Raleigh, NC, USA
e-mail: Roland.Kays@Naturalsciences.org

M. Wikelski
Max Planck Institute for Ornithology, Radolfzell, Germany

North Carolina Museum of Natural Sciences, Raleigh, NC, USA
e-mail: martin@orn.mpg.de

M. Lowman
North Carolina Museum of Natural Sciences, North Carolina State University,
121 W. Jones Street, Raleigh, NC 27603, USA
e-mail: canopymeg@gmail.com

M. Lowman et al. (eds.), *Treetops at Risk: Challenges of Global Canopy Ecology and Conservation*, DOI 10.1007/978-1-4614-7161-5_19,

notorious name, which is one of the "seven deadly sins," accurately describes their average day. With a basal metabolism nearly half of what would be expected for a mammal of its size, they often move less than a dozen meters in a day (Sunquist and Montgomery 1973). But this lazy lifestyle is not a sign of weakness or ineptitude, but rather a survival adaptation resulting from millions of years of selective adaptation and evolution (Goffart 1971).

Daily energy expenditure for three-toed sloths has been estimated at just 147 kJ (per kg of body weight), meaning that a typical adult 4-kg sloth would need to eat only 60 g of leaves each day (Montgomery and Sunquist 1975). Their sedentary lifestyle has led to a unique body design, with only 25–30 % of the total body mass being muscle, which is lower than other mammal species (typically 40–45 % muscle). Instead, up to 30 % of a sloth's body weight consists of food in the stomach and urine. Additionally, unlike other mammals that are homeothermic, sloths spend less energy on energetically expensive thermoregulatory process. As a result, their low body temperature varies, over 10 °C, along with the ambient temperature and exposure to sunlight (Nagy and Montgomery 1980).

Sloths have evolved to fill a highly specialized niche in the tropics. Their home is high up in the canopy, where they are surrounded on all sides by their food source—leaves. Their diet consists of plant leaves and buds—a diet too low in vital nutrients for most mammals to eat. Sloths possess multichambered stomachs that are almost large enough to fill the entire abdominal cavity, suggesting evolutionarily convergence with ruminants. They engage in pre-gastric fermentation in their massive four-chambered forestomach, relying on a specific diversity of gut bacteria to break down food and aid in digestion. Their gut flora is so vital that sloths have developed special stomach lysozymes to protect their gut bacteria from being altered by foreign bacteria. Food may remain in the stomach for up to 90 h and can take a week to move through the alimentary canal to the anus (Pacheco et al. 2007).

Three-toed sloths are strictly folivores and are known to eat leaves from 28 different species of trees and lianas (Montgomery and Sunquist 1978). On Barro Colorado Island, a research site in Panama, three-toed sloths crop leaves at a rate of 38 kg per hectare, suggesting that they eat 2 % of the total annual leaf production of the island (McNab 1978; Montgomery and Sunquist 1975; Leigh 1999). Two-toed sloths are known to occasionally also eat fruits and flowers, but the majority of their food includes leaves. In order to survive on leaves alone, sloths have had to greatly curtail their daily energy expenditure through millions of years of evolution of their special energy–conscience existence.

The average day in the life of a sloth is rather uneventful. They sleep in periods of less than 1 h throughout the day and night, interrupted by occasional bouts of feeding, scratching, and crawling (Sunquist and Montgomery 1973). Two-toed sloths tend to be more mobile than three-toed sloths, the latter of which can often be seen on the same branch for days on end. But overall, sloths are one of the least mobile mammal species on Earth.

Surprisingly though, sloths do not spend most of their time asleep. In the wild they sleep around 9.5 h per day, significantly less that was originally thought (Rattenborg et al. 2008). The majority of their day (and night) is spent quietly hanging, motionless, and awake, in the trees. Whether they are in deep thought, meditating, or just staring blankly into space, this "wake but inactive" state comprises the majority of their existence and is crucial for energy conservation and their ultimate survival.

With so much food available in the canopy, the key to sloth survival is to avoid winding up as someone else's dinner. Prey species across the animal kingdom have developed a plethora of ways to combat predation. Sloths have forgone the modes of defensive teeth, claws, or speedy retreat and instead are completely reliant of crypsis for their survival. Remaining motionless and well camouflaged has proven successful for sloths. Their movement around the canopy, climbing from branch to branch, is slow and methodical enough not to draw attention. Even to the trained human eye, a sloth in the canopy will appear to be nothing more than a dark clump of leaves (Beebe 1926). While catching and radio-tagging sloths in Panama, we have often climbed to the top of a tree only to discover the sloth we thought we spotted from the ground was actually a dead leaf.

Up in the canopy, the main predator risk for sloths is Harpy Eagles, the largest predatory bird in the world (Fowler and Cope 1964). This magnificent raptor feeds predominately on two- and three-toed sloths, with mounds of their bones piled up beneath their nests (Galetti and Carvalho 2000). When hunting, the Harpy Eagle spots sloths while soaring or perched in the canopy. Sloths hiding in thick tangles of lianas and branches are unreachable. But when the eagle spots an exposed sloth, it swoops through the canopy, inverts its body so the claws are skyward, and grabs the back of the sloth with its talons. The sloth is then ripped out of the canopy and carried to the Harpy's nest to consume (Rettig 1978). This dramatic predation event has become relatively uncommon across most of Central America, as the Harpy Eagle population have decreased in recent years (Touchton et al. 2002).

If caught on the ground, however, sloths face predation from an array of species. Ocelots, jaguars, and pumas are their main threats, but other less common species will also take advantage of a slow moving sloth on the forest floor. Tayras, spectacled owls, and even coatis have been observed killing defenseless sloths on the ground (Voirin 2009). Clearly, sloths are vulnerable on the ground. So why do they choose to risk their survival with a bathroom visit on terra firma?

G.L. Buffon first described the curious toilet behavior of sloths in 1848, and since then, it has been the source of jokes and much ridicule for the species. But mortality from going to the bathroom is no joking matter for sloths. Here is an arboreal mammal that is born in the trees, mates in the trees, and dies (when not predated) in the trees as well. Yet, once a week they climb out of the canopy, down to the forest floor, and defecate. Three-toed sloths dig a small hole with their tail, sitting in an awkward squatting position. Once finished, they partially cover the dung with soil, much like a dog, and climb back into the canopy. Two-toed sloths, lacking any tail, simply squat and go on the exposed soil (Montgomery and Sunquist 1975).

For such an energetically costly behavior to evolve in both genera of sloths despite such obvious predation, risks suggest that it must have some strong adaptive value. Several theories have been put forth to explain this, but none is fully satisfactory. The most prominent theory is related to nutrient cycling, where both species of sloths defecate at the base of their preferred trees, effectively fertilizing the tree's root system and helping the tree. The idea that this fertilizer offering would help the tree survive and produce more leaves for the hungry sloth seems far-fetched.

Sloth feces are extremely hard and dry and are largely underutilized by dung eaters in the forest. They often remain untouched on the forest floor for many days (Young 1981). Furthermore, it is not clear how climbing down the tree to defecate would help fertilize the tree any more than dropping the scat from within the tree's canopy. Moreover, given that sloths move about their home range throughout the year, there may be months that this particular tree is not visited.

Another hypothesis put forth in the mid-twentieth century is that a mutualistic relationship exists between sloth moths (*Bradypodicola hahneli* and *Cryptoses choloepi*) and their respective sloth hosts. These moths live in the fur of sloths, leaving the animal only to lay their eggs in sloth dung. The moth's survival is probably reliant on the sloth's ground defecation; if the sloth were to defecate up in the tree, the moths would have a more difficult time finding the dung. However, any mutualistic benefit to the sloth from having moths is unknown. If the sloths defecate on the ground solely for the benefit of the moth, there must be a strong benefit of the moth to the sloth for this behavior to evolve.

In the days leading up to a defecation event, increasing numbers of sloth moths congregate around the rear of the sloth. Just prior to actual event, as the sloth climbs down towards the ground, the moths become extremely active and swarm around the sloth. Once the sloth dung has been deposited onto the forest floor, female moths fly from the sloth and deposit their eggs in the sloth dung. Once hatched, the larva spins silk threads between dung pellets to help them feed. After the larvae pupate, the adult moths will fly through the canopy searching for a resident sloth (Waage and Montgomery 1976). It is possible these moths take in nutrients through the tear ducts or sebaceous glands at the base of hair follicles, as many other moth species do. However, such behavior has yet to be observed.

Our favorite hypothesis focuses on the sloth's diet of leaves, which is devoid of certain important trace elements. Many leaf-eating mammals ingest dirt or soil (aka geophagy) to supplement their dietary needs not met by traditional food. Golden-faced Saki monkeys are known to eat dirt from termite nests, even in periods of plentiful foods, suggesting that their geophagy is related to uptake of trace elements not available in their arboreal diets. Ingesting soil has been shown to function as a mineral supplement, a moderator of pH levels in the stomach, and as a catalyst for breaking down tannins in leaves.

We suspect that sloths are also dependent on soil and that their strictly arboreal diet is limited in nutrients, perhaps lacking key elements such as iron and sodium. Baby two-toed sloths have been seen ingesting soil while the mother is on the ground pooping. Two-toed sloths were seen eating human feces from a latrine in Costa Rica, and a similar behavior was recorded for three-toed sloths in Amazonian Peru (Lowman, personal communication). Such decomposing matter is rich in salts and nutrients that the sloths crave. In captivity, two-toed sloths will fight over access to salt licks, which likely aid them in water retention.

Although not directly observed in the wild, we hypothesize that sloths take up nutrients and trace elements from their ground defecation behavior. While on the ground, sloths crawl across the soil with significant effort, digging their claws into the ground and dragging their bodies along. Their claws have cupped ridges on their interior, which is often caked with soil. Sloths subsequently lick their claws, and this could provide soil nutrients that supplement their diet in important ways.

Amazingly, the two existing families of sloths evolved from distinctly separate ground sloths, independently taking to the trees. Thus, their terrestrial toilet behavior represents convergent evolution and must be under strong selective pressure. We hypothesize that during their adaptation to a life in the canopy, sloths continued to visit the ground to defecate by habit. This bathroom ritual aided them in their nutrient balance, which in turn allowed them to continue their diet of abundant green leaves.

Unfortunately, no rigorous data sets exit to test any of these three hypotheses (tree fertilizer, moth coevolution, or trace nutrients), leaving the canopy sloth's terrestrial toilet behaviors an unsolved mystery of the rainforest.

References

Beebe W (1926) The three-toed sloth. Bradypus cucullinger cucullinger Wagler. Zoologica 7:1–67

Buffon BGEL (1848) Oeuvres completes de Buffon. Volume 4. Au Bureaudes publications illustrees, Paris, France

Fowler JM, Cope JB (1964) Notes on Harpy Eagle in British Guiana. Auk 81:257–273

Galetti M, Carvalho O Jr (2000) Sloths in the diet of a harpy eagle nestling in Eastern Amazon. Wilson Bull 112(4):535–536

Goffart M (1971) Function and form in the sloth. Pergamon Press, Oxford/New York/Sydney/Toronto/Braunschweig

Leigh E (1999) Tropical forest ecology: a view from Barro Colorado Island. Oxford University Press, Oxford

McNab BK (1978) Energetics of arboreal folivores: physiological problems and ecological consequences of feeding on an ubiquitous food supply. In: Montgomery GG (ed) The ecology of arboreal folivores. Smithsonian Institution Press, Washington, DC

Montgomery GG, Sunquist ME (1975) Impact of sloths on neotropical forest energy flow and nutrient cycling. In: Golley FB, Medina E (eds) Tropical ecological systems, Trends in terrestrial and aquatic research. Springer, New York, 293 pp

Montgomery GG, Sunquist ME (1978) Habitat selection and use by two- and three-toed sloths. In: Montgomery, GG (ed). The ecology of arboreal folivores. Washington, DC : Smithsonian Institution Press

Nagy A, Montgomery GG (1980) Field metabolic rate, water flux, and food consumption in three-toed sloths (Bradypus variegatus). J Mammal 61:465–472

Pacheco MA, Concepci6n JL, Rosales Rangel JD, Ruiz MC, Michelangeli F, Domínguez–Bello MG (2007) Stomach lysozymes of the three-toed sloth (Bradypus variegatus), an arboreal folivore from the Neotropics. Comp Biochem Physiol 147(3):808–819

Rattenborg NC, Voirin JB, Vyssotski AL, Kays RW, Spoelstra K, Kuemmeth F, Heidrich W, Wikelski MC (2008) Sleeping outside the box: electroencephalographic measures of sleep in sloths inhabiting a rainforest. Biol Lett 4:402–405

Rettig N (1978) Breeding behavior of the Harpy Eagle (Harpia harpyja). The Auk 95:629–643

Sunquist ME, Montgomery GG (1973) Activity patterns and rates of movement of two-toed and three-toed sloths (*Choloepus hoffmanni* and *Bradypus infuscatus*). J Mammal 54:946–954

Touchton J, Hsu YC, Palleroni A (2002) Foraging ecology of reintroduced captive–bred subadult Harpy Eagles (*Harpia harpyja*) on Barro Colorado Island, Panama. Ornithol Neotrop 13:365–379

Voirin B, Kays R, Lowman M, Wikelski M (2009) Evidence of three-toed sloth predation by spectacled owl. Edentata 10:15–20

Waage JK, Montgomery GG (1976) Cryptoses choloepi: a coprophagous moth that lives on a sloth. Science 193(4248):157–158

Young OP (1981) The utilization of sloth dung in a neotropical forest. Coleopterists Bull 35(4):427–430

Chapter 20
Birds of the "Canopy": Historical Perspective, Current Trends, and Future Directions

Vivek Ramachandran and T. Ganesh

Keywords Canopy • Sampling • Conservation • Birds • Crane • Walkways • History

1 Why Canopy?

In structurally complex tropical forests, an understanding of forest ecosystems must address biodiversity and community interactions at all levels (Lowman and Rinker 2004). The structural complexity, species diversity, fluctuations in microclimate, and resource availability separate the canopy from other strata. Canopy bird communities include important functional groups, such as seed dispersers, pollinators, and predators (Howe 1977; Nadkarni and Matelson 1989; Holbrook and Smith 2000). In addition, many long-distance and elevational migrants use the canopies, and an understanding of their ecology, distribution, and abundance is necessary for their conservation (Loiselle 1987; Anderson 2009).

Ground-based studies have disadvantages when the habitat is structurally complex, such as tall mature canopies. In tropical habitats, the ability of an observer to detect birds within the dense foliage depends on the distance from the observer to the canopy and the characteristics of the species observed in

V. Ramachandran (✉)
Ashoka Trust for Research in Ecology and the Environment (ATREE),
Royal Enclave, Srirampura, Jakkur Post, Bangalore, KA 560064, India

Manipal University, Manipal.edu, Madhav Nagar, Manipal 576104, India
e-mail: vivekr@atree.org

T. Ganesh
Ashoka Trust for Research in Ecology and the Environment (ATREE),
Royal Enclave, Srirampura, Jakkur Post, Bangalore, KA 560064, India
e-mail: tganesh@atree.org

M. Lowman et al. (eds.), *Treetops at Risk: Challenges of Global Canopy Ecology and Conservation*, DOI 10.1007/978-1-4614-7161-5_20,

terms of coloration, vocalization, and movement (Anderson 2009). But studying the canopy avifauna is critical to having a complete picture of tropical-forest bird communities (Naka 2004).

2 Do Canopies Host a Unique Set of Species?

Studies show that frugivores and some omnivores represent the "core" canopy avifauna of the neotropics (Greenberg 1981; Loiselle 1988; Levey and Stiles 1992) and display narrow individual diet spectra (Cohn-Haft and Sherry 1994). The patchy distribution of resources over time and space in the canopy (Greenberg 1981; Loiselle 1988; Levey and Stiles 1994) has led to long-distance movements of many canopy species (Karr and James 1975), creating fluctuations in local populations (Winkler and Preleuthner 2001). Birds inhabiting the forest canopy or the forest edge are claimed to be more ecologically and phylogenetically similar, than with those of the forest understory (Orians 1969; Pearson 1971; Cohn-Haft 1995). A major difference between canopy and understory bird assemblages is thought to be their trophic organization (Sherry 1984).

Despite their importance, a majority of studies on tropical-forest canopy birds are conducted from the ground, primarily due to the difficulty of canopy access. Within the last few decades, advances in canopy access techniques enable scientists to safely access the canopy (Lowman and Wittman 1996; Sutton 2001). Avian studies greatly benefit from these new techniques, and our knowledge of the ecology and behavior of rain forest birds has improved (Munn and Loiselle 1995; Winkler and Preleuthner 2001). To get more quantifiable and reliable data on canopy-dwelling birds, mist netting in the canopy and observations from platforms, towers, cranes, and walkways are being employed.

3 Methods to Study Canopy Birds

3.1 Mist Netting

Early attempts to study birds in the canopy used mist netting (Greenlaw and Swinebroad 1967; Humphrey et al. 1968). These studies describe techniques to capture birds in the forest canopy at heights of 30 m or more using mist nets constructed with low-cost materials and operable with minimum manpower using pulleys. Subsequently, the Humphrey et al. method was modified by Webber (1975) and Whitaker (1972) and successfully used to capture canopy birds. Beehler (1983) and McClure (1984) used mist nets that were slipped up and down tall poles to capture canopy birds, but this method did not gain popularity because it is too labor intensive and complicated. Munn (1985, 1986) used the modified Humphrey et al. method to study mixed species flocks in the Amazon. Finding this method laborious and destructive to vegetation, Munn (1991) described an efficient method to get lines into tall trees and a modification of the Greenlaw and Swinebroad method to successfully capture birds at heights of 40–60 m.

Meyers and Pardieck (1993) developed three lightweight, low-canopy mist nets and evaluated them in dry tropical scrub, mangrove, and forest habitats. They found that the telescopic aluminum pole system, in spite of the initial costs, was the most efficient in capturing psittacines, columbids, passerines, and possibly chiropterans in habitats with canopies <10 m or in the forest sub-canopy. Stokes and Schultz (1995) and Stokes et al. (2000) describe a method of setting up mist nets horizontally using canopy platforms paralleling conventional netting to allow comparisons between strata (useful for studies that require simultaneous mist netting in forest canopy and understory levels). In a study by Derlindati and Caziani (2005) in the Chaco forests of Argentina, a comparison between canopy mist netting and point counts showed similar patterns of bird diversity and distribution. Using mist netting from three canopy walkways in Malaysia, Rehman (2002) suggests that deployment from canopy walkways was useful in avifauna surveys to detect species commonly missed by other techniques.

All the above studies showed that with a little technical innovation, the upper strata of structurally complex habitats could be sampled repeatedly and reliably. They also encountered species not captured or detected in previous studies that relied on ground-based sampling. In addition, vital information on the vertical distribution of birds in these habitats was obtained.

3.2 *Towers, Platforms, and Cranes*

One of the earliest observational studies carried out from canopy towers was of Lovejoy (1975), who conducted censuses from a tower in the Brazilian Amazon at Manaus (but did not separate canopy census results in his published analysis). Greenberg (1981) carried out censuses from a canopy tower in Barro Colorado Island, Panama, and found that canopy and scrubby second growth have many species in common.

The abundance and seasonality of migrants in the canopy was studied by Loiselle (1987, 1988) in the lowland rain forests of Costa Rica who found that canopy use by migrants was seasonal with abundances highest during the fall and lowest during spring migration. The avifauna of the forest canopy, with few exceptions, was distinct from the understory avifauna; few of her common understory species were recorded in the canopy.

Portable canopy platforms were used to observe foraging behavior of Costa Rican birds (Nadkarni 1988; Nadkarni and Matelson 1989). Cohn-Haft et al. (1997) demonstrated how canopy surveys from a single canopy tower served to detect many species that were not encountered from the ground. Conducting censuses from three canopy towers in Manaus, Brazil, Naka (2004) showed that the canopy avifauna of Manaus was mostly composed of typical canopy bird species, as was the case with Costa Rica.

Since the establishment of canopy crane sites in the early 1990s in tropical and temperate forests around the world, many ornithological studies have taken advantage of these facilities. The Surumoni crane at Venezuela hosted studies of frugivorous bird assemblages and their foraging behavior. A study by Walther (2000) found that degree of frugivory did not correlate with specialization for particular fruits. Fruit size was dependent on abundance of alternate fruits, fruit accessibility, and secondary metabolite content (Schaefer et al. 2003). A study on vocalization by Nemeth et al. (2001) found that factors that most influence bird song are reverberation and ground attenuation. Optical communication was studied in manakins (*Pipridae*), which displayed where ambient light increased their color signal against the background vegetation (Heindl and Winkler 2003). Vertical stratification of the avian community and effects of light habitats and height on foraging breath of species were studied by Walther (2002a, b).

Shaw and Flick (1999) and Shaw et al. (2002), at the Wind River canopy crane in Oregon, USA, found that birds shifted to the upper canopy during winter, and small bird assemblages of the old-growth forest stand were seasonally and vertically patterned to reflect the patterns of vertical forest structure of the upper canopy. Van Bael et al. (2003) studied the effect of birds on herbivory in a Panamanian neotropical-forest canopy crane. In a similar study, Kalka et al. (2008) experimentally partitioned bird predation from bat predation and found that bats had dramatic ecological effects on herbivory that were previously overlooked.

These studies shed light on processes and patterns often missed from ground-based studies. Insights gained from these studies further underline the importance of canopy-based efforts in avian biology.

4 Trends in Canopy Bird Research

Despite the improved methods and increase in ease and safety of canopy access, the number of research publications on canopy birds has not exponentially increased (Fig. 20.1). Mist-netting techniques contributed to the early studies followed by studies from observational towers. Rope-based techniques were used in the 1980s and the establishment of the canopy cranes led to a spurt in canopy studies in the early 1990s and 2000s. The second half of the last decade has seen a slowdown in canopy bird studies. Reasons for such slowdown include a lack of trained personnel to access canopies, especially in the old world tropics, and paucity of infrastructure to access the canopies.

Access to the canopies for ornithological studies has provided valuable data in regard to the structure and functioning of avian assemblages in tropical and temperate forests. Walther (2003) used observations made from a canopy crane located in the Amazonian rain forest (Surumoni Crane Project, southern Venezuela) to illustrate how canopy access may enhance our understanding of canopy birds. His observations from the crane demonstrated that some species that are portrayed as edge loving are more often found in closed canopy than was obvious from ground-based observations.

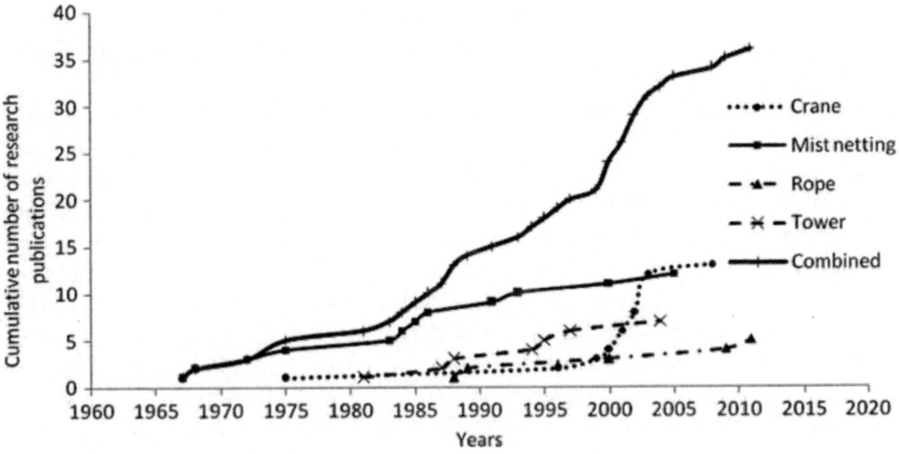

Fig. 20.1 Trend in peer-reviewed research publications on canopy birds using various access techniques

A comparison on ground-based sampling with canopy sampling by Anderson (2009) found ground methods significantly underestimated species and familial richness as well as abundance of individuals in the canopy stratum and highlights the risk of using ground-based methods for bird studies in structurally complex tropical forests.

The obvious disadvantage of using cranes is the n = 1 caveat as the focus on a single forest stand. Despite this obvious limitation, no other method can offer access to all parts of the vertical column of the forest and quality of observation provided by a crane. Cranes allow flexible and continuous access to a large area with minimal effect on birds. They also facilitate deployment of canopy nets in places inaccessible for ground-operated canopy nets.

Similar to the crane, platforms or canopy towers can provide localized permanent access to locations in the canopy and facilitate good opportunities for observation, but with slightly less expanse (Walther 2003). With cranes and single-point towers, it is important to consider the high spatiotemporal variability of food resources in the canopy (Leigh et al. 1996), which could influence the number of species and individual birds available to the observer. In our studies (Vivek and Ganesh 2012), we worked around these problems by installing several canopy platforms in an area of about 12 km^2, accessed using the Single Rope Technique (SRT). These platforms are spread across a habitat mosaic of primary, selection-felled, and clear-felled regenerating forests.

5 Implications for Conservation

With increasing pressure from human development and various climate change scenarios, conservation of biologically important sites is a global priority. Since canopies are at the interface between the atmosphere and earth, canopy avifauna could be used as indicators of potential change. Under current rates of warming, montane bird communities of Australasia, the Afrotropics, and the Nearctic are exceptionally threatened (La Sorte and Jetz 2010).

The determination of conservation priorities is currently based on ground-based methods. With the inclusion of canopy-based techniques, species richness and abundance of biodiversity including birds can be more accurately estimated without bias and will help in prioritization of the conservation potential of sites (Anderson 2009). Better estimates of abundance and densities can also be obtained by applying distance sampling to canopy-based samples. Our results have shown that detection probabilities for small canopy nectarivores were much higher when sampling was occurred in the canopy (Vivek and Ganesh 2012). In addition, our study also facilitated research in other disciplines that benefit from such distributed canopy access like the pollination ecology of *Loranthaceae* that are pollinated and dispersed by canopy birds (Giby and Devy, pers comm). Our canopy-based studies have also shown that the change in habitat structure after disturbance led to expatriation of mid-story species and the resultant assemblage are dissimilar to the primary forest assemblages.

There is a need for the ornithological community to sustain and encourage such studies. Similar to canopy arthropod surveys done under IBISCA (Basset et al. 2007), canopy bird studies spanning across tropical and temperate regions could facilitate intertropical and tropical vs. temperate comparisons. The additional effort and expense in canopy research is easily justified by its added accuracy and holistic view of the bird assemblages. Existing large ecological monitoring such as NEON and CTFS should also incorporate canopy-based methods to study canopy bird assemblages. This additional canopy dimension would benefit studies of not only birds but also other lesser known taxa such as lichens, epiphytes, amphibians, reptiles, and fungi, often overlooked in whole-forest initiatives.

Acknowledgments We thank the Department of Science and Technology, Government of India for funding, Tamil Nadu Forest Department for permits, and Tamil Nadu Electricity Board for logistics. Noragric, Norwegian University of Life Sciences, and Sir Dorabji Tata Trust provided a fellowship to RV during the preparation of this chapter.

References

Anderson DL (2009) Ground versus canopy methods for the study of birds in tropical forest canopies: implications for ecology and conservation. Condor 111(2):226–237

Basset Y, Corbara B, Barrios H, Cuénoud P, Leponce M, Aberlenc HP et al (2007) IBISCA-Panama, a large-scale study of arthropod beta-diversity and vertical stratification in a lowland rainforest: rationale, study sites and field protocols. Bulletin de l'Institut Royal des Sciences Naturelles de Belgique, Entomologie 77:39–69

Cohn-Haft M (1995) Dietary specialization by birds forest interior versus canopy and edge habitats. M.S. thesis, Tulane University, New Orleans

Cohn-Haft M, Sherry TW (1994) Evolution of avian foraging stereotypes in tropical rain forest habitats. J Ornithol 135:481

Cohn-Haft M, Whittaker A, Stouffer PC (1997) A new look at the "species-poor" Central Amazon: the avifauna north of Manaus, Brazil. Ornithol Monogr 48:205–235
Derlindati EJ, Caziani SM (2005) Using canopy and understory mist nets and point counts to study bird assemblages in Chaco forests. Wilson Bull 117:92–99
Greenberg R (1981) The abundance and seasonality of forests canopy birds on Barro Colorado Island, Panama. Biotropica 13:241–251
Greenlaw JS, Swinebroad J (1967) A method for constructing and erecting aerial-nets in a forest. Bird Band 38:114–119
Heindl M, Winkler H (2003) Interacting effects of ambient light and plumage colour patterns in displaying wire-tailed manikins (Aves, Pipridae). Behav Ecol Sociobiol 53:153–162
Holbrook KM, Smith TB (2000) Seed dispersal and movement patterns in two species of *Ceratogymna* hornbills in a West African tropical lowland forest. Oecologia 125:249–257
Howe HF (1977) Bird activity and seed dispersal of a tropical wet forest tree. Ecology 58:539–550
Humphrey PS, Bridge D, Lovejoy T (1968) A technique for mist-netting in the Forest canopy. Bird Band 39:43–50
Kalka MB, Adam RS, Kalko EKV (2008) Bats limit arthropods and herbivory in a tropical forest. Science 320(5872):71
Karr JR, James FC (1975) Eco-morphological configurations and convergent evolution in species and communities. Ecol Evol Commun 258–291
La Sorte FA, Jetz W (2010) Projected range contractions of montane biodiversity under global warming. Proc R Soc B 277:3401–3410
Leigh EG Jr, Rand AS, Windsor DW (eds) (1996) The ecology of a tropical forest: seasonal rhythms and long-term changes. Smithsonian Institution Press, Washington, DC
Levey DJ, Stiles FG (1992) Evolutionary precursors of long distance migration: resource availability and movement patterns in Neotropical landbirds. Am Nat 140:447–476
Levey DJ, Stiles FG (1994) Birds: ecology, behavior, and taxonomic affinities. In: McDade L, Bawa KS, Hespenheide HA, Hartshorn GS (eds) La Selva, ecology and natural history of a Neotropical rain forest. University of Chicago Press, Chicago
Loiselle BA (1987) Migrant abundance in a Costa Rican lowland forest canopy. J Trop Ecol 3:163–168
Loiselle BA (1988) Bird abundance and seasonality in a Costa Rican lowland forest canopy. Condor 90:761–772
Lovejoy T (1975) Bird diversity and abundance in Amazonian forest communities. Living Bird 14:127–191
Lowman MD, Rinker HB (2004) Forest canopies, 2nd edn. Elsevier, Amsterdam
Lowman MD, Wittman PK (1996) Forest canopies: methods, hypotheses, and future directions. Annu Rev Ecol Syst 27:55–81
McClure E (1984) Bird banding. The Boxwood Press, Pacific Grove
Beehler B McP (1983) The behavioral ecology of four birds of paradise. Ph.D. thesis, Princeton University, Princeton
Meyers JM, Pardieck KL (1993) Evaluation of three elevated mist net systems for sampling birds. J Field Ornithol 64:270–277
Munn CA (1985) Permanent canopy and understory flocks in Amazonia: species composition and population density. Ornithol Monogr 36:683–712
Munn CA (1986) Birds that 'cry wolf'. Nature 319:143–145
Munn CA (1991) Tropical canopy netting and shooting lines over tall trees. J Field Ornithol 62:454–463
Munn CA, Loiselle BA (1995) Canopy access techniques and their importance for the study of tropical forest canopy birds. In: Lowman MD, Nadkarni NM (eds) Forest canopies. Academic, San Diego
Nadkarni NM (1988) Use of a portable platform for observations of tropical forest canopy animals. Biotropica 20:350–351

Nadkarni NM, Matelson TJ (1989) Bird use of epiphyte resources in neotropical trees. Condor 91:891–907

Naka LN (2004) Structure and organization of canopy bird assemblages in central Amazonia. Auk 121:88–102

Nemeth J, Winkler H, Dabelsteen T (2001) Differential degradation of Antbird songs in a Neotropical rainforest: adaptations to perch height? J Acoust Soc Am 110:3263–3274

Orians GH (1969) The number of bird species in some tropical forests. Ecology 50:783–801

Pearson DL (1971) Vertical stratification of birds in a tropical dry forest. Condor 73:46–55

Rehman MA (2002) Using mist-nets on canopy walkways in Malaysia to study canopy avifauna. Raff Bull Zool 50(2):499–506

Schaefer HM, Schmidt V, Winkler H (2003) Testing the defence trade-off hypothesis: how contents of nutrients and secondary compounds affects fruit removal. Oikos 102:318–328

Shaw D, Flick C (1999) Are resident songbirds stratified within the canopy of a coniferous old-growth forest? Selbyana 20:324–331

Shaw DC, Freeman EA, Flick C (2002) The vertical occurrence of small birds in an old-growth Douglas-fir-western hemlock forest stand. Northwest Sci 76(4):322–334

Sherry TW (1984) Comparative dietary ecology of sympatric, insectivorous neotropical flycatchers (Tyrannidae). Ecol Monogr 54:313–338

Stokes AE, Schultz BB (1995) Mist netting birds from canopy platforms. Selbyana 16:144–146

Stokes AE, Schultz BB, Degraaf RM, Griffin CR (2000) Setting mist nets from platforms in the forest canopy. J Field Ornithol 71:57–65

Sutton SL (2001) Alice grows up: canopy science in transition from wonderland to reality. Plant Ecol 153:13–21

Van Bael S, Brawn J, Robinson S (2003) Birds defend trees from insect herbivores in a Neotropical forest canopy. Proc Natl Acad Sci 100:8304–8307

Vivek R, Ganesh T (2012) Habitat structure and its effects on bird assemblages in the Kalakad-Mundanthurai Tiger Reserve, India. JBNHS 109(1&2):87–95

Walther BA (2000) Fruit size and frugivore species richness: additional evidence from observations at a large Ficus tree. Ecotropica 6:197–201

Walther BA (2002a) Vertical stratification and use of vegetation and light habitats by Neotropical forest birds. J Ornithol 143:64–81

Walther BA (2002b) Grounded ground birds and surfing canopy birds: variation of foraging stratum breadth observed in Neotropical forest birds and tested with simulation models using boundary constraints. Auk 119:658–675

Walther BA (2003) Why canopy access is essential to understand Canopy birds: four examples from the Surumoni Crane project. Ornithol Neotrop 15:41–52

Webber MI (1975) Some aspects of the nonbreeding population dynamics of the Great Tit (Parus major), Appendix 1. Ph.D thesis, University of Oxford, Oxford

Whitaker AH (1972) An improved mist net rig for use in forests. Bird Band 43:1–8

Winkler H, Preleuthner M (2001) Behaviour and ecology of birds in tropical rain forest canopies. Plant Ecol 153:193–202

Chapter 21
Functional Roles of Lianas for Forest Canopy Animals

Stephen P. Yanoviak and Stefan A. Schnitzer

Keywords Ants • Connectivity • Formicidae • Panama • Resources • Structure • Tropics

Bullet Points

1. Lianas provide nutritional and structural resources for consumers and are a key source of physical connectivity within the canopy.
2. Ongoing experiments in Panama suggest that physical connectivity is an important determinant of local arboreal ant species richness.
3. Understanding the functional ecology of lianas in the canopy will provide a basis for predicting the broader effects of increasing liana abundance.

Summary

Lianas (woody vines) are key components of tropical forest canopies. They represent a large fraction of total stem density and upper canopy foliage, and have important effects on forest dynamics. Liana abundance is increasing in neotropical forests, which could have far-reaching community- and ecosystem-level effects. Accurately predicting the consequences of this change depends on a clear understanding of the functional roles of lianas. Clearly lianas are detrimental to tropical trees via mechanical loading and competition, but less is known regarding their effects on arthropod

S.P. Yanoviak (✉)
Department of Biology, University of Louisville, Louisville, KY 40292, USA
e-mail: steve.yanoviak@louisville.edu

S.A. Schnitzer
Department of Biological Sciences, University of Wisconsin – Milwaukee, Milwaukee, WI 53201, USA
e-mail: S1@uwm.edu

M. Lowman et al. (eds.), *Treetops at Risk: Challenges of Global Canopy Ecology and Conservation*, DOI 10.1007/978-1-4614-7161-5_21,

diversity in the canopy. This chapter briefly reviews the functional ecology of lianas and summarizes evidence that the inter-tree physical connectivity provided by lianas is an important determinant of local canopy ant diversity. Quantifying ecological associations between lianas and other canopy inhabitants is essential for understanding how ongoing changes in forest structure will affect tropical forest canopy biodiversity, and community and ecosystem processes.

1 Introduction

Lianas are climbing plants with relatively long, slender, woody stems rooted in soil and extending to the forest canopy, where they produce abundant foliage. Like "tree" or "shrub," "liana" refers to a polyphyletic functional group that exhibits considerable structural diversity among taxa (Schnitzer and Bongers 2002). Hundreds of species of lianas exist worldwide, and the liana growth form is represented in nearly all major plant families (Putz and Mooney 1991).

Lianas are among the most conspicuous structural elements of tropical forests. Mature individuals range in length from a few meters to more than half a kilometer and in diameter from a few millimeters to more than half a meter. In lowland tropical forests, lianas commonly represent > 25 % of the rooted woody stems, 35 % of the woody species, and up to 40 % of the foliage area of the upper canopy (Putz and Mooney 1991; Schnitzer and Bongers 2002; Schnitzer et al. 2012). Recent and ongoing studies indicate that liana abundance is increasing in neotropical forests due to multiple, potentially interrelated factors that are partly associated with climatic change (Schnitzer and Bongers 2011; Schnitzer et al. 2011). Consequently, lianas are rapidly emerging as key elements of present and future forest dynamics (Schnitzer et al. 2014).

2 Functional Roles of Lianas

The large, rapidly expanding literature regarding the biology of lianas and their ecological relevance in forest ecosystems is comprehensively reviewed elsewhere (Putz and Mooney 1991; Schnitzer et al. 2014). Briefly, lianas are agents of tree mortality; they influence forest dynamics and patterns of forest regeneration via competition and mechanical parasitism on trees (Schnitzer and Bongers 2002, 2011). At the ecosystem scale, lianas affect gross transpiration rates, forest productivity, and carbon budgets (Schnitzer and Bongers 2002, 2011). Whereas most studies view the functional roles of lianas in forests from the perspective of their conspicuous negative effects on trees, lianas also provide important resources for animals and likely play a role in shaping the evolution of arboreal lifestyles

(Emmons and Gentry 1983; Kilgore et al. 2010). However, very little information exists concerning interactions between lianas and animals.

2.1 Lianas as Food

Liana foliage is an important food source for large herbivores in forest canopies (e.g., Dunn et al. 2012), and the seeds and fruits of some taxa (e.g., *Tontelea, Strychnos*) are opportunistically consumed by primates and other vertebrates both in the canopy and on the ground (e.g., Croat 1978). In contrast, information regarding herbivory and frugivory on lianas by arthropods is relatively limited. This is a significant gap in our understanding of trophic dynamics in tropical forests, given that arthropods are the most diverse and widespread herbivores in tropical forests (Basset et al. 2012). The best studied examples of arthropod herbivory on lianas include *Heliconius* butterflies feeding on *Passiflora* and ithomiine butterflies feeding on lianas in the family Solanaceae (see Ødegaard 2000 and references therein). Arthropod inventories conducted at various tropical locations provide strong correlative evidence that lianas partly determine local arthropod community structure in forest canopies (e.g., Dial et al. 2006). The mechanisms for this pattern remain largely unexplored, with the exception of Ødegaard's (2000) results showing that host specificity for lianas among phytophagous beetles increases local species richness in Panama.

Apart from leaf, fruit, and seed resources, lianas also provide pollen, nectar, and extrafloral nectar that are attractive to a broad range of consumers. Extrafloral nectar is a particularly important food source for aggressive, behaviorally dominant arboreal ants, and the distribution of carbohydrates derived from lianas and their herbivores may determine local ant community structure (reviewed in Rico-Gray and Oliveira 2007). Despite a wealth of literature pertaining to ant-plant associations and ant ecology in general (e.g., Hölldobler and Wilson 1990), many fundamental questions regarding ant-liana trophic interactions remain unanswered.

2.2 Lianas as Structure

Lianas contribute to the structural heterogeneity of tropical forests and expand the availability of physical resources for animals. Specifically, liana foliage provides shelter for a wide range of arboreal taxa, liana stems provide perches and grips for volant and brachiating species, and all liana surfaces provide foraging substrates for insect gleaners. The ecological importance of these functional contributions is revealed by differences in vertebrate assemblages between forests with and without abundant lianas (e.g., Emmons and Gentry 1983). Structural contributions of lianas appear to be particularly important to the maintenance of understory bird diversity and are likely to facilitate the presence of phyllostomids and other bats that feed by gleaning.

Many of the nutritional and structural resources described above are also provided by trees and epiphytes, but lianas are the only source of widespread, persistent connections between tree crowns in the forest canopy. Despite the complex, highly interconnected appearance of tropical forest canopies, continuous physical contact between leaves and branches of neighboring canopy trees is rare. Trees generally maintain a narrow, leaf-free gap between neighboring crowns, commonly called "crown shyness" (Ng 1977). Crown shyness is most evident in even-aged monotypic stands, but is equally common in dense primary forest. The connectivity provided by lianas overcomes crown shyness and influences the foraging activities and general mobility of nonvolant canopy animals, including mammals (Emmons and Gentry 1983). As with many other aspects of forest structure, measuring connectivity is logistically and quantitatively challenging. However, advances in remote-sensing technology (e.g., LiDAR) are quickly making such measurements feasible at ecologically relevant scales.

2.3 *Lianas and Canopy Ants*

Among invertebrates, ants are the most conspicuous users of liana structure; they commonly incorporate lianas into their foraging trails (Clay et al. 2010; Yanoviak et al. 2012) and often nest in hollow liana stems (e.g., Ward 1989). Some Asian and African ants have quasi-mutualistic associations with lianas, including *Cladomyrma* on *Spatholobus* and *Tetraponera* on *Vitex* (Djiéto-Lordon et al. 2005). In contrast, occupancy of lianas by neotropical twig-nesting ants appears to be entirely opportunistic. Foraging ants may preferentially use lianas to access patchy resources (Clay et al. 2010), and field observations suggest that the narrow, often smooth structure of liana stems enhances ant locomotion and predator avoidance relative to tree surfaces (Yanoviak, unpublished).

Inter-crown connectivity should be particularly important to arboreal ants. In the absence of lianas, access to resources in multiple tree crowns can only be accomplished by descending to the forest floor and traversing the leaf litter, which is hazardous and inefficient (e.g., Yanoviak et al. 2011). Thus, the connectivity provided by lianas is likely to be a key structural resource for ants. We are testing this hypothesis via a large, ongoing liana removal experiment in Panama. We surveyed canopy ants in sixteen 80 × 80 m experimental plots before and after liana extermination, then added structural connectivity in the form of used climbing ropes suspended between tree crowns. Ants began using the added connectivity almost immediately (Fig. 21.1). Preliminary results show that liana removal reduced average ant species richness in tree crowns by 22 % relative to controls (i.e., plots with intact lianas). The addition of connectivity partly counteracted this effect and increased local (per tree) ant species richness in control plots by ca. 25 %. Collectively, these results indicate that the physical connectivity provided by lianas is an important determinant of local ant diversity in the canopy. Our ongoing studies are exploring the mechanisms for these patterns.

Fig. 21.1 A worker ant, *Camponotus senex*, uses an artificial connection (old climbing rope) to travel between isolated tree crowns. (Photo by S.P. Yanoviak)

3 Conclusions

Increased liana abundance (Schnitzer and Bongers 2011) is one of many factors that will affect the structure of tropical forests over the next few decades (e.g., Wright 2005). Predicting the effects of such changes on forest ecosystems requires a clear understanding of the specific contributions of lianas to canopy biodiversity and other ecological parameters. Lianas are functionally redundant with trees and epiphytes in some respects, but their ecological role as inter-tree connectors clearly distinguishes them from other canopy components. Preliminary results from large-scale experiments show that connectivity is a key resource for small, cursorial organisms like arboreal ants. Similar studies are needed to determine the specific contributions of lianas to the diversity of other major taxa, especially arboreal vertebrates and herbivorous arthropods.

References

Basset Y, Cizek L, Cuénoud P, Didham RK, Guilhaumon F, Missa O, Novotny V, Ødegaard F, Roslin T, Schmidl J, Tishechkin AK, Winchester NN, Roubik DW, Aberlenc HP, Bail J, Barrios H, Bridle JR, Castaño-Meneses G, Corbara B, Curletti G, da Rocha WD, Bakker DD, Delabie JHC, Dejean A, Fagan LL, Floren A, Kitching RL, Medianero E, Miller SE, Oliveira EG, Orivel O, Pollet M, Rapp M, Ribeiro SP, Roisin Y, Schmidt JB, Sørensen L, Leponce M (2012) Arthropod diversity in a tropical forest. Science 338:1481–1484

Clay NA, Bauer M, Solis M, Yanoviak SP (2010) Arboreal substrates influence foraging in tropical ants. Ecol Entomol 35:417–423

Croat TB (1978) Flora of Barro Colorado Island. Stanford University Press, Stanford

Dial RJ, Ellwood MDF, Turner EC, Foster WA (2006) Arthropod abundance, canopy structure, and microclimate in a Bornean lowland tropical rain forest. Biotropica 38:643–652

Djiéto-Lordon C, Dejean A, Ring RA, Nkongmeneck BA, Lauga J, McKey D (2005) Ecology of an improbable association: the pseudomyrmecine plant-ant *Tetraponera tessmanni* and the myrmecophytic liana *Vitex thyrsiflora* (Lamiaceae) in Cameroon. Biotropica 37:421–430

Dunn JC, Asensio N, Arroyo-Rodríguez V, Schnitzer S, Cristóbal-Azkarate J (2012) The ranging costs of a fallback food: liana consumption supplements diet but increases foraging effort in howler monkeys. Biotropica 44:705–714

Emmons LH, Gentry AH (1983) Tropical forest structure and the distribution of gliding and prehensile-tailed vertebrates. Am Nat 121:513–524

Hölldobler B, Wilson EO (1990) The ants. Harvard University Press, Cambridge, MA

Kilgore A, Lambert TD, Adler GH (2010) Lianas influence fruit and seed use by rodents in a tropical forest. Trop Ecol 51:265–271

Ng FSP (1977) Shyness in trees. Nat Malays 2:35–37

Ødegaard F (2000) The relative importance of trees versus lianas as hosts for phytophagous beetles (Coleoptera) in tropical forests. J Biogeogr 27:283–296

Putz FE, Mooney HA (eds) (1991) The biology of vines. Cambridge University Press, Cambridge, UK

Rico-Gray V, Oliveira PS (2007) The ecology and evolution of ant-plant interactions. University of Chicago Press, Chicago

Schnitzer SA, Bongers F (2002) The ecology of lianas and their role in forests. Trends Ecol Evol 17:223–230

Schnitzer SA, Bongers F (2011) Increasing liana abundance and biomass in tropical forests: emerging patterns and putative mechanisms. Ecol Lett 14:397–406

Schnitzer SA, Bongers F, Wright SJ (2011) Community and ecosystem ramifications of increasing lianas in neotropical forests. Plant Signal Behav 6:598–600

Schnitzer SA, Mangan SA, Dalling JW, Baldeck C, Hubbell SP, Ledo A, Muller-Landau H, Tobin M, Aguilar S, Brassfield D, Hernandez A, Lao S, Perez R, Valdez O, Yorke SR (2012) Liana abundance, diversity, and distribution on Barro Colorado Island, Panama. PLoS One 7(12): e52114

Schnitzer SA, Bongers F, Burnham R, Putz FE (eds) (2014) The ecology of lianas. Wiley-Blackwell. Oxford, UK

Ward PS (1989) Systematic studies on pseudomyrmecine ants: revision of the *Pseudomyrmex oculatus* and *P. subtilissimus* species groups, with taxonomic comments on other species. Quaest Entomol 25:393–468

Wright SJ (2005) Tropical forests in a changing environment. Trends Ecol Evol 20:553–560

Yanoviak SP, Munk Y, Dudley R (2011) Evolution and ecology of directed aerial descent in arboreal ants. Integr Comp Biol 51:944–956

Yanoviak SP, Silveri C, Hamm CA, Solis M (2012) Stem characteristics and ant body size in a Costa Rican rain forest. J Trop Ecol 28:199–204

Chapter 22
Islands in a Sea of Foliage: Mistletoes as Discrete Components of Forest Canopies

Anna E. Burns and David M. Watson

Keywords Mistletoe • Host-plant • Insects • Spiders • Ecosystem health • Habitat fragmentation

Bullet Points

1. Mistletoes and their dependent insects, such as butterflies, moths, and psyllids, are sensitive indicators of disturbance regime and overall forest integrity.
2. The health of the whole forest ecosystem is important for maintaining populations of insects in the canopy because different insect life stages depend on different types of plants.
3. Herbivorous insects and mammals, along with fire, help to regulate mistletoe abundance.

Summary

This chapter focuses on the ecological interactions between mistletoe plants and their invertebrate inhabitants (insects and spiders), drawing on our own empirical research in southeastern Australia and several other studies conducted worldwide. Our research on box mistletoes and their host eucalypt trees in remnant woodlands was the first direct comparison of invertebrate fauna inhabiting mistletoes and their host plants (Burns 2009; Burns et al. 2011). We discovered the occurrence of

A.E. Burns (✉)
Department of Zoology, La Trobe University,
Bundoora, VIC 3086, Australia
e-mail: aeburns78@gmail.com

D.M. Watson
School of Environmental Sciences, and Institute for Land, Water and Society,
Charles Sturt University, PO BOX 789, 2640, NSW, Australia
e-mail: dwatson@csu.edu.au

M. Lowman et al. (eds.), *Treetops at Risk: Challenges of Global Canopy Ecology and Conservation*, DOI 10.1007/978-1-4614-7161-5_22,

distinct assemblages of insects on mistletoes, which are dependent on mistletoes for their entire lives. Spiders were found to be generalists on mistletoes and their host trees. These invertebrates represent important components of the food web in forest and woodland canopies.

1 Introduction

Mistletoes are a functional category of plants, defined as shrubby, aerial hemiparasites which depend on their host plants for water and mineral nutrition. Although three aberrant species (all in monotypic genera) are root parasitic, mistletoes are stem parasites, attaching to their hosts via specialized organs called haustoria. Mistletoes belong to the order Santalales and are arranged in the families Loranthaceae, Viscaceae, Misodendraceae, and Santalaceae, with the great majority of species belonging to the first two families (Nickrent 2001; Nickrent et al. 2010). Although some species regularly parasitize lianas and other mistletoes, most species are dependent on trees and shrubs as principal hosts. The more than 1,500 species worldwide live in diverse habitats from rainforest to semiarid woodlands and are absent only from habitats devoid of woody hosts (e.g., polar, alpine and desert environments). Mistletoes are an important food source and nesting site for many birds and mammals (Cooney et al. 2006; Mathiasen et al. 2008; Watson 2001) and have a positive effect on the diversity and distribution of vertebrate animals in a range of habitats (Mathiasen et al. 2008; Watson 2002). By comparison, the ecological interactions between mistletoes and invertebrates—particularly arthropods—are poorly known. Unlike birds and mammals—long-lived and highly mobile animals that visit mistletoes periodically—many insects live their entire lives within mistletoe clumps, completely dependent on them for food and shelter. In this contribution, we summarize recent research on mistletoe-dependent arthropods, contrasting the extreme specialism exhibited by herbivorous groups, with lower substrate specificity (and greater dependence on structural complexity) displayed by predatory taxa. Rather than simply a subset of the biota found in the host canopy, we demonstrate that arthropods in mistletoes represent discrete and complementary assemblages, hitherto overlooked islands within a sea of forest and woodland treetops. Finally, we consider these findings in terms of the threats facing many forested systems, demonstrating the utility of mistletoe-dependent arthropods as sensitive indicators of overall forest health and ecosystem integrity.

2 Diversity of Invertebrates on Mistletoes

The recorded diversity of invertebrates occurring on mistletoes includes more than ten orders of arthropods including beetles (Coleoptera); bugs (Hemiptera); wasps, ants, and bees (Hymenoptera); butterflies and moths (Lepidoptera); thrips (Thysanoptera); flies (Diptera); bark lice (Psocoptera); cockroaches (Blattodea); praying mantis

(Mantodea); grasshoppers (Orthoptera); lace wings (Neuroptera); mites (Acari); and spiders (Araneae) (Anderson and Braby 2009; Baloch and Mohyuddin 1969; Burns et al. 2011; French 2004; Robertson et al. 2005; Room 1972; Tassone and Majer 1997; Whittaker 1982). Some of the animals in these groups are herbivorous, feeding on the foliage, flowers, fruit, or stems; others are predators of insects or scavengers that feed on fungi or dead plant matter. Thus, some invertebrates on mistletoes are specialists that particularly seek out these plants and are dependent on them for their survival, while others are canopy generalists, occurring on mistletoes haphazardly or opportunistically for food and/or shelter. Some examples of specialist and generalist invertebrates inhabiting mistletoes are examined in detail below.

2.1 Specialists: Psyllids

Also known as lerp insects, these sap-sucking bugs (in the order Hemiptera) resemble small cicadas but are more similar in size to aphids (1–10 mm in length). The name "lerp" refers to the shell-like covering produced by the developing insects of some species (Fig. 22.1a), which acts as a shelter and may help maintain a humid microenvironment. These structures consist of sugar compounds excreted by the larvae and come in many shapes and sizes, some with elaborate decorations, which can be diagnostic tools for identification of species (Hollis 2004). Both the lerps and larvae inside them are a food source for birds (Lockwood and Gilroy 2004; te Marvelde et al. 2009). These insects can occur in outbreak proportions and cause severe foliar damage, which is particularly detrimental to the host plants' health in drought conditions (Clark and Dallwitz 1974; Semple and Koen 2007).

These insects are also known as jumping plant lice or psyllids that belong to the superfamily Psylloidea (suborder Sternorrhyncha). More than 3,000 species have been identified to date, worldwide, and they are very host-plant specific; most occur on one or a few closely related plant species and can only complete their whole life cycle on plants of those species (Hollis 2004). Accordingly, the species that occur on mistletoes are different from those that live on the mistletoes' host plants. Three psyllid species have been identified inhabiting species of *Amyema* mistletoes in Australia (Taylor 1999). In our study comparing the insect communities on box mistletoe, *Amyema miquelii*, and some of its host *Eucalyptus* species (Burns 2009; Burns et al. 2011), we found completely different assemblages of psyllids on the two plant types. While two species (*Acizzia loranthaceae* and *A. amyemae;* see Fig. 22.1b) were found to inhabit box mistletoe across the whole study area, 17 species inhabited the three eucalypt species.

Psyllid species that inhabit mistletoes are more closely related to those inhabiting *Acacia* than *Eucalyptus* species (Taylor 1999; Yen 2002). This may indicate an ancient radiation of psyllids from *Acacia*, which are mostly understory plants, to mistletoes on *Acacia* and thence to mistletoes in the canopy of *Eucalyptus* trees (Taylor 1999), potentially coinciding with the spread of eucalypts throughout Australia as rainforests retreated (White 1994).

Fig. 22.1 (**a**) Lerp of a psyllid species on a *Eucalyptus* leaf. (**b**) An adult psyllid of *Acizzia amyemae* on a box mistletoe (*Amyema miquelii*) leaf (Images: Anna Burns)

2.2 Generalists: Spiders

Spiders are one of the most abundant and speciose groups of invertebrate animals that occur on mistletoes (Anderson and Braby 2009; Burns et al. 2011; Room 1972), and their occurrence is seemingly opportunistic. Both our study of box mistletoe and its host eucalypts and that by Jennings et al. (1989) of dwarf mistletoes (*Arceuthobium* spp.) and their conifer hosts showed that similar species of spiders inhabit mistletoes and their hosts. Specifically, 50 % of the 42 species of spiders found on box mistletoes and the eucalypts occurred on both mistletoes and their host trees. Most of the spiders in our study were less than 10 mm in size and belonged to six families, including orb weavers (Araneidae), jumping spiders (Salticidae), crab spiders (Thomisidae), comb-footed spiders (Theridiidae), and lynx spiders (Oxypidae; see Fig. 22.2). The spiders and insects were both more abundant on the eucalypts than on mistletoes, indicating that the distribution of spiders was likely influenced by the abundance of potential prey. Habitat structure

Fig. 22.2 A lynx spider protecting its egg sac on box mistletoe (Image: Anna Burns)

is also an important factor that influences the distribution and composition of spider communities (Foelix 1982). Examination of the microclimate and specific substrates within mistletoe clumps, such as the haustorium (connection with the host plant) and folded leaves or other leaf structures, could reveal habitat-specific spiders within mistletoe clumps, similar to the assemblages of spiders that inhabit bromeliads (Romero 2006).

3 Indicators of Forest Health: Butterflies, Moths, and Mistletoes

Butterflies and moths (order Lepidoptera) form associations with mistletoes for food and shelter. Their larvae feed on mistletoe foliage and often live in colonies on mistletoe or its host plant until completion of pupation (Braby 2000, 2004; De Baar 1985). In Australia, 25 species of butterflies (in families Pieridae and Lycaenidae) and at least four species of moths (in families Saturniidae, Agaristidae, Lymantriidae, and Noctuidae; see Fig. 22.3) feed on a range of mistletoe species, either exclusively or as a large proportion of their diet (Braby 2000; De Baar 1985). In the Asian region, 20 species of *Delias* (Pieridae) have been recorded feeding on mistletoes in the families Loranthaceae and Viscaceae (Braby 2006); in Africa, 14 species of *Mylothris* (Pieridae) feed on mistletoes (Braby 2005); and five Lepidopteran species

Fig. 22.3 Caterpillar of mistletoe day moth, *Comocrus behri*, on box mistletoe (Image: Catherine Bach)

feed on dwarf mistletoes in the USA, including the rare Johnson's (or mistletoe) hairstreak (McCorkle 1962; Mooney 2003). Many of these caterpillars are completely dependent on mistletoe as a food source, and thus, as important herbivores, they help regulate mistletoe abundance. Due to their holometabolous life cycle (i.e., complete metamorphosis, with larvae and adults dependent on different food sources), the occurrence and abundance of these insects can act as sensitive indicators of overall habitat health and integrity.

Mammalian herbivores also contribute to regulation of mistletoe abundance, along with fire in some habitats (Kelly et al. 1997; Mathiasen et al. 2008; Parker et al. 2006). It is when these natural disturbances, along with insect herbivores, are eliminated from or reduced in ecosystems that mistletoes can become overabundant and impair the health of their host trees (Mathiasen et al. 2008; Reid et al. 1995). Maintenance of the habitat requirements of herbivores is thus important for sustainable abundance of mistletoes in forests and woodlands (Watson 2011). For example, an adequate supply of the food source of adult moths and butterflies (e.g., understory nectar-bearing plants) is required to maintain healthy populations of hungry caterpillars. However, in modified and degraded landscapes, understory plants often decline. Habitat degradation and fragmentation also affect populations of mammalian herbivores (e.g., possums, koalas, and deer), rendering them less effective control agents, and can alter mistletoe dispersal patterns, leading to overabundance in isolated trees (Reid 1997). On the other hand, when mistletoes become too scarce, many of their dependent animals are affected, often leading to local extinction (McCorkle 1962; Hawksworth and Wiens 1996). Considering these factors, we propose that mistletoes and their dependent arthropods are sensitive indicators of disturbance regime and overall forest health.

References

Anderson SJ, Braby MF (2009) Invertebrate diversity associated with tropical mistletoe in a suburban landscape from northern Australia. North Territ Nat 21:2–23

Baloch GM, Mohyuddin AI (1969) The phytophagous fauna of a mistletoe (*Loranthus longiflorus* Desr.: Loranthaceae) in West Pakistan. Weed Res 9:62–64

Braby MF (2000) Butterflies of Australia: their identification, biology and distribution. CSIRO, Melbourne

Braby MF (2004) The complete field guide to butterflies of Australia. CSIRO, Melbourne

Braby MF (2005) Afrotropical mistletoe butterflies: larval food plant relationships of *Mylothris* Hubner (Lepidoptera: Pieridae). J Nat Hist 39:499–513

Braby MF (2006) Evolution of larval food plant associations in *Delias* Hubner butterflies (Lepidoptera: Pieridae). Entomol Sci 9:383–398

Burns AE (2009) Diversity and dynamics of the arthropod assemblages inhabiting mistletoe in eucalypt woodlands. Ph.D. thesis, Charles Sturt University, Australia

Burns AE, Cunningham SA, Watson DM (2011) Arthropod assemblages in tree canopies: a comparison of orders on box mistletoe (*Amyema miquelii*) and its host eucalypts. Aust J Entomol 50:221–230

Clark LR, Dallwitz MJ (1974) On the relative abundance of some Australian Psyllidae that coexist on *Eucalyptus blakelyi*. Aust J Zool 22:387–415

Cooney SJN, Watson DM, Young J (2006) Mistletoe nesting in Australian birds: a review. Emu 106:1–12

De Baar M (1985) The complex mistletoe-insect community. Entomol Soc Qld Bull 13:100–102

Foelix RF (1982) Biology of spiders. Harvard University Press, Cambridge

French JA (2004) Ecological interactions between western hemlock dwarf mistletoe (*Arceuthobium tsugense* subsp. *tsugense*) and insects within an old-growth forest. Master of Science. University of Washington, Washington, DC

Hawksworth FG, Wiens D (1996) Dwarf mistletoes: biology, pathology, and systematics. Handbook, vol 709. Department of Agriculture, Forest Services and Agriculture, Washington, DC

Hollis D (2004) Australian Psylloidea: jumping plantlice and lerp insects. Australian Biological Resources Study, Canberra

Jennings DT, Penfield FB, Stevens RE, Hawksworth FG (1989) Spiders (Araneae) associated with dwarf mistletoes (*Arceuthobium* sp.) in Colorado. Southwest Nat 34:349–355

Kelly P, Reid N, Davies I (1997) Effects of experimental burning, defoliation, and pruning on survival and vegetative resprouting in mistletoes (*Amyema miquelii and Amyema pendula*). Int J Plant Sci 158:856–861

Lockwood JL, Gilroy JJ (2004) The portability of foodweb dynamics: reassembling an Australian eucalypt-psyllid-bird association within California. Glob Ecol Biogeogr 13:445–450

Mathiasen RL, Nickrent DL, Shaw DC, Watson DM (2008) Mistletoes: pathology, systematics, ecology, and management. Plant Dis 92:988–1006

McCorkle DV (1962) Notes on the life history of *Callophrys* (Mitoura) *johnsoni* Skinner (Lepidoptera, Lycaenidae). Proc Wash State Entomol Soc 14:103–105

Mooney KA (2003) *Promylea lunigerella glendella* Dyar (Pyralidae) feeds on both conifers and parasitic dwarf mistletoe (*Arceuthobium* spp.): one example of food plant shifting between parasitic plants and their hosts. J Lepidopter Soc 57:47–53
Nickrent DL (2001) Santalales (Mistletoe). In: Encyclopedia of Life Sciences. Article A3714. Macmillan, New York
Nickrent DL, Malecot V, Vidal-Russell R, Der JP (2010) A revised classification of Santalales. Taxon 59(2):538–558
Parker TJ, Clancy KM, Mathiasen RL (2006) Interactions among fire, insects and pathogens in coniferous forests of the interior western United States and Canada. Agric Forest Entomol 8:167–189
Reid N (1997) Control of mistletoes by possums and fire: a review of the evidence. Vic Nat 114:149–158
Reid N, Smith SMM, Yan Z (1995) Ecology and population biology of mistletoes. In: Lowman MD, Nadkarni NM (eds) Forest canopies. Academic, San Diego
Robertson AW, Ladley JJ, Kelly D (2005) Effectiveness of short-tongued bees as pollinators of apparently ornithophilous New Zealand mistletoes. Aust Ecol 30:298–309
Romero GQ (2006) Geographic range, habitats, and host plants of bromeliad-living jumping spiders (Salticidae). Biotropica 38:522–530
Room PM (1972) The constitution and natural history of the fauna of the mistletoe *Tapinanthus bangwensis* (Engl. & K. Krause) growing on cocoa in Ghana. J Anim Ecol 41:519–535
Semple WS, Koen TB (2007) Observations of insect damage to leaves of woodland eucalypts on the central western slopes of New South Wales: 1990 to 2004. Proc Linn Soc NSW 128:99–110
Tassone RA, Majer JD (1997) Abundance of arthropods in tree canopies of *Banksia* woodland on the Swan Coastal Plain. J R Soc West Aust 80:281–286
Taylor GS (1999) New species of Acizzia Heslop-Harrison (Hemiptera: Psyllidae) from Australian mistletoe (Loranthaceae). Aust J Entomol 38:66–71
te Marvelde L, McDonald PG, Kazem AJN, Wright J (2009) Do helpers really help? Provisioning biomass and prey type effects on nestling growth in the cooperative bell miner. Anim Behav 77(3):727–735
Watson DM (2001) Mistletoe – a keystone resource in forests and woodlands worldwide. Annu Rev Ecol Syst 32:219–249
Watson DM (2002) Effects of mistletoe on diversity: a case-study from southern New South Wales. Emu 102:275–281
Watson DM (2011) Mistletoes of Southern Australia. CSIRO, Collingwood
White ME (1994) The greening of Gondwana, 2nd edn. Reed, Chatswood
Whittaker PL (1982) Community ecology of *Phoradendron tomentosum* in southern Texas. Ph.D. thesis, University of Texas, Austin
Yen AL (2002) Short-range endemism and Australian Psylloidea (Insecta: Hemiptera) in the genera *Glycaspis* and *Acizzia* (Psyllidae). Invert Syst 16:631–636

Chapter 23
Nonvascular Epiphytes: Functions and Risks at the Tree Canopy

Michael Lakatos and Alexandra Fischer-Pardow

Keywords Cryptogams • Poikilohydry • Microclimate • Desiccation tolerance • Canopy access • Research tower • Bryophytes

1 Research on Nonvascular Epiphytes

In the frame of current discussions on the value of forest canopies, a short introduction and recent advances of nonvascular epiphyte (NVE) research are presented here regarding a novel canopy access facility to study mechanisms and functions of carbon and water exchange as well as impacts of climate and land-use change on NVE.

Sometimes inconspicuous, but highly diverse and omnipresent in tree tops are microorganisms and nonvascular plants occupying plant surfaces (Lakatos 2011). The communities are composed of bacteria, fungi, cyanobacteria, lichens, algae, and bryophytes in variable proportions. The first four groups are generally categorized as microbes, whereas the latter are regarded as cryptogams and often subsumed as nonvascular epiphytes (NVE) due to the dominance of photoautotrophic organisms and the lack of lignified vascular water conduction system. They cover almost every spot of bark (corticolous) and develop from thin green biofilms comprising mainly cyanobacteria, algae, and fungi with a thickness of less than one millimeter to well visible and colorful epiphytic cryptogamic crusts or mats of several centimeters, dominated by lichens and bryophytes (Fig. 23.1). Biofilms also occur as epiphylls on the surface of long-living and evergreen leaves in the phyllosphere (Coley et al. 1993; Furnkranz et al. 2008; Sonnleitner et al. 2009; Rigonato et al. 2012). As corticolous and epiphyllic photoautotrophic communities, they influence important ecological processes

M. Lakatos (✉) • A. Fischer-Pardow
Department of Plant Ecology and Systematics, University of Kaiserslautern, 3049, Kaiserslautern 67653, Germany
e-mail: lakatos@rhrk.uni-kl.de; apardow@rhrk.uni-kl.de

M. Lowman et al. (eds.), *Treetops at Risk: Challenges of Global Canopy Ecology and Conservation*, DOI 10.1007/978-1-4614-7161-5_23,

Fig. 23.1 Colorful lichen assemblage in the Mata Atlantica, Brazil (*left*), and canopy studies on bryophytes in the lowland cloud forest of French Guiana

such as carbon, nitrogen, and water cycles (reviewed: Rhoades 1995; Sillett and Antoine 2004; Elbert et al. 2012); represent a large pool of species diversity (e.g., Mandl et al. 2010; Sporn et al. 2010; Ellis 2012); and effect trophic cascades as a food resource and as hosts for microbes, protists, and invertebrates in the phytosphere and phyllosphere. Despite their omnipresence and numerous roles occupied in the canopy, researchers have focused on macroscopic vascular plants and animals, whereas the direct and indirect impacts of NVE remain relatively understudied.

Most investigations on NVE are restricted to the tree base or lower stem of trees although several studies indicate that this limitation may underestimate species diversity by 30 % (for lichens in temperate forests; John and Schröck 2001) or even 50 % (for bryophytes in tropical forests; Cornelissen and Gradstein 1990b; Wolf 1995; Costa 1999; Gradstein et al. 2001). In entire vertical assessments of NVE assemblages, fallen trees were often examined (Jarman and Kantvilas 1995; Fritz 2009). However, epiphytic surveys for cyanobacteria are scarce (12 species in the tropical lowland forest, Singapore, Neustupa and Škaloud 2010) as well as for green algae, which range between 20 species (temperate forest, Germany, Freystein et al. 2008) and 40 species (tropical: Neustupa and Škaloud 2010; Lemes-Da-Silva et al. 2010) mainly from the families of Chlorophyceae, Trebouxiophyceae, and Ulvophyceae. Diatoms are common as epiphytes on epiphytic lichens (18 species in the tropical *Coenogonium linkii*; Lakatos et al. 2004) and also on mosses (Foerster 1971). Lichen diversity ranges from 36 to 76 species and bryophyte diversity from 28 to 55 species in Australia (Jarman and Kantvilas 1995; Milne and Louwhoff 1999), but with values of 88–100 bryophyte species reported in tropical lowland forests of Guiana and French Guiana (Cornelissen and Gradstein 1990; Gradstein 1995) and 153–190 bryophyte species in tropical montane forests (Wolf 1993; Gradstein et al. 2001). One single tree (*Elaeocarpus*) in Papua New Guinea boasted 173 lichen species (Aptroot 1997). Such high species diversity is one of the reasons that most studies on NVE in the canopy occur in tropical canopies, using single rope techniques (e.g., Nadkarni 1984; Wolf 1995; Freiberg and Freiberg 2000;

Nadkarni et al. 2004; Holz and Gradstein 2005; Gehrig-Downie et al. 2011), walkways, aluminum towers (e.g., Zotz and Winter 1994), and construction cranes (e.g., Komposch and Hafellner 2000). These studies mainly documented species distribution, cover, and biomass of NVE. In temperate and boreal regions, several studies documented biomass and species distribution on canopies largely consisting of conifers (e.g., McCune et al. 1997; Clement and Shaw 1999; Ellyson and Sillett 2003; Williams and Sillett 2007). One constraint of canopy research is the limitation of noninvasive accessibility that permits studying organisms in their natural undisturbed environment. This might explain why ecophysiological and long-term studies on NVE in the canopy are infrequent (but see Renhorn et al. 1997; Zotz and Schleicher 2003; Zotz et al. 2003; Antoine 2004; Lakatos et al. 2006; Romero et al. 2006; Gauslaa et al. 2012; Pardow and Lakatos 2013).

2 The "Biosphere Tower": A Novel Canopy Access for Long-Term Research on Microbes and Nonvascular Epiphytes

One critical issue of investigations on microbes and NVE is that most canopy access methods are invasive and harm the ensembles of bark-dwelling (corticolous) organisms, particularly for long-term studies that require repeated access. To reduce the destructive impact of access, a new canopy tower was developed to conduct long-term studies. The "Biosphere Tower" (Fig. 23.2) is a wooden canopy access

Fig. 23.2 A novel canopy access: the wooden 36 m tall Biosphere Tower – the Biosphere Reserve Palatinate Forest–Northern Vosges, Germany

tower with mobile cantilevers, installed in a representative old-growth (170 years) and commercial oak forest in the Biosphere Reserve Palatinate Forest–Northern Vosges, Germany. The construction has a basal area of 3×3 m plus 6 platforms each ascending 6 m reaching a total heigh of 36 m height. The structure surmounts the treetops and the cantilevers of 5–10 m in length reach each strata of adjacent oaks (*Quercus robur*) and beeches (*Fagus sylvatica*), which comprise economically the two most important broadleaf tree species in Europe. To emphasize abiotic canopy processes, a micrometeorological monitoring system was installed at the ground level, and both within and above the canopy, providing continuous measurements of abiotic conditions (e.g., light, temperature, humidity, turbulences), including UV and global radiation. All impregnation of the wood was avoided, because volatile substances may affect the target organisms of the epiphytic community. This interdisciplinary project is a partnership of the University of Kaiserslautern (Departments of Plant Ecology and Systematics, and Architecture), the Kaiserslautern forestry office, and the Rhineland-Palatinate Foundation for Innovation.

The Biosphere Tower facilitates long-term research on forest canopy biodiversity, ecology, and global climate change with the focus on microbial ecosystems. NVE are not often studied in silvicultural forests, despite their economic importance for wood production and water balance. Moreover, noninvasive long-term canopy research offers new insights into seasonality, succession, and long-term changes. Owing to its innovative design (renewable primary products, low-invasive construction method, microbial canopy research), the Tower will expand the novel field of microbial canopy ecology, promote ecologically and economically relevant long-term-research in the "silvicultural canopy biosphere," and provide an attractive platform for environmental education.

3 Mechanisms and Functions of Nonvascular Epiphytes

The mechanisms and ecological impacts of NVE are still poorly understood (Cornelissen et al. 2007). In some ecosystems, they contribute a substantial proportion to primary production (Rhoades 1995; Lakatos 2011; Elbert et al. 2012), influence nutrient fluxes (Forman 1975; Knops et al. 1996; Sillett and Antoine 2004; Clark et al. 2005), and promote animal life (Pettersson et al. 1995; Richardson et al. 2000; van der Wal 2006). Two main aspects involved in understanding the NVE mechanisms and functions are the exchange of carbon and water.

Carbon exchange of NVE is closely bound with water availability. NVE are desiccation-tolerant organisms suspending metabolism when dry and recover after hydration by liquid water or high humidity (Proctor and Tuba 2002). They are capable of efficiently exploiting several liquid water sources such as rain, fog, and dew which are absorbed by their whole "plant" surface to activate metabolism and photosynthesis. This independence on permanent water supply enables them to photosynthesize and grow at sites inhospitable for vascular plants. Given NVEs' poikilohydric strategy, they successfully inhabit almost all terrestrial habitats from the tropics to cold and hot deserts.

The success of their poikilohydric was catalogued by biomass distribution and by their role in ecosystem carbon fluxes. In general, a clear pattern of biomass distribution occurs in tropical forests: biomass increases with altitude ranging from 10 gm^{-2} (relative to branch surface area) at lowland forests to almost 3,000 gm^{-2} of bryophytes in upper montane forests at 3,700 m a.s.l. (Hofstede et al. 1993; Freiberg and Freiberg 2000). In contrast, the biomass of vascular epiphytes (usually between 400 and 900 gm^{-2}) seems not to increase with altitude (Freiberg and Freiberg 2000). Also, patterns of biomass distribution change within the tree – the biomass of NVE (dominated by bryophytes) decreases from the branches of the inner crown to the periphery of the canopy (Freiberg and Freiberg 2000; Romanski et al. 2011), whereas lichens and biofilms increase (Werner et al. 2012; M. Lakatos unpublished data). On a regional level, local studies revealed maximal NVE biomass of around 2,000–16,000 $kgha^{-1}$ relative to ground surface area in tropical montane forests (Coxson et al. 1992; Köhler et al. 2007; Werner et al. 2012) compared to boreal and coastal forests where NVE biomass of up to 4,220 $kgha^{-1}$ was reported (Rhoades 1981). On an ecosystem level, calculated NVE biomass per biome was calculated to range from 470 to 2,120 $kgha^{-1}$ relative to ground surface area (Elbert et al. 2012). A global extrapolation of NVE approximates ~3.1 Pg (1.4–6.5 Pg) dry biomass on the basis of an global epiphytic area of 30.57×10^{12} m^2 (Friedl et al. 2002). The calculated carbon uptake fluxes of NVE range from 5.8 gm^{-2} $year^{-1}$ in extratropical forests to 10 $gm^{-2}year^{-1}$ in tropical forests and correspond to an annual carbon net uptake of 0.45–0.95 $Pgyr^{-1}$ (Elbert et al. 2012). In an ecosystem comparison, the total global carbon net uptake for NVE of 1.4 $Pgyr^{-1}$ would be in the same range as that for tundra (1.83 $Pgyr^{-1}$) and desert ecosystem (1.72 $Pgyr^{-1}$) (Cleveland et al. 1999), accounting for 2.5 % of the total net primary production of terrestrial vegetation (56 $Pgyr^{-1}$; Zhao et al. 2005). By these rough calculations, NVE could compensate for almost half of the global annual carbon release from biomass burning (3.6 $Pgyr^{-1}$). Although these calculations are based on very few data, the function of NVE as part of global carbon fluxes nevertheless has been underestimated until now.

The contribution of NVE to the hydrological cycle is also often overlooked (Hölscher et al. 2004; Köhler et al. 2007). Poikilohydric NVE exploit nearly every available water sources and absorb rain, fog, dew, and vapor over their entire "plant" surface to activate metabolism and photosynthesis. Many lichens and bryophytes significantly alter water contents (WC) by as much as 250–400 % of dry weight (DW) in green algal lichens (Blum 1973; Rundel 1988), 600–2,000 % DW in cyanolichens (Lange et al. 1993), and 2,500 % DW in bryophytes (Proctor et al. 1998). Optimal photosynthetic capacity is achieved between full turgor and turgor loss of the cells, varying between life-forms and species from 100 % to 2,000 % DW (e.g., Proctor et al. 1998; Hajek and Beckett 2008). During desiccation, photosynthesis decreases until cessation is reached at low water potential (−22 MPa) with a critical water content of 30–70 % DW for bryophytes or even down at −38 MPa for lichens (Nash III et al. 1990) where WC ranges between 10 % and 20 % DW. For the reactivation of photosynthesis, green algae (Edlich 1936; Bertsch 1966) and green algal lichens require only relative humidity between 75 % and 85 % RH, corresponding to water potential of −37 to −22 MPa (Lakatos 2011). In contrast, it was assumed that cyanobacteria (Lange et al. 1994) and bryophytes need liquid water for photosynthetic reactivation (Green et al. 1994). However, recent studies on bryophytes demonstrated for almost all investigated tropical mosses and liverworts the recovery of PSII already at around 85 % relative humidity (RH) (−22 MPa; Pardow and Lakatos 2013). The exchange of vapor with the organism occurs quite fast in NVE. Equilibration experiments during desiccation at 60 % RH and traced by the stable isotope ^{18}O demonstrated total exchange of thallus water with vapor within only 2 h for the fruticose *Usnea filipendula* and within 3–4 h for other lichen growth forms (Hartard et al. 2009). In comparison, the epiphytic vascular plant *Tillandsia usneoides* needed days for total equilibrium and only under RH above 95 % (Helliker and Griffiths 2007). Due to this rapid equilibration, lichens and other NVE may serve as prospective long-term proxies for water sources (Hartard et al. 2009). Moreover, their use as indicators for relative humidity and precipitation is important (Pardow et al. 2010; Obregon et al. 2011; Karger et al. 2012; Pardow and Lakatos 2013). The exploitation of vapor as a water source has several advantages: (i) the loss of CO_2 by respiration can be compensated to some degree; (ii) when liquid water becomes available, total recovery of metabolism happens faster; and (iii) some organisms perform photosynthesis with increasing efficiency at equilibrium with vapor because the absorption of liquid water reduces CO_2 diffusion and leads to higher respiration rates of the fungal partner (Pintado and Sancho 2002). Besides vapor, recent insights indicate that dew is another important water source. In particular, NVE tightly attached to the bark benefit from dew condensation on the tree surface supplying up to 0.7 mm H_2O day^{-1} or *c.* 176 mm year^{-1} (Lakatos et al. 2012). This is comparable with dew rates in the midlatitudes of *c.* 0.5 mm per 10-h night (Jacobs et al. 2008) and much more than stem water flow calculated for this area of c. 19 mm year^{-1} (Jetten 1996). Due to delayed radiative loss and heat storage of the tree stem, dew formation occurs on stem and NVE surfaces until midday. This novel phenomenon was studied recently in the perhumid tropics and predicted to occur worldwide in forests (Lakatos et al. 2012).

Both water exchange processes and biomass in NVE impacts the hydrology not only in tropical forests but also in boreal, temperate, and coastal (rain) forests. Water (as rain, fog, or dew) is intercepted by the entire tree canopy (made up of foliage, stem, branches, and epiphytes) and slowly drops or flows to the ground (stem flow and throughfall), where it is absorbed, retained, and evaporated by the NVE. In tropical ecosystems, 273 and 724 mm year^{-1} are intercepted by canopy epiphytes in a submontane and in a cloud forest of Tanzania, representing 10 % and 18 % of annual precipitation (Pocs 1980, 1982). Even 34 % annual interception rates were maintained by NVE in a submontane rain forest of Uganda (Hopkins 1960). The absorbed amount ranges from 2 % to 61 % of the total precipitation at the Central Cordillera of Panamá (Cavelier et al. 1996) and from 10 % to 93 % in elfin cloud forests of Venezuela and Colombia (Cavelier and Goldstein 1989). The equivalent of approximately 0.5 mm of cloud water droplets may be sufficient to recharge the water-holding capacity of green algal lichens and pendent mosses (Leon-Vargas et al. 2006). NVE can absorb twice to twentyfold of their dry weight (or fourfold to fivefold as community ensemble; Pocs 1982; Hölscher et al. 2004), providing an essential "canopy water pool." In the temperate forest of Germany, a single 200-year-old oak tree (*Quercus robur*) harbored 8–25 kg of NVE dry biomass that retained 100–400 l of intercepted water; in the tropical mountain forest, this amount could be at least one magnitude higher. Thus, NVE function as "capacitors" influencing the discharge of precipitation, thereby ameliorating erosion and floods as well as water storage (Still et al. 1999; Weathers 1999). Intercepted water is discharged by NVE long after precipitation has stopped and contributes to high humidity within the canopy and understory (e.g. Perry 1984; Veneklaas et al. 1990). For example, epiphytic bryophytes in a Costa Rican montane cloud forest evaporated up to 2.5-fold of their dry weight biomass in 3 days, exceeding evaporation of canopy humus (Köhler et al. 2007). The direct ecological influence of this humidity input on stomatal opening of canopy leaves and thereby photosynthesis and transpiration in forests remains unknown.

The mechanism and function of NVE significantly contribute to canopy ecology, especially micro- and meso-fluxes of carbon and water. Ecophysiological studies of NVE, nonetheless, will likely reveal that secrets and surprises still remain to be discovered in the frontier of forest canopies.

4 Risks of Land-Use Change and Global Warming

Climate change is altering the amount and distribution of precipitation and cloud water in many forests (Still et al. 1999; Solomon et al. 2007). Further impacts of global change promote ongoing deforestation, fragmentation, and forest disturbance (Sala et al. 2000). These processes affect microhabitats of remaining forest patches and trees by increasing penetration of light and wind (Murcia 1995; Laurance and Williamson 2001; Pohlman et al. 2007). As a consequence, microhabitats in forest canopies are becoming progressively drier, with moist, shaded microhabitats confined to lower parts of the tree and interior of forest patches. Epiphytes are sensitive to these environmental changes since they live at the interface between the terrestrial and atmospheric environment (Benzing 1998).

Changes in NVE communities are expected to be most pronounced in areas of high moisture availability (Zotz and Bader 2009), where organisms are poorly adapted to droughts. Sensitive species will respond to increasing dryness by migrating to moister microhabitats further down the tree (Acebey et al. 2003; Alvarenga et al. 2010; Gradstein and Sporn 2010) or shifting to higher altitudes (Nadkarni and Solano 2002; Zotz and Bader 2009). In an undisturbed evergreen lowland moist rain forest in French Guiana, desiccation tolerance of bryophytes indicated that one-third of the canopy species would not tolerate extended drought periods and are expected to migrate to more humid understory microhabitats (Pardow & Lakatos 2013). Species from the understory are regarded as particularly vulnerable to forest disturbance since they have no refugia to migrate into (Gradstein 1992) and are highly desiccation sensitive (Pardow and Lakatos 2013). Zotz and Bader (2009) predict that an increase in temperature in tropical lowland forests by only a few degrees Celsius would increase metabolism and nocturnal respiration rates of epiphytic bryophytes and lichens to such a degree that compensation by photosynthesis during the day is highly challenged and the risk of mass extinctions in this habitat quite likely.

Rising temperatures, on the other hand, can also have a positive effect on species distribution by extending the range of suitable habitats. Ranges of subtropical epiphytic lichens and Atlantic and Mediterranean bryophytes, for example, are currently expanding as they migrate towards Central Europe (Frahm 2001; Aptroot and van Herk 2007).

In sum, NVE are sensitive to environmental change and apparently respond by migrating to more suitable microhabitats within the tree or along elevational gradients where possible. While the change in NVE communities and potential loss of biodiversity are documented (Aptroot and van Herk 2007; Zotz and Bader 2009; Tuba et al. 2011), less attention is given to functional consequences of these changes. There is yet little understanding of how NVE contribute to functional diversity and resource provision in forest canopies. For example, can NVE influence host tree health and pathogen defense through secondary compounds? Will changes in NVE affect the abiotic and biotic canopy environment? Figure 23.3 shows how small-scale differences in moisture supply affect the functional diversity of epiphytic bryophytes in a tropical lowland forest (Pardow et al. 2012). The distribution of seven life-forms is shown across microhabitats within the tree for two adjacent forests, tropical lowland rain forest and tropical lowland cloud forest (Gradstein et al. 2010). These forests share the same meso-climate and differ only in small-scale moisture supply (i.e., cloud forests have frequent early morning fog events (Obregon et al. 2011)). If this represents a model scenario of drier environmental conditions transitioning from lowland cloud forest to lowland rain forest, then life-form composition will change, particularly in the canopy strata. Some life-forms will dissappear or respond by migration to lower – more humid – microhbaitats. This leads to a loss of biocomplexity in strata of drier canopies by harboring fewer and less structurally diverse life-forms (mainly mats). The loss of certain "bryophyte structures" is likely to correlate with a loss of certain microhabitats and other resources confined to a particular life-form. Future research should be directed towards the ecological impact of these changes in functional diversity. The newly designed Biosphere Tower is ideal for noninvasive studies of the NVE community with respect to climate change.

5 Conclusion

This overview of recent outcomes in canopy research on NVE shows knowledge gaps on three topics: (i) for the most part, long-term investigations on seasonality, succession, and long-term changes are missing, (ii) the research in silvicultural tree canopies is underrepresented, and (iii) ecological and physiological measurements in situ are scarce. The consequences of climate and land-use change for NVE and thereby their influence on forest structural diversity, interactions, and carbon and water balance remain unknown.

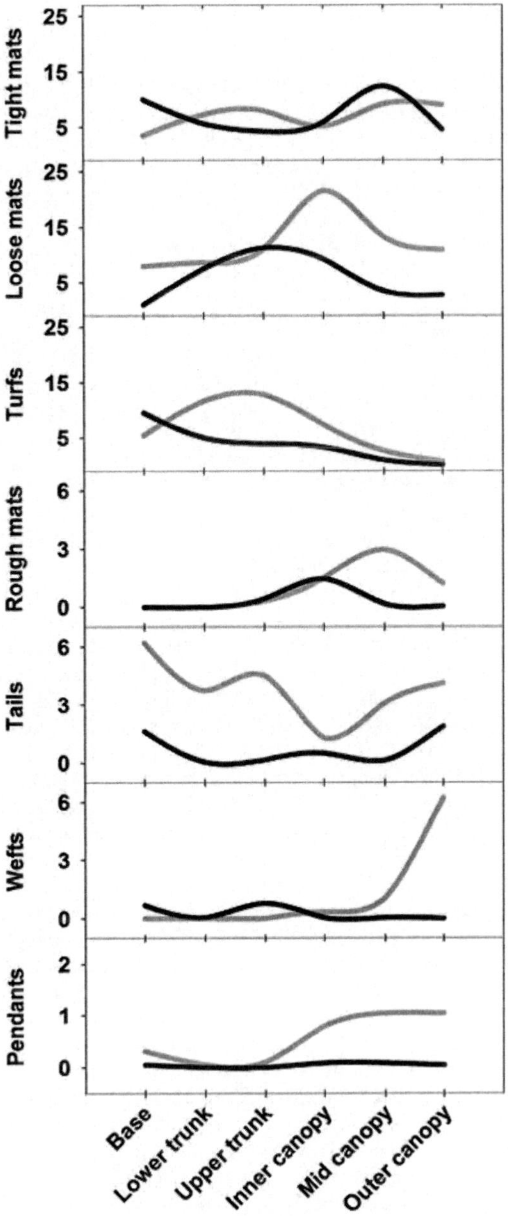

Fig. 23.3 Percent cover of bryophyte life-forms at each height zone in a lowland cloud forest (*gray line*) and lowland rain forest (*black line*) at French Guiana (Modified after Pardow et al. 2012)

References

Acebey A, Gradstein SR, Kromer T (2003) Species richness and habitat diversification of bryophytes in submontane rain forest and fallows of Bolivia. J Trop Ecol 19:9–18

Alvarenga LDP, Porto KC, de Oliveira JRDM (2010) Habitat loss effects on spatial distribution of non-vascular epiphytes in a Brazilian Atlantic forest. Biodivers Conserv 19:619–635

Andreae MO, Merlet P (2001) Emission of trace gases and aerosols from biomass burning. Global Biogeochem Cycles 15:955–966

Antoine ME (2004) An ecophysiological approach to quantifying nitrogen fixation by *Lobaria oregana*. Bryologist 107:82–87

Aptroot A (1997) Lichen biodiversity in Papua New Guinea, with the report of 173 species on one tree. J Cramer Gebr. Borntraeger, Berlin

Aptroot A, van Herk CM (2007) Further evidence of the effects of global warming on lichens, particularly those with *Trentepohlia phycobionts*. Environ Pollut 146:293–298

Benzing DH (1998) Vulnerabilities of tropical forests to climate change: the significance of resident epiphytes. Clim Change 39:519–540

Bertsch A (1966) CO_2-Gaswechsel und Wasserhaushalt der Aerophilen Grünalge *Apatococcus lobatus*. Planta 70:46–72

Blum OB (1973) Water relation. In: Ahmadjian V, Hale ME (eds) The lichens. Academic, New York

Cavelier J, Goldstein G (1989) Mist and fog interception in elfin cloud forests in Colombia and Venezuela. J Trop Ecol 5:309–322

Cavelier J, Solis D, Jaramillo MA (1996) Fog interception in montane forest across the Central Cordillera of Panama. J Trop Ecol 12:357–369

Clark KL, Nadkarni NM, Gholz HL (2005) Retention of inorganic nitrogen by epiphytic bryophytes in a tropical montane forest. Biotropica 37:328–336

Clement JP, Shaw DC (1999) Crown structure and the distribution of epiphyte functional group biomass in old-growth *Pseudotsuga menziesii* trees. Ecoscience 6:243–254

Cleveland CC, Townsend AR, Schimel DS, Fisher H, Howarth RW, Hedin LO, Perakis SS, Latty EF, Von Fischer JC, Elseroad A, Wasson MF (1999) Global patterns of terrestrial biological nitrogen (N-2) fixation in natural ecosystems. Global Biogeochem Cycles 13:623–645

Coley PD, Kursar TA, Machado JL (1993) Colonization of tropical rain forest leaves by epiphylls: effects of site and host plant leaf lifetime. Ecology 74:619–623

Cornelissen JHC, Gradstein SR (1990) On the occurrence of bryophytes and macrolichens in different lowland rain forest types at Mabura Hill, Guyana. Trop Bryol 3:29–35

Cornelissen JHC, Lang SI, Soudzilovskaia NA, During HJ (2007) Comparative cryptogam ecology: a review of bryophyte and lichen traits that drive biogeochemistry. Ann Bot 99:987–1001

Costa DP (1999) Epiphytic bryophyte diversity in primary and secondary lowland rainforests in southeastern Brazil. Bryologist 102:320–326

Coxson DS, Mcintyre DD, Vogel HJ (1992) Pulse release of sugars and polyols from canopy bryophytes in tropical montane rain forest (Guadeloupe, French West Indies). Biotropica 24:121–133

Edlich F (1936) Einwirkung von Temperatur und Wasser auf aerophile Algen. Arch Microbiol 7:62–109

Elbert W, Weber B, Burrows S, Steinkamp J, Budel B, Andreae MO & Poschl U (2012) Contribution of cryptogamic covers to the global cycles of carbon and nitrogen. Nat Geosci 5:459–462

Ellis CJ (2012) Lichen epiphyte diversity: a species, community and trait-based review. Perspect Plant Ecol Evol Syst 14:131–152

Ellyson WJT, Sillett SC (2003) Epiphyte communities on Sitka spruce in an old-growth redwood forest. Bryologist 106:197–211

Foerster JW (1971) The ecology of an elfin forest in Puerto Rico, 14. The algae of Pico del Oeste. J Arnold Arbor 52:86–109

Forman RTT (1975) Canopy lichens with blue-green-algae – nitrogen-source in a Colombian rain-forest. Ecology 56:1176–1184

Frahm JP (2001) Bryophytes as indicators of recent climate fluctuations in central Europe. Lindbergia 26:97–104

Freiberg M, Freiberg E (2000) Epiphyte diversity and biomass in the canopy of lowland and montane forests in Ecuador. J Trop Ecol 16:673–688

Freystein K, Salisch M, Reisser W (2008) Algal biofilms on tree bark to monitor airborne pollutants. Biologia 63:866–872

Friedl MA, McIver DK, Hodges JCF, Zhang XY, Muchoney D, Strahler AH, Woodcock CE, Gopal S, Schneider A, Cooper A, Baccini A, Gao F, Schaaf C (2002) Global land cover mapping from MODIS: algorithms and early results. Remote Sens Environ 83:287–302

Fritz O (2009) Vertical distribution of epiphytic bryophytes and lichens emphasizes the importance of old beeches in conservation. Biodivers Conserv 18:289–304

Furnkranz M, Wanek W, Richter A, Abell G, Rasche F, Sessitsch A (2008) Nitrogen fixation by phyllosphere bacteria associated with higher plants and their colonizing epiphytes of a tropical lowland rainforest of Costa Rica. ISME J 2:561–570

Gauslaa Y, Coxson DS, Solhaug KA (2012) The paradox of higher light tolerance during desiccation in rare old forest cyanolichens than in more widespread co-occurring chloro- and cephalolichens. New Phytol 195:812–822

Gehrig-Downie C, Obregón A, Bendix J, Gradstein SR (2011) Epiphyte biomass and canopy microclimate in the tropical lowland cloud forest of French Guiana. Biotropica 43:591–596

Gradstein SR (1992) The vanishing tropical rainforest as an environment for bryophytes and lichens. In: Bates JW, Farmer AM (eds) Bryophytes and lichens in a changing environment. Clarendon, Oxford

Gradstein S (1995) Bryophyte diversity of the tropical rainforest. Arch Sci [Société de physique et d'histoire naturelle de Genève] 48:91–96

Gradstein SR, Sporn SG (2010) Land-use change and epiphytic bryophyte diversity in the tropics. Nova Hedwig 138:311–323

Gradstein S, Griffin D, Morales M, Nadkarni N (2001) Diversity and habitat differentiation of mosses and liverworts in the cloud forest of Monteverde, Costa Rica. Caldasia 23: 203–212

Gradstein SR, Obregón A, Gehrig C, Bendix J (2010) Tropical lowland cloud forest – a neglected forest type. In: Bruijnzeel LA, Scatena FN, Hamilton LS (eds) Tropical montane cloud forests – science for conservation and management. Cambridge University Press, Cambridge, UK

Green TGA, Lange OL, Cowan IRC (1994) Ecophysiology of lichen photosynthesis: the role of water status and thallus diffusion resistances. Crypt Bot 4:166–178

Hajek T, Beckett RP (2008) Effect of water content components on desiccation and recovery in *Sphagnum mosses*. Ann Bot 101:165–173

Hartard B, Cuntz M, Máguas C, Lakatos M (2009) Water isotopes in desiccating lichens. Planta 231:179–193

Helliker BR, Griffiths H (2007) Toward a plant-based proxy for the isotope ratio of atmospheric water vapor. Glob Chang Biol 13:723–733

Hofstede R, Wolf J, Benzing D (1993) Epiphytic biomass and nutrient status of a Colombian upper montane rain forest. Selbyana 14:37–45

Hölscher D, Köhler L, Van Dijk AIJM, Bruijnzeel LA (2004) The importance of epiphytes to total rainfall interception by a tropical montane rain forest in Costa Rica. J Hydrol 292:308–322

Holz I, Gradstein RS (2005) Cryptogamic epiphytes in primary and recovering upper montane oak forests of Costa Rica – species richness, community composition and ecology. Plant Ecol 178:89–109

Hopkins B (1960) Observations on rainfall interception by a tropical forest in Uganda. E Afr Agric Forest J 25:255–258

Jacobs AFG, Heusinkveld BG, Berkowicz SM (2008) Passive dew collection in a grassland area, the Netherlands. Atmos Res 87:377–385

Jarman SJ, Kantvilas G (1995) Epiphytes on an old Huon pine tree (Lagarostrobos-Franklinii) in Tasmanian rain-forest. N Z J Bot 33:65–78

Jetten VG (1996) Interception of tropical rain forest: performance of a canopy water balance model. Hydrol Process 10:671–685

John V, Schröck HW (2001) Flechten im Kronen- und Stammbereich geschlossener Waldbestände in Rheinland-Pfalz (SW-Deutschland). Fauna und Flora in Rheinland-Pfalz 9:727–750

Karger DN, Kluge J, Abrahamczyk S, Salazar L, Homeier J, Lehnert M, Amoroso VB, Kessler M (2012) Bryophyte cover on trees as proxy for air humidity in the tropics. Ecol Indic 20:277–281

Knops JMH, Nash TH III, Schlesinger WH (1996) The influence of epiphytic lichens on the nutrient cycling of an oak woodland. Ecol Monogr 66:159–179

Köhler L, Tobón C, Frumau K, Bruijnzeel L (2007) Biomass and water storage dynamics of epiphytes in old-growth and secondary montane cloud forest stands in Costa Rica. Plant Ecol 193:171–184

Komposch H, Hafellner J (2000) Diversity and vertical distribution of lichens in a Venezuelan tropical lowland forest. Selbyana 21:11–24

Lakatos M (2011) Lichens and bryophytes: habitats and species. In: Lüttge U, Beck E, Bartels D (eds) Plant desiccation tolerance. Springer, Berlin/Heidelberg, p. 65–87

Lakatos M, Lange-Bertalot H, Büdel B (2004) Diatoms living inside the thallus of the green algal lichen *Coenogonium linkii* in neotropical lowland rain forests. J Phycol 40:70–73

Lakatos M, Rascher U, Büdel B (2006) Functional characteristics of corticolous lichens in the understory of a tropical lowland rain forest. New Phytol 172:679–695

Lakatos M, Obregón A, Büdel B, Bendix J (2012) Midday dew – an overlooked factor enhancing photosynthetic activity of corticolous epiphytes in a wet tropical rain forest. New Phytol 194:245–253

Lange OL, Büdel B, Meyer A, Kilian E (1993) Further evidence that activation of net photosynthesis by dry cyanobacterial lichens requires liquid water. Lichenologist 25:175–189

Lange OL, Meyer A, Büdel A (1994) Net photosynthesis activation of a desiccated cyanobacterium without liquid water in high air humidity alone - experiments with *Microcoleus sociatus* isolated from a desert soil crust. Funct Ecol 8:52–57

Laurance WF, Williamson GB (2001) Positive feedbacks among forest fragmentation, drought, and climate change in the Amazon. Conserv Biol 15:1529–1535

Lemes-Da-Silva NM, Branco LHZ, Necchi-Júnior O (2010) Corticolous green algae from tropical forest remnants in the northwest region of São Paulo State, Brazil. Rev Bras Bot 33:215–226

Leon-Vargas Y, Engwald S, Proctor MCF (2006) Microclimate, light adaptation and desiccation tolerance of epiphytic bryophytes in two Venezuelan cloud forests. J Biogeogr 33:901–913

Mandl N, Lehnert M, Kessler M, Gradstein S (2010) A comparison of alpha and beta diversity patterns of ferns, bryophytes and macrolichens in tropical montane forests of southern Ecuador. Biodivers Conserv 19:2359–2369

McCune B, Amsberry KA, Camacho FJ, Clery S, Cole C, Emerson C, Felder G, French P, Greene D, Harris R, Hutten M, Larson B, Lesko M, Majors S, Markwell T, Parker GG, Pendergrass K, Peterson EB, Peterson ET, Platt J, Proctor J, Rambo T, Rosso A, Shaw D, Turner R, Widmer M (1997) Vertical profile of epiphytes in a Pacific Northwest old-growth forest. Northwest Sci 71:145–152

Milne J, Louwhoff S (1999) Vertical distribution of bryophytes and lichens on a Myrtle Beech, *Nothofagus cunninghamii* (Hook.) Oerst. Hikobia 13:23–30

Murcia C (1995) Edge effects in fragmented forests – implications for conservation. Trends Ecol Evol 10:58–62

Nadkarni NM (1984) Epiphyte biomass and nutrient capital of a Neotropical Elfin forest. Biotropica 16:249–256

Nadkarni NM, Solano R (2002) Potential effects of climate change on canopy communities in a tropical cloud forest: an experimental approach. Oecologia 131:580–586

Nadkarni NM, Schaefer D, Matelson TJ, Solano R (2004) Biomass and nutrient pools of canopy and terrestrial components in a primary and a secondary montane cloud forest, Costa Rica. For Ecol Manag 198:223–236

Nash TH III, Reiner A, Demmig-Adams B, Kilian E, Kaiser WM, Lange OL (1990) The effect of atmospheric desiccation and osmotic water stress on photosynthesis and dark respiration of lichens. New Phytol 116:269–276

Neustupa J, Škaloud P (2010) Diversity of subaerial algae and cyanobacteria growing on bark and wood in the lowland tropical forests of Singapore. Plant Ecol Evol 143:51–62

Obregon A, Gehrig-Downie C, Gradstein SR, Rollenbeck R, Bendix J (2011) Canopy level fog occurrence in a tropical lowland forest of French Guiana as a prerequisite for high epiphyte diversity. Agric For Meteorol 151:290–300

Pardow A, Lakatos M (2013) Desiccation Tolerance and Global Change: Implications for Tropical Bryophytes in Lowland Forests. Biotropica 45(1):27–36

Pardow A, Hartard B, Lakatos M (2010) Morphological, photosynthetic and water relations traits underpin the contrasting success of two tropical lichen groups at the interior and edge of forest fragments. Ann Bot Plant 2010:plq004

Pardow A, Gehrig-Downie C, Gradstein SR, Lakatos M (2012) Functional diversity of epiphytes in two tropical lowland rainforests, French Guiana: Using bryophyte life-forms to detect areas of high biodiversity. Biodivers Conserv 21:3637–3655

Perry DR (1984) The canopy of the tropical rain-forest. Sci Am 251:138–147

Pettersson R, Ball J, Renhorn K, Esseen P, Sjöberg K (1995) Invertebrate communities in boreal forest canopies as influenced by forestry and lichens with implications for passerine birds. Biol Conserv 74:57–63

Pintado A, Sancho LG (2002) Ecological significance of net photosynthesis activation by water vapour uptake in *Ramalina capitata* from rain-protected habitats in central Spain. Lichenologist 34:403–413

Pocs T (1980) The epiphytic biomass and its effect on the water balance of 2 rain forest types in the Uluguru Mountains (Tanzania, East-Africa). Acta Botanica Academiae Scientiarum Hungaricae 26:143–167

Pocs T (1982) Tropical forest bryophytes. In: Smith AJE (ed) Bryophyte ecology. Chapman & Hall, London

Pohlman CL, Turton SM, Goosem M (2007) Edge effects of linear canopy openings on tropical rain forest understory microclimate. Biotropica 39:62–71

Proctor MCF, Tuba Z (2002) Poikilohydry and homoihydry: antithesis or spectrum of possibilities? New Phytol 156:327–349

Proctor MC, Nagy Z, Csintalan Z, Takacs Z (1998) Water-content components in bryophytes: analysis of pressure-volume relationships. J Exp Bot 49:1845–1854

Renhorn KE, Esseen PA, Palmqvist K, Sundberg B (1997) Growth and vitality of epiphytic lichens.1. Responses to microclimate along a forest edge-interior gradient. Oecologia 109:1–9

Rhoades FM (1981) Biomass of epiphytic lichens and bryophytes on *Abies lasiocarpa* on a Mt. Baker lava flow, Washington. Bryologist 84:39–47

Rhoades FM (1995) Nonvascular epiphytes in forest canopies: worldwide distribution, abundance, and ecological roles. In: Lowman MD, Nadkarni NM (eds). Academic press, San Diego

Richardson B, Rogers C, Richardson M (2000) Nutrients, diversity, and community structure of two phytotelm systems in a lower montane forest, Puerto Rico. Ecol Entomol 25: 348–356

Rigonato J, Alvarenga DO, Andreote FD, Dias ACF, Melo IS, Kent A, Fiore MF (2012) Cyanobacterial diversity in the phyllosphere of a mangrove forest. FEMS Microbiol Ecol 80:312–322

Romanski J, Pharo EJ, Kirkpatrick JB (2011) Epiphytic bryophytes and habitat variation in montane rainforest, Peru. Bryologist 114:720–731

Romero C, Putz FE, Kitajima K (2006) Ecophysiology in relation to exposure of pendant epiphytic bryophytes in the canopy of a tropical montane oak forest. Biotropica 38:35–41

Rundel PW (1988) Water relation. In: Galum M (ed) Handbook of lichenology II. CRC Press, Boca Raton

Sala OE, Chapin FS III, Armesto JJ, Berlow E, Bloomfield J, Dirzo R, Huber-Sanwald E, Huenneke LF, Jackson RB, Kinzig A, Leemans R, Lodge DM, Mooney MA, Oesterheld M, Poff NL, Sykes MT, Walker BH, Walker M, Wall DH (2000) Global biodiversity scenarios for the year 2100. Science 287:1770–1774

Sillett SC, Antoine ME (2004) Lichens and bryophytes in forest canopies. In: Lowman MD, Rinker HB (eds) Forest canopies. Elsevier Academic, New York

Solomon S, Qin D, Manning M, Chen ZZ, Marquis M, Averyt KB, Tignor M, Miller HLE (2007) Climate Change 2007: the physical science basis. Contribution of working group I to the Fourth assessment report of the intergovernmental panel on climate change. Cambridge University Press, Cambridge

Sonnleitner M, Dullinger S, Wanek W, Zechmeister H (2009) Microclimatic patterns correlate with the distribution of epiphyllous bryophytes in a tropical lowland rain forest in Costa Rica. J Trop Ecol 25:321–330

Sporn SG, Bos MM, Kessler M, Gradstein SR (2010) Vertical distribution of epiphytic bryophytes in an Indonesian rainforest. Biodivers Conserv 19:745–760

Still C, Foster P, Schneider S (1999) Simulating the effects of climate change on tropical montane cloud forests. Nature 398:608–610

Tuba Z, Slack NG, Stark LR (2011) Bryophyte ecology and climate change. Cambridge University Press, Cambridge, UK

Van der Wal R (2006) Do herbivores cause habitat degradation or vegetation state transition? Evidence from the tundra. Oikos 114:177

Veneklaas EJ, Zagt RJ, Van Leerdam A, Van Ek R, Broekhoven AJ, Van Genderen M (1990) Hydrological properties of the epiphyte mass of a montane tropical rain forest Colombia. Vegetatio 89:183–192

Weathers KC (1999) The importance of cloud and fog in the maintenance of ecosystems. Trends Ecol Evol 14:214–215

Werner FA, Homeier J, Oesker M, Boy J (2012) Epiphytic biomass of a tropical montane forest varies with topography. J Trop Ecol 28:23–31

Williams C, Sillett S (2007) Epiphyte communities on redwood (*Sequoia sempervirens*) in northwestern California. Bryologist 110:420–452

Wolf JHD (1993) Diversity patterns and biomass of epiphytic bryophytes and lichens along an altitudinal gradient in the northern Andes. Ann Mo Bot Gard 80:928–960

Wolf JHD (1995) Non-vascular epiphyte diversity patterns in the canopy of an upper montane rain forest (2550–3670 m), central Cordillera, Colombia. Selbyana 16:185–195

Zhao MS, Heinsch FA, Nemani RR, Running SW (2005) Improvements of the MODIS terrestrial gross and net primary production global data set. Remote Sens Environ 95:164–176

Zotz G, Bader MY (2009) Epiphytic plants in a changing world-global: change effects on vascular and non-vascular epiphytes. Prog Bot 70:147–170

Zotz G, Schleicher T (2003) Growth and survival of the foliose lichen Parmotrema endosulphureum in the lowland tropics of Panama. Ecotropica 9:39–44

Zotz G, Winter K (1994) Photosynthesis and carbon gain of the lichen, *Leptogium azureum*, in a lowland tropical forest. Flora 189:179–186

Zotz G, Schultz S, Rottenberger S (2003) Are tropical lowlands a marginal habitat for macrolichens? Evidence from a field study with *Parmotrema endosulphureum* in Panama. Flora 198:71–77

Chapter 24
Canopy Texture Analysis for Large-Scale Assessments of Tropical Forest Stand Structure and Biomass

Pierre Ploton, Raphaël Pélissier, N. Barbier, Christophe Proisy, B.R. Ramesh, and P. Couteron

Keywords Canopy texture • Fourier spectra • High-resolution images • Forest structure • Above-ground biomass

Bullet Points

1. Detecting carbon stock variations throughout the entire range of high-biomass tropical forests remains technically challenging with conventional remote sensing techniques and requires alternative detection methods and newly available data types (e.g. very-high-resolution optical imagery, LiDAR).
2. FOTO characterizes the canopy grain properties or texture (i.e. a combination of mean size and density of tree crowns), which link stand structure parameters through allometric relationships.

P. Ploton (✉)
Ecology Department, French Institute of Pondicherry, UMIFRE 21 MAEE-CNRS, Pondicherry 605001, India

IRD, UMR AMAP, University of Yaounde I, Yaounde, Cameroon
e-mail: pierre.ploton@ird.fr

R. Pélissier (✉)
Ecology Department, French Institute of Pondicherry, UMIFRE 21 MAEE-CNRS, Pondicherry 605001, India

IRD, UMR AMAP, Montpellier F- 34000, France
e-mail: raphael.pelissier@ird.fr

N. Barbier • C. Proisy • P. Couteron
IRD, UMR AMAP, Montpellier F- 34000, France
e-mail: nicolas.barbier@ird.fr; christophe.proisy@ird.fr; pierre.couteron@ird.fr

B.R. Ramesh
Ecology Department, French Institute of Pondicherry, UMIFRE 21 MAEE-CNRS, Pondicherry 605001, India
e-mail: ramesh.br@ifpindia.org

M. Lowman et al. (eds.), *Treetops at Risk: Challenges of Global Canopy Ecology and Conservation*, DOI 10.1007/978-1-4614-7161-5_24,

3. At local scale, FOTO has been compared to field data in different contexts across the tropics and consistently provides promising results in the distinction of forest types or stand ages and in the inversion of various structural parameters, notably mean tree diameter and standing biomass.
4. Canopy grain measurements are potentially powerful for large-scale forest structure assessments.

Summary

The structural organization of a forest canopy is an important descriptor that may provide information for vegetation mapping and management planning. We present a new approach of canopy texture analysis from diverse very-high remotely sensed optical image types, such as digitized aerial photographs, very-high-resolution satellite scenes or Google Earth extractions. Based on the multivariate ordination of Fourier spectra, the FOTO method allows us to ordinate canopy images with respect to canopy grain, i.e. a combination of mean size and density of tree crowns per sampling window. Compared to field data in different contexts across the tropics, in mangroves, evergreen to semievergreen lowland and mountain forests, FOTO-derived indices proved powerful for consistently retrieving certain stand structure parameters, notably aboveground biomass up to the highest levels observed. We illustrate the potential of the texture-structure model inversion for predicting stand structure parameters over vast poorly documented forest areas in India, Amazonia and central Africa and discuss limitations of this method, such as instrumental and topography-induced biases.

1 Introduction

The structural organization of a forest canopy is an important descriptor that may provide spatial information for vegetation mapping and management planning, such as attributes of plant species distributions, intensity of disturbances, aboveground biomass or carbon stock. A variety of airborne and satellite images characterize forest stands from above the canopy, providing the advantage of a rapid exploration of extensive and sometimes inaccessible zones. Unfortunately, this approach has limited applicability in wet tropical regions, because most optical and radar signals that deliver medium to high spatial resolution data will saturate at intermediate levels of biomass ranges (ca. 150–200 $t.ha^{-1}$) or leaf area index values (Gibbs et al. 2007). As a consequence, while forest vs. non-forest classifications are nowadays routinely performed from such data, variations in stand structure and biomass within forests of fairly closed canopy remain almost undetectable with classical techniques, and the forest treetops seen from above appear as a homogeneously undulating green carpet. However, rainforest structure varies substantially from place to place either naturally (as the soil, composition or forest dynamics vary) or from anthropogenic degradations. Detecting, characterizing and mapping these variations over vast areas are critical to emerging policies, such as the REDD+ agenda (Maniatis and Mollicone 2010), whereby participating countries will monitor their carbon stock variation. Although promising technology such as LiDAR (light detection and ranging) has great potential, they remain very expensive to systematically assess large expanses of tropical forests (but see Asner et al. 2010). We propose as a cost-effective alternative "canopy grain texture analysis" from very-high-resolution air- or space-borne images, which proved efficient for retrieving and mapping stand structure parameters including aboveground biomass over vast poorly documented areas of tropical forest.

2 Methodological Background and Rationale

Given allometric relationships that exist between individual tree dimensions (such as trunk diameter, height and crow size), our approach stems from the idea that the number and size of tree crowns visible from above the canopy should inform on some other forest structural parameters. The reasoning is straightforward when considering for instance stand basal area (BA) or aboveground biomass (AGB) in closed forest conditions since in this case the largest trees that reach the canopy can account for up to 70–80 % of stand level values. However recent developments of the metabolic theory of ecology, which links tree dimensions and their frequency in a stand by scaling up relationships, provide a more general perspective for inferring stand properties, such as spacing relations, mortality rates or stand dynamics, from canopy characteristics (Enquist et al. 2009). While most previous attempts were based on visual or automated delineations of individual tree crowns from very-high resolution (VHR) canopy images, we present hereafter a more holistic characterization of canopy geometrical properties through canopy grain texture analysis by two-dimensional Fourier power spectrum.

Our method is called Fourier textural ordination (FOTO) (Fig. 24.1). From a digital VHR panchromatic canopy image, optionally masked for non-forested areas (clouds, water bodies, bare soils, etc.), a set of square windows with size set to include several repetitions of the largest textural pattern that compose canopy grain is first extracted (Fig. 24.1a). For closed-canopy scenes, canopy grain results from the shape, size and spatial arrangement of dominant tree crowns, so that square windows of about 1 ha proved a good option. A Fourier radial power spectrum (or r-spectrum) is then computed for each window, which features how image grey levels' variance partitions into increasing spatial frequency bins (in cycles.km^{-1}, i.e. the number of repetitions over 1 km) or equivalently into wavelengths (pattern sizes in m) (see Couteron 2002 for further details). In other words, an r-spectrum represents the frequency distribution of pattern sizes in the canopy grain: while coarse canopy grain yields r-spectra significantly skewed towards small frequencies (large wavelengths), fine canopy grain yields r-spectra significantly skewed towards large frequencies (small wavelengths; Fig. 24.1b). All windows' r-spectra computed from a VHR canopy image can then be stacked into a single matrix with the canopy windows (observations) as rows and the spatial frequencies (variables) as columns (Fig. 24.1c). Such a matrix may be submitted to a standardized PCA that systematically

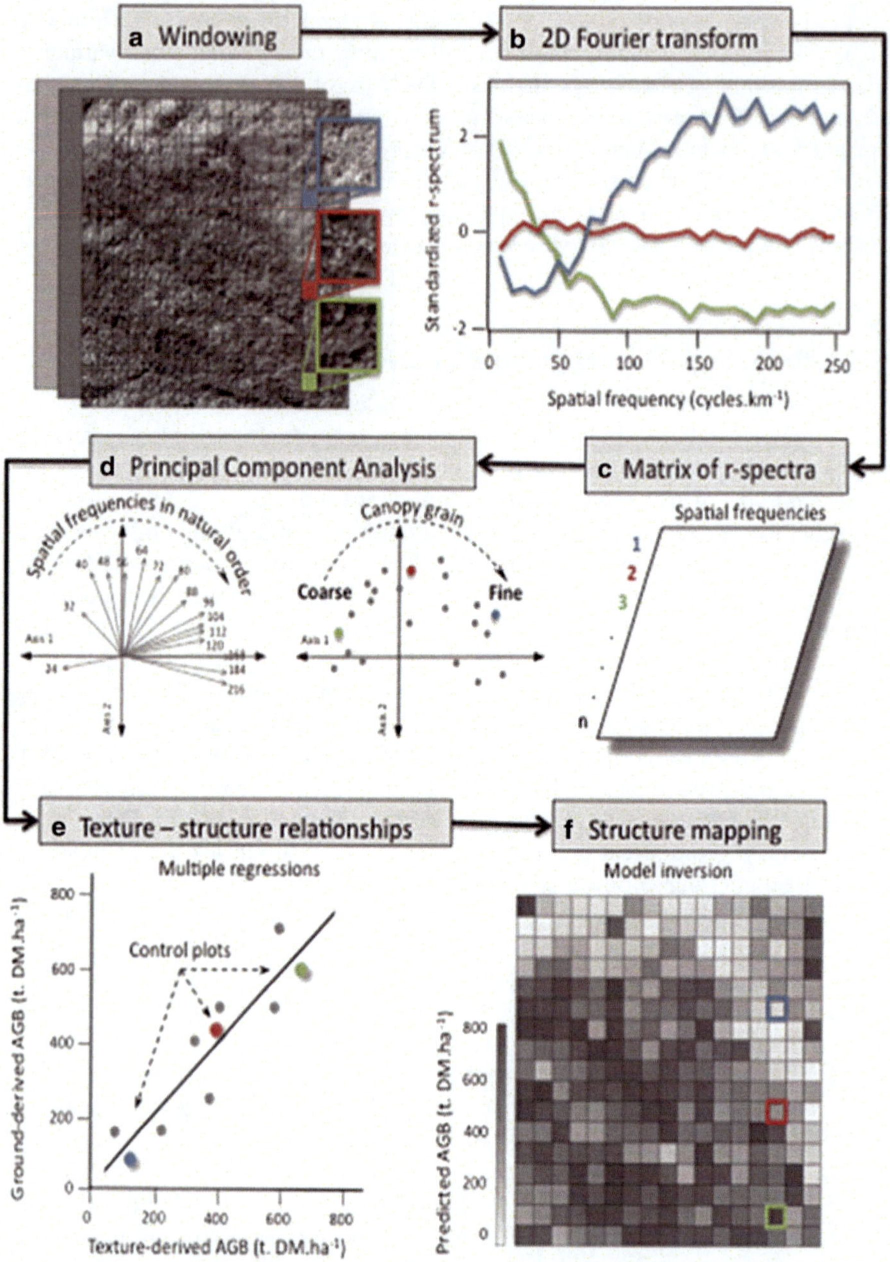

Fig. 24.1 Flow of operations of the Fourier textural ordination (*FOTO*) method. R-spectrum corresponds to azimuthally averaged 2D Fourier periodograms

compares the canopy windows with respect to the relative importance of spatial frequencies in their respective spectra. The typical clockwise distribution of the spatial frequency variables in the first PCA plane straightforwardly creates a canopy texture gradient which ordinates canopy windows, from coarse- to fine-grained patterns (Fig. 24.1d). Windows' scores against the main PCA axes thus represent canopy texture indices (so-called FOTO indices), which can be related to control plots data using multivariate linear models and to calibrate and assess the indices' ability to infer stand structure parameters (Fig. 24.1e). Once calibrated, texture-structure relationships can be inverted for predicting and mapping stand structure parameters over the whole area covered by the initial VHR image (Fig. 24.1f).

3 Results from Some Case Studies

The method was tested on a variety of tropical forest types (evergreen, semievergreen and mangrove forests) in various regions (Amazonia, India and central Africa) using different VHR image data (digitized aerial photographs and commercial and freely accessible satellite images). We also used virtual closed-forest canopy scenes simulated using allometric 3D forest mock-ups and a discrete anisotropic radiative transfer model to benchmark FOTO indices against controlled stand structures (Barbier et al. 2012). Such virtual images were used by Barbier et al. (2010) as a preliminary step to a basin-wide analysis of Amazonian *terra firme* lowland forest canopy images, to evidence a strong relationship between FOTO indices and stand mean apparent crown sizes in simulated images ($R^2=0.96$). These results illustrate the theoretical backbone underlying the allometric assumption on which the approach relies: canopy grain information as captured by FOTO indices mostly pertains to pseudo-periodic patterns of crown diameters' repetitions within the scene. Though pseudo-periodic canopy patterns are generally noisy in real forest types as a result of interacting endogenous (e.g. stand composition and dynamics) and exogenous (e.g. topography, degradation intensity) factors, we were able to consistently discriminate observed forest scenes on the basis of their canopy grain features and to reveal in a number of situations strong correlations with structural parameters measured in ground-truth field plots (Table 24.1).

In even-aged mangrove stands where most trees occupy the canopy layer, FOTO yielded very good predictions on AGB ($R^2=0.92$, with no apparent maximal biomass limitation; Proisy et al. 2007). As expected, weaker though strong relationships were obtained on uneven, hyper-diverse forests displaying more complex canopy patterns such as lowland evergreen forests of French Guiana, India and Cameroon (Table 24.1). While some structure parameters showed fairly stable relationships with canopy texture, such as the mean quadratic diameter (Dg) in Indian and Guianan *terra firme* forests ($R^2=0.68$ and 0.71, respectively) or AGB in Indian forests and Guianan mangroves ($R^2=0.78$ and 0.92, respectively), other parameters showed contrasted relationships that reveal differences in the local variation of stand structural characteristics. For instance, the gradient of sampled forest structures in French Guiana *terra firme* plots relates to old-growth forests on contrasted soil conditions, yielding a strong correlation between canopy grain and tree density (N)

Table 24.1 FOTO indices' predictive power on several classical forest stand structure attributes over a variety of tropical forest types. The quality of the prediction is characterized by the coefficient of determination (R^2), the associated P-value (ns: >0.05) and the relative root mean square error Rrmse (in %). Forest attributes: *N* density of trees more than 10 cm dbh (trees.ha^{-1}), N_{30} density of trees more than 30 cm dbh (trees.ha^{-1}), N_{100} density of trees more than 100 cm dbh (trees.ha^{-1}), D_{max} maximum tree dbh (cm), *Dg* quadratic mean dbh (cm), *BA* basal area (m^2.ha^{-1}), *AGB* aboveground biomass (t.ha^{-1} dry matter), *Cd* mean crown diameter (m), *H* dominant tree height (m)

Study site		Data type		Forest attribute		Texture-structure		
Area	Forest type	Image	Control plot	Parameter	Range	R^2	Rrmse (%)	Sources
French Guiana	Evergreen (*terra firme*)	Aerial photographs	Field plots (12 1 ha)	N	455–861	0.8	–	Couteron et al. (2005)
				BA	28.41–42.38	0.007 ns	–	
				Dg	20.6–34.2	0.71	–	
				H	21.1–30.4	0.57	–	
	Mangroves	IKONOS	Field plots (26 1 ha)	AGB	80–436	0.92	17.3	Proisy et al. (2007)
Amazonia	*terra firme*	Google Earth	Simulations (330 2.25 ha)	Cd	7–25	0.96	–	Barbier et al. (2010)
India	Evergreen	IKONOS/Google Earth	Field plots (15 1 ha)	N	371–733	0.109 ns/0.132 ns	14.4/13.8	Ploton et al. (2012)
				N_{30}		0.773/0.770	14.6/14.3	
				BA	10.3–54.4	0.779/0.741	13.1/13.5	
				Dg	17.9–36.4	0.657/0.687	7.4/7.9	
				AGB	124.1–683.6	0.779/0.741	13.1/13.5	
			Simulations (10 1 ha)	Cd	7–25	0.96/0.93	1.17/1.64	
Cameroon	Evergreen to semievergreen	GeoEye	Field plots (10 1 ha)	D_{max}	89–144	0.93		Unpublished results
				N_{100}	0–10	0.64		
				N	225–525	0.34		
				BA	13–43	0.38		
				Dg	22–33	0.30		

or mean quadratic diameter (Dg), but no relationship with basal area (BA) because the denser stands are located on poor soils that do not support many large trees (Couteron et al. 2005).

Conversely, in India the area studied encompassed a gradient of forest succession stages from highly degraded secondary formations recovering from burning to old-growth undisturbed forests. In this case, FOTO clearly detected the gradual increase in density of the largest trees (N_{30}), BA and AGB throughout forest successions, while total tree density (N) did not lead to a predictable relationship (Ploton et al. 2012). In a different context, in southeastern Cameroon forests that mix monodominant *Gilbertiodendron dewevrei* forests, along with degraded *Maranthaceae* and old-growth mixed-forest patches, canopy texture mostly correlates with maximum tree diameter (D_{max}) and density of very large trees (N_{100}), while overall stand structure parameters exhibit only weak relationships with FOTO indices (see Table 24.1). On the one hand, high-biomass *G. dewevrei* stands display a very fine canopy grain, possibly because of the high evenness of tree heights and strong imbrications of their crowns. On the other hand, the varying degree of canopy closeness in *Maranthaceae* forests (that may contain very few emergent trees dominating a low understory) results in bimodal r-spectrum dominated by both large and fine textural patterns. Interestingly, canopy texture allows the segmentation of these different forest types, therefore potentially permitting separate, type-wise inversions. The FOTO method not only offers good prospects for saturation-free tropical forest biomass assessments but also appears valuable to draw ecological insights into forest stand structure variation.

Our case studies were conducted from image types as different as digitized panchromatic aerial photographs (French Guiana *terra firme*), panchromatic IKONOS or GeoEye satellite data (India, Cameroon and French Guiana mangroves) or RGB true-colour bands average of Google Earth (GE) extractions (Amazonian *terra firme* and India) that nonetheless provided consistent results as long as images were in the optical domain with a spatial resolution of metric order (i.e. VHR). This underlines a promising feature of FOTO with respect to inter-operating image data types and provides a cost-free alternative to commercial data (GE) for large-scale assessments (for instance, within the REDD+ framework) or retrospective analyses from old aerial photographs in the pre-satellite period.

4 Limits and Perspectives

Further case studies and simulation works are ongoing to validate the FOTO method in various tropical forest contexts. One of the limits of the canopy grain approach arises when canopy texture properties deviate too much from a pseudo-periodicity so that the main canopy pattern does not result from repetitions of crown diameters. Such heterogeneity may result from the presence of canopy gaps or treefalls of varying sizes or of contrasted illumination patterns due to abrupt relief variations (e.g. two sides of a ridge line or a deep thalweg). Typically, r-spectra of windows displaying a high degree of spatial heterogeneity are skewed towards low frequencies due to the contribution of heterogeneity to the largest size patterns. It follows that such windows can be erroneously interpreted as containing large tree crowns and must be removed from the analysis to eliminate bias of the texture-structure relationship (Ploton et al. 2012). Similarly, when shifting from pure evergreen to semievergreen or mixed deciduous tropical forest types, textural information may be influenced by seasonal changes in the canopy that still require to be investigated.

Though the FOTO method offers good prospects for large-scale implementation, further difficulties arise when several canopy images with different acquisition parameters are placed into a mosaic. Indeed, the sun-scene-sensor geometry influences the size and proportion of tree shadows in the canopy scene and thus modifies the textural properties as detected by Fourier r-spectrum. To ensure a consistent comparison between different canopy images, one must either use similar acquisition conditions or correct for the influences of changing acquisition conditions on canopy textural properties. To this end, Barbier et al. (2011) introduced a bidirectional texture function (BTF) that corrects for instrumental bias based on a partitioned standardization of the r-spectrum prior to PCA. A similar problem arises with relief variations, which modify the proportion of sun-lighted vs. shadowed crowns. In mountainous regions, this effect has the potential for FOTO to detect a finer grain on illuminated hillsides and coarser grain on shaded ones regardless of forest structure. A partitioned standardization of the r-spectrum according to hillshade classes may correct for such a bias (Ploton 2010).

References

Asner GP, Powell GVN, Mascaro J, Knapp DE, Clark JK, Jacobson J, Kennedy-Bowdoin T, Balaji A, Paez-Acosta G, Victoria E (2010) High-resolution forest carbon stocks and emissions in the Amazon. Proc Natl Acad Sci 107:16738

Barbier N, Couteron P, Proisy C, Malhi Y, Gastellu Etchegorry J-P (2010) The variation of apparent crown size and canopy heterogeneity across lowland Amazonian forests. Glob Ecol Biogeogr 19:72–84

Barbier N, Proisy C, Vega C, Sabatier D, Couteron P (2011) Bidirectional texture function of high resolution optical images of tropical forest: an approach using LiDAR hillshade simulations. Remote Sens Environ 115:167–179

Barbier N, Couteron P, Gastelly-Etchegorry JP, Proisy C (2012) Linking canopy images to forest structural parameters: potential of a modeling framework. Ann For Sci 69(2):305–311

Couteron P (2002) Quantifying change in patterned semi-arid vegetation by Fourier analysis of digitized aerial photographs. Int J Remote Sens 23(17):3407–3425

Couteron P, Pélissier R, Nicolini E, Paget D (2005) Predicting tropical forest stand structure parameters from Fourier transform of very high-resolution remotely sensed canopy images. J Appl Ecol 42:1121–1128

Enquist BJ, West GB, Brown JH (2009) Extensions and evaluations of a general quantitative theory of forest structure and dynamics. Proc Natl Acad Sci 106:7046

Gibbs HK, Brown S, Niles JO, Foley JA (2007) Monitoring and estimating tropical forest carbon stocks: making REDD a reality. Environ Res Lett 2:1–13

Maniatis D, Mollicone D (2010) Options for sampling and stratification for national forest inventories to implement REDD+ under the UNFCCC. Carbon Balance Manag 5:1–14

Ploton P (2010) Analyzing canopy heterogeneity of the tropical forests by texture analysis of very-high resolution images – a case study in the Western Ghats of India. Pondy Pap Ecol 10:1–71

Ploton P, Pélissier R, Proisy C, Flavenot T, Barbier N, Rai SN, Couteron P (2012) Assessing aboveground tropical forest biomass using Google Earth canopy images. Ecol Appl 22(3): 993–1003

Proisy C, Couteron P, Fromard F (2007) Predicting and mapping mangrove biomass from canopy grain analysis using Fourier-based textural ordination of IKONOS images. Remote Sens Environ 109:379–392

Chapter 25
Changing Tropical Forest Dynamics and Their Effects on Canopy Geometry and Tropical Forest Biodiversity

David B. Greenberg and Simon L. Lewis

Keywords Changing forest dynamics • Canopy structure • Canopy gap • Forest microclimate • Forest light • Forest biodiversity

Bullet Points

1. Trees within Amazon forests are recruiting, growing, and dying faster, and such changes may be occurring within tropical forests worldwide.
2. We have poor understanding of the drivers of the changing forest dynamics, which are likely complex and interacting; but the data suggest that rising atmospheric CO_2 concentrations increasing rates of photosynthesis and therefore rates of tree growth may be a dominant driver of change, leading to knock-on effects of accelerated tree population dynamics.
3. We hypothesize that these changes are altering the physical structure (i.e., geometry) of tropical forest canopies, in that canopy gaps are likely to have become larger and/or more numerous. A review of potential drivers of the demography of stems in tropical forest stands finds that all have either increased tree growth (which should eventually to lead to higher mortality) or have increased mortality, which would lead to canopy gaps becoming more numerous and/or larger.

D.B. Greenberg (✉)
Environmental Change Institute, Oxford University Centre for the Environment, School of Geography and the Environment, University of Oxford, South Parks Road, Oxford OX1 3QY, United Kingdom

School of Geography, University of Leeds, Leeds, LS2 9JT, United Kingdom
e-mail: david.greenberg@ouce.ox.ac.uk

S.L. Lewis
School of Geography, University of Leeds, Leeds LS2 9JT, United Kingdom

Department of Geography, University College London, London, WC1E, United Kingdom
e-mail: s.l.lewis@leeds.ac.uk

M. Lowman et al. (eds.), *Treetops at Risk: Challenges of Global Canopy Ecology and Conservation*, DOI 10.1007/978-1-4614-7161-5_25,

4. Changes in the geometry of forest canopies should concomitantly alter the life histories of plant and animal species within these forests, in complex ways that we understand only poorly. Careful study is required in order to predict how tropical forest biodiversity, dynamics, and function might be influenced by changes to forest canopy structure.

Summary

Risks to forest canopies are the result of risks to the trees that comprise them. Whether a canopy degrades depends on the relative rates of recruitment, growth, reproduction, and mortality among trees within the forest, and assessing that risk means studying how those dynamics are affected by environmental changes. We focus here on (1) the recent changes in the dynamics of tropical forest trees, particularly in Amazonia, as evidenced by long-term forest inventories, (2) the potential drivers of these changing dynamics, (3) how the changing dynamics may be altering the physical structure of forest canopies, and (4) how such changes in canopy structure might influence tropical forest biodiversity. We review evidence that in Amazonia, and likely elsewhere, tropical forest turnover has increased over recent decades (i.e., that the mean time a stem, or a unit of basal area, or unit of biomass, remains within a forest, has decreased), concomitant with an acceleration of tree recruitment, growth, and mortality. We suggest that this is increasing the physical heterogeneity of tropical forest canopies, as more and larger treefall gaps are formed (and filled in faster), and that these changes will have consequences for tropical forest biodiversity, dynamics, and function that demand investigation.

1 Introduction

Tropical forests are changing, in complex ways, for reasons we are only beginning to understand and with consequences that are hard to predict (Lewis 2006; Lewis et al. 2004a, b, 2009a; Wright 2010). Long-term field studies indicate that rates of tree recruitment, growth, and mortality are increasing within Amazon forests (Phillips et al. 2008) and possibly within tropical forests worldwide (Lewis et al. 2009a). The net impact of these changes has been an increase in aboveground biomass (Phillips et al. 2009; Lewis et al. 2009a, b; Chave et al. 2008). Atmospheric CO_2 measurements indicate increased uptake of carbon in tropical regions (after accounting for emissions from deforestation and degradation), also suggesting that tropical forest biomass has increased (Stephens et al. 2007; Deng and Chen 2011), and satellite-based studies suggest significant changes in tropical forests globally (Nemani et al. 2003; Weishampel et al. 2001). There are hints that tropical forest species composition and phenology are changing too, for example, that lianas are becoming more abundant across Amazonia and in one part of Central America (Phillips et al. 2002; Wright and Calderon 2006), and phenology is changing in Eastern Central Africa (Chapman et al. 2005).

How might the changing dynamics of tropical forests affect their canopies? This is worth considering, given that tropical forests are estimated to house over half of the world's terrestrial species (Groombridge and Jenkins 2003). Furthermore, because canopies are the interface across which carbon cycles between the terrestrial land surface and the atmosphere, they influence the nature and rate of climate change globally. Indeed, the tropics are the main source of interannual variability in atmospheric CO_2 concentrations (Gurney et al. 2008). Since the abundance and composition of trees and lianas within tropical forests are changing, we should expect change in the physical architecture, range of microclimates, chemistry, and physiological operation of tropical forest canopies along with concomitant effects on both tree and non-tree biodiversity. However, the nature and mechanisms of these changes remain unclear.

In this chapter we firstly review the observed changes in tropical forest tree growth, recruitment, and mortality, interpreting these in terms of possible changes in the geometry of forest canopies. Secondly, we review the likely drivers that may be causing these changes. Thirdly, we conclude with a more speculative section on what the changes in and likely drivers of these forest dynamics implies for tropical forest canopies and their continued functioning as biodiversity reservoirs.

2 Observed Changes Within Tropical Forests

Studies of long-term forest inventory plots reveal two changes in the dynamics and physical structure of tropical forests, both in the Neotropics and in the Paleotropics: (i) an overall increase in stem turnover (Phillips and Gentry 1994; Phillips 1996; Phillips et al. 2004; Lewis et al. 2004c) and (ii) an overall increase in aboveground biomass (Phillips et al. 1998, 2009; Lewis et al. 2009a, b; Chave et al. 2008). The increase in stem turnover – the mean of the ratio of stem recruitment to stem mortality, a measure of forest dynamism – was first documented using data from 40 plots across the tropics, though these were largely in Amazonia and SE Asia (Phillips and Gentry 1994), and was documented again with data from 67 plots again mostly spanning Amazonia and SE Asia (Phillips 1996); these studies suggested an overall doubling of stem turnover rates from the 1950s to 1990s. For South America, Phillips et al. (2004) showed a similar pattern when analyzing plots occurring on richer soils, poorer soils, aseasonal climates, and seasonal climates suggesting a widespread pattern (Phillips et al. 2004). Similarly, disaggregating turnover into component stem recruitment and stem mortality, both showed increases over time (Phillips 1996; Phillips et al. 2004). Among plots from published studies reviewed in a more recent synthesis (Lewis et al. 2009b), 78 % showed increases in stem recruitment and 71 % showed increases in stem mortality. For forest canopies, this implies a greater frequency of gap formation caused by the higher tree mortality. The greater stem recruitment may imply that canopy gaps are filling in more rapidly than in the past, or it may reflect changing size-frequency distributions of canopy gaps being formed over recent decades.

The overall increase in aboveground biomass was first documented in plots from Amazonia (Phillips et al. 1998) and later using plots from Africa (Lewis et al. 2009a),

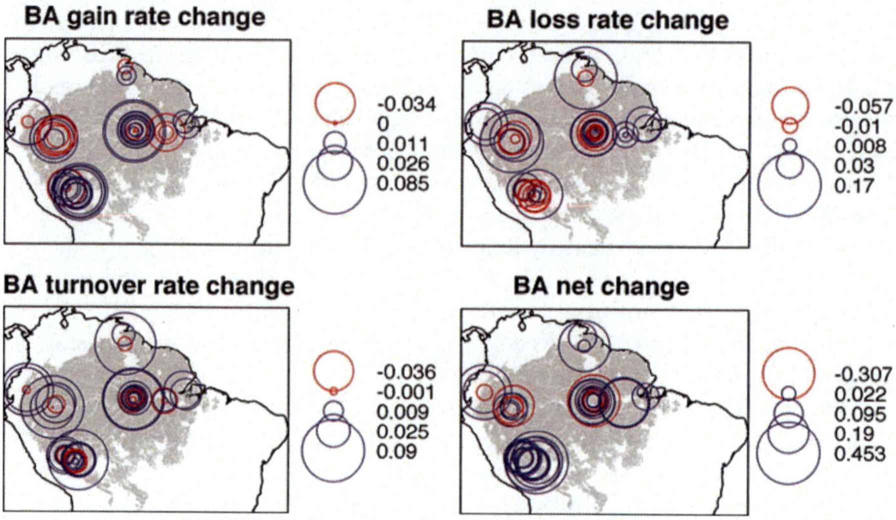

Fig. 25.1 Changes in basal area (BA, sum of the cross-sectional areas of trees in a plot), gain rate (sum of all the increments of BA of surviving stems and newly recruited stems over a census interval), BA loss rate (sum of BA of all stems that died over a census interval), BA turnover (the mean of the changes in BA gain and BA losses), and the net change in basal area (end BA minus initial BA scaled by the census interval), for 50 long-term forest plots with three census intervals, all in the units m^2 ha^{-1} year^{-1}. Mean initial census year was 1985; mean final census year was 1999. Gray background indicates continuous forest cover across the Amazon basin

with some evidence too that this is a pantropical phenomenon (Chave et al. 2008; Lewis et al. 2009a, b). These results imply that intact tropical forests, on average, are increasing in total biomass, year on year, by ~0.5 Mg C ha^{-1} year^{-1} or ~0.3 %, based on a review of published studies reporting inventory plot data on changes in tropical forest biomass (Lewis et al. 2009b). Given that biomass has increased, yet increases in the density of stems per unit area have increased only marginally (Lewis et al. 2004a, 2009b), then the mean size of trees has increased over recent decades. In terms of the physical structure of forest canopies, this implies that the mean size of tree crowns may be increasing. Furthermore, all else being equal, if trees are larger on average, they will leave larger canopy gaps when they die. Given that stem turnover has also been increasing, we predict that canopy gaps may have increased in both frequency and size over recent decades. Furthermore, given that stem mortality has increased, the documented increase in biomass is therefore not caused by a decrease in mortality but rather by increases in the rate of tree growth, providing some evidence that while gaps may be forming more rapidly, they may also be closing more rapidly too.

For 50 plots spanning Amazonia, the simultaneous changes in forest growth, mortality, and dynamics are shown in Fig. 25.1. We show changes in (1) the additions to a forest plot, as basal area (BA, the summed cross-sectional area of trees within a defined area, in m^2 ha^{-1} year^{-1}) gain (the sum of all the increments of surviving trees and newly recruited trees over a census interval); (2) the subtractions from a forest plot, as in basal area loss (the sum of the BA of all the trees dying over a census interval); (3) the dynamics of the system, as BA turnover (the average of the means of BA gains and BA losses); and (4) the overall impacts of these change, as net changes in BA. The maps show overall pan-Amazon increases in all four variables over recent decades at most locations, i.e., growth, mortality, and dynamics increased, while the forests added mass. Hence, the results suggest increases in the heterogeneity of the forest canopy and likely larger canopy gaps as well as more frequently forming canopy gaps.

Satellite data can also reveal changes in tropical forest canopies, but few studies have used them to attempt to detect widespread changes over time. Two types of relevant studies have been published. First, 1 km^2 scale AVHRR data have been used to parameterize models of change in forest net primary productivity (NPP), with two major studies of changes in NPP showing increases of 0.4 % $year^{-1}$ (Nemani et al. 2003) and 0.5 % $year^{-1}$ (Cao et al. 2004) from the early 1980s to the late 1990s. These results are consistent with the increases in biomass seen in plot studies. Second, Weishampel et al. (2001) studied textural metrics derived from Landsat images covering Africa, South America, and SE Asia and reported highly significant increases in the average size of aggregations (groupings of similar texture) and in entropy (the degree of inverse correlation between neighboring pixels within a raster image) from the mid-1970s to the mid-1980s. The authors interpret this as forests becoming more dynamic and gaining biomass, with more emergent trees and thus more shadows, and as them having larger canopy gaps (Weishampel et al. 2001). However, these claims should be treated cautiously as the features identified were considerably larger than most canopy gaps.

Other observed (but less widely documented) changes in forest dynamics may affect forest canopies. Firstly, species composition may be changing, though the limited available data do not suggest clear patterns (Lewis et al. 2009b). One consistent trend has been found, however, in the abundance of lianas – structural parasites that can compose a significant fraction of a forest canopy. Phillips et al. (2002) documented increase across Western Amazonia, in the density of large lianas $\geq$100 mm in diameter (0.22 $\pm$ 0.11 lianas ha^{-1} $year^{-1}$), in their basal area (0.0037 m^2 ha^{-1} $year^{-1}$), and in their rates of recruitment and mortality, across plots in a network spanning 1979–2002 (Phillips et al. 2002). In addition, at a site in Central America, litterfall records over 18 years show that the production of liana leaves has increased (Wright et al. 2004), as did liana flower production though not seed production (Wright and Calderon 2006). Both the Western Amazon and Central America studies showed an increase in the relative dominance of lianas compared to trees. In contrast, the only study in African forest to document changes in liana density and basal area over time showed a significant decrease at Makokou, Gabon (Caballe and Martin 2001).

Tree species phenology may be changing within tropical forests as well. However, only two studies have assessed changes in tree reproduction over time. A 32-year record of change was compiled from a forest in Uganda, but the sampling protocol changed over time, making robust conclusions difficult (Chapman et al. 2005). Over the 1990s the proportion of trees fruiting increased dramatically (Chapman et al. 2005). Meanwhile in Central America, flower production was monitored from 1987 to 2003 for 81 species using consistent methods, also showing large increases over time for both tree and liana species (Wright and Calderon 2006). Such increases in productivity and the ratio of lianas to trees would probably be reflected in changes in the physical structure of forest canopies as well.

3 The Potential Drivers of Changing Tropical Forest Dynamics

The physical, chemical, and biological environments of tropical forests are changing (Lewis et al. 2004b). For example, the physical environment has changed as air temperatures have increased and will do so for the coming decades, while the chemical environmental has altered as atmospheric CO_2 concentrations have increased and continue to do so (IPCC 2007). Similarly, the biological environment is changing, with widespread increases in hunting pressure and land-use changes leading to remaining forests being nearer to populations of pioneer tree species, meaning alterations of seed inputs (less transport of big seeds, greater numbers of small seeds available for transport) and altered competitive interactions among tree species in forests. Furthermore, each driver of change likely has a unique spatiotemporal pattern and initially acts on different processes (i.e., growth or mortality or recruitment) and via different mechanisms (e.g., photosynthesis or leaf respiration). Thus, each potential driver of change would leave a unique dominant "fingerprint" in datasets (Lewis et al. 2004b; Parmesan and Yohe 2003).

A recent survey identified ten potential drivers: (a) increasing air temperatures (altering growth), (b) changes in rainfall (altering growth), (c) changes in incoming solar radiation (altering growth), (d) changes in the frequency and/or intensity of extreme climatic (including El Niño-Southern Oscillation) events (altering mortality), (e) rising atmospheric CO_2 concentrations (stimulating growth), (f) increasing nutrient deposition (nitrogen and phosphorus; stimulating growth), (g) ozone and acid deposition (reducing growth), (h) hunting pressure (altering recruitment), (i) land-use change (altering recruitment), and (j) secondary biological changes such as the abundance of lianas (can alter growth, recruitment, and mortality; Lewis et al. 2009b). Three of these drivers could potentially alter multiple processes that affect forest ecosystem functioning: rising atmospheric CO_2 concentrations can affect photosynthesis and water-use and nutrient-use efficiency of plants, increasing air temperatures can affect rates of photosynthetis and respiration and may alter soil-nutrient availability by increasing soil temperatures, and land-use changes can affect plant mortality rates via edge effects or by increasing seed rain of disturbance-adapted plants from nearby degraded lands. This gives 17 potential mechanisms of changing forest dynamics, plus potential historical effects (e.g., recovery from past disturbance) that may be influencing tropical forests today (Lewis et al. 2004a, 2009b).

As described, the limited available data suggest that tree growth, recruitment, and mortality, as well as forest biomass have increased; this implies a driver that increases resource supply to tropical trees over large areas. Thus, we can tentatively exclude those changes that first drive either mortality or recruitment, as neither of these can increase biomass via increased tree growth (d, h, i, j). Similarly, those factors that drive tree growth but negatively so (g), or where the driver has decreased over recent decades, can also be excluded (b), as the pantropical trend in rainfall is negative (Malhi and Wright 2004). Of the remaining possibilities, average air temperatures have increased over the tropics, but such increases are usually hypothesized to decrease tree growth, not to increase it (Clark et al. 2003; Feeley et al. 2007). Yet, if increasing soil temperatures increase soil-nutrient mineralization rates and other chemical reactions, then it is possible that air temperatures could actually increase tree growth and forest biomass, although most evidence points to lower growth with rising temperatures due to higher allocation of scarce carbon to respiration, all else being equal. Changes in incoming solar radiation and ratios of direct to diffuse light (c) are complex and poorly understood, and it is unclear whether trees in tropical forests have experienced increased or decreased photosynthetically active radiation on average over recent decades or whether direct/diffuse ratios of photosynthetically active radiation have altered significantly.

Rising atmospheric CO_2 concentrations, being relatively uniform globally and with CO_2 being a key substrate for photosynthesis, are a leading candidate cause of the widespread changes in tropical forests (Lloyd and Farquhar 1996). The evidence that nutrient deposition has increased, largely from biomass burning, is not strong, and there is evidence that it may in fact have decreased (Marlon et al. 2008), and rates of tropical deforestation have not changed markedly over recent decades (Achard et al. 2002; Ramankutty et al. 2007). However, it is conceivable that nutrient inputs into remaining forests from biomass burning are increasing as the rate at which the deforestation occurs near once-remote forests is increasing (Laurance 2005). In addition, long-range inputs of Saharan dust to both African and the Latin American forests may have increased over recent decades, possibly in response to climate change (Engelstaedter et al. 2006). Overall, the observed large-scale directional changes in tropical forests have most likely been caused by an increase in resource availability, with leading candidates being increasing atmospheric CO_2 concentrations, possibly in concert with increases in incoming solar radiation, direct/diffuse light ratios, and perhaps nutrient deposition rates and higher air temperatures increasing nutrient mineralization rates. Other interpretations of both the data and possible drivers are possible (summarized in Lewis et al. 2009b, and see Wright 2010 for some alternative views).

4 Effects of Changes in the Physical Structure of Tropical Forest Canopies on Biodiversity

With changes in the geometry of canopy vegetation (i.e., in the locations, sizes, shapes, and orientations of leaves, stems, branches, flowers, and fruits) come changes in the diversity, abundance, and distribution of within- and below-canopy habitats suitable for plants and animals. When even a small canopy gap forms, it may change local microclimates beyond the tolerance of species that normally occupy an area. If canopy gaps are becoming more frequent and larger, owing to trees growing faster and dying younger, and if these more frequent, larger gaps are filling faster as stems recruit and grow more quickly, then within- and below-canopy habitats are changing rapidly and the diversity and relative abundance of tropical forest species will likely be changing in complex ways.

Because canopy vegetation limits the transmission of incoming solar radiation, forest light environments are spatially and temporally complex (Lieberman et al. 1989). Light attenuation can be greater than 90 % over the first 5 m of the canopy (Johnson and Atwood 1970), and light intensities of less than two percent of full sunlight have been reported at ground level (Bazzaz and Pickett 1980; Lewis and Tanner 2000). Resulting effects on where plants can and cannot photosynthesize sufficiently, and concomitant effects on tree species recruitment, growth, and mortality, have been studied at length (e.g., Denslow 1980, 1987; Augspurger 1984; Canham et al. 1990; Hubbell et al. 1999). We will not review this here, except to say that many tree species are so photosensitive that they grow entirely on the light they receive from sunflecks (Chazdon and Fetcher 1984a, b), which means that where and at what rates they recruit, grow, and survive will change if canopy gaps are increasing in frequency, size, and the rate at which they close. We should see pervasive alteration in the diversity, abundances, and distributions of tree species in tropical forests if canopy gap dynamics are changing.

With most research on how light availability within forests influences tree species focusing on effects related to photosynthesis, the emphasis has been on the total *intensity* of light within the 400–700 nm spectrum; these are the wavelengths that constitute photosynthetically active radiation. However, the spectral composition of forest light (i.e., its *color*) – the relative intensities of different wavelengths both within and outside 400–700 nm – also influences plant physiology, along with visual perception in animals and thereby many aspects of their ecology (Endler 1993, 1997; Théry 2001). There are few data on how these effects may influence the richness, abundance, and distributions of species or on how they can affect ecosystem function or dynamics; to our knowledge, these topics have only been explicitly investigated twice (Théry 2001; Proulx and Parrot 2008). This is an area ripe for investigation.

Endler (1993, 1997) reviewed variation in the color of forest light and its effects on the ecology of forest plants and animals. The color of light influences plant growth and phenology as well as the ability of animals to communicate visually with each other and to see potential food, mates, competitors, predators, oviposition sites, and refugia. Endler (1993, 1997) describes four major forest light environments, each defined by the geometry of local vegetation and its effects on the relative intensities of direct versus diffuse light (the latter radiated by blue sky, clouds, and vegetation) and each dominated by differently colored ambient light. They include *forest shade,* under closed canopy and rich in yellow-green wavelengths; *small gaps* under a mix of closed canopy and small openings, which are rich in yellow-orange wavelengths; *woodland shade* under more open canopy and rich in bluish wavelengths; and *large gaps* which are similar to wide open areas in being largely whitish (Endler 1993, 1997). If the observed changing dynamics of tropical forests are causing canopy gaps to increase in frequency, size, and the rate at which they close, we should see pervasive changes in the relative frequencies of these four light environments within them. There will be less forest shade and woodland shade, and more large- and small-gap light environments.

One result of canopies developing larger gaps more frequently would be tropical forests becoming dominated by yellowish-orange and whitish ambient light, with fewer areas containing yellow-green or bluish light. This should have profound effects on the growth, survival, and reproduction of forest species, and because this will influence their abundances and distributions we should see considerable change in tropical forest biodiversity we can only theorize about the mechanisms to be involved and specific changes that will occur, however, as there has been little research on the topic.

Consider how the color of light affects growth and morphogenesis in plants. Blue and ultraviolet light influence germination, stem and branch elongation and expansion, pigment production, chloroplast development, stomatal opening, and other physiological processes (Endler 1993, and references therein). Because tropical forests should lose forest shade and woodland shade, becoming dominated by small- and large-gap light environments, tree species adapted to and normally present within bluish environments (the woodland shade) should become less abundant. Their distributions in the forest should become more restricted, and some could even be extirpated locally. How this may indirectly affect organisms that depend on them, e.g., certain herbivorous animals, begs investigation.

Changes in the color of forest light can also directly influence survival and reproduction in animal species. For example, in French Guiana the ability of certain species of lekking birds to attract mates but also avoid predators may depend on the availability of the forest shade light environment (Endler and Théry 1996). Leks are sites where groups of males display themselves to females, often showing brightly colored plumage or other elaborated characteristics, and are then selected as mates by the females based on the quality of their individual displays (Bradbury 1981). In French Guiana, the color patterns of male cocks-of-the-rock (*Rupicola rupicola*) and white-throated manakins (*Corapipo gutturalis*) make them cryptic in forest shade, where they remain still and quiet when at the lek but not displaying, but for their displays they move partially into a sunfleck which makes them highly visible (Endler and Théry 1996). If forests become dominated by large-gap light environments, males of these species will not only become highly conspicuous to predators at their leks even when they are not displaying, but they will also lose the ability to court females as their visibility when displaying requires a combination of the forest shade, sunflecks, and reflected light from surrounding objects. In *C. gutturalis*, for example, males display by putting their white front and chest in the sunfleck, where the direct sunlight plus reflected light from the green mossy log at which they display, in contrast with their black back and sides in adjacent forest shade, make them most conspicuous (Endler and Théry 1996). Selective logging in the study area apparently disrupted the placement of suitable logs, in which case the reflected light from green moss was lost and their ability to court females was diminished, causing the males to move 1 km to new display sites (Théry, unpublished, cited in Endler and Théry 1996). Furthermore, Théry (2001) reports that along a gradient from primary to disturbed forest there is a positive correlation between hue range (which should be lower in disturbed forests with many large gaps, which will be dominated by whitish light) and the species diversity of manakins; this is consistent with the hypothesis that by reducing complexity in the color of forest light, forest disturbance may limit the availability of suitable courtship sites for lekking birds, limiting their species' potential distributions.

By attenuating incoming solar radiation, canopy vegetation also creates vertical gradients in temperature, such that the ground typically remains cooler than the upper forest boundary (Madigosky 2004). Vertical gradients develop concomitantly in relative humidity, which tends to be higher at the ground because the temperature is lower (Madigosky 2004). When a canopy gap forms, however, these gradients may become inverted each day once the sun passes overhead because the ground and understory vegetation will slowly radiate the solar energy that they have absorbed back upwards and the higher temperature would cause evaporation from the soil which would lower relative humidity at the ground. We would then expect species with low tolerance for desiccation to become less abundant on the ground and within the understory beneath the canopy gap.

Some of the clearest examples of vertical variation in the diversity and abundance of desiccation-prone species within tropical forests come from studies of arthropods, and some of these specifically illustrate the effects of vertical variation in relative humidity. Bassett et al. (2001) report that within a tropical forest in Gabon, herbivorous insects were three times as abundant, and twice as diverse, in the upper canopy than in the understory; furthermore, in one genus (*Agrilus*), only three of 26 species collected were found in both the upper canopy and the understory, suggesting low overlap among arthropod species between the upper and lower forest. Dial et al. (2006) measured microclimatic variables at sites where they sampled arthropods within a Bornean tropical forest and found arthropod abundance to be highest in mid-canopy where dense foliage kept the air moist. Moreover, they found nonuniform *horizontal* distributions of arthropod density within the canopy itself, which suggests that geometric variation even within and among crowns may determine the availability of suitable habitats for tropical forest arthropod species.

We do not know how increasing dynamism in canopy structure – i.e., with gaps presumably forming more frequently, becoming larger, and filling faster because rates of tree growth and mortality are increasing within tropical forests – is influencing the abundances, diversity, and distributions of tropical forest arthropod species. This begs investigation, given that these species comprise much or most of the terrestrial biodiversity on Earth (Wilson 1988; Godfray et al. 1999). We should especially study potential effects on the ecosystem dynamics of tropical forests - for example, trophic cascades that may follow changes in insect herbivore abundance, or changes in insect pollinator behavior that may influence tree reproduction and thereby patterns of forest regeneration (e.g., forest bees changing the heights at which they forage for nectar, as temperature, relative humidity, and wind speed can constrain their ability to shed metabolic heat during flight [Roubik 1993]).

Wood-decaying fungi present a particularly salient case of organisms whose response to changing canopy geometry may influence forest ecosystem dynamics. Wood-decaying fungi are ecosystem engineers – the decayed wood that they create provides ecological niches for a variety of other organisms (Odling-Smee et al. 2003), and wood decomposition is fundamental to nutrient recycling, soil formation, and carbon cycling within forests (Lonsdale et al. 2008) which are essential to forest regeneration. Regeneration often requires the decay of fallen logs in situ, and the survival of seedlings may be enhanced when they are growing on decayed wood (Lonsdale et al. 2008). Because the distribution of wood-decaying fungi depends on temperature and moisture, at whole-forest scales (e.g., the canopy may support different species than the understory because deadwood desiccates faster under the higher temperatures there) as well as across short distances within the canopy itself (Unterseher and Tal 2006, and references therein), pervasive changes in canopy geometry will likely have cascading effects on the ecosystem dynamics and function of tropical forests.

5 Conclusions

In summary, we have evidence that the dynamics of tropical forests are undergoing pervasive change, in the Amazon and quite possibly worldwide, and given the nature of these changes, the geometry of their canopies should be changing concomitantly. The drivers of these pervasive changes are not well understood, though this is an area of active research and it seems we are on our way to achieving some clarity. What we know almost nothing about is how subsequent effects on the physical structure of canopies within tropical forests may be affecting their biodiversity, and we believe that this topic demands scientific investigation. Global change research frequently attempts to predict how changing climate will influence the abundance and distribution of tropical forests and the carbon sinks that they provide, but the modeling involved does not consider the effects of change in tropical forest biodiversity and the feedbacks that must influence rates at which different tree species with different growth forms, wood densities, and carbon sequestering capacities will be successfully pollinated or have their fruits dispersed by the animal species whose diversity, abundance, and distributions must be influenced by changing canopy geometry. We agree with Endler (1997) and Théry (2001): through its influence on the intensity and color of light, as well as on temperature and relative humidity, changes in canopy geometry will have nontrivial effects on the biodiversity, dynamics, and function of tropical forests; without carefully investigating these phenomena, we are necessarily limiting our capacity to conserve (or even to restore) these ecosystems.

Acknowledgments We thank the editors for inviting DBG to participate in the 2009 Canopy Conference in Bangalore, India, and for their invitation to contribute to this volume. DBG was supported by a grant to the RAINFOR project from the Gordon and Betty Moore Foundation. SLL is supported by a Royal Society University Research Fellowship.

References

Achard F, Eva HD, Stibig HJ, Mayaux P, Gallego J et al (2002) Determination of deforestation rates of the world's humid tropical forests. Science 297:999–1002

Augspurger CK (1984) Light requirements of neotropical tree seedlings: a comparative study of growth and survival. J Ecol 72:777–795

Bassett Y, Aberlenc HP, Barrios H, Curletti G, Berenger JM, Vesco JP, Causse P, Haug A, Hennion AS, Lesobre L, Marques F, O'meara R (2001) Stratification and diel activity of arthropods in a lowland rainforest in Gabon. Biol J Linn Soc 72:585–607

Bazzaz FA, Pickett SAT (1980) Physiological ecology of a tropical succession: a comparative review. Annu Rev Ecol Syst 11:287–310

Bradbury JW (1981) The evolution of leks. In: Alexander RD, Tinkle DW (eds) Natural selection and social behavior: recent research and new theory. Chiron Press, New York

Caballe G, Martin A (2001) Thirteen years of change in trees and lianas in a Gabonese rainforest. Plant Ecol 152:167–173

Canham CD, Denslow JS, Platt WJ, Runkle JR, Speices TA, White PS (1990) Light regimes beneath closed canopies and tree-fall gaps in temperate and tropical forests. Can J For Res 20:620–631

Cao MK, Prince SD, Small J, Goetz SJ (2004) Remotely sensed interannual variations and trends in terrestrial net primary productivity 1981–2000. Ecosystems 7:233–242

Chapman CA, Chapman LJ, Struhsaker TT, Zanne AE, Clark CJ, Poulsen JR (2005) A long-term evaluation of fruiting phenology: importance of climate change. J Trop Ecol 21:31–45

Chave J, Condit R, Muller-Landau HC, Thomas SC, Ashton PS et al (2008) Assessing evidence for a pervasive alteration in tropical tree communities. PLOS Biology 6:455–462

Chazdon RL, Fletcher N (1984a) Photosynthetic light environments in a lowland tropical rainforest in Costa Rica. J Ecol 72:553–564

Chazdon RL, Fletcher N (1984b) Light environments of tropical forests. In: Medina E, Mooney HA, Vázquez-Yánez C (eds) Physiological ecology of plants of the wet tropics. Dr W Junk, The Hague

Clark DA, Piper SC, Keeling CD, Clark DB (2003) Tropical rain forest tree growth and atmospheric carbon dynamics linked to interannual temperature variation during 1984–2000. Proc Natl Acad Sci 100:5852–5857

Deng F, Chen JM (2011) Recent global CO_2 flux inferred from atmospheric CO_2 observations and its regional analyses. Biogeosciences 8:3263–3281

Denslow JS (1980) Gap Partitioning among tropical rainforest trees. Biotropica 12(suppl):47–55

Denslow JS (1987) Tropical rain-forest gaps and tree species diversity. Annu Rev Ecol Syst 18:431–451

Dial RJ, Ellwood MDF, Turner EC, Foster W (2006) Arthropod abundance, canopy structure, and microclimate in a Bornean lowland tropical rain forest. Biotropica 38:643–652

Endler JA (1993) The color of light in forests and its implications. Ecol Monogr 63:1–27

Endler JA (1997) Light, behavior and conservation of forest-dwelling organisms. In: Clemmons JR, Buchholz R (eds) Behavioral approaches to conservation in the wild. Cambridge University Press, Cambridge, UK

Endler JA, Théry M (1996) Interacting effects of lek placement, display behavior, ambient light, and color patterns in three Neotropical forest-dwelling birds. Am Nat 148:421–452

Engelstaedter S, Tigen I, Washington R (2006) North African dust emissions and transport. Earth Sci Rev 79:73–100

Feeley KJ, Wright SJ, Supardi MNN, Kassim AR, Davies SJ (2007) Decelerating growth in tropical forest trees. Ecol Lett 10:461–469

Godfray HCJ, Lewis OT, Memmott J (1999) Studying insect diversity in the tropics. Philos Trans R Soc Lond B Biol Sci 354:1811–1824

Groombridge B, Jenkins MD (2003) World atlas of biodiversity. University of California Press, Berkeley

Gurney KR, Baker D et al (2008) Interannual variations in continental-scale net carbon exchange and sensitivity to observing networks estimated from atmospheric CO_2 inversions for the period 1980 to 2005. Glob Biogeochem Cycle 22:3

Hubbell SP, Foster RB, O'Brien S, Wechsler B, Condit R, Harms K, Wright SJ, Loo de Lau S (1999) Light gaps, recruitment limitation and tree diversity in a Neotropical forest. Science 283:554–557

IPCC (2007) Summary for policymakers. In: Solomon S, Qin D, Manning M, Chen Z, Marquis M, Averyt KB, Tignor M, Miller HL (eds) Climate change 2007: the physical science basis. Contribution of working group I to the Fourth Assessment Report of the Intergovernmental Panel on Climate Change. Cambridge University Press, Cambridge, UK/New York

Johnson PL, Atwood DM (1970) Aerial sensing and photographic study of the El Verde rain forest. In: Odum HT, Pigeon RF (eds) A tropical rain forest; a study of irradiation and ecology at El Verde, Puerto Rico. Division of Technical Information, U.S. Atomic Energy Commission, Oak Ridge

Laurance WF (2005) Forest-climate interactions in fragmented tropical landscapes. In: Malhi Y, Phillips OL (eds) Tropical forests and global atmospheric change. Oxford University Press, Oxford

Lewis SL (2006) Tropical forests and the changing earth system. Philos Trans R Soc Lond B Biol Sci 261:195–210

Lewis SL, Tanner EVJ (2000) Effects of above- and belowground competition on growth and survival of rain forest tree seedlings. Ecology 81:2525–2538

Lewis SL, Malhi Y, Phillips OL (2004a) Fingerprinting the impacts of global change on tropical forests. Philos Trans R Soc Lond B Biol Sci 359:437–462

Lewis SL, Phillips OL, Baker TR, Lloyd J, Malhi Y et al (2004b) Concerted changes in tropical forest structure and dynamics: evidence from 50 South American long-term plots. Philos Trans R Soc Lond B Biol Sci 359:421–436

Lewis SL, Phillips OL, Sheil D, Vinceti B, Baker TR et al (2004c) Tropical forest tree mortality, recruitment and turnover rates: calculation, interpretation and comparison when census intervals vary. J Ecol 92:929–944

Lewis SL, Lopez-Gonzalez G, Sonke B, Affum-Baffoe K, Baker TR et al (2009a) Increasing carbon storage in intact African tropical forests. Nature 457:1003–1007

Lewis SL, Lloyd J, Sitch S, Mitchard ETA, Laurance WF (2009b) Changing ecology of tropical forests: evidence and drivers. Annu Rev Ecol Evol Syst 40:529–549

Lieberman M, Lieberman D, Peralta R (1989) Forests are not just swiss cheese: canopy stereogeometry of non-gaps in tropical forests. Ecology 70:550–552

Lloyd J, Farquhar GD (1996) The CO_2 dependence of photosynthesis, plant growth responses to elevated atmospheric CO_2 concentrations and their interaction with soil nutrient status. General principles and forest ecosystems. Funct Ecol 10:4–32

Lonsdale D, Pautasso M, Holderieder O (2008) Wood-decaying fungi in the forest: conservation needs and management options. Eur J For Res 127:1–22

Madigosky SR (2004) Tropical microclimate considerations. In: Lowman MD, Rinker HB (eds) Forest canopies, 2nd edn. Elsevier, Burlington

Malhi Y, Wright J (2004) Spatial patterns and recent trends in the climate of tropical rainforest regions. Philos Trans R Soc London B Biol Sci 359:311–329

Marlon JR, Bartlein PJ, Carcaillet C, Gavin DG, Harrison SP et al (2008) Climate and human influences on global biomass burning over the past two millennia. Nat Geosci 1:697–702

Nemani RR, Keeling CD, Hashimoto H, Jolly WM, Piper SC et al (2003) Climate-driven increases in global terrestrial net primary production from 1982 to 1999. Science 300:1560–1563

Odling-Smee FJ, Laland KN, Feldman MW (2003) Niche construction—the neglected process in evolution. Princeton University Press, Princeton

Parmesan C, Yohe G (2003) A globally coherent fingerprint of climate change impacts across natural systems. Nature 421:37–42

Phillips OL (1996) Long-term environmental change in tropical forests: increasing tree turnover. Environ Conserv 23:235–248

Phillips OL, Gentry AH (1994) Increasing turnover through time in tropical forests. Science 263:954–958

Phillips OL, Malhi Y, Higuchi N, Laurance WF, Nunez PV et al (1998) Changes in the carbon balance of tropical forests: evidence from long-term plots. Science 282:439–442

Phillips OL, Martinez RV, Arroyo L, Baker TR, Killeen T et al (2002) Increasing dominance of large lianas in Amazonian forests. Nature 418:770–774

Phillips OL, Baker TR, Arroyo L, Higuchi N, Killeen TJ et al (2004) Pattern and process in Amazon tree turnover, 1976–2001. Philos Trans R Soc Lond B Biol Sci 359:381–407

Phillips OL, Lewis SL, Baker TR, Chao KJ, Higuchi N (2008) The changing Amazon forest. Philos Trans R Soc Lond B Biol Sci 363:1819–1827

Phillips OL, Aragao L, Lewis SL, Fisher JB, Lloyd J et al (2009) Drought sensitivity of the Amazon rainforest. Science 323:1344–1347

Proulx R, Parrot L (2008) Measures of structural complexity in digital images for monitoring the ecological signature of an old-growth forest ecosystem. Ecol Indic 8:270–284

Ramankutty N, Gibbs HK, Achard F, Defries R, Foley JA, Houghton RA (2007) Challenges to estimating carbon emissions from tropical deforestation. Glob Change Biol 13:51–66

Roubik DW (1993) Tropical pollinators in the canopy and understory: field data and theory for stratum "preferences". J Insect Behav 6:659–673

Stephens BB, Gurney KR, Tans PP, Sweeney C, Peters W et al (2007) Weak northern and strong tropical land carbon uptake from vertical profiles of atmospheric CO_2. Science 316:1732–1735

Théry M (2001) Forest light and its influence on habitat selection. Plant Ecol 153:251–261

Unterseher M, Tal O (2006) Influence of small scale conditions on the diversity of wood decay fungi in a temperate, mixed deciduous forest canopy. Mycol Res 110:169–178

Weishampel JF, Godin JR, Henebry GM (2001) Pantropical dynamics of 'intact' rain forest canopy texture. Glob Ecol Biogeogr 10:389–397

Wilson EO (1988) Biodiversity. National Academies Press, Washington, DC

Wright SJ (2010) The future of tropical forests. Ann N Y Acad Sci 125:1–27

Wright SJ, Calderón O, Hernandéz A, Paton S (2004) Are lianas increasing in importance in tropical forests? A 17- year record from Panama. Ecology 85:484–489

Wright SJ, Calderon O (2006) Seasonal El Nino and longer term changes in flower and seed production in a moist tropical forest. Ecol Lett 9:35–44

Chapter 26
Reproductive Biology and Population Genetics of Some Canopy- and Understorey-Dominant Tree Species of Sri Lanka: Implications for Conservation Management in a Fragmented Landscape

Nimal Gunatilleke and Savitri Gunatilleke

Keywords Sri Lanka • Sinharaja rain forest • Forest fragmentation • Plant reproductive biology • Population genetics

Bullet Points

1. While all the canopy-dominant tree species examined are predominantly outbreeding and pollinated by *Apis* spp., the understorey-dominant tree *Xylopia championii* of Annonaceae is pollinated by a curculionid beetle species.
2. Forest fragmentation, degradation and eventual isolation by distance may lead to enhanced levels of inbreeding in these naturally outbreeding species.
3. Population-level genetic diversity estimates in some of the canopy-dominant species was found to be higher in forests outside the current protected area system, and consequently, such estimates could serve as useful criteria in selecting additional conservation areas.

Summary

Reproductive biological features of several tree species, including closely related canopy-dominant dipterocarp species, in the fragmented lowland rain forests of Sri Lanka in the context of their genetic diversity conservation are reviewed in this chapter. The members of the partially sympatric endemic dipterocarp clade of *Doona* exhibit distinct differences in their leaf morphology, patterns of flowering phenology, and microhabitat requirements for seedling growth. While nearly half the *Shorea* (*Doona*) species displayed asynchronous annual flowering and fruiting patterns, the remainder exhibited a supra-annual episodic flowering and mast fruiting at irregular multi-year intervals at the population level.

N. Gunatilleke (✉) • S. Gunatilleke
University of Peradeniya, Peradeniya, Sri Lanka
e-mail: savnim@gmail.com

M. Lowman et al. (eds.), *Treetops at Risk: Challenges of Global Canopy Ecology and Conservation*, DOI 10.1007/978-1-4614-7161-5_26,

Studies on pollination ecology and breeding systems of these species were facilitated by canopy platforms and aluminium ladder systems erected on each of the trees sampled. Selective logging of these bee-pollinated and thus predominantly outcrossing *Shorea* (*Doona*) species has shown significantly increased levels of inbreeding. Population-level genetic variation estimated for the *Shorea* (*Doona*) and *Syzygium* species indicate a strong potential for biparental inbreeding depression and partial reproductive isolation among adult trees in isolated small forest fragments leading to genetic erosion over time. In *S. cordifolia*, fruit set increased steadily with outcrossing distance, peaking at intermediate distance in within-forest crosses and then declined in cross-fertility (i.e. outbreeding depression) in the between-forest crosses. This may be due to genetic divergence of separate forest populations through drift and/or the influence of a selectively heterogeneous environment.

The understorey-dominant trees of *Xylopia championii* of Annonaceae examined from canopy platforms showed typical adaptations for a cantherophilic pollination syndrome and a facultatively xenogamous breeding system. Population-level genetic diversity estimates in some of the canopy-dominant species were found to be higher in forests outside the current protected area system, and consequently, such estimates could serve as useful criteria in selecting additional conservation areas. Studies of this nature would serve as benchmarks in estimating genetic differentiation with time in future generations of adult tree populations in these degraded and fragmented forests.

1 Introduction

Fragmentation of the continuous tropical forest-scape into isolated remnant forest patches among seas of human influenced alternative landscapes is a widespread pantropical phenomenon compounded by the emerging impacts of global climate change events (Loo et al. 2011). Sri Lanka, with exceptionally rich biodiversity, particularly in its aseasonal south-western sector, is no exception to this alarming trend, acknowledged as such by declaring the island together with Western Ghats of India as one of 34 global hotspots of biodiversity (Mittermeier et al. 2004).

Fragmentation leads to a decrease in size and often an increase in spatial isolation of a population, while its effects on the dynamics of gene flow may directly influence genetic structure within a forest fragment. When forest fragments are isolated and mating is restricted largely to neighbouring genetically related trees, biparental inbreeding depression is a predicted consequence (Stacy 2001). Genetic isolation, by restricted gene flow among fragments, may ultimately have detrimental consequences on the evolutionary viability of a population, through increased levels of inbreeding, random genetic drift and founder effects (White et al. 2002). Consequently, there is a growing and urgent need for information on reproductive and genetic responses of tree species to disturbance, resulting from fragmentation accompanying degradation due to selective logging, in order to manage such forest-scapes in a sustainable manner (Nason and Hamrick 1997).

Sound scientific information on how species survive and maintain their genetic diversity in fragmented landscapes will help decisions on conservation management of species and their habitats. This is of particular importance in today's context in redesigning and demarcating networks of protected areas in a milieu of rapid deforestation, in implementing targets set by the Global Strategic Plan for Biodiversity 2011–2020 (www.cbd/int/sp). These networks include forests of varying sizes, shapes and neighborhoods as well as varying proximities to one another resulting from changes in land use. In this respect, a better understanding of reproductive phenology and ecology, breeding and mating systems leading to historic genetic differentiation, contemporary gene flow patterns and genetic diversity of tree species among the remaining large continuous and isolated small populations are of critical importance in developing strategies for potential adaptations to climate change and anthropogenic disturbance (Finger et al. 2011).

In response to this critical situation in Sri Lanka, our research team (including several overseas universities) conducted a series of long-term studies on reproductive biology and genetic diversity of rain forest plant species, particularly of canopy-dominant dipterocarps and several other tree species (Murawski et al. 1994a, b; Dayanandan et al. 1990; Dayanandan 1996; Stacy 2001; Stacy et al. 2001; Ratnayake et al. 2007). An overview of our major findings is given in this chapter.

2 Reproductive Phenology

Phenology is the study of the timing of recurrent biological events of the life cycles of a species, the causes for timing of such events with regard to biotic and abiotic factors and the interaction among phases of the same or different species (Sakai 2001). Reproductive phenological patterns indicate the periodicity and levels of gene flow via pollen and seeds, both within and among populations, of a given species. Long-term phenological observations made over 12 years have revealed that there are two general population-level patterns of reproductive phenology in the Sri Lankan *Shorea (Doona)* species, the most dominant group of canopy species in lowland and lower montane rain forests of Sri Lanka (Figs. 26.1, 26.2, and 26.3):

(i) The light hardwood *Thiniya-Yakahalu Dun* group of *Shorea* species flower and fruit annually at population level, sometimes with more than one reproductive episode per year in a more or less asynchronous manner.
(ii) The flowering events of the medium hardwood Beraliya group of *Shorea* species, on the other hand, are more tightly synchronized and exhibit supra-annual general flowering and mast fruiting. Nevertheless, they seem to show some degree of sequential flowering episodes leading to a partial overlap in flower availability (Dayanandan et al. 1990). Sequential flowering among these *Shorea* species seems to be maintained by extremely poor reproductive success of individuals flowering outside of their conspecific flowering periods (Stacy 2001).

All four species within the second group appear to be responding to a common supra-annual environmental stimulus for their reproductive phenology. Consequently,

Fig. 26.1 Crowns of the annually flowering *Shorea (Doona) trapezifolia* (Dipterocarpaceae) dominating the canopy of lowland and mid-elevational rain forests in Sri Lanka

Fig. 26.2 Lower montane rain forest canopy co-dominated by *Shorea (Doona) trapezifolia* and *Shorea (Doona) gardneri* (in flower) in SW Sri Lanka

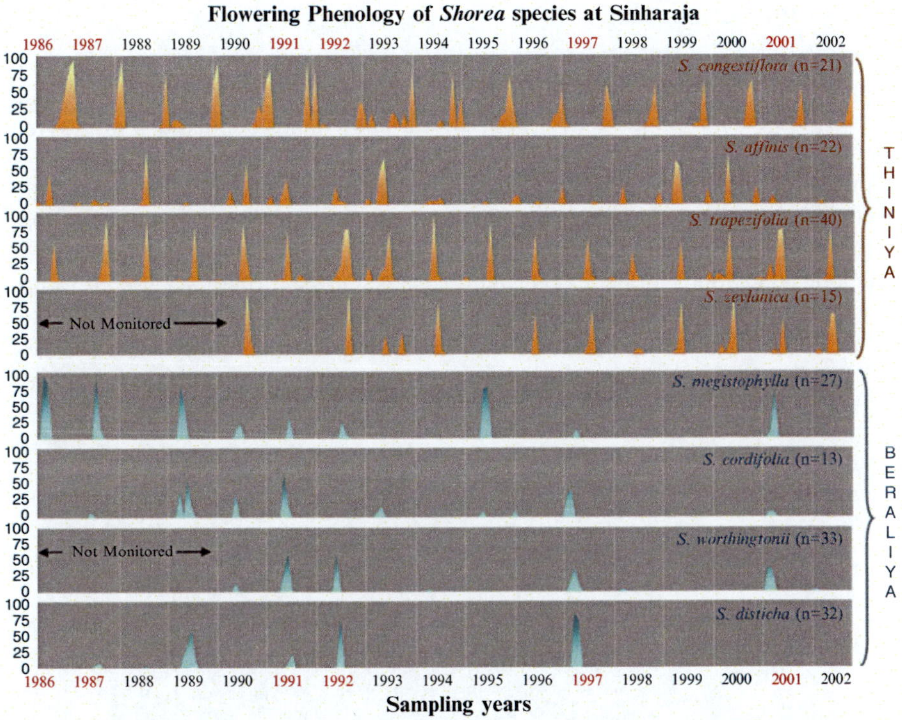

Fig. 26.3 Flowering phenology of the *Thiniya-Yakahalu Dun* and *Beraliya* groups of *Shorea (Doona)* species in NW Sinharaja rain forest, Sri Lanka. Years in *red* colour on the X-axis are *El Nino* years. Y-axis denotes percent of trees flowering in sample observed (n) in each species

unlike the *Thiniya-Yakahalu Dun* group, the *Beraliya* group exhibits episodic flowering and mast fruiting at irregular, multi-year intervals at the population level with some degree of coincidence with *El Nino*-Southern Oscillation events (Gunatilleke 1999).

Members of both groups, all endemic to the island, exhibit partial sympatry in Sri Lankan rainforests yet differ markedly in terms of flowering phenology, flower and leaf morphology and microhabitats for growth. The correlation between species-specific seedling performance and microhabitat conditions is so strong in these two groups that speciation is believed to have occurred in relation to topography-dependent variation in soil nutrition and light regimes at the ground level (Dayanandan et al. 1990; Ashton et al. 1995; Gunatilleke et al. 1997; Ashton et al. 2001; Gunatilleke et al. 2006).

Population-level phenological studies conducted over a three-year period (2002–2004) in one of the understorey-dominant annonaceous tree species, *Xylopia championii* endemic to lowland rain forests of Sri Lanka, revealed that its peak flowering generally occurs during the moderately rainy inter-monsoonal season from December to April (Ratnayake et al. 2007). Unlike in canopy dipterocarps, the flowering and fruiting intensities are consistently low at both the individual and population level in this understorey tree species.

3 Pollination Ecology

Plant-animal interactions related to both pollen and seed dispersal are also important factors that determine gene flow both within and among populations of a given species. While invertebrates like bees and butterflies usually have short flight ranges with limited gene flow, mammals such as bats, which pollinate wild durians, fly over great distances facilitating gene exchange over a much larger area (Bawa 1990; Marshall 1983). In pollination ecology, we examined plant-pollinator interactions and floral adaptations for pollen transfer (Dayanandan et al. 1990; Stacy et al. 1999). Close-range observations of pollination ecology of canopy giants which are often 30 m tall were made by accessing the tree crowns with aluminium ladder systems and wooden platforms (Figs. 26.4 and 26.5).

Fig. 26.4 Erecting aluminium ladder and rope systems to reach the canopy of a *Shorea (Doona) trapezifolia* tree in NW Sinharaja lowland rain forest, Sri Lanka

Fig. 26.5 Climbing up the ladder system to the canopy platform for pollination ecological studies of *Shorea* (*Doona*) sp. in NW Sinharaja lowland rain forest, Sri Lanka

Over 20 species of insect visitors have been observed to visit *Shorea* flowers (Dayanandan et al. 1990; Stacy 2001). However, the rock bee (*Apis dorsata*) and the Asian honey bee (*Apis cerana* var. *indica*) were found to be the most frequent visitors as well as the most effective pollinators based on close-range observations made from canopy platforms (Fig. 26.6). The other social, stingless bee species were seen moving slowly on stamens or petals but rarely touching the stigma while collecting pollen.

Another subcanopy-dominant tree species in SW Sri Lanka, *Syzygium rubicundum* (Myrtaceae), similarly attracted a broad array of floral visitors during its annual flowering event in January/February, as observed from close-range canopy platforms. In this species too, the most effective pollinators were found to be the medium-sized bees (*Apis* spp.) as in dipterocarps (Stacy et al. 1999).

On the other hand, one of the most widespread understorey tree species in lowland rain forests of Sri Lanka, *Xylopia championii*, showed typical adaptations for cantherophilic pollination syndrome, viz., creamy-yellow protogynous flowers with a floral chamber, thermogenesis and scent production during the functionally active pistillate and staminate phases of a two-day rhythm of anthesis (Ratnayake et al. 2007). From observations made on crown platforms on this understorey tree, its most likely pollinator was observed to be a crepuscular and nocturnal coleopteran, *Endaeus* sp.

Fig. 26.6 A honey bee (*Apis cerana var indica*) collecting pollen from a *Shorea megistophylla* (*Doona macrophylla*) flower (Picture taken from a canopy platform)

4 Breeding Systems

Breeding system studies examine the morphological attributes of plant species that facilitate either self-pollination or cross-pollination, e.g. bisexuality, monoecy, dioecy and their variants.

The breeding systems of *Shorea* species were examined by a series of controlled hand-pollination experiments. These studies revealed that the *Shorea* species favor outcrossing to selfing, although they have bisexual flowers (Dayanandan et al. 1990). The exceptionally low fruit set in selfed flowers of almost all of the species examined may be due to strong inbreeding depression, pollinator limitation or to genetic self-incompatibility (Bawa 1992; Stacy 2001).

Fruit set in *Shorea* spp. in artificially cross-pollinated treatments was higher than in natural open-pollinated controls (Dayanandan et al. 1990). This suggests the limitation of pollinator availability for lower fruit set in open-pollinated flowers. Stacy (2001) also observed a significant relationship between outcrossing distance and fruit set in *S. cordifolia*, where fruit set increased steadily with outcrossing distance, peaking at intermediate distance in within-forest crosses (1–10 km) and then declining in the between-forest crosses. These studies coupled with seed germination and seedling growth resulting from these crosses indicated a strong potential for biparental inbreeding depression within-forest tree populations, resulting from outcrossing among genetically related individuals. They also showed partial reproductive isolation among trees in different forest fragments. The dramatic drop in cross-fertility (i.e. outbreeding depression) observed for between-forest crosses relative to within-forest crosses may be due to genetic divergence of separate forest populations through drift and/or the influence of a selectively heterogeneous environment (Stacy et al. 2001). *Xylopia championi* was found to possess a facultative xenogamous breeding system.

5 Mating Systems

Mating system studies examine the genetic consequences of the products of the fertilization events, by observing the polymorphic variations in genetic markers in the form of allozymes or DNA fragments (microsatellites, single sequence repeats, etc.) in progeny arrays. Using these, the mating systems in several *Shorea* species (*S. trapezifolia, S. congestiflora, S. cordifolia* and *S. megistophylla*) and one *Stemonoporus* species (*St. oblongifolius*) have been quantified (Murawski et al. 1994a, b; Dayanandan 1996; Stacy 2001).

Segregating patterns of these genetic markers among progeny seeds (half-sib families) were compared with that of 20–40 individual mother trees. From these results, the relative proportions of seeds that developed from inbreeding and outbreeding were calculated for single as well as multiple loci, using the mixed-mating model of Ritland and Jain (1981).

The results revealed that *S. congestiflora* is a predominantly outcrossing species with a multi-locus outcrossing rate (t_m) of 87 % (Murawski et al. 1994a). Both *S. megistophylla* and *S. trapezifolia,* on the other hand, has a mixed-mating system with a t_m value ranging between 54 % and 64 % (Murawski et al. 1994b). In *S. megistophylla,* outcrossing was more prevalent in the unlogged primary forest than in the selectively logged forest. In this species, the single-locus outcrossing rate was significantly lower than the multi-locus outcrossing rate in the logged forest, but not in the unlogged forest. This significant difference in the logged forest may be attributed to greater inbreeding as a result of removal of adult trees during logging.

The fixation index, yet another tool for estimating the level of inbreeding, in these three *Shorea* species is lower in adults than that of their respective progeny. This suggests that their adult trees may be subject to considerable selection against selfed progeny, compared to those that are outcrossed. Inbreeding appears to be a consequence of both selfing and mating among close relatives (kin mating), which is more common in logged forests leading to genetic isolation and enhanced inbreeding (Murawski et al. 1994b). Nearly 10-fold decrease in the density of adults of *S. megistophylla* in the logged forest would have restricted tree-to-tree movement of pollinators, due either to the increased distance among conspecific flowering individuals or greater intensity of flowering within each crown resulting from greater incidence of light, or a combination of both. A similar pattern of inbreeding depression was observed for both *Syzygium rubicundum* and *Shorea cordifolia* (Stacy 2001).

6 Genetic Diversity

Genetic diversity is the foundation of evolutionary processes and therefore its understanding is essential for adaptation under changing environments. The study of genetic structure and diversity reveals the organization and distribution of genetic variability within as well as among populations of a species (Hamrick 1994). Using several genetic parameters, Dayanandan (1996) estimated the genetic variation of *S. trapezifolia* both within and outside the currently protected Sinharaja National Heritage Wilderness Area. The reserves outside the protected area included both large and small relict forest patches resulting from forest fragmentation and isolation. Furthermore, the genetic variation in *S. megistophylla* populations in logged and unlogged forests within Sinharaja rain forest was also compared (Murawski et al. 1994b).

The proportion of polymorphic loci, effective number of alleles per locus and observed and expected heterozygosities of *S. trapezifolia* were lowest in a small relict forest fragment in Guruluwana near Ratnapura. This may be a consequence of forest fragmentation, degradation and genetic isolation. The genetic substructuring in this population could be a result of fragmentation and process of isolation by distance. Forest fragmentation leads to decrease in population sizes and accentuates the effects of genetic drift, thus enhancing inbreeding depression.

Genetic variation of *S. trapezifolia* was high in Walankanda, Suriyakanda and Balangoda-Rassagala populations, which are at a higher altitude than those in the middle elevations in SW Sri Lanka. This species has a wider ecological amplitude, but its greatest abundance is towards the upper end of the lowland rain forest and in lower montane regions at an altitudinal range of 700–1,100 m. As such, the genetic diversity of this species may be at its highest at these altitudes (Dayanandan 1996). Genetic diversity studies also revealed that 89 % of the genetic variation is found within populations and only 11 % of it is among the ten populations of *S. trapezifolia* sampled, suggesting that there are only a few alleles and genotypes restricted to each of the populations sampled. In genetic terminology, this is referred to as low genetic differentiation or low genetic substructuring (low G_{ST}).

A phenogram of genetic distances constructed using UPGMA cluster analysis failed to recognize any geographical pattern of population differentiation in *S. trapezifolia* (Dayanandan 1996). The results also indicated that the Nei's genetic distance did not correlate with geographic distance between pairs (Spearman's rho=0.214; p>0.05). Using seven polymorphic microsatellite markers developed, *S. cordifolia*, too, showed that the genetic differentiation among populations was low (F_{ST}=0.031) and did not fit a strict isolation by the distance model (Stacy 2001). It suggests that there is still widespread gene flow and little genetic structuring in the original populations represented by adult trees in these forest fragments.

These findings suggest that the populations are still linked by cohesive forces through higher levels of gene flow among populations or that forest fragmentation in this region is relatively more recent compared to the expected life spans of forest canopy trees (Stacy 2001). These studies would serve as a benchmark in estimating genetic differentiation with time, in the future generations of adult trees in these fragmented forests.

7 Discussions and Conclusions

The above research reveals that forest fragmentation and degradation lead to accentuation of genetic drift resulting from inbreeding and kin mating causing a decrease in heterozygosity which may, in the longer term, lead to accelerated genetic erosion in Sri Lankan rain forests. Although genetic substructuring is still not widespread among present-day adult individuals, there are indications of a strong potential for biparental inbreeding depression within-forest tree populations and a partial reproductive isolation among trees occupying the remaining fragmented forest reserves in SW Sri Lanka (Stacy 2001). These factors can have a negative influence on average individual fitness. The loss of genetic variability can diminish future adaptability to changing environments. Forest fragmentation could reduce the reproductive success, through reductions in germination rate and eventual fruit production resulting from reduced fitness. This may consequently affect the resilience of a given species to adapt and survive in changing environments.

This genetic drift in fragmented forests could be reversed by establishing 'gene corridors' or 'gene bridges' among these forest fragments. We have developed a restoration model which could be used for this purpose by converting a monoculture *Pinus caribaea* stand into a mixed species native forest stand that include *S. trapezifolia, S. megistophylla* and *S. cordifolia* among other indigenous species (Ashton et al. 1997, 1998, 2001). If such monoculture tree plantations raised among natural forest fragments could be enriched with suitable indigenous species, they have the potential to serve as gene corridors or bridges linking nearby natural forest fragments.

The Global Strategic Plan for Biodiversity developed for implantation in the decade of Biodiversity from 2011 to 2020 has set global conservation targets for establishing 17 % of terrestrial and inland water areas and restoring at least 15 % of degraded areas through conservation and restoration activities, thereby contributing to climate change mitigation and adaptation and to combating desertification. In this context, a sound understanding of how species survive and the extent to which their genetic diversity is maintained in fragmented landscapes and in fragments connected through gene corridors is imperative. This has become all the more relevant and propitious at a time when Sri Lanka is moving towards adopting strategies for Reducing Emissions from Deforestation and Degradation ($REDD^{+}$), while ensuring co-benefits which naturally include gene pool conservation of both native and naturalized species.

Acknowledgements This chapter is based on research carried out by a number of colleagues whose contributions to enhance our understanding on reproductive biology and Population genetics of Sri Lankan forest tree species are gratefully acknowledged. Thanks are due to Profs. Peter Ashton, Kamaljith Bawa, Mark Ashton and Richard Saunders for generating funds to carry out research and provide training. The facilities provided by the University of Peradeniya, Forest Department and the National Science Foundation of Sri Lanka are also appreciated.

References

Ashton PMS, Gunatilleke CVS, Gunatilleke IAUN (1995) Seedling survival and growth of four *Shorea* species in a Sri Lankan rainforest. J Trop Ecol 11:263–279

Ashton PMS, Gamage S, Gunatilleke CVS, Gunatilleke IAUN (1997) Restoration of a Sri Lankan rain forest: using Caribbean pine, *Pinus caribaea* as a nurse for establishing late successional tree species. J Appl Ecol 34:915–925

Ashton PMS, Gamage S, Gunatilleke IAUN, Gunatilleke CVS (1998) Using Caribbean pine to establish a mixed plantation: testing effects of pine canopy removal on plantings of rain forest tree species. For Ecol Manage 106(2):211–222

Ashton MS, Gunatilleke CVS, Singhakumara BMP, Gunatilleke IAUN (2001) Restoration pathways for rain forests in southwest Sri Lanka: a review of concepts and models. For Ecol Manage 154:409–430

Bawa KS (1990) Plant-pollinator interactions in tropical rain forests. Ann Rev Ecol Syst 21:399–422

Bawa KS (1992) Mating systems, genetic differentiation and speciation in tropical rain forest plants. Biotropica 24:250–255

Dayanandan S (1996) Genetic diversity and mating system of *Shorea trapezifolia* (Dipterocarpaceae). M.Sc. thesis, University of Massachusetts, Boston

Dayanandan S, Attygalle DNC, Abeygunasekera AWWL, Gunatilleke IAUN, Gunatilleke CVS (1990) Phenology and floral morphology in relation to pollination of some Sri Lankan dipterocarps. In: Bawa KS, Hadley M (eds) Reproductive ecology of tropical forest plants, Man & biosphere series. The Parthenon Publishing Group/UNESCO, Carnforth/Paris

Finger A, Kaiser-Bunbury CN, Kettle CJ (2011) Development of polymorphic microsatellite markers of the Seychelles endemic tree *Gleonnetia sericea* (Rubiaceae). Conserv Genet Resour. doi:10.1007/s12686-011-9515

Gunatilleke IAUN (1999) Reproductive biology and genetics of selected rain forest plant species of Sri Lanka: Implications for conservation. (Presidential address-Section D, 1998). In: Proceedings of the 54th annual session, Part II. Sri Lanka association for the advancement of science, Colombo, p. 95–122

Gunatilleke CVS, Gunatilleke IAUN, Perera GAD, Burslem DFRP, Ashton PMS (1997) Responses to nutrient addition among seedlings of eight closely related species of *Shorea* in Sri Lanka. J Ecol 85(3):301–311

Gunatilleke CVS, Gunatilleke IAUN, Esufali S, Harms KE, Ashton PMS, Burslem DFRP, Ashton PMS (2006) Species-habitat associations in a Sri Lankan dipterocarp forest. J Trop Ecol 22:371–384

Hamrick JL (1994) Genetic diversity and conservation in tropical forests. In: Drysdale RM, John SET, Yopa AC (eds) Proceedings of the international symposium on genetic conservation and production of tropical forest tree seed, ASEAN Canada Forest Seed Center Project, Muak Lek

Loo J, Bruno F, Dawson I, Vinceti B, Baldinelli G (2011) Climate change and forest genetic resources – state of knowledge, risks and opportunities. Background study paper no.56 prepared for the Commission on Genetic Resources for Food and Agriculture, Rome

Marshall AG (1983) Bats, flowers and fruit: evolutionary relationships in the old world. Biol J Linn Soc 20(1):115–135

Mittermeier RA, Gil PR, Hoffmann M, Pilgrim J, Brooks T, Mittermeier CG, Lamoreux J, da Fonseca GAB (2004) Hotspots revisited. Cemex, Mexico

Murawski DA, Dayanandan S, Bawa KS (1994a) Outcrossing rates of two endemic *Shorea* species from Sri Lankan tropical rain forests. Biotropica 26:23–29

Murawski DA, Gunatilleke IAUN, Bawa KS (1994b) The effects of selective logging on inbreeding in *Shorea megistophylla* (Dipterocarpaceae) from Sri Lanka. Conserv Biol 8:997–1002

Nason JD, Hamrick JL (1997) Reproductive and genetic consequences of forest fragmentation: two case studies of neotropical canopy trees. J Hered 8:264–276

Ratnayake RMCS, Gunatilleke IAUN, Wijesundera DSA, Saunders RMK (2007) Pollination ecology and breeding system of *Xylopia championii* (Annonaceae): curculionid beetle pollination promoted by floral scents and elevated floral temperatures. Int J Plant Sci 168(9):1255–1268

Ritland K Jain SK (1981) A model for the estimation of outcrossing rate and gene frequencies based on n independent loci. Heredity 47:37–54

Sakai S (2001) Phenological diversity in tropical forests. Popul Ecol 43:77–86

Stacy EA (2001) Relatedness-dependent cross-fertility in tropical rain forest trees and the nature of species boundaries. Ph.D. thesis, Boston University

Stacy EA, Harischandran S, Gunatilleke IAUN (1999) Reproductive biology of *Syzygium rubicundum* wight and arn. (Myrtaceae) in selectively logged forest in Sinharaja world heritage site. Sri Lanka Forester XXIII(3 & 4):18–29

Stacy EA, Gunatilleke IAUN, Gunatilleke CVS, Dayanandan S, Schneider CJ, Khasa PD (2001) Crossing barriers within rainforest tree species can arise over small geographic scales. In: Ganeshaiah KN, Uma Shanker R, Bawa KS (eds) Structure, diversity and human welfare. Proceedings of the international conference on tropical ecosystems.,Oxford-IBH, New Delhi

Strategic plan for biodiversity 2011–2020 and the Aichi Targets. www.cbd/int/sp

White GM, Boshier DH, Powell W (2002) Increased pollen flow counteracts fragmentation in a tropical dry forest: an example from *Swietenia humilis* Zuccarini. PNAS 99(4):2038–2042

Chapter 27
The Importance of Flowers for Beetle Biodiversity and Abundance

Carl W. Wardhaugh

Keywords Flower-visitors • Host specialization • Rainforest canopy • Resource differentiation • Species richness

Bullet Points

1. The tropical rainforest canopy supports a large proportion of extant biodiversity, yet we have little knowledge of the spatial distribution of canopy organisms or what resources most species are utilizing.
2. Distinct microhabitats (mature leaves, new leaves, flowers) supported their own unique beetle communities in terms of abundance, diversity, and species composition.
3. Flowers were especially important microhabitats and supported a disproportionately large number of species, while beetle densities (per unit biomass of microhabitat) were 10–10,000 times greater on flowers than on adjacent foliage.
4. Accounting for flower-visiting species in conservation planning may be very important for protecting biodiversity and maintaining ecosystem function under a changing climate.

Summary

Tropical rainforest canopies support a large proportion of all terrestrial invertebrate species. Estimates of total biodiversity, energy flow and nutrient cycles, and food web analyses in tropical rainforests are based in part on our current understanding of the distribution of canopy invertebrate abundance. However, knowledge of canopy invertebrate distribution patterns have been based almost exclusively on species

C.W. Wardhaugh (✉)
School of Marine and Tropical Biology, James Cook University,
Cairns Campus, McGregor Road, Smithfield, Cairns, QLD 4870, Australia
e-mail: carl.wardhaugh@my.jcu.edu.au

M. Lowman et al. (eds.), *Treetops at Risk: Challenges of Global Canopy Ecology and Conservation*, DOI 10.1007/978-1-4614-7161-5_27,

inhabiting the foliage, under the assumption that foliage samples are representative of the entire canopy. I tested this assumption by comparing the density, species richness, and specialization of beetles (Coleoptera) on three microhabitats (mature leaves, new leaves, and flowers) in an Australian tropical rainforest. There was marked differentiation between canopy microhabitats in the density and composition of associated beetle species. Flowers supported densities (per unit resource biomass) that were several orders of magnitude greater than those on adjacent foliage. Furthermore, microhabitat specificity was very high and flowers in particular supported a disproportionately rich fauna compared to the leaves. I therefore reject the hypothesis that the beetle fauna on mature foliage is representative of other microhabitats or the wider canopy community. The identification of flowers as important resources for a large proportion of canopy beetle species in an Australian rainforest, and potentially in other rainforests, could alter our understanding of the evolution of plant-herbivore interactions and food web dynamics. This information could provide a better platform for protecting biodiversity and functioning ecosystems from human-induced disturbances.

1 Introduction

The rainforest canopy supports a large, but as yet uncertain proportion of global biodiversity (Price 2002; Ozanne et al. 2003; Grimbacher and Stork 2007). The high species richness of plants and animals in the canopy and their interactions have been shown to be strongly influential in determining food web dynamics (Novotny et al. 2010) and form the basis for many estimates of global species richness (Erwin 1982; Stork 1993; Ødegaard 2000a; Novotny et al. 2002; Hamilton et al. 2010). However, while the high diversity of invertebrates in rainforest canopies is acknowledged (Southwood 1961; Erwin 1982; Moran and Southwood 1982; Stork 1988), the logistical difficulties in accessing the canopy have placed sampling limitations on previous biodiversity and ecological studies. Consequently, most studies to date have used sampling techniques that indiscriminately sample many arboreal microhabitats together, such as insecticide fogging (Erwin 1982; Moran and Southwood 1982; Stork 1988) or flight interception/Malaise traps (Stork and Grimbacher 2006), or focus only on sampling species from leaves as the dominant habitat (Novotny and Basset 2005).

But rainforest canopies contain a range of other resources that may be exploited by invertebrates, including flowers, fruits, bark, and living and dead wood, and resource differentiation and niche-based theories predict specialization by invertebrates on different microhabitats (e.g., Condon et al. 2008). The practical result of the previous sample bias is that it remains unknown whether samples taken from foliage accurately reflect the spatial and temporal patterns of invertebrate abundance and diversity in the canopy as a whole. Without information on how species are distributed in time and space or with respect to microhabitat differentiation, it is difficult to make any generalizations about ecological processes such as nutrient cycling and energy flow.

Apart from the recent work by me and my collaborators (Wardhaugh et al. 2012a, b, c, 2013), the only other study of which I am aware that has simultaneously examined the insect assemblages on more than one microhabitat from tropical rainforest canopy trees is that of Ødegaard (2000b, 2004) using the canopy cranes in Panama. He focussed on the host specificity of the herbivorous beetle (Buprestoidea, Chrysomeloidea, and Curculionoidea) assemblages collected from leaves, flowers, and suspended dead wood. Approximately 20 % of all the beetle species he collected were flower-visitors, and each microhabitat type supported a unique beetle community (Ødegaard 2000b). Furthermore, a greater number of beetle species were collected from suspended dead wood than the leaves on *Brosimum utile* (Moraceae), with little overlap between each assemblage (Ødegaard 2004).

Although an accurate understanding of the distribution and diversity of canopy insects is important for quantifying food web dynamics and ecosystem function, apart from our own work, and that of Ødegaard (2000b, 2004), no published studies have quantified microhabitat differentiation in the rainforest canopy. In this study, I examine microhabitat differentiation among the beetle community by comparing density per unit dry weight, species richness, and microhabitat specialization on discrete canopy microhabitats. Specifically, I examine the assemblages on mature leaves, new leaves, and flowers from 23 species of rainforest canopy plants to assess the spatial distribution of arboreal beetles and determine the relative contribution of each microhabitat to overall canopy beetle diversity. I tested two hypotheses: (1) canopy beetle density and species richness are directly proportional to the amount of resource available and (2) canopy microhabitats represent discrete resources that are utilized by their own specialized beetle assemblages. This protocol allowed for an assessment of the validity of using the foliage-inhabiting beetle assemblage to represent canopy-wide patterns in abundance, density, and species richness. This chapter integrates prior research that looked specifically at microhabitat differentiation and the spatial distribution and specialization of rainforest canopy invertebrates, with a special emphasis on the beetle community (Wardhaugh et al. 2012a, b, c, 2013)

2 Methods

2.1 *Study Site*

All fieldwork was conducted using the Australian Canopy Crane (www.jcu.edu.au/canopycrane/) at the Daintree Rainforest Observatory (a Long-Term Ecological Research site), near Cape Tribulation (16°17′S, 145°29′E) Queensland, Australia (Stork 2007). The crane is situated approximately 40 m a.s.l. and >300 m from the forest edge in complex mesophyll vine forest (Tracey 1982) that is contiguous with the extensive lowland and upland rainforests of the Daintree National Park and Wet Tropics World Heritage Area (0 m a.s.l. to >1,300 m a.s.l.). Approximately 1 ha of rainforest containing 745 individual trees (>10 cm d.b.h) from 82 species and 34 families is accessible from the crane gondola (based on a recent (2009) survey at the crane site which updates previously published data (Laidlaw et al. 2007)). The canopy is noticeably uneven in height, varying from 10 to 35 m. Although some rain does fall each month (the lowest average monthly rainfall occurs in August; 80 mm), there is a distinct wet season from November to April (the highest average monthly rainfall occurs in March; 550 mm). The 50-year average annual precipitation at Cape Tribulation is 3,996 mm (Hopkins et al. 1996).

2.2 Sampling

Beetles (Coleoptera) were sampled from 23 canopy plant species in three microhabitats: mature leaves, new leaves, and flowers. Beetles were also collected from fruit and suspended dead wood, but these microhabitats were scarce at the crane site, and their associated beetle communities were small and subsequently omitted from this study. The beetle fauna was chosen because of its ecological diversity and high species richness (Grove and Stork 2000), which allowed for the comparisons necessary to test the hypotheses. All adult beetles were pinned or pointed and sorted to morphospecies (hereafter referred to as species) for use in species richness and specialization analyses. Species were compared with previous collections from the site (Stork and Grimbacher 2006).

The host tree species selected represent a broad range of taxonomic relatedness, growth pattern, phenology, distribution, size, and abundance. In addition to woody trees (19 species), two species of palms and two species of lianas were sampled (Table 27.1). These species comprise 435/745 individuals and >70 % of the basal area of all trees >10 cm d.b.h. in the 1 ha area of forest directly under the crane (Laidlaw et al. 2007). One to three individuals of each host species were sampled each month for 1 year (May 2008–May 2009). Sampling was carried out by beating the microhabitat over a beating sheet to dislodge beetles (Fig. 27.1). Each microhabitat on each replicate tree was sampled for 10 min. In general, trees that were flowering and/or leaf flushing were selected wherever possible, to maximize the number and temporal distribution of samples from these more ephemeral microhabitats. Cross contamination between microhabitat samples was kept to a minimum by only sampling microhabitats that were discretely partitioned on host trees.

Microhabitat specialization was calculated for each beetle species using *Sm* (specificity to microhabitat *m*, analogous to the host specialization (*HS*) measure of Novotny et al. (2004) which is based on an earlier measure by Thomas (1990)). This technique produces similar results to the commonly used Lloyd's index as they both account for differences in abundance between samples, which reduce bias caused by increasing numbers of rare records that inevitably accumulate from large sample sizes. Indeed, the *Sm* measure and Lloyd's index for my data were closely correlated ($r=0.98$). However, Lloyd's index is a relative measure of specialization for each species in a community, which means that it can only show that species *a* is more or less specialized than species *b*. The *Sm* method was therefore chosen as it allowed

Table 27.1 The canopy plant species sampled, including the number of individuals of each species that is accessible from the crane. Lianas have not been mapped on the crane site

Habit	Family	Species	Trees on site
Trees	Lauraceae	*Endiandra microneura*	22
		Cryptocarya mackinnoniana	16
		Cryptocarya grandis	7
		Cryptocarya hypospodia	1
	Myrtaceae	*Acmena graveolens*	16
		Syzygium sayeri	9
		Syzygium gustavioides	8
	Meliaceae	*Dysoxylum papuanum*	12
		Dysoxylum pettigrewianum	9
	Euphorbiaceae	*Cleistanthus myrianthus*	90
	Apocynaceae	*Alstonia scholaris*	61
	Elaeocarpaceae	*Elaeocarpus angustifolius*	7
		Elaeocarpus bancrofti	1
	Cunoniaceae	*Gillbeea whypallana*	5
	Proteaceae	*Cardwellia sublimis*	14
		Musgravia heterophylla	7
	Sterculiaceae	*Argyrodendron peralatum*	17
	Myristicaceae	*Myristica globosa*	59
	Fabaceae	*Castanospermum australe*	8
Lianas		*Entada phaseoloides*	
	Convolvulaceae	*Merremia peltata*	
Palms	Arecaceae	*Normanbya normanbyi*	59
		Archontophoenix alexandrae	7

for the identification of host specificity for each beetle species (e.g., species *a* is a specialist while species *b* is a generalist).

The *Sm* method involved assigning each beetle species to one of three groups based on the proportion of the total number of individuals collected from the microhabitat that supported the highest number of individuals. The categories were:

(a) Specialists: species where $Sm > 0.9$
(b) Preferences (or oligophages): species where $0.5 < Sm < 0.9$, since most individuals were collected from a single microhabitat, indicating that they have a preference for it, but are not necessarily specialized
(c) Generalists: species where $0.33 < Sm < 0.5$, since no microhabitat supported more than half of all individuals

Assigning specialization in this way is sensitive to absolute number of records per species. Specialization analyses were therefore restricted to the 75 beetle species where at least 12 individuals were collected from the focal microhabitats. The limit of 12 individuals was chosen as a compromise between including a maximum number of species and reducing errors arising from potential assignation of specialization when none actually exists.

Fig. 27.1 The author using the Australian Canopy Crane to sample beetles from the foliage of a tree using a beating sheet suspended in a frame

It should be noted that since mature leaf biomass constitutes >90 % of the combined biomass of the focal microhabitats, a randomly distributed beetle species will be assigned as a "mature leaf specialist" since >90 % of its population should be found on mature leaves. It is therefore not possible to discern mature leaf specialists from randomly distributed microhabitat generalists, since both should be found predominantly on mature foliage. However, for the sake of clarity, I refer to all beetles where $Sm > 0.9$ on mature leaves as specialists. This is not the case for flowers and new leaves, however. The spatially and temporally restricted distribution of flowers and new leaves means a random distribution of individuals across microhabitats should produce (on average) less than 10 % of all records for each species on these resources. Defining microhabitat specialization using cutoff values of >90 % and >50 % as employed by the *Sm* method is therefore considered robust in determining preference for flower and new leaf beetles. Differences in microhabitat specialization between assemblages collected from each microhabitat were assessed using Student's t tests. The Chao 1 biodiversity indicator was used to estimate the number of beetle species that utilize each microhabitat on the tree species sampled using Estimates 8.20 (Colwell 2009).

2.3 Microhabitat Biomass Estimation

Different microhabitats vary considerably in size and biomass both between tree species and within individual trees. As such, a time-based measure of collecting effort, where it is assumed that an equal amount of each microhabitat will be sampled during a set time period, is inappropriate to estimate beetle density as a function of biomass available. Furthermore, an attempt to sample an equal amount (weight, surface area, or volume) of each microhabitat on each tree was unfeasible, due to the large differences in biomass between microhabitats. Therefore, I combined the time-based sampling protocol (each microhabitat was sampled on each tree for 10 min), with an estimate of the biomass of each microhabitat in each sample to produce densities of beetles/kg of resource.

To calculate the biomass of a unit of microhabitat (i.e., a single leaf or flower), mature leaves and flowers were collected from each plant species and dried at 60 °C for 48 h. Mature leaves (n = 9–40/species, mean 30.7) and flowers (n = 1–10, mean 8.2) were weighed and the mean used in subsequent calculations of biomass. New leaves were distinguished from mature leaves on the basis of color and texture. Many new leaves on a flushing tree are still expanding and will therefore weigh much less than fully expanded new foliage. Nevertheless, measurement of all new leaves to account for the wide range of sizes is logistically impossible. Samples of fully expanded, but not yet toughened, new leaves weighed just 56.5 % (±6.7 %) of conspecific mature leaves. I therefore estimated the biomass of a single new leaf to be 50 % of the biomass of a conspecific mature leaf.

The amount of each microhabitat present on each tree was calculated following Chapman et al. (1992) and was based on visually estimating the number of units (leaves, flowers) of each microhabitat within tree crowns. Chapman et al. (1992) demonstrated this technique to be both quick and accurate and was considered ideal for assessing varying amounts of available resources on different trees. Specifically, the number of resource units (i.e., leaves, flowers/inflorescences) within five, randomly located, 1 m^3 samples of tree crown were counted and extrapolated to the total estimated volume (m^3) of tree crown sampled (Chapman et al. 1992). The number of resource units sampled was then multiplied by the measured biomass of that particular resource unit to generate an estimated amount (kg) of microhabitat sampled. Conversely, in cases where there were few flowers or new leaves on a tree, the total number of resource units sampled would be manually counted rather than extrapolated. This provided a basis for a calculation of the density of individuals and number of species per kilogram of resource within each tree species, making between- and within-microhabitat comparisons possible. Densities on each microhabitat were weighted for biomass/tree species each month, to avoid potential bias produced by high densities or high microhabitat biomass on single tree species. Differences between microhabitats in mean density of individuals and number of species per kg of microhabitat sampled were examined using ANOVA and $Log_{10}(x + 1)$ transformations.

3 Results

Over 1 year, a total of 10,185 beetles from 358 species were collected from mature leaves, new leaves, and flowers (Table 27.2). Monthly assessments of the biomass of flowers and new leaves showed that these microhabitats constitute a mean (±95 % CI) of just 0.06 % ±0.05 and 1.8 % ±0.52, respectively, of mature foliage biomass/ha. Expressed per unit biomass, a disproportionately large number of individuals were associated with new leaves and especially flowers, where densities were one to four orders of magnitude greater than on the foliage, a pattern consistent across all 18 canopy plant species that flowered during the study. The density of beetles per unit biomass of resource varied significantly between microhabitats ($F_{2,56}=181.27$, $P<0.0001$), with flowers supporting 4,440.3 ± 1,020.1 (weighted mean ± SE) individuals per kilogram and 14.0 ± 5.0/kg on new leaves compared to just 1.5 ± 0.1/kg on mature foliage.

The number of species per kg of microhabitat varied significantly between microhabitats ($F_{2,33}=214.6$, $P<0.0001$), with flowers supporting 404.7 ± 89.9 (weighted mean ± SE) beetle species for every kilogram sampled, while new leaves supported 4.76 ± 0.84 species/kg and mature leaves just 0.57 ± 0.02 species/kg. Note that the number of species/kg on flowers is greater than the total number of species collected from flowers, as this figure is the mean across all temporal samples and less than 1 kg of flowers was sampled during most sampling periods.

Table 27.2 The number of species and individuals within each beetle family collected

Family	No. of species	No. of individuals	Family	No. of species	No. of individuals
Aderidae	7	18	Languriidae	4	93
Anobiidae	4	9	Latridiidae	2	139
Anthicidae	1	2	Leiodidae	1	1
Anthribidae	11	24	Melyridae	10	67
Attelabidae	1	1	Monotomidae	3	7
Boganiidae	2	4	Mordellidae	3	13
Brentidae	7	35	Mycetophagidae	1	27
Buprestidae	2	2	Nitidulidae	20	932
Cantharidae	1	19	Phalacridae	19	213
Carabidae	1	2	Psephenidae	1	1
Cerambycidae	13	19	Ptiliidae	2	2
Chrysomelidae	32	855	Rhipiphoridae	2	7
Ciidae	1	1	Salpingidae	1	1
Cleridae	2	22	Scarabaeidae	7	38
Coccinellidae	24	449	Scirtidae	3	4
Corylophidae	14	401	Scraptiidae	1	6
Curculionidae	111	4,607	Silvanidae	1	422
Dermestidae	6	22	Staphylinidae	16	1,675
Elateridae	8	26	Tenebrionidae	7	9
Hydrophilidae	2	3	Zopheridae	3	5
Laemophloeidae	1	2			

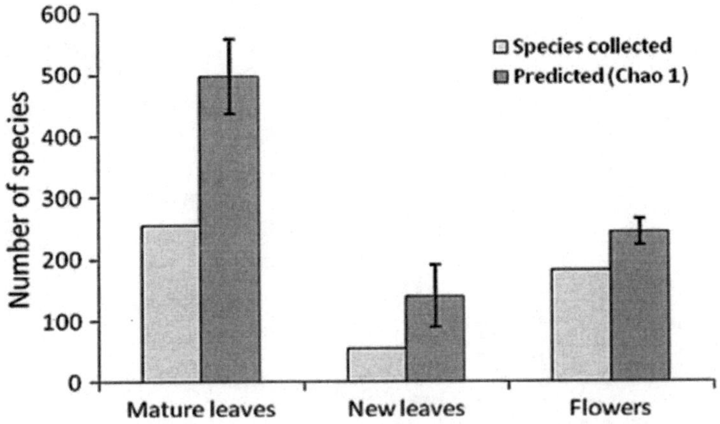

Fig. 27.2 The total number of beetle species collected from and the total estimated number of beetle species (Chao 1 (± SD) species richness estimator) utilizing mature leaves, new leaves, and flowers (Modified from Wardhaugh et al. 2012a)

A disproportionately high concentration of species was found on flowers and, to a lesser extent, new leaves. The majority of the estimated number of beetle species was found on mature leaves, reflecting the large proportion of canopy biomass this microhabitat constitutes (Fig. 27.2). However, the Chao 1 biodiversity indicator showed that 41 % of beetle species utilize flowers and 23 % utilize new leaves (Fig. 27.2), percentages far greater than the relative contributions of these microhabitats to total canopy biomass. None of the species accumulation curves calculated for each microhabitat reached an asymptote, indicating that none of the assemblages on the focal microhabitats were completely sampled (Fig. 27.3).

Microhabitat specialization was high, with only one species showing no preference for a particular microhabitat, although it did show a preference for foliage (mature and new leaves combined). Mean *Sm* (± SE) for beetle species collected predominantly ($Sm > 0.5$) from flowers was 0.92 (±0.014), which is above the 0.9 threshold for categorization as a specialist. Mean *Sm* for beetles collected predominantly from mature leaves was significantly lower than that of flower-visitors ($t_{70} = 3.13$, $P = 0.0026$) but was still relatively high (0.83 ± 0.027). Only three species displayed a preference for new leaves. When mature leaf and new leaf beetle assemblages were pooled into a single foliage group, *Sm* increased to 0.9 (±0.018), which did not differ significantly from the *Sm* of flower-visitors ($t_{74} = 1.08$, $P = 0.28$).

4 Discussion

Overall I found strong support for microhabitat differentiation between beetle communities inhabiting different canopy microhabitats. In particular, flowers were identified as important sites for very high densities of individuals and a

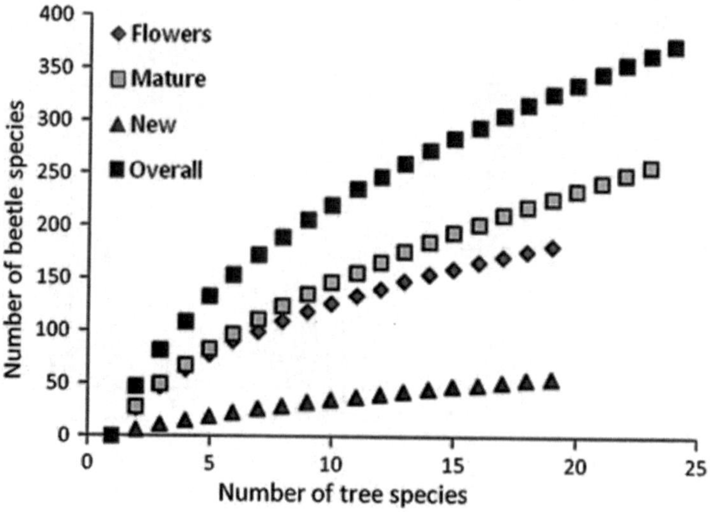

Fig. 27.3 Species accumulation curves (Sobs – Mao Tau) of beetle species associated with mature leaves, new leaves, and flowers, as well as for the pooled sample. Accumulation curves for each microhabitat and the combined total are the products of 100 randomizations (Modified from Wardhaugh et al. 2012b)

disproportionately large number of species of rainforest canopy beetles, despite flowers constituting a tiny fraction of the biomass of mature foliage. Microhabitat specialization was also very high, indicating that flowers support a complementary fauna to that on leaves. Consequently, the hypothesis that beetle abundance and species richness in the canopy is proportional to microhabitat biomass is rejected, and the hypothesis that each microhabitat is utilized by a relatively unique and specialized community is accepted. The inherent assumption that the abundance, diversity, and spatial distribution of the foliage-inhabiting beetle assemblage can be applied to assemblages inhabiting other canopy microhabitats is also rejected. Based on these results, I suggest that flower-associated insects may constitute a neglected component of biodiversity in tropical rainforests.

The identification of flower-visiting insects as a potential missing or underrepresented component of rainforest biodiversity means our current theories that are either directly or indirectly related to the spatial and temporal distribution of insects in the canopy may be inaccurate. In particular, microhabitat differentiation may need to be accounted for when addressing conservation issues such as restoration projects and plans to mitigate the effects of climate change or other anthropogenic disturbances. For instance, flowering is often under the influence of climatic cues (Chapman et al. 2005), especially in mass-flowering species such as the Dipterocarpaceae (Brearley et al. 2007). These climatic cues can also be tracked by flower-visiting insects, resulting in temporal synchronization of floral resources and flower-visitors (Hegland et al. 2009). Similar tracking can also occur for insect species that utilize fruit or flush leaves (van Asch and Visser 2007). Alterations in the intensity, frequency, and timing of phenological events and insect abundance patterns are therefore likely under a changing climate (Graham et al. 2003; Chapman et al. 2005). These changes could potentially alter pollination, herbivory, seed predation, and seed dispersal (Coley 1998; Memmott et al. 2007; van Asch and Visser 2007; Hegland et al. 2009; Potts et al. 2010). Tropical rainforests may be especially vulnerable to climate change-induced phenological decoupling of mutualistic partners since over 90 % of rainforest trees are pollinated by insects (Bawa et al. 1985; Ollerton et al. 2011). Moreover, many non-pollinating insect species are also likely to depend on floral resources (Frame 2003), as it is doubtful that all of the species that rely on flowers are active pollinators of their host trees. These species are just as vulnerable to phenological decoupling from their respective host plants as pollinators yet are often omitted from pollination studies (Wäckers et al. 2007) and unaccounted for in biodiversity studies (see Novotny et al. 2010). More information is required on the biodiversity of flower-associated insects and how they interact with the host plants to plan for future disturbances and to maximize biodiversity conservation and protect functioning ecosystems in the long term.

This study showed that microhabitat differentiation among the beetle community in a rainforest canopy in Australia is pronounced, which may have important implications for a wide range of ecological, biodiversity, and conservation applications. But how general are these results likely to be? In particular, are flowers likely to be as important to canopy diversity and abundance among beetles in other rainforests? First, there is evidence that a disproportionately large number of beetle species may specialize on flowers in other rainforest systems (Ødegaard 2004). Next the feeding biology of beetles is relatively conservative at the family/subfamily level (Lawrence et al. 2000). Indeed, most flower-visiting beetles, including those I collected, belong to a discrete number of families that have been repeatedly recorded as flower-visitors or pollinators in many parts of the world (Gottsberger 1989; Boulter et al. 2005; Teichert et al. 2011). Moreover, these families almost all have a pantropical distribution and display similar patterns of rank species richness across different rainforests (Stork 1993; Hammond et al. 1996). Lastly, 11 of the 13 plant families sampled have a pantropical distribution: Arecaceae, Myristicaceae, Lauraceae, Proteaceae, Euphorbiaceae, Fabaceae, Myrtaceae, Sterculiaceae, Meliaceae, Apocynaceae, and Convolvulaceae. The remaining two families, Elaeocarpaceae and Cunonaceae, are also distributed beyond Australia. All of these factors reduce the likelihood that my results are the product of local host tree phylogeny or beetle assemblage composition. Rather, they suggest that my findings may be indicative of tropical rainforests in general.

Although it can be argued that the results of this study may be applicable to beetles in other rainforests, it is unlikely that similar species richness patterns will be found among most other invertebrate groups, since beetles include many anthophilous species, and most other groups do not. In a concurrent study (Wardhaugh et al. 2012a), it was shown that the density patterns prevalent among the beetle fauna on each microhabitat (i.e., very high on flowers, very low on mature leaves, with new leaves intermediate) are also shown by the total invertebrate fauna. However, flowers supported a distinct fauna in terms of feeding guild and taxonomic composition to that on the leaves. In particular, the flower-visiting community was dominated by high abundances of herbivorous groups (Lepidoptera caterpillars, Hemiptera, Thysanoptera, and Coleoptera), whereas the leaves supported a broader range of taxonomic groups at more equitable relative abundances (C. Wardhaugh unpublished data). For instance, beetles constituted 37.7 % of the total number of invertebrates collected from flowers, compared to just 11.9 % of all invertebrates collected from mature leaves. But beetles, and the other highly diverse insect orders (e.g., Diptera, Lepidoptera, Hymenoptera, Hemiptera), constitute the majority of insect diversity and the majority of flower-visiting species (Bawa 1990). As a result, it is feasible that flowers support a large proportion of insect species in the rainforest canopy because they are utilized predominantly by the most diverse taxonomic groups.

The logistical difficulties in accessing the canopy is largely responsible for the previously exclusion of flower-visiting species from insect biodiversity studies in tropical rainforest canopies. Although canopy access has greatly improved in recent years, the small biomass of flowers compared to the foliage could also be used to justify the omission of flower-visiting species. However, I have shown that a relatively small biomass of flowers can support a disproportionately large number of individuals and species compared to the leaves. Consequently, measures of the spatial and temporal distribution of canopy insect abundance and diversity based only on samples obtained from the foliage are unlikely to be representative of the entire canopy fauna, and the disproportionately large number of species that specialize on other microhabitats, especially the flowers, may be missed altogether. The potential for the flower-visiting invertebrate fauna to contribute significantly to local, regional, and even global biodiversity, food web dynamics, and ecosystem function emphasizes the need to consider this assemblage a conservation priority.

Acknowledgments I thank Nigel Stork, Will Edwards, Peter Grimbacher, and Katherine Wardhaugh for their help during this project, and Cassandra Nichols, Andrew Thompson, Shane Kelly, and Russell Holmes at the Australian Canopy Crane for allowing access to the canopy. Financial support was provided by an MTSRF grant, the Skyrail Rainforest Foundation, and an Australian Postgraduate Award.

References

Bawa KS (1990) Plant-pollinator interactions in tropical rain forests. Annu Rev Ecol Syst 21:399–422

Bawa KS, Bullock SH, Perry DR, Coville RE, Grayum MH (1985) Reproductive biology of tropical lowland rain forest trees. II. Pollination systems. Am J Bot 72:346–356

Boulter SL, Kitching RL, Howlett BG, Goodall K (2005) Any which way will do – the pollination biology of a northern Australian rainforest canopy tree (*Syzygium sayeri*; Myrtaceae). Bot J Linn Soc 149:69–84

Brearley FQ, Proctor J, Suriantata NL, Dalrymple G, Voysey BC (2007) Reproductive phenology over a 10-year period in a lowland evergreen rain forest of central Borneo. J Ecol 95:828–839

Chapman CA, Chapman LJ, Wangham R, Hunt K, Gebo D, Gardner L (1992) Estimators of fruit abundance of tropical trees. Biotropica 24:527–531

Chapman CA, Chapman LJ, Struhsaker TT, Zanne AE, Clark CJ, Poulsen JR (2005) A long-term evaluation of fruiting phenology: importance of climate change. J Trop Ecol 21:31–45

Coley PD (1998) Possible effects of climate change on plant/herbivore interactions in moist tropical forests. Clim Change 39:455–472

Colwell RK (2009) EstimateS: statistical estimation of species richness and shared species from samples. http://viceroy.eeb.uconn.edu/EstimateS

Condon MA, Scheffer SJ, Lewis ML, Swensen SM (2008) Hidden neotropical diversity: greater than the sum of its parts. Science 320:928–931

Erwin TL (1982) Tropical forests: their richness in Coleoptera and other species. Coleopt Bull 36:74–75

Frame D (2003) Generalist flowers, biodiversity and florivory: implications for angiosperm origins. Taxon 52:681–685

Gottsberger G (1989) Comments on flower evolution and beetle pollination in the genera Annona and Rollinia (Annonaceae). Plant Syst Evol 167:189–194

Graham EA, Mulkey SS, Kitajima K, Phillips NG, Wright SJ (2003) Cloud cover limits net CO_2 uptake and growth of a rainforest tree during tropical rainy seasons. Proc Natl Acad Sci USA 100:572–576

Grimbacher PS, Stork NE (2007) Vertical stratification of feeding guilds and body size in beetle assemblages from as Australian tropical rainforest. Austral Ecol 32:77–85

Grove SJ, Stork NE (2000) An inordinate fondness for beetles. Invert Taxon 14:733–739

Hamilton AJ, Basset Y, Benke KK, Grimbacher PS, Miller SE, Novotny V, Samuelson GA, Stork NE, Weiblen GD, Yen JDL (2010) Quantifying uncertainty in estimation of tropical arthropod species richness. Am Nat 176:90–95

Hammond PM, Kitching RL, Stork NE (1996) The composition and richness of the tree-crown Coleoptera assemblage in an Australian subtropical forest. Ecotropica 2:99–108

Hegland SJ, Nielson A, Lázaro A, Bjerknes A-L, Totland Ø (2009) How does climate warming affect plant-pollinator interactions? Ecol Lett 12:184–195

Hopkins MS, Head J, Ash JE, Hewett RK, Graham AW (1996) Evidence of a Holocene and continuing recent expansion of lowland rain forest in humid, tropical north Queensland. J Biogeogr 23:737–745

Laidlaw M, Kitching RL, Goodall K, Small A, Stork NE (2007) Temporal and spatial variation in an Australian tropical rainforest. Austral Ecol 32:10–20

Lawrence JF, Hastings AM, Dallwitz MJ, Paine TA, Zurcher EJ (2000) Beetles of the World. CSIRO, Collingwood

Memmott J, Craze PG, Waser NM, Price MV (2007) Global warming and the disruption of plant-pollinator interactions. Ecol Lett 10:710–717

Moran VC, Southwood TRE (1982) The guild composition of arthropod communities in trees. J Anim Ecol 51:289–306

Novotny V, Basset Y (2005) Host specificity of insect herbivores in tropical forests. Proc R Soc Lond B 272:1083–1090

Novotny V, Basset Y, Miller SE, Weiblen GD, Bremer B, Cizek L, Drozd P (2002) Low host specificity of herbivorous insects in a tropical forest. Nature 416:841–844

Novotny V, Miller SE, Leps J, Basset Y, Bito D, Janda M, Hulcr J, Damas K, Weiblen GD (2004) No tree an island: the plant-caterpillar food web of a secondary rain forest in New Guinea. Ecol Lett 7:1090–1100

Novotny V, Miller SE, Baje L, Balagawi S, Basset Y, Cizek L, Craft KJ, Dem F, Drew RAI, Hulcr J, Leps J, Lewis OT, Pokon R, Stewart AJA, Samuelson GA, Weiblen GA (2010) Guild-specific patterns of species richness and host specialization in plant-herbivore food webs from a tropical forest. J Anim Ecol 79:1193–1203

Ødegaard F (2000a) How many species of arthropods? Erwin's estimate revised. Biol J Linn Soc 71:583–597

Ødegaard F (2000b) The relative importance of trees versus lianas as hosts for phytophagous beetles (Coleoptera) in tropical forests. J Biogeogr 27:283–296

Ødegaard F (2004) Species richness of phytophagous beetles in the tropical tree *Brosimum utile* (Moraceae): the effects of sampling strategy and the problem of tourists. Ecol Entomol 29:76–88

Ollerton J, Winfree R, Tarrant S (2011) How many flowering plants are pollinated by animals? Oikos 120:321–326

Ozanne CMP, Anhuf D, Boulter SL, Keller M, Kitching RL, Körner C, Meinzer FC, Mitchell AW, Nakashizuka T, Silva Dias PL, Stork NE, Wright SJ, Yoshimura M (2003) Biodiversity meets the atmosphere: a global view of forest canopies. Science 301:183–186

Potts SG, Biesmeijer JC, Kremen C, Neumann P, Schweiger O, Kunin WE (2010) Global pollinator declines: trends, impacts and drivers. Trends Ecol Evol 25:345–353

Price PW (2002) Resource-driven terrestrial interaction webs. Ecol Res 17:241–247

Southwood TRE (1961) The number of species of insect associated with various trees. J Anim Ecol 30:1–8

Stork NE (1988) Insect diversity: facts, fiction and speculation. Biol J Linn Soc 35:321–337

Stork NE (1993) How many species are there? Biodivers Conserv 2:215–232

Stork NE (2007) Australian tropical forest canopy crane: new tools for new frontiers. Austral Ecol 32:4–9

Stork NE, Grimbacher PS (2006) Beetle assemblages from an Australian tropical rainforest show that the canopy and the ground strata contribute equally to biodiversity. Proc R Soc Lond B 273:1969–1975

Teichert H, Dötterl S, Gottsberger G (2011) Heterodichogamy and nitidulid beetle pollination in *Anaxagorea prinoides*, an early divergent Annonaceae. Plant Syst Evol 291:25–33

Thomas CD (1990) Herbivore diets, herbivore colonization, and the escape hypothesis. Ecology 71:610–615

Tracey JG (1982) The vegetation of the humid tropical region of North Queensland. CSIRO, Melbourne

van Asch M, Visser ME (2007) Phenology of forest caterpillars and their host trees: the importance of synchrony. Annu Rev Entomol 52:37–55

Wäckers FL, Romeis J, van Rijn P (2007) Nectar and pollen feeding by insect herbivores and implications for multitrophic interactions. Annu Rev Entomol 52:301–323

Wardhaugh CW, Stork NE, Edwards W, Grimbacher PS (2012a) The overlooked biodiversity of flower-visiting invertebrates. PLoS ONE 7:e45796

Wardhaugh CW, Edwards W, Stork NE (2012b)Variation in beetle community structure across five microhabitats in Australian tropical rainforest trees. Insect Conserv Div. doi: 10.1111/icad.12001

Wardhaugh CW, Stork NE, Edwards W (2012c) Feeding guild structure of beetles on Australian tropical rainforest trees reflects microhabitat resource availability. J Anim Ecol 81:1086–1094

Wardhaugh CW, Stork NE, Edwards, W (2013) Specialization of rainforest canopy beetles to host trees and microhabitats: not all specialists are leaf-feeding herbivores. Biol J Linn Soc 109: 215–228

Chapter 28
Assessing Canopy Processes at Large Landscape Scales in the Western Ghats Using Remote Sensing

Jagdish Krishnaswamy

Keywords Evapotranspiration • Plant-available moisture • NDVI • Western Ghats

Bullet Points

1. We used NDVI (Normalized Difference Vegetation Index) to assess the landscape of the Western Ghats, India.
2. Remote sensing data, in combination with ground truthing, can be utilized to assess variation in canopy foliar status.

Summary

Forest canopy processes such as seasonal foliar phenology, interannual changes in photosynthetic activity, and leaf area index in tropical seasonal forests are linked with changes in associated processes such as transpiration and are often responses to changes in availability of soil and groundwater.

Direct measurements of these processes are expensive and unfeasible except at a few sites in a large landscape. We demonstrate how time series of remotely sensed data in combination with ground data from reference sites can be utilized to estimate seasonal and interannual variability in canopy foliar status in response to environmental drivers. This is suggested as the basis for short- and longer-term monitoring of canopy processes in relation to natural and anthropogenic disturbance and climatic perturbations and cycles.

J. Krishnaswamy (✉)
Ashoka Trust for Research in Ecology and the Environment (ATREE), Royal Enclave, Sriramapura, Jakkur Post, Bangalore, KA 560064, India
e-mail: jagdish@atree.org

M. Lowman et al. (eds.), *Treetops at Risk: Challenges of Global Canopy Ecology and Conservation*, DOI 10.1007/978-1-4614-7161-5_28,

1 Introduction

Forest canopy processes such as seasonal foliar phenology, interannual changes in photosynthetic activity, and leaf area index in tropical seasonal forests are linked with changes in associated processes such as transpiration and are often responses to changes in availability of soil and groundwater (Iturbe and Porporato 2004). Furthermore, there is now clear evidence that trees may redistribute moisture through hydraulic redistribution both laterally and vertically in the soil profile from relatively higher soil moisture potential sites to sites with lower soil water potential (Neumann and Cardon 2012).

Direct measurements of these processes are expensive and unfeasible except at a few sites in a large landscape. Remotely sensed indices especially Normalized Difference Vegetation Index (NDVI) have been known to respond to plant canopy processes at various spatial and temporal scales (Glenn et al. 2008). These canopy processes that can be detected through time series of remotely sensed data at large landscape scales are linked to carbon and hydrologic services (Krishnaswamy et al. 2009). Furthermore, these remotely sensed indices can be generated at various spatial and temporal scales, enabling monitoring of canopy processes at large spatial scales that would otherwise be impossible.

We now demonstrate the utility of NDVI-based measures in quantifying and mapping canopy processes at large spatial scales in the Western Ghats. NDVI is correlated with photosynthetically active canopy and in seasonally dry forests can help track the changes in plant available moisture at the landscape scale. As trees shed their leaves through the dry season in response to moisture limitations, both NDVI and transpiration are expected to decline (Fig. 28.1).

This suggests that multi-date NDVI can be the basis for an ecoclimatic distance measure for every pixel using a reference class that is moisture rich and maintains high NDVI through the dry season, which in the case of the Western Ghats would be the wet-evergreen forests (Krishnaswamy et al. 2009). On this continuous scale, the higher the value, the greater the ecoclimatic distance from the reference evergreen class and less would be the evapotranspiration. Such a measure can be generated for every pixel over large landscape at regular intervals of time (Fig. 28.2). Furthermore, these multi-date NDVI indices which act as proxy for plant available moisture can be calibrated with known sites with specific moisture regimes and can be the basis of approximate estimates of evapotranspiration (Green water flux) using semiempirical models such as the Zhang and Walker model which only requires rainfall and potential evapotranspiration (PET) as inputs.

The Zhang, Walker, and Dawes model (Zhang et al. 2001) predicts evapotranspiration (ET) as a function of precipitation (P), E_0 (potential evapotranspiration), and a plant available moisture index, w:

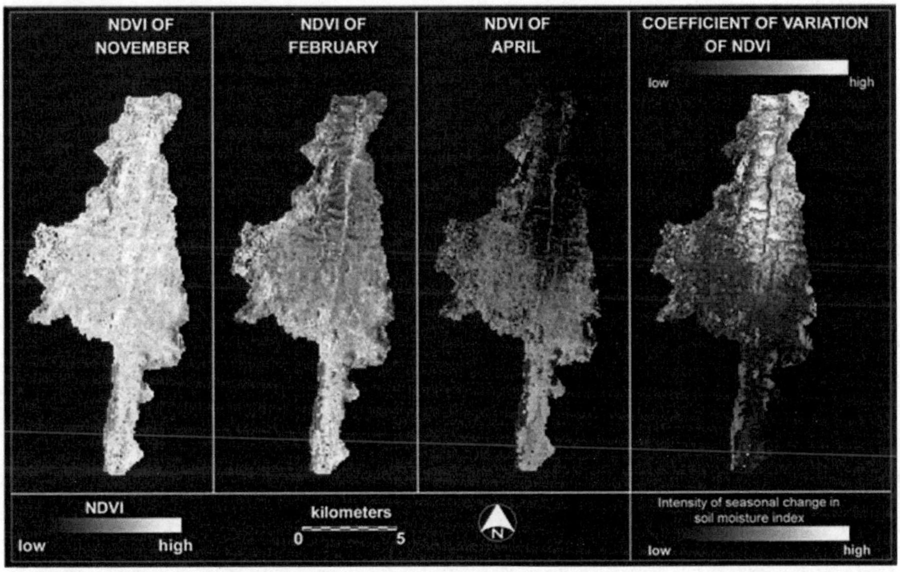

Fig. 28.1 NDVI change through the dry season in Gundal catchment and the potential of using coefficient of variation of multi-date NDVI for mapping spatial variability of plant available moisture. Note that the riparian forest is largely leafless by April

$$\mathrm{ET} = \frac{1 + w\dfrac{E_0}{P}}{1 + w\dfrac{E_0}{P} + \left(\dfrac{E_0}{P}\right)^{-1}}$$

As annual rainfall and PET are easily available from global and regional data sets, only w is the main variable to be estimated. W ranges from 0.5 (moisture-stressed grassland) to 2 (wet-evergreen forest) in the model. The multi-date NDVI-based indices can be calibrated against the range of w and can be the basis for estimating blue and green water fluxes for large landscapes (Fig. 28.3).

2 Conclusion

The potential applications of monitoring forest canopy status over large landscapes using carefully calibrated remotely sensed indices are very diverse. These range from quantifying hydrologic and carbon functions and services to tracking

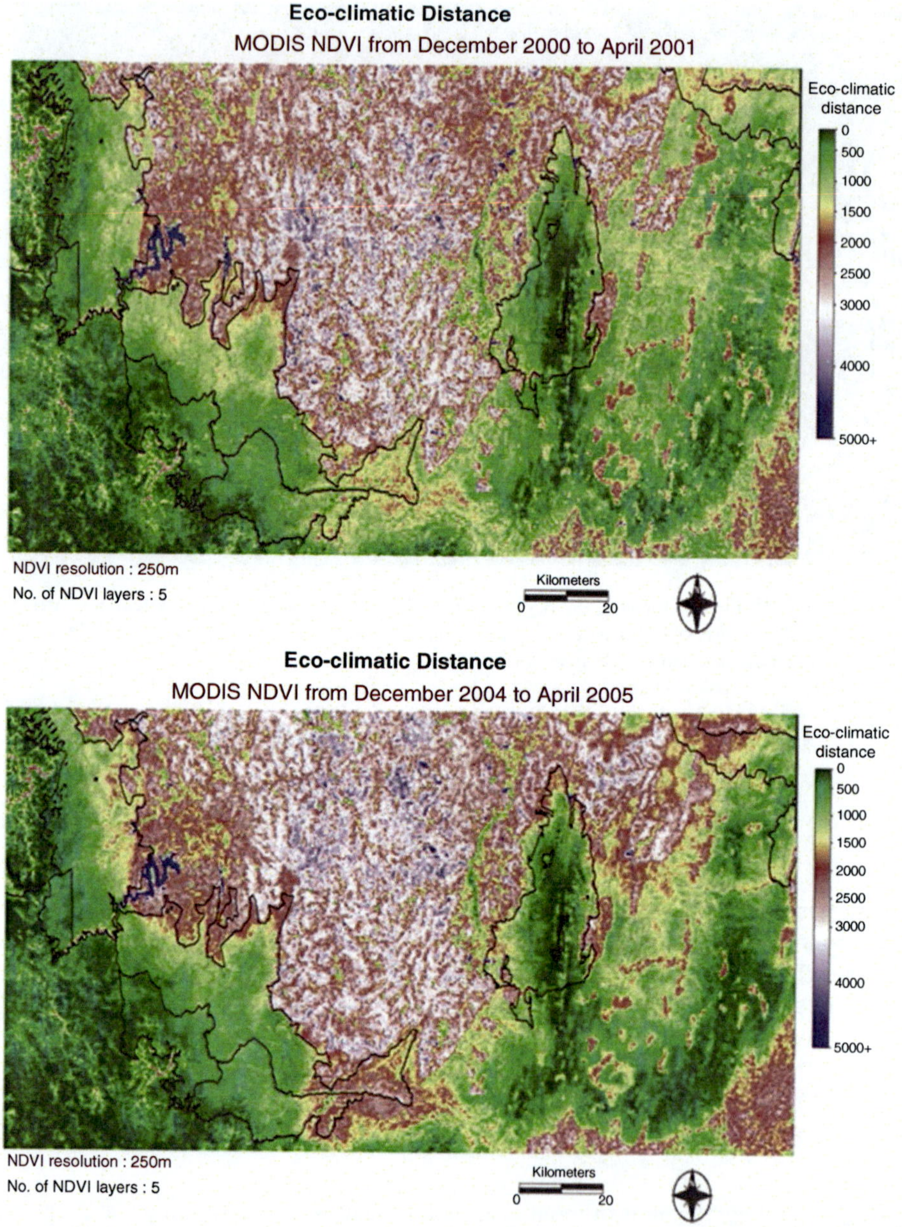

Fig. 28.2 Temporal variability in ecoclimatic distance for large landscapes over two time frames

influences of droughts and forest fire to encroachments, forest degradation, and natural disturbances such as tree-fall gaps and elephant activity. These same indices can be used to measure the downstream effects of dams and water diversion on riparian forests and effects of extreme climatic perturbations and climate change.

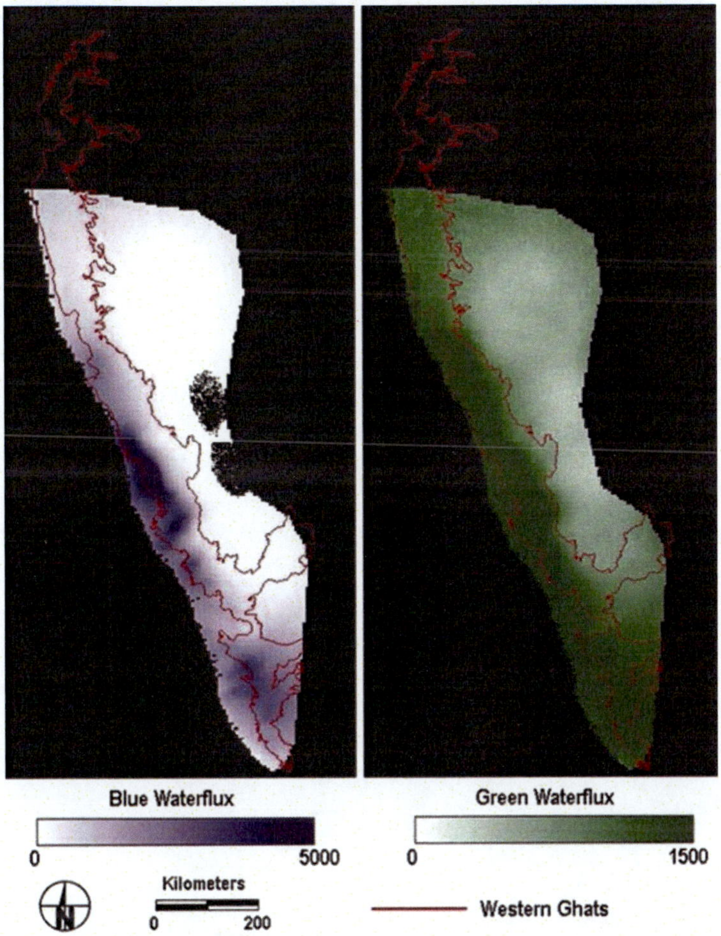

Fig. 28.3 NDVI indices can form the basis for estimating *blue* and *green* water fluxes in the Western Ghats landscape, India

Acknowledgments M.C. Kiran of the Eco-informatic laboratory at the Ashoka Trust for Research in Ecology and the Environment (ATREE) helped generate the figures.

References

Glenn EP, Huete AR, Nagler PL, Nelson SG (2008) Relationship between remotely-sensed vegetation indices, canopy attributes and plant physiological processes: what vegetation indices can and cannot tell us about the landscape. Sensors 8:2136–2160

Iturbe IR, Porporato A (2004) Ecohydrology of water controlled eco systems. Soil moisture and plant dynamics. Cambridge University Press, New York

Krishnaswamy J, Kiran MC, Bawa KS, Ganeshaiah KN (2009) Quantifying and mapping biodiversity and ecosystem services using a multi-date NDVI based Mahalanobis distance measure. Remote Sens Environ 113:857–867

Neumann RB, Cardon ZG (2012) The magnitude of hydraulic redistribution by plant roots: a review and synthesis of empirical and modeling studies. New Phytol 194:337–352. doi:10.1111/j.1469-8137.2012.04088.x

Zhang L, Dawes WR, Walker GR (2001) Response of mean annual evapotranspiration to vegetation changes at catchment scale. Water Resour Res 37(3):701–708. doi:10.1029/2000WR900325

Chapter 29
Ontogeny of Herbivory on Leaves in a Tropical Rain Forest in Madagascar

Harold Heatwole, Sybille Unsicker, and Margaret Lowman

Keywords Canopy • Epiphyllae • Galls • Herbivory • Madagascar • Ontogeny

Summary

This study was undertaken to ascertain the extent to which damage inflicted by different agents to foliage in a tropical rain forest in Madagascar varied during the flushing and maturation of leaves. There was an ontogenetic sequence in which agents attacked leaves. New, tender foliage was attacked primarily by grazing insects, and by the time the leaf had hardened, it had already suffered a high proportion of the total loss of leaf area it would sustain from grazers throughout its life. As grazing waned, other agents, such as fungi, skeletonizing insects, and galls, came into play. The last agent was epiphyllae; it started late and gradually accumulated over the life of the leaf.

1 Introduction

Leaves undergo ontogenetic changes in their toughness, chemical composition, and texture, all of which affect their palatability to herbivores and pathogens and their resistance to physical agents (Lowman and Box 1983; Lowman 1995; Coley and

H. Heatwole (✉)
Department of Zoology, North Carolina State University, Raleigh, NC 27695-7617, USA
e-mail: harold_heatwole@ncsu.edu

S. Unsicker
Institute of Ecology, University of Jena, Dornburgerstrasse 159, Jena 07743, Germany
e-mail: sybille.unsicker@uni-jena.de

M. Lowman
North Carolina Museum of Natural Sciences, North Carolina State University,
121 W. Jones Street, Raleigh, NC 27603, USA
e-mail: canopymeg@gmail.com

M. Lowman et al. (eds.), *Treetops at Risk: Challenges of Global Canopy Ecology and Conservation*, DOI 10.1007/978-1-4614-7161-5_29,

Kursar 1996; Coley et al. 2006). Consequently, rates of herbivory and other damage would be expected to change during the life of the leaf, and hence, there may be temporal as well as spatial dimensions to the vicissitudes of leaves. The present study was carried out to ascertain the changes in vulnerability of leaves to damage by various agents as the leaves undergo ontogeny.

2 The Study Area

Fieldwork was conducted from 31 October to 13 November 2001 in the Masoala National Park on Masoala Peninsula in northeastern Madagascar. The park includes an elevational gradient spanning habitats from lowland, humid, evergreen rain forest near sea level to montane thicket and cloud forest at >1,200 m. With an area of 211,230 ha, this park is larger than all other forest reserves in Madagascar combined (Kremen et al. 1999). Since the park was gazetted in 1997, disturbance by humans had been minimal and, at the time of the study, consisted mainly of hand felling a few trees and processing them by pit sawing (Kremen et al. 1999). At one time, cinnamon was grown in the vicinity of the research encampment, and there was some incursion by cinnamon trees (*Canella winterana*) into the edges of the forest, but otherwise the forest appeared relatively undisturbed.

The study was carried out in lowland to mid-elevation humid evergreen forest from a base of operations at the Tampolo Field Station (15° 43′ 48″ S; 49° 57′ 34″ E) (Monte-Alegre et al. 2005). There was a closed upper canopy at 15–22 m and a less complete mid-canopy at 10–13 m. Scattered "emergent" trees, up to 37 m tall, protruded above the upper canopy. Branches and trunks at all levels supported a profuse growth of epiphytes, lianas, vines, and parasitic plants. The ground layer consisted of a sparse cover of seedlings, saplings, and shrubs.

3 Materials and Methods

The study consisted of three parts: (1) a comparison of damage to mature leaves of different relative ages by a variety of agents, including fungi, grazing insects, skeletonizing insects, leaf miners, galls, mechanical forces, and epiphyllae; (2) a comparison of damage by grazing and skeletonizing insects to recently flushed versus mature leaves on the same twigs; and (3) measurement of the weekly rate of damage by various agents to newly flushed leaves. Because the data obtained were not normally distributed, nonparametric tests were used throughout.

3.1 Damage by Various Agents to Hardened, Mature Leaves of Different Relative Ages

The effect of relative age of mature leaves upon the extent of grazing by insects was ascertained for one canopy liana and for branches from 36 saplings or shrubs from the ground layer. Samples from the ground layer were taken by walking a service path through the forest, and at 10-m intervals, the shrub or sapling nearest the path was selected for sampling, choosing the plant to the right of the transect at odd-numbered stations and to the left at even-numbered stations. For the liana, a given branch of a shrub, or for an entire sapling, leaves were numbered, beginning apically (relatively youngest leaves) and progressing toward the base (oldest leaves). All leaves were mature (hardened and deep green).

Method of assessment of damage varied depending on the agent. Leaves were individually scored for mechanical breakage, incrustation by epiphyllae, attack by fungi, and attack by leaf miners, as follows: nil (no damage), + (damage present, but inconsequential), and slight, moderate, or heavy. The number of galls on each leaf was recorded. In comparisons of incidences by chi-square analyses, the actual values (rather than percentages), including the "nil" category, were used. When expected values fell below five, categories were lumped until this minimum criterion was met.

For assessing damage by insect grazers and skeletonizers, the outline of each leaf was traced on paper and the extent of damage traced within the original (before damage) outline of the leaf, separately for these two trophic groups. The tracings were cut out with manicure scissors and the total undamaged area of each leaf was measured to the nearest 0.01 cm^2 by a Li-Cor area meter, model LI-3100. Then, the area eaten by insect grazers was excised and a second reading taken. Finally, the area eaten by skeletonizers was excised and a final measurement made. Subtraction of the various leaf areas allowed separate calculation of the percentage of the total leaf surface eaten by grazers and skeletonizers. Insects, bite marks were easily distinguished from mechanical damage by the nature of the broken surfaces.

The prediction of a direct, positive relationship between leaf age and extent of grazing damage was tested by the Spearman Rank Correlation test; since the direction of the result could be predicted a priori, the test was considered one-tailed, with a rejection level of 2.5 %.

3.2 Damage to Recently Flushed Versus Mature Foliage by Grazing and Skeletonizing Insects

Recently flushed leaves were distinguished from mature leaves by color, sclerophylly, and texture (Lowman 1985). New leaves were flimsy and light green, red, or light brown, depending on the species, compared to the deep green of older, thicker, and less flexible leaves. Nine twigs from the crown of a *Uapaca* sp. (Euphorbiaceae) canopy tree containing an apical cohort of newly flushed leaves and a more basal one of mature ones were collected by single-rope climbing techniques

(Montgomery 1977). Damage by grazing and skeletonizing insects to these leaves was assessed as described above, and the rejection level of statistical testing was 2.5 %.

3.3 *Weekly Rate of Damage to Newly Flushed Leaves*

Using a motor-powered, hot-air dirigible (Montgomery 1977; Hallé et al. 2000; Hallé 2002; Mitchell et al. 2002; Heatwole et al. 2009), an 800-m rope transect was laid across the upper canopy of the forest and fixed at intervals to tree limbs. A helium-filled balloon, harnessed to a single investigator and attached by jumars to the transect rope, was used to follow this transect to find branches with newly flushed leaves. Whenever such a branch was encountered, it was flagged and leaves already suffering damage were pinched off and discarded. One week later, the transect was revisited and the marked branches collected for measuring the damage sustained since the original visit, using the methods described for assessing damage to mature leaves of different ages. Even if leaves continued to grow during the interval of study, proportional extent of grazing can be directly compared as Lowman (1987) showed that the holes in leaves caused by herbivores increase in size in proportion to the growth of the leaf itself.

4 Results

4.1 *Damage by Various Agents to Hardened, Mature Leaves of Different Relative Ages*

It was predicted that among mature leaves, the oldest, most basal, ones would have suffered a greater incidence of damage or infestation than would have the youngest, more apical, ones. This was true only for coverage by epiphyllae (Table 29.1). Only 15 % of the youngest leaves had epiphyllae, whereas 47 % of the oldest leaves were infested. The difference was significant, and thus, even after maturing, leaves clearly continued to accumulate epiphyllae. All the other agents had slightly higher incidences for the two oldest leaves, as opposed to the two youngest ones, except for incidence of damage by skeletonizing insects in which the incidence on older leaves was slightly lower than on younger ones; none of these differences, however, were significant (Table 29.1). It would appear, then, that the incidence of these agents does not increase markedly with further ageing once the leaf has matured, hardened, and changed color.

The extent of damage by grazing insects could be treated in more detail than could incidence. Extent of grazing damage was positively correlated with relative leaf age, as determined by its location on the apical-to-basal scale (leaf number), in 19 (51 %) of the 37 tests, negatively correlated in 17 (46 %), and identical in one (2.7 %) (Table 29.2). These correlations were significant in only eight samples

Table 29.1 Comparison of incidences of different kinds of damage to the two youngest and two oldest of mature leaves on 36 saplings, shrubs, and vines from the ground layer of a Madagascaran rain forest. All tests are one-tailed; **boldface** indicates significance at a rejection level of 2.5 %. For taxonomic identification of these plants, see Table 29.2

	Number of leaves			
Kind of damage	Damaged	Undamaged	Chi-squared	P
Fungus				
Youngest	34	38	0.02	0.45>P>0.40
Oldest	42	30		
Leaf miners				
Youngest	21	51	0.11	0.40>P>0.35
Oldest	24	48		
Epiphyllae				
Youngest	11	61	15.64	**P<0.0005**
Oldest	34	38		
Mechanical damage				
Youngest	9	63	0.06	0.40>P>0.35
Oldest	11	61		
Galls				
Youngest	1	71	0.83	0.25>P>0.15
Oldest	4	68		
Insect grazing				
Youngest	49	23	0.03	0.99>P>0.98
Oldest	50	22		
Insect skeletonizing				
Youngest	4	68	0.15	0.95>P>0.90
Oldest	3	69		

(six positive, two negative). There did not seem to be any taxonomic consistency in the occurrence of significant positive correlations except that perhaps they tended to be more prevalent in the family Rubiaceae than in the Euphorbiaceae. In summary, there is seldom a significant increase in grazing damage with age in leaves after they are already hardened and mature.

4.2 *Damage to Recently Flushed Versus Mature Foliage by Grazing and Skeletonizing Insects*

Newly flushed leaves had lower incidence of attack and less damage by grazing insects than did mature leaves (Table 29.3). Wilcoxon Matched-Pairs Signed-Ranks tests, pairing by twig, gave overall significance to both measures (in both cases, $T=0$; $P<0.005$). The same was true for skeletonizers (Wilcoxon test: $T=0$; $P<0.005$). Despite these expected differences, the most striking result was the amount of grazing already sustained by new leaves. The incidence of attack on newly flushing leaves by grazing insects averaged 14 % that on mature, hardened leaves, with values for individual twigs reaching as high as 47 %. The proportion of the area consumed

Table 29.2 Relation of relative age of mature, hardened leaves to their extent of damage by grazing insects. Values in **boldface** indicate significance at the 2.5 % rejection level (one-tailed Spearman Rank Correlation test). *N* number of leaves

Plant type/taxon/plant no.	N	r	P
Canopy liana			
Menispermaceae	26	−0.033	0.873
Ground-layer plants (saplings, shrubs, vines)			
Arecaceae, *Dypsis* sp.	5	0.900	0.072
Canellaceae, *Canella* sp.[a]	16	−0.563	**0.029**
Clusiaceae sp. 1	7	0.083	0.838
Clusiaceae, *Garcinia* sp.			
Tree 1	13	−0.630	**0.029**
Tree 2	10	−0.124	0.709
Dioscoreaceae, *Dioscorea* sp.	6	0.507	0.257
Ebenaceae, *Diospiros* sp.			
Tree 1	14	−0.064	0.818
Tree 2	15	0.269	0.315
Euphorbiaceae, *Uapaca* sp. 1			
Tree 1	10	−0.220	0.510
Tree 2	16	−0.381	0.141
Tree 3	4	0.316	0.584
Tree 4	18	0.244	0.315
Tree 5	7	−0.793	0.052
Tree 6	11	−0.364	0.250
Tree 7	8	−0.383	0.311
Tree 8	10	−0.267	0.422
Euphorbiaceae, *Uapaca* sp. 2	26	−0.322	0.108
Menispermaceae			
Tree 1	14	0.607	**0.029**
Tree 2	21	0.369	0.099
Monimiaceae, *Tambourissa* sp.	10	−0.203	0.542
Moraceae, *Bosqueia* sp.	8	0.082	0.827
Myrsinaceae	14	−0.020	0.943
Myrsinaceae, *Oncostemum* sp.	10	0.365	0.273
Oleaceae	11	−0.073	0.818
Rubiaceae			
Tree 1	10	0.038	0.909
Tree 2	4	0	>0.999
Tree 3	12	0.387	0.199
Tree 4	10	0.766	**0.022**
Tree 5	10	0.730	**0.029**
Tree 6	5	0.264	0.598
Tree 7	13	0.699	**0.015**
Tree 8	7	0.808	**0.048**
Tree 9	7	−0.506	0.215
Rubiaceae, *Coffea* sp.	8	−0.172	0.649
Rubiaceae, *Ixora* sp.	9	0.693	**0.050**
Violaceae, *Rinorea* sp.	6	0.698	0.118
All ground-layer plants	385	0.058	0.256

[a]Escaped domestic tree

Table 29.3 Comparison of damage by grazing and skeletonizing insects between new and mature foliage on twigs of a *Uapaca* sp. tree (Euphorbiaceae). Note that means are presented for descriptive purposes, but all statistical tests are nonparametric (Mann–Whitney *U* test). Leaves per twig=2–18 for new foliage and 5–39 for mature foliage. **Boldface** indicates significance at the 2.5 % rejection level

	Mean number of attacks per leaf				Mean percent of total leaf area consumed			
Twig no.	Old foliage	New foliage (% of old)	Z	P	Old foliage	New foliage (% of old)	Z	P
Grazing insects								
1	1.29	0.60 (47)	−0.46	**0.048**	7.42	2.02 (27)	−0.27	0.787
2	1.00	0.29 (29)	−1.45	0.146	18.30	0.54 (3)	−1.84	0.066
3	2.17	0.50 (23)	−0.86	0.390	8.05	2.95 (37)	−1.03	0.306
4	2.00	0 (0)	−3.31	**0.0009**	12.48	0 (0)	−3.30	**0.001**
5	4.00	0.14 (4)	−3.93	**0.001**	12.37	1.63 (13)	−3.44	**0.0006**
6	2.00	0 (0)	−3.04	**0.002**	1.70	0 (0)	−3.03	**0.003**
7	1.76	0 (0)	−3.70	**0.002**	7.27	0 (0)	−3.69	**0.002**
8	3.79	0.50 (13)	4.47	**<0.0001**	6.06	4.78 (79)	−3.02	**0.003**
9	1.60	0.14 (9)	−3.13	**0.0018**	10.23	0.61 (6)	−2.76	**0.0058**
Skeletonizing insects								
1	0.14	0	−0.85	0.398	0.14	0	−0.85	0.398
2	0.20	0	−1.18	0.237	0.20	0	−1.18	0.237
3	0.17	0	−0.58	0.564	0.17	0	−0.58	0.564
4	0.11	0	−1.05	0.292	0.11	0	−1.05	0.292
5	0.17	0	−1.53	0.127	0.17	0	−1.53	0.127
6	0	0	–	–	0	0	–	–
7	0	0	–	–	0	0	–	–
8	0	0	–	–	0	0	–	–
9	0	0	–	–	0	0	–	–

of leaves still in the unhardened state had already reached as high as 79 % (mean = 18 %) of the cumulative levels in mature foliage on the same branch.

Incidences and damages by skeletonizers were much lower; none of the new foliage had suffered any damage from skeletonizers, whereas 56 % of the older leaves had. Thus, grazing occurs on leaves earlier than does skeletonizing.

4.3 *Weekly Rate of Damage to Newly Flushed Leaves*

Neither time nor resources were available for tracking the damage by insect herbivores throughout the entire lifetime of leaves as was done by Lowman and Heatwole (1992). The harvesting of new foliage from individual branches a week after previously damaged leaves had been excised did, however, allow a short-term assessment of rates of herbivory on new, succulent foliage.

On average, the newly flushing leaves of all tree species collectively lost 1.6 % of their total area to insect grazers per week, whereas the accumulated damage to

Table 29.4 Comparison of damage caused by one week of grazing by insects on new growth of the canopy with cumulative grazing damage to mature leaves throughout the canopy of a Madagascaran rain forest. Statistical testing by Mann–Whitney *U* test, one-tailed. **Boldface** indicates significance at a rejection level of 2.5 %

Family (N)	Mean % of leaf area grazed (New as % of old)	Z	P
Trees			
Clusiaceae			
New growth (6)	0 (0)	−2.05	**0.041**
Mature leaves (90)	5.0		
Euphorbiaceae			
New growth	1.9 (65)	−3.55	**0.0004**
Mature leaves	2.9		
Meliaceae			
New growth	3.2 (17)	−8.71	**<0.0001**
Mature leaves	18.5		
Myrtaceae			
New growth	2.3 (54)	−5.12	**<0.0001**
Mature leaves	4.3		
Rubiaceae			
New growth	0.8 (16)	−0.66	0.503
Mature leaves	4.9		
Mean for all tree families			
New growth	1.6 (23)	–	–
Mature leaves	7.1		
Epiphytes			
Asclepiadaceae			
New growth (13)	0 (0)	−0.86	0.123
Mature leaves (30)	3.9		
Loranthaceae			
New growth	3.4	–	–
Mature leaves	No data on old growth		

mature foliage was 7.1 % (Table 29.4). The mean damage by grazing insects to new leaves was 23 % of the mean accumulated damage to mature leaves, and values for individual branches reached as high as 65 %. After one week, all the new leaves were still in the immature state and did not appear to be beginning to harden. Consequently, it is likely they would sustain considerably more damage before assuming mature texture and color.

No skeletonization was observed in leaves that were still tender and with their young coloration and no leaf miners, epiphyllae, or galls were seen; only a few small of spots of fungi were detected. Thus, there seems to be an ontogenetic sequence with grazing occurring first, followed by fungal attack and then by other kinds of damage.

5 Discussion

Although a tropical rain forest is commonly perceived as providing an equable and benign environment, in reality it presents a kaleidoscope of spatial and temporal challenges, emanating especially from the biotic component of the ecosystem. Damage inflicted by various agents on leaves in a tropical rain forest in Madagascar is vertically stratified (Heatwole et al. 2009). Attack by grazing insects and fungi is a major challenge for seedlings at ground level, but with increasing height of saplings above the ground as the tree grows, fungal attack on leaves becomes more prevalent, while attack by grazing insects diminishes. Thus, the optimal adaptive strategy for a tree may change throughout its life. It not only needs to balance its defenses to meet multiple threats at a particular time within a given stratum, but it also must adapt to a series of different suites of challenges as it develops from a seedling into a sapling and finally into a mature tree while growing upward through different strata.

At least part of such adaptation occurs via physiological adjustments to immediate conditions. Attack may induce a response in the plant that counters that particular stress. For example, Haukioja and Niemelä (1979) and Haukioja and Neuvonen (1985) showed that birch trees respond to mechanical simulations of light grazing by elevating the levels of defensive chemicals in their leaves. Such responses may exhibit spatial resilience and selectively direct the response to local sites where grazing is greatest, rather than wasting resources by mobilizing antidotes in parts of the tree not subject to high intensities of grazing. For example, Lowman and Heatwole (1987) found that branches of eucalypt saplings subject to attack by grazing insects produce lower levels of autochthonous defensive chemicals than do branches on the same tree and at the same time that had been sprayed with insecticides and thus were protected from attack by grazers.

In addition to temporal shifts in responses by an entire tree, or some part thereof, in adjusting to its changing milieu as it grows through different strata, individual leaves in a given stratum face sequential shifts in challenges during their own ontogeny. They are first subject mainly to grazing by insects, then to attack by fungi, then skeletonizers and leaf miners, and finally to a gradual accumulation of epiphyllae. During flushing, young leaves suffer a more rapid and extensive damage from grazing insects than they do for the rest of their lives (Lowman 1985; Selman and Lowman 1983; Coley and Kursar 1996). Supporting evidences from the present study are as follows: (1) on most branches of ground-layer plants, the relatively older, mature leaves seldom showed a significant increase in grazing damage over that suffered by younger mature leaves, and (2) newly flushed leaves had high rates of grazing upon them, in only 1 week suffering an average loss of leaf area of 23 % of the cumulative loss by mature, hardened leaves. In contrast to their relative immunity to attack by grazing insects, hardened leaves are more subject to fungal infections and the blocking of their leaves' photosynthetic surfaces by epiphyllae.

The interaction of challenges and adaptive response is complex, with the severity of damage serving both as a cause and a consequence of a leaf's attributes. As noted above, the intensity of grazing by insects can induce adaptive defensive responses. It also transpires that a leaf's character, independent of induced responses, has an influence on intensity of attack. For example, young leaves are selectively targeted by grazing insects probably because flushing leaves are mechanically easier to chew than are hardened ones (Lowman 1985). Thus, the pattern of grazing is a consequence of the vulnerability of young, tender leaves to grazing by insects. One would expect, then, that tender leaves would (1) maximize growth rate as a means of shortening the time spent in flushing, (2) channel available resources into production of chemical defenses of young leaves, while (3) deferring use of resources for other functions, such as photosynthesis.

Young tropical leaves are usually light brown, light green, or red; only later, after they have developed mechanical protection, do the leaves become deep green as metabolic resources are redirected from growth and chemical protection into producing chlorophyll and assembling the photosynthetic apparatus (Coley and Aide 1989). In this way resources are not wasted by producing expensive photosynthetic materials, part of which would be harvested by insects rather than benefit the plant.

In conclusion, there are spatial and ontogenetic components of a complex interaction of assaults upon leaves and of trees' physiological and evolutionary responses to them. The nature of this complexity is only beginning to be understood and constitutes an important challenge for research into canopy dynamics.

Acknowledgments We thank Pro-Natura International and Operation Canopée for financial and logistic support; Judith Thomas for use of the leaf meter; Francis Hallé for sharing his expertise; the dirigible pilots, Dany Cleyet-Marrel and Jean Drouault, for their expert flying; Giles Ebersolt, Thierry Aubert, and François Idiené, for technical support and teaching us to fly the helium balloon; the late Audry Heatwole and the late Shirley Waters for laboratory assistance; and Susan King for critical comment on the manuscript.

References

Coley PD, Aide TM (1989) Red coloration of tropical young leaves: a possible antifungal defense? J Trop Ecol 5:293–300

Coley PD, Kursar T (1996) Anti-herbivore defenses of young tropical leaves: physiological constraints and ecological trade-offs. In: Mulkey SS, Chazdon LR, Smith AP (eds) Tropical forest plant physiology. Chapman & Hall, New York

Coley PD, Bateman ML, Kursar TA (2006) The effects of plant quality on caterpillar growth and defense against natural enemies. Oikos 115:219–228

Hallé EF (2002) The canopy raft. In: Mitchell AW, Secoy K, Jackson T (eds) The global canopy handbook, techniques of access and study in the forest roof. Global Canopy Programme, Oxford

Hallé EF, Cleyet-Marrel D, Ebersolt G (2000) Le Radeau des Cimes, exploration des canopées forestières. Editions Jean-Claude Lattès, Paris

Haukioja E, Neuvonen S (1985) Induced long-term resistance of birch foliage against defoliators: defensive or incidental? Ecology 66:1303–1308

Haukioja E, Niemelä P (1979) Birch leaves as a resource for herbivores: Seasonal occurrence of increased resistance in foliage after mechanical damage of adjacent leaves. Oecologia 39:151–159

Heatwole H, Unsicker S, Andriamiarisoa LR, Lowman MD (2009) Vicissitudes of leaves in a tropical rain forest in Madagascar. J Trop Ecol 25:615–624

Kremen C, Razafimahatratra V, Guillery RP, Rakotomalala J, Weiss A, Ratsisompatrarivo JS (1999) Designing the Masoala National Park in Madagascar based on biological and socioeconomic data. Conserv Biol 13:1055–1068

Lowman MD (1985) Temporal and spatial variability in insect grazing of the canopies of five Australian rain forest tree species. Aust J Ecol 10:7–24

Lowman MD (1987) Relationships between leaf growth and holes caused by herbivores. Aust J Ecol 12:189–191

Lowman MD (1995) Herbivory as a canopy process in rain forest trees. In: Lowman MD, Nadkarni N (eds) Forest canopies. Academic, San Diego

Lowman MD, Box JD (1983) Variation in leaf toughness and phenolic content among five species of Australian rain forest trees. Aust J Ecol 8:17–25

Lowman MD, Heatwole H (1987) The impact of defoliating insects on the growth of eucalypt saplings. Aust J Ecol 12:175–181

Lowman MD, Heatwole H (1992) Spatial and temporal variability in defoliation of Australian eucalypts. Ecology 73:129–142

Mitchell AW, Secoy K, Jackson T (2002) Global canopy handbook, techniques of access and study in the forest roof. Global Canopy Programme, Oxford

Monte-Alegre A, Rakotomalala Z, Levedgle M, Coulier F, Faure E (2005) Ecological, vertical and elevational distributions of Heteropsis (Lepidoptera, Satyrinae) in the rainforest of Masoala (NE Madagascar). J Zool Syst Evol Res 43:214–222

Montgomery N (1977) Single rope techniques, a guide for vertical cavers. Sydney Speleogical Society Occasional Papers no. 7, p. 1–122

Selman BJ, Lowman MD (1983) The biology and herbivory rates of Novacastria nothofagi Selman (Coleoptera: Chrysomelidae), a new genus and species on *Nothofagus moorei* in Australian temperate rain forests. Aust J Zool 31:179–91

Chapter 30
Do Water Bears Climb Trees Too?

William R. Miller, Logan Gallardo, and Tiffany Clark

Keywords Tardigrade • Vertical distribution • Diversity

Bullet Points

1. Tardigrades are a ubiquitous, but relatively unknown, phylum of invertebrates.
2. Tardigrades represent excellent subjects for undergraduate research in forest canopies.

Summary

Tardigrades or water bears remain a little known and little studied phylum of invertebrates that inhabit the mosses and lichens that grow on the trees of the world. Tardigrades are renowned for cryptobiosis, the process by which they desiccate with their environment and reconstitute later when moisture is available. This physiological adaptation gives them the ability to survive the extreme environmental conditions found in the Antarctic, in the deep oceans, on high mountains, in the deserts, and in the canopies of the world. Tardigrades eat bacteria, algae, protozoans, and other microscopic organisms. They are in turn eaten by rotifers, nematodes, mites, and insect larva. They are part of an unseen trophic level in the web of life found on in the canopy. Yet, we know little about the limits to their normal living.

Modern canopy research has challenged our assumption of low biodiversity in trees but to date there is no data to support tardigrade usage of the canopy. To investigate, we conducted a vertical transect in a 25-m white pine (*Pinus strobus*) on the campus of Baker University in Kansas. Samples of tardigrade habitat (moss and

W.R. Miller (✉) • L. Gallardo • T. Clark
Department of Biology, Baker University, Baldwin City, KS 66006, USA

North Carolina Museum of Natural Sciences, Raleigh, NC, USA
e-mail: william.miller@bakeru.edu; Logan.gallardo@bakeru.edu; Tiffany.clark@bakeru.edu

M. Lowman et al. (eds.), *Treetops at Risk: Challenges of Global Canopy Ecology and Conservation*, DOI 10.1007/978-1-4614-7161-5_30,

lichens) were collected at four levels 3 m apart and 183 tardigrades from five genera were extracted and classified. More than half of the animals were collected at the highest level. The most common genus (*Milnesium*) with 68 % of the specimens occurred only at the upper levels with their greatest abundance at the highest level. Two genera (*Minibiotus* and *Paramacrobiotus*) with 30 % of the specimens occurred at all four levels but in greater numbers higher in the tree. The remaining two genera (*Echiniscus* and *Isohypsibius*) with only 2 % of the specimens were found at the lower levels. There were more water bears higher in the tree. These findings document diversity in the vertical distribution of microscopic animals on trees. We feel a three-dimensional model for microinvertebrate distribution, and abundance should be developed as a tool to monitor change of this unseen but critical trophic level in canopy health.

1 Introduction

Microinvertebrates are those animals on the edge of invisibility, generally requiring a microscope to see. They include mites, rotifers, nematodes, and tardigrades and form an unseen part of the food web. These animals eat smaller organisms (bacteria, algae, and protozoans,) and are eaten by larger ones (mites, insect larva, insects, or each other). In turn, they are consumed by the animals that graze on their habitat of moss and lichen. But their habitat is threatened by logging, burning, air pollution, and global warming. Edge effect and forest islands are increasing, as are average temperatures. It is clear that the underlying habitat for microinvertebrates (moss, lichen, and algae on trees) must also be declining and changing and the formula for the survival of these invisible components of the canopy ecosystem is being altered. It is unclear if the microinvertebrates can or will be able to change as the paradigm for their survival shifts. If not, they might be the weak link in the productivity of forest ecosystems.

For example, water bears or tardigrades remain a little known and little studied phylum of invertebrates. They can be found in the interstitial spaces in mosses, lichens, algae, and leaf litter habitats and enjoy worldwide distribution (Kinchin 1994). Tardigrades are renowned for cryptobiosis, the process by which they desiccate with their environment and reconstitute later when moisture is available (Miller 1997). This physiological adaptation gives them the ability to survive the extreme environmental conditions found in the Antarctic, on high mountains, in deserts, and in the canopies of the world. Despite the recent success of water bears in becoming the first multi-celled life form to survive exposure to space (Jönnson et al. 2008), we know little about the limits to their normal living.

Canopy knowledge is also still in its infancy (Lowman and Rinker 2004) but has evolved from the assumption of "not much up there" to Wilson's (2005) eighth continent with the revelation of the great biodiversity in trees (Erwin 1982). Researchers have found different diversities of different organisms at different heights in the canopy (Erwin 1982; Lowman and Rinker 2004). To date there is no data to support tardigrade usage of the canopy (Miller 2004). When mentioned, tardigrades fall into the "also found" category (Voegtlin 1982; Counts et al. 2001).

Fig. 30.1 Collecting tardigrade habitat in white pine tree, (a) taking sample, (b) ascending safely

Mitchell et al. (2009) tested the hypothesis that substrate (species of tree) influences the habitat (moss or lichen) which in turn influences the distribution, density, and diversity of microinvertebrates, especially the tardigrade. They also noted some variation in height in the tree.

Because water bears are aquatic, they are only active during periods of moisture in their habitat. Because tardigrades are microscopic in size and passively distributed by the winds, they do not seek or select their habitat. Arriving as "Tardigrade Rain," water bears accept or reject the habitat into which they are deposited by the success of their population. Thus, if tardigrades are not present, then one or more of the elements for successful life is missing. Because this process goes on continuously, any sample is a statement in time and an expression of the success or failure of tardigrade habitation.

Nelson (1975) who collected only at ground level suggested height from the ground might be a contributing factor in tardigrade success. Mitchell et al. (2009) documented a relationship between water bears and tree species (substrates) and pH, but her study was limited to ground-reachable height.

To gain insight into tardigrade distribution and diversity within the canopy, we chose to conduct a vertical transect in a 25-m white pine (*Pinus strobus*) on the campus of Baker University (N 38.76684, W 95.18342), Baldwin City, KS. We borrowed a safety harness and a climbing rope from a local tree trimmer and one of the authors climbed the limbs of the tree while the other belayed her (Fig. 30.1a). Sites were established at 1, 3, 7, and 9 m above the ground. At each site, moss and lichen habitat samples were scraped into brown paper bags with a pocket knife (Miller 1997).

In the lab, samples were soaked in a cup of water for 24 h. Subsamples were examined under a 20× dissecting microscope. Tardigrades were transferred with an Irwin loop into a drop of PVA media on glass slides (Miller 1997). Tardigrades were identified based on visible characteristics such as cuticle, claw structure, and mouth parts using the keys of Ramazzotti and Maucci (1983), Nelson and McInnes (2002), and Pilato and Binda (2010). All data was recorded in an excel spreadsheet and chi-square analysis applied to the difference between the observed and an expected value of equal distribution (Zar 1999).

2 Results

Thirty lichen and moss samples were recovered from the tree. One hundred eighty-three water bears representing two classes, three orders, and five genera were identified (Table 30.1). Three common genera (*Milnesium, Paramacrobiotus*, and *Minibiotus*) made up 98 % of the specimens, while the other two (*Echiniscus* and *Isohypsibius*) were represented by only three specimens (Table 30.1).

More than 50 % of the tardigrades were found at the highest level in the tree and more than 75 % were found on the two upper levels (Table 30.1). The most common genus, *Milnesium* 125 of 183 (68 %), was found only at the three upper levels. Both of the other common genera (*Paramacrobiotus* and *Minibiotus*) were found at all four levels but occurred more frequently at the three higher levels (89 % and 81 %, respectively), while 66 % of the two uncommon genera were found at lower levels (Table 30.1). Based on chi-square analysis ($X^2(3, n=183)=84.48, p<0.001$), we must conclude that the distribution of tardigrades in this tree is not a random pattern nor equal dispersion. There are significantly more tardigrades higher in the tree.

3 Discussion

Little is known about the tardigrade or any microinvertebrate relationship to substrates or habitats or heights in a tree. Habitats (moss and lichen) respond slowly to environmental change (positive or negative) and the animals that live within them

Table 30.1 Height distribution of genera of tardigrades in a white pine tree in Kansas

Height	*Milnesium*	*Paramacrobiotus*	*Minibiotus*	*Echiniscus*	*Isohypsibius*	Totals	
9.5 m	70	17	7			94	51 %
7.3 m	39	2	7	1		49	27 %
3.3 m	16	6	8			30	16 %
1.0 m		3	5	1	1	10	6 %
Total	125	28	27	2	1	183	

are likewise slow to change; thus, they are useable monitors of change. Using a statistically designed sampling plan, it would be possible to use tardigrades to measure environmental change in the canopy.

The presence or absence of microorganisms such as tardigrades, rotifers, or nematodes is an expression of the health of their habitat. If the habitat changes, then its acceptability for microscopic life may change and the survivability of animals like tardigrade may be affected.

As the macroenvironmental forces (temperature, moisture, structure) act on the forest, it is unclear if the microinvertebrates can adjust as the paradigm for their survival also shifts. If animals like water bears cannot adjust, they might be the weak link in the productivity of forest ecosystems. We feel a three-dimensional model for microinvertebrate distribution and abundance should be developed as a tool to monitor change of this unseen but critical trophic level in canopy health. Undergraduate students are capable of not only exploring tardigrades in the canopy but also discovering new species and making meaningful contributions to our understanding of canopy biodiversity.

References

Counts JW, Henley L, Skrabal M, Keller HW (2001) Biological jewels in tree canopies. Trans Missouri Acad Sci 35(2):69

Erwin TL (1982) Tropical forests: their richness in Coleoptera and other arthropod species. Coleopt Bull 36:74–75

Jönnson KI, Rabbow E, Schill RO, Harms-Ringdahl M, Rettberg P (2008) Tardigrades survive exposure to space in low earth orbit. Curr Biol 18(17):729–731

Kinchin IM (1994) The biology of tardigrades. Portland Press, London, pp. 186

Lowman MD, Rinker HB (eds) (2004) Forest Canopies. Elsevier Academic Press, San Diego, pp. 1–544

Miller WR (1997) Tardigrades: bears of the moss. Kans Sch Nat 43(3):1–16

Miller WR (2004) Tardigrades: bears of the canopy. In: Lowman MD, Rinker HB (eds) Forest Canopies. Elsevier AcademicPress, San Diego, pp. 251–258

Mitchell CR, Miller WR, Davis B (2009) Tardigrades of North America: influence of substrate on habitat selection. J Pa Acad Sci 83(1):10–16

Nelson DR (1975) Ecological Distribution of Tardigrades on Roan Mountain, Tennessee-North Carolina. In: Higgins RP (ed) International Symposium on Tardigrades, 1974. Memorie dell'Istituto Italiano di Idrobiologia, Suppl. 32:225–276

Nelson D, McInnes SJ (2002) Tardigrada. In: Rundle SD, Robertson AL, Schmid-Araya JM (eds) Freshwater meiofaunal biology and ecology. Backhuys, Leiden, pp. 177–215

Pilato G, Binda MG (2010) Definition of families, subfamilies, genera and sub genera of the Eutardigrada and keys to their identification. Zootaxa 2402:1–54

Ramazzotti G, Maucci W (1983) Il Phylum Tardigrada. III edizione riveduta e aggiornata, Memorie dell'Istituto Italiano di Idrobiologia, vol 42. p. 1–1011

Voegtlin DJ (1982) Invertebrates of H.J. Andrews Experimental forest, western Cascade Mountains, Oregon: a survey of arthropods associated with the canopy of old growth *Pseudotsuga menziesii*. Forest Research Laboratory, School of Forestry, Oregon State University, Corvallis

Wilson EO (2005) The future of life. Knopf, Borzoi Books, Random House, New York, pp. 230

Zar JH (1999) Biostatistical analysis, 4th edn. Prentice Hall, Upper Saddle River

Chapter 31
From Leaf Litter to Canopy: Noninvasive and Reliable Sampling in a Tropical Rainforest

Manjari Jain and Rohini Balakrishnan

Keywords Forest structure • Microhabitat selection • Acoustic monitoring • Horizontal canopy access • Crickets • Kudremukh

Summary

- An indirect, inexpensive, and noninvasive method to estimate foliage density in different strata of a rainforest is described.
- Foliage density-height profile of the rainforest reveals that the canopy has the highest foliage density among all strata of the rainforest.
- Study on microhabitat selection for calling sites shows that ensiferan species are microhabitat specialists with some calling exclusively from the abundant canopy, while others call from rare microhabitats such as dead logs.
- Lab and field-based tests reveal the potential of acoustic monitoring as a noninvasive and reliable method to quantify and monitor biodiversity of acoustically active animals such as insects and bats.
- A nondestructive method to move horizontally within the canopy is described that can potentially serve as a low-cost alternative to canopy walkways.

M. Jain (✉)
Institute of Evolutionary Biology and Environmental Studies, University of Zurich, Zurich, Switzerland

Centre for Ecological Sciences, Indian Institute of Science, Bangalore, India
e-mail: drmanjarijain@gmail.com

R. Balakrishnan
Centre for Ecological Sciences, Indian Institute of Science, Bangalore, India
e-mail: rohini@ces.iisc.ernet.in

M. Lowman et al. (eds.), *Treetops at Risk: Challenges of Global Canopy Ecology and Conservation*, DOI 10.1007/978-1-4614-7161-5_31,

1 Introduction

Precise estimates of vegetation structure variables are important for understanding trophic relationships, ecosystem processes such as energy fluxes, transpiration, and photosynthesis. Tropical rainforests are known for their heterogeneous and stratified architecture as vegetation structure varies greatly between layers of the forest. However, quantitative studies examining vegetation structure in tropical forests through vertical strata are scarce. The heterogeneous vegetation structure of rainforests gives rise to diverse microhabitats that differ in microclimatic conditions, species composition, and resource availability. These microhabitats are, however, poorly characterized in terms of their relative availability and the diversity of species they accommodate. Animals choose certain microhabitats over others in response to the availability of resources (such as sitting space, food, or mates) or to avoid predators. Thus, microhabitat selection has important fitness consequences for animals. Yet, few studies have examined microhabitat selection in a quantitative fashion, especially in insects. This problem is even more acute in tropical rainforests. It is known that most of the earth's terrestrial diversity is concentrated in the tropics and tropical rainforests are home to about half of this diversity. Yet a large fraction of this biodiversity, especially that of invertebrate species, is still unknown to science. This immense tropical biodiversity is under threat because of habitat loss and destruction due to anthropogenic activities. As a result inventorying of biodiversity becomes crucial in order to concentrate conservation efforts to species-rich areas (hotspots). However, field surveys are usually time consuming, logistically difficult, and often invasive in nature. The magnitude and urgency of the task of inventorying and monitoring biodiversity in tropical rainforests makes it imperative to use easy, noninvasive, and reliable techniques. The task of biodiversity inventorying and characterization of forest structure, however, becomes harder as one goes towards the higher layers of the rainforest such as the canopy.

In this chapter we describe an easy, inexpensive method to estimate foliage density in different layers of tropical rainforest, a method to quantify microhabitat selection and discuss the use of acoustic monitoring as a technique to monitor orthopteran species diversity. We briefly describe the main findings of our research carried out using these methods. The inaccessibility of forest canopies has severely impeded research in forest canopies that remain largely unexplored. When sampling or recording in the canopy, one needs to move both vertically and horizontally. While several techniques exist to move vertically in and out of the canopy, most people rely on using a canopy walkway to move horizontally in the canopy. While canopy walkways are important for reliable and easy access into the canopy, they cannot be established at large scale. In this chapter we describe an innovative method to move horizontally in the canopy of rainforests in areas devoid of canopy walkways. All methods have been developed keeping in mind the complex architecture of tropical rainforests and the inaccessibility of the rainforest canopy.

2 Study Site

The work was carried out in a tropical wet evergreen forest in Kudremukh National Park (lat. 13°01′–13°29′N, long. 75°01′–75°25′E), which is situated in a biodiversity hotspot, the Western Ghats, in southern India.

3 Methods and Results

3.1 Forest Structure

We developed an indirect photographic method to estimate foliage density in different heights of the forest including the canopy (Jain et al. 2010). The method involves taking a picture of the foliage (of a given height class) from a fixed distance D, against the backdrop of a background of known dimension. The photographs are then analyzed using open access software ImageJ (version 1.32j, Wayne Rasband, National Institutes of Health, USA) to measure the total area of the background that is obscured by foliage. Foliage density in each height class is then calculated using the method described by MacArthur and MacArthur (1961):

$$k = \ln(1/X)/D$$

where k is the foliage density, X the proportion of board unobscured, and D the height class interval.

Using this method we examined the foliage density-height profile of the forest. We estimated foliage density in three distinct layers of the understory (0–1 m, 1–5 m, 5–8 m) and the canopy (8–30 m) of the forest in six large quadrats ($20\times30\times30$ m^3). We found foliage density to decrease with height in the understory, with a significant drop in foliage density above 5 m (Fig. 31.1). Among the different strata of the forest, the canopy had the highest mean foliage density (Fig. 31.1).

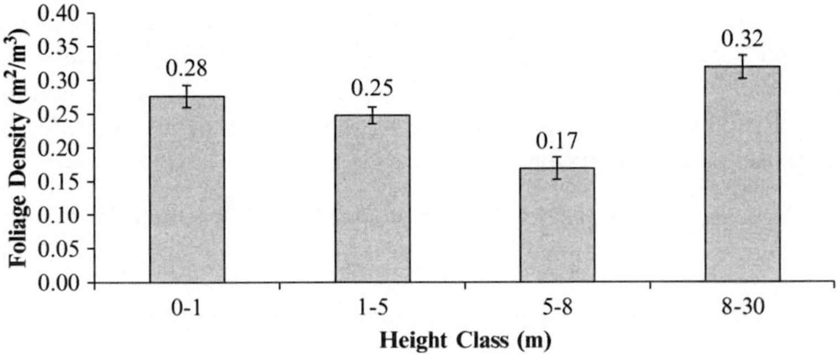

Fig. 31.1 Foliage density-height profile – mean foliage density in different strata of the forest (averaged across six $20\times30\times30$ m^3 quadrats). Error bars indicate SE

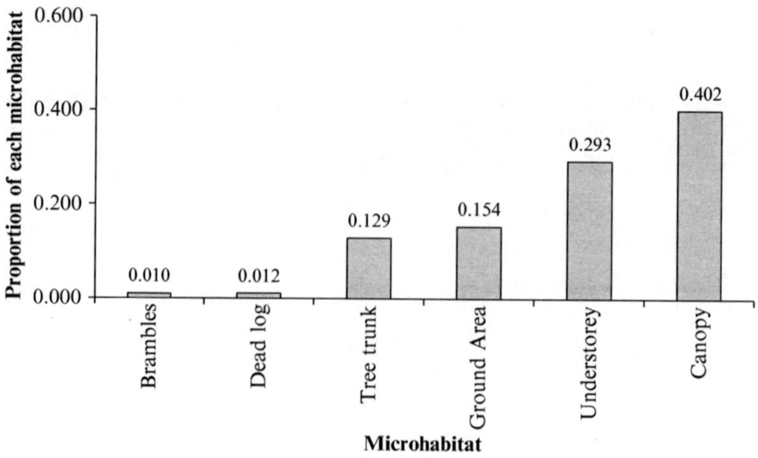

Fig. 31.2 Proportional availability of the six microhabitats

We then quantified the amount of sitting space provided by six different microhabitats of a tropical rainforest, namely, leaf-litter-covered ground, brambles, dead log, tree trunks, understory, and canopy. We found that of the different microhabitats examined, the canopy contributes approximately 40 % of the total sitting space offered across all microhabitats (Fig. 31.2; Jain and Balakrishnan 2011).

3.2 *Microhabitat Selection*

Selection for a resource occurs only when the use of the resource is disproportionately more than its availability. Using psychoacoustic sampling we quantified the differential use of six microhabitats (leaf litter-covered ground, tree trunk, brambles, dead logs, understory, and canopy) by 13 of the 20 known species of crickets and katydids (Diwakar and Balakrishnan 2007) inhabiting the rainforests of Kudremukh National Park. We used the data on proportional availability and use of different microhabitats to calculate Resource Selection Functions (RSF; McLoughlin et al. 2002) to quantify microhabitat selection as described below:

Step 1. At first we calculated the selection function ($w_{h,s}$) of a microhabitat (h) by a species (s):

$$\mathbf{w_{h,s} = proportion\ used_h\ /\ proportion\ available_h}$$

Step 2. We then use a standardization protocol to estimate the likelihood that an individual of species (s) will choose microhabitat (h) over all other microhabitats that are equally available to it. The standardization is carried out as follows:

$$b_{h,s} = w_{h,s} \,/\, \sum_{h=1}^{H} w_{h,s}$$

for s = 1 to S, where S is the total number of species and H is the total number of microhabitats.

Using RSF we found that of the 13 species examined, 10 showed 100 % selection for a specific microhabitat. For example, a species of weta, *Gryllacropsis* sp., called exclusively from tree trunks of large trees (mean girth at breast height > 1 m), while a gryllid species, *Landreva* sp., always called from dead rotten logs, a rare microhabitat providing less than 1 % of total available space. Of the six katydid species belonging to family Tettigoniidae, two species (*Onomarchus* sp. and *Phyllomimus* sp.) were found to be canopy specialists, three others (*Brochopeplus* sp., *Pirmeda rosetta*, and "*15 kHz*") called exclusively from the understory, while one species (*Mecopoda "Two-part"*) showed close to 100 % selection for brambles, a rare habitat that provides less than 2 % of the total sitting space in the forest (Jain and Balakrishnan 2011).

3.3 Biodiversity Estimation and Monitoring

The ensiferan community of Kudremukh was characterized using acoustic monitoring (Diwakar and Balakrishnan 2007). Acoustic monitoring is typically carried out in two ways, (a) by manual call surveys and (b) through ambient noise recordings using an automated system. We wanted to assess the reliability of both these methods in biodiversity monitoring of orthopteran species. For this we carried out lab-based psychophysical tests on a trained listener and compared field-based psychoacoustic spot sampling with simultaneous ambient noise recordings in the field. We also compared the hearing threshold (or sensitivity for different frequencies of sound) of a human listener and that of a sound recorder to examine the detection efficiency for sounds of different frequencies. The reliability of correct species identification by the trained listener was 100 % for 16 out of 20 species and that for identifying the numbers of individuals correctly was 100 % for 13 out of 20 species (Diwakar et al. 2007). This demonstrates the reliability of human listeners in differentiating between species and numbers of calling individuals based on hearing calls. We also found that the human listener performed better than the instrument in detecting low-frequency and broadband calls in the field, whereas the recorder detected high-frequency calls with greater probability. We suggest that trained listener-based sampling is preferable for low-frequency calls, whereas broadband recorders are preferable for high-frequency calls for accurate estimation of orthopteran species richness and relative abundance (Diwakar et al. 2007). In a different endeavor

Fig. 31.3 Setup for horizontal movement through the canopy. Mr. Gaurav (*left*) and Mr. Xavier (*right*) making recordings in the canopy of an evergreen forest in Kudremukh

we recorded the echolocation calls of 13 microchiropteran bat species in Kudremukh that were captured using mist-netting. Of these, 7 species have been recorded for the first time. Discriminant function analyses (DFA) of the calls of 11 species provided 91.7 % correct classification of individuals to their respective species. The accuracy of DFA improves the prospects of using calls of species as a proxy for the presence of these animals in an area. The echolocation calls recorded will serve as baseline data for noninvasive acoustic surveys and monitoring of bat species in the future.

3.4 Horizontal Movement Through the Canopy

Canopy science is greatly hindered by poor methods of horizontal sampling in the canopy. With the help of two professional climbers Mr. Gaurav J and Mr. C.B. Xavier, we developed a safe and low-cost method to move horizontally between the crowns of trees in the canopy of a rainforest in order to make sound recordings in the canopy. The steps involved in setting up the access system include (1) setting up anchor points and a vertical line at starting node, (2) establishing two horizontal lines between the crown of trees using a pulley system, (3) setting up another vertical line at the end node, (4) ascending into the canopy from node 1 using Single Rope Technique, (5) transferring to the horizontal lines to dangle midair till sampling/ recording is done, (6) doing a monkey crawl to approach the next sampling location along the horizontal line, and finally, (7) descending back to the ground from the end node (Fig. 31.3).

4 Discussion

Our study reveals the forest canopy as being the layer with the highest foliage density, highlighting its importance in ecosystem processes such as transpiration, photosynthesis, and carbon fluxes. The fact that the canopy accounted for approximately 40 % of the total available sitting space emphasizes its role in providing refuge for animals.

Our findings on microhabitat selection in ensiferan species of Kudremukh have important conservation implications. The tight association between insect species and different microhabitats highlights the need to preserve all components of the forest structure, including the abundant canopy and rare microhabitats such as dead logs that could be of critical survival value to invertebrate species. It must be noted that this selection for microhabitats is unlikely to be driven by phylogenetic constraints since species of the same family selected different microhabitats for calling. However, other factors such as avoidance of eavesdropping predators or improving efficiency of communication could be potential drivers of microhabitat selection for calling sites in these acoustically communicating insects.

It is known that forest canopies harbor a staggering diversity of species, a large fraction of which is waiting to be documented. We propose that acoustic monitoring through human listener-based sampling and automated recordings can be used for rapid, easy, reliable, and noninvasive biodiversity inventorying and monitoring of acoustically communicating animal species such as frogs, orthopteran insects, as well as birds having species-specific calls. Our research demonstrates the potential of acoustic monitoring as a noninvasive and reliable method to monitor biodiversity of acoustically active animals. This improves the prospects of using remote acoustic monitoring for sampling biodiversity in inaccessible parts of the forests such as the canopy. With advances in technology it is now possible to set up recording and networking devices in the canopy and carry out passive recording and monitoring of biodiversity without physically being present. Yet, several other studies require researchers to gain direct access to the canopy. The canopy access method described here provides a low-cost, nondestructive, and safe method to access and sample in the canopy in areas lacking a canopy walkway. Several methods have been developed to move horizontally in the canopy (Dial et al. 2004) and the choice of method depends on the gear available and the location of sampling. One must, however, bear in mind that professional training and sufficient skill is required before attempting such methods of canopy access. Given that tropical rainforests are fast disappearing and the remaining ones are found in developing or underdeveloped nations, such methods could complement other conventional methods of canopy access in order to unveil the mysteries of the last remaining biological frontiers.

References

Dial RJ, Sillett SC, Antoine ME, Spickler JC (2004) Methods for horizontal movement through forest canopies. Selbyana 25:151–163

Diwakar S, Balakrishnan R (2007) The assemblage of acoustically communicating crickets of a tropical evergreen forest in Southern India: call diversity and diel calling patterns. Bioacoustics 16:113–135

Diwakar S, Jain M, Balakrishnan R (2007) Psychoacoustic sampling as a reliable, non- invasive method to monitor orthopteran species diversity in tropical forests. Biodivers Conserv 16:4081–4093

Jain M, Balakrishnan R (2011) Microhabitat selection in an assemblage of crickets (Orthoptera: Ensifera) of a tropical evergreen forest in Southern India. Insect Conserv Divers 4:152–158

Jain M, Kuriakose G, Balakrishnan R (2010) Evaluation of methods to estimate understorey foliage density in a tropical evergreen forest. Curr Sci 98:508–515

MacArthur RH, MacArthur JW (1961) On bird species diversity. Ecology 42:594–598

McLoughlin PD et al (2002) Hierarchical habitat selection by barren-ground grizzly bears in the central Canadian Arctic. Oecologia 132:102–108

Part IV
Education and Outreach

Introduction

Research on the forest canopy constantly shows the importance of conserving and understanding this unique habitat. Scholars continue to push for projects and regulations to aid in conservation efforts. However, the future of the canopy relies largely on the education of the youth and enthusiasm by the public for these conservation efforts. The articles in this chapter display numerous ways to cultivate public interest in the canopy – from learning about its biodiversity firsthand to integrating new technology with teaching techniques in the classroom. Research demonstrates the role the forest canopy can have on society and the need for conservation, but without public support, many of these goals will not be achieved. There is an increase in programs allowing "citizen scientists" and youth to go into the field to interact with nature and experience research firsthand. The canopy provides numerous educational tools linking the field to the classroom and bringing scientists and educators together. This chapter demonstrates how effective application of groundbreaking research to the global community can inspire canopy education and conservation.

Chapter 32
Win-Win for Scientists and Citizen Scientists Who Engage in Amazon Canopy Expeditions

D.C. Randle and Frances Gatz

Keywords Amazonia • Rainforests • Canopy research • Ecotourism • Experiential education • Citizen science • Service learning • Student travel

Summary and Bullet Points

The future protection of biodiversity and forest canopies depends on our youth becoming concerned scientists, responsible policy makers, and informed global citizens. With rainforest education marginalized in K-12 classrooms in the USA, how do we nurture this next generation?

Environmental Expeditions' Amazon Rainforest Workshops serve as an example of how informal education can address this concern by engaging teachers, nature lovers, and students in an experiential citizen science program focusing on:

1. Biodiversity: Students discover a rich diversity of plant and animal species, an intricate world of jungle relationships at each level of the rainforest, and why biodiversity matters.
2. Research and Conservation: Scientists provide hands-on experiences of what canopy research is about and how research is essential to canopy preservation.
3. Cultural Immersion: A service project in a local village offers students a personal connection with local people and an understanding of the critical role of local people in the conservation equation.

D.C. Randle (✉)
St. Francis High School, St. Francis, MN 55070, USA
e-mail: drandle55@msn.com

F. Gatz
Frances Gatz, Environmental Expeditions, Silver Spring, MD 20910, USA
e-mail: frances@ee-link.net

M. Lowman et al. (eds.), *Treetops at Risk: Challenges of Global Canopy Ecology and Conservation*, DOI 10.1007/978-1-4614-7161-5_32,

4. Global Perspective: Experiences to broaden each participant's perspective of (1) the role of natural and human resources in our global economy, (2) our interconnectedness and the impacts that we as global citizens make through our choices, and (3) our role in addressing climate change and alleviating poverty.

1 Introduction

The only source of knowledge is experience. —Albert Einstein

The future protection of rainforest biodiversity and forest canopies depends on our youth becoming concerned scientists, responsible policy makers, and informed global citizens. With rainforest education marginalized in K-12 classrooms around the world, how do we nurture this next generation to address this concern? Environmental Expeditions' Amazon Rainforest Workshops rely on the leadership of scientists who can address research and conservation issues by involving teachers, nature enthusiasts, and students in an experiential travel program. St. Francis High School in Minneapolis, Minnesota, is an example of a school whose students first traveled to the Amazon in 2005. While returning again and again, they continue their education outreach at home (Fig. 32.1).

Fig. 32.1 St. Francis High School students present scientist Meg Lowman with a handmade plaque to celebrate her canopy research

How can this be a win-win experience to all involved? Teachers are mentored by the scientists, who share their personal experiences in the classroom. Students benefit from the teachers' heightened enthusiasm and direct contact with researchers and carry the message to other students and to the community. Lastly, the canopy researcher, by conducting outreach in this way, fulfills a requirement often mandated by research grant funding, to engage in public outreach. These workshop programs are also offered by scientists themselves collaborating with other university faculty as part of university field courses.

The workshop experience stimulates awareness, compassion, and connection. On expedition there is no "teaching to the test." Participants are guided by curiosity and discovery. Unlike a typical classroom course of study, this Amazon experience offers the opportunity to overcome fears, build confidence, recharge enthusiasm, learn team building, see the value of reflection, and develop a connection with an otherwise remote part of the planet. It is often stated that "my experience in the Amazon changed my life!" The intent is that these transformed travelers will become part of our informed and active global citizenry. The Amazon Rainforest Workshops are an eye-opening introduction to the environmental and cultural aspects of the Amazon Rainforest and river system. Canopy scientists play a vital role in these programs. Participants get a real-time immersion in a scientific quest for learning and discovery, serving as citizen scientists. They witness a biologist's passion in the field and get a close-up view of how insects, birds, and plants depend upon one another to spin the web of life.

> The plane sweeps over lush rainforest on approach to Iquitos, Peru. A huge, rambling river courses below, stretching to the horizon. What awaits me on this journey? A tribe that hunts with blowguns in this dense jungle? A troupe of cackling monkeys? Tarantulas, anacondas, or flesh-eating piranha? Walking out onto the tarmac, the reality of this remote jungle city thousands of miles from home begins to take hold. There are no tall buildings, no stoplights, and the chaotic streets are filled with three-wheeled mototaxis. Soon, you board a boat and head down river, passing children paddling dugout canoes and barges carrying timber.

Each intriguing element of this new environment is a personal discovery and an opportunity to fuel a passion for learning. Every step of the way, participants are engaged by scientists who work hand-in-hand with naturalist guides and teacher leaders to take advantage of every teachable moment.

Workshop scientists collaborate with the local guides, offering the opportunity to contribute to ongoing research, providing a wider context for what is discovered along the way and exploring significant questions as: How did such enormous biodiversity develop? Why is the Amazon called the "lungs of the planet"?

2 Experiencing Biodiversity

Most students know the word "biodiversity" but have not experienced it in Amazon living color. The diversity of birds, though astounding with a list of nearly 600 bird species compiled for the Iquitos region, pales in comparison to that of Amazon insects, with hundreds of thousands of species. Millions of years in the making,

Fig. 32.2 A never to be forgotten experience with a red-tailed boa gives biodiversity personal meaning

Amazon creatures who adapt and survive the test of immense competition are unusual, diverse, and fascinating (Fig. 32.2). Every shade of green is represented here, the individual patterns, lines, curves, forms, and textures. Adaptations to the environment become apparent: prop roots, buttress trunks, large leaf surfaces, drip tips, waxy leaves, and smooth bark. With the lack of wind, fleshy fruit seeds rely on animal dispersal and flower structures accommodate specific animal pollinators. Through these firsthand experiences, students begin to understand in a new way how evolution has honed these characteristics over the eons and see the natural world with new curiosity.

As the week unfolds, other exciting sessions include the excavation of a leafcutter ant colony accompanied by a detailed explanation of this remarkable creature's social structure and reproductive cycle; viewing bioluminescent fungus deep in the forest at night; cataloguing the diversity of tree frogs, butterflies, and insects; an excursion to a blackwater swamp to see the large, prehistoric-looking Hoatzin bird; and encounters with mammals including three-toed sloth, capybara, bats, tapir, and at least seven species of monkeys.

3 Research and Conservation

Conservation decisions rely on the outcomes of rainforest research. It is important for students to have a glimpse of *how* we know *what* we know about the rainforest. It becomes clear that we actually know relatively little about the canopy layer where most tropical forest life exists. Barely 30 years old, the field of canopy research is a fledging science in which researchers access this still largely unexplored part of the rainforest via ropes, walkways, or cranes. One of the high points (literally!) of the workshop is the visit to Amazon Conservatory of Tropical Studies (ACTS) to experience a one-quarter mile canopy walkway—an extensive, multilevel system of aerial platforms and pathways at a height of over 100 ft. ACTS is a science, education, and tourism facility located on a wilderness area encompassing 250,000 acres of rainforest (Fig. 32.3).

> Dawn on the canopy walkway overlooking the rainforest is one of the most memorable parts of the expedition experience. The vastness of the forest is revealed as the fog gradually lifts. Embracing this seemingly endless canopy is truly a contemplative, spiritual experience.

Field study on the canopy walkway is exciting and can generate a lifelong interest in rainforests. Field sessions are inquiry oriented, demonstrating interconnections

Fig. 32.3 Over 100 ft above the forest floor, a workshop participant enjoys the extraordinary Amazon canopy views from the ACTS canopy walkway

Fig. 32.4 St. Francis High School students complete their canopy research and share a quiet moment on the canopy walkway at dusk

among living organisms and environmental conditions. The workshops give participants a window on how a scientist thinks, questions, and problem-solves (Fig. 32.4). There are millions of research questions waiting to be answered. Peer into a bromeliad tank to observe a microclimate, where the overlapping leaves of the plant absorb nutrients from the water collected in its basin, frogs and insects lay their eggs, tadpoles and damselfly nymphs mature, and other aquatic microorganisms live. One could spend a lifetime pursuing questions of this microhabitat alone.

4 Cultural Immersion

The indigenous people of the rainforest are essential to its preservation and are an integral part of Amazon Rainforest Workshops.

> Along the riverbanks, you first see thatch-roofed dwellings built on stilts, and feel the joy of the children who offer their exuberant waves to passersby. The river is the life blood of the Ribereños, or "river people." It is their highway, bathtub, laundry, and source for water and food.

The program is an example of ecotourism—responsible travel with the purpose of education, conservation, and improving the well-being of local people. During our Village Service Project Day, there is a chance to work alongside local people in a Ribereños village. The community welcomes everyone with a ceremony.

Songs and praises are exchanged, and then work begins to paint a school, plant a garden, or mend a fence or roof. Food is shared, as is a game of soccer, and most leave humbled by how these local people survive in the simplest of settings, grateful for what they have. Students are particularly moved by the experience, having seen the stark contrast with their own culture.

5 Global Perspective

Conservation of forest canopies is fostered by a long-term perspective where natural and human resources are both part of the economic equation. The Amazon Rainforest Workshops enable travelers to see firsthand how indigenous peoples depend on the preservation of the rainforest and how villagers might find more sustainable ways to use their resources.

The Amazon rainforest has global significance because of its impact on climate stability—through oxygen production and carbon retention. It is a vital resource for plant- and animal-derived pharmaceuticals, natural pesticides, crop alternatives, and much more.

Workshop participants take home a personal understanding of the global interdependence of ecosystems and how their personal choices at home impact the Amazon rainforest. They have a grasp of the many perspectives and considerations involved as governments address international environmental problems. Many educational projects, public expositions, articles, and media coverage result from inspired travelers returning home from a workshop.

The preservation of rainforests requires immediate and heroic efforts worldwide; an educated, caring, and motivated younger generation, combined with the passionate research efforts of scientists, will continue to provide the energy and enthusiasm needed to promote public awareness and changes in policy. To this end, Amazon Rainforest Workshops continue to make an important contribution.

References

Lowman MD (2004) Ecotourism and the treetops: complement, not conflict, for an emerging science. In: Lowman MD, Rinker HB (eds) Forest canopies. Elsevier Press, San Diego

Lowman MD, Randle DC (2009) Ecological mentoring: inspiring future scientists. Front Ecol Environ 7:119

Nadkarni N, Parker G, Lowman MD (2011) Forest canopy studies as an emerging field of science. Ann For Sci 68:217–224

Rinker HB, Jarzen DM (2004) Reintegration of wonder into the emerging science of canopy ecology. In: Lowman MD, Rinker HB (eds) Forest canopies. Elsevier Press, San Diego

Chapter 33
In the Canopy with Wheelchairs: A Model for Teaching Field Biology

William R. Miller, Margaret Lowman, and E. McCord

Keywords Mobility-limited students • Tardigrades • Wheelchairs

Bullet Points

1. Mobility-limited students deserve opportunities to seek careers in field biology.
2. The vertical (not horizontal) ambulatory actions of canopy research are accessible for wheelchair-bound students.

Summary

We have created a model whereby students with disabilities can explore, study, and discover ecological truths via canopy research. The authors make the case that physical limitations such as crutches or a wheelchair do not preclude a career in field biology. This chapter discusses several research projects that can be conducted with simple vertical transects up different tree canopies and subsequently inspire career paths in field biology for underserved students with mobility limitations.

W.R. Miller (✉)
Department of Biology, Baker University, Baldwin City, KS 66006, USA

North Carolina Museum of Natural Sciences, Raleigh, NC, USA
e-mail: william.miller@bakeru.edu

M. Lowman
North Carolina Museum of Natural Sciences, North Carolina State University, 121 W. Jones Street, Raleigh, NC 27603, USA
e-mail: canopymeg@gmail.com

E. McCord
Department of Biology and Environmental Studies, New College of Florida, Sarasota, FL, USA
e-mail: mccord@ncf.edu

M. Lowman et al. (eds.), *Treetops at Risk: Challenges of Global Canopy Ecology and Conservation*, DOI 10.1007/978-1-4614-7161-5_33,

1 Introduction

Forest canopies are often called the "eighth continent of planet Earth" (Wilson 2005) due to their complexity and enormous biodiversity. With the explosion of canopy exploration over the last two decades (Lowman 1999; Lowman and Rinker 2004a), made possible by the simultaneous development of safe single- and doubled-rope tree climbing techniques (SRT/DRT), canopy researchers have confirmed that a large portion of global biodiversity inhabits the treetops (Lowman and Rinker 2004a). Ascent into the treetops is critical to answering scientific questions about forests, biodiversity, and ecology.

Historically, the vast preponderance of canopy exploration has prioritized tropical rain forests, with almost no quantitative surveys in temperate forests (but see Preisser et al. 1999; Shaw et al. 2006). A need exists to address this disparity by exploring the diversity, density, and vertical distribution of invertebrates in four North American forest types using protocols developed in tropical forest canopies to facilitate tropical-subtropical-temperate comparisons. This paucity of data and knowledge can also be used to train students to survey diverse groups of invertebrates (such as tardigrades and beetles) to answer ecological questions about habitat selection, herbivory, diversity, vertical stratification, and evolution. Most aspects of stratification, abundance, and responses to environmental change are unexplored in the American temperate forest canopies; thus, all of our students will be on the cutting-edge of discovery (Bergelson et al. 1993).

2 Students with Disabilities as Field Biologists

As equal opportunity employers and educators, we are dedicated to support the commitment to broadening opportunities for women, minorities, and persons with disabilities. We feel that students in underrepresented groups should be encouraged to pursue careers in field biology if that is their passion.

The authors have worked with students and colleagues with physical disabilities and recognize that being ambulation-restricted is not a condition that precludes good science. Burgstahler (2009) observed that any student with a research opportunity can observe, sample, and understand ecology. Students with disabilities have the same ability to think, question, hypothesize, analyze, and write as any other student. In the laboratory, students in or out of a wheelchair can use microscopes, analytical instruments, computers, and books to classify invertebrates, analyze environmental samples, and interpret temporal and spatial data (Burgstahler 2009).

"How can a student in a wheelchair climb a tree?" is often the first question, to which the response is that canopy researchers climb ropes, not trees, and that a person with average arm strength can hoist themselves into the canopy. During the Fifth International Canopy Conference, in Bangalore, India (October, 2009), we consulted Tim Kovar, a professional tree climbing instructor from New Tribe, Inc., and tested a new DRT climbing rig. One of us (WM)—who is older, overweight, and recently

Fig. 33.1 Miller in India

underwent knee replacement surgery—quickly and easily lifted himself into a tree without the use of his legs (Fig. 33.1).

Gathright et al. (2008) recently wrote of the value of tree climbing in the therapeutic treatment of disabilities. His *Treehab Program* (treeclimbing.jp/Treehab.html) aims to help physically challenged persons climb trees, and although not ecological research, his work is testimony to the fact that students with an ambulatory disability need not be excluded from canopy exploration.

The TV news Magazine, *CBS Sunday Morning* (CBS 2002) aired the story of Patty Jenkins, a crutches-dependent polio survivor in her late 50s, as she and her husband, Peter, ascended a 167-ft Sitka Spruce in the Hoh rain forest and spent the night in its branches. Patty sees no reasons why students with similar physical disabilities cannot explore scientific questions in the canopy (personal communication).

Dan House, Becki Peterson, and Stan Stalnaker of Tree Climbers Kansas City (treeclimberskc.com) have considerable experience with teaching students of all ages to climb trees (Figs. 33.2 and 33.3). Students with ambulatory disabilities need only the tools, mentorship, and enthusiasm to pursue science, including canopy ecology.

3 Suggested Opportunities for Mobility-Limited Students to Pursue Canopy Research

Student research can be based on simple vertical transects with sampling stations established at standard elevations, say each 3 or 5 m. At each station, a set of protocols could be carried out to collect and sample the desired data. Quickly the data set

Fig. 33.2 Tree Climbers KC, Dan House teaching climbing to student

Fig. 33.3 Dan and student success

from each species of tree (substrate) becomes comparable to other tree species. The results at different elevations are another valid data set as is the matrix of habitats found within the canopy (moss, lichen, algae, and plants). Differences between forest types can be studied and analyzed both temporally and spatially.

Fig. 33.4 Tardigrade (water bear)

3.1 Tardigrades

Tardigrades (Fig. 33.4) are a little studied phylum of aquatic invertebrates found in the interstitial moisture that collects within the leaves of mosses and thalli of lichens that inhabit the trunks and branches of trees all over the world. Tardigrades are known for cryptobiosis, the process of desiccating with its habitat and reconstituting with the return of moisture (Miller 1997). Recently, cryptobiotic tardigrades became the first multicelled animal to survive exposure to the vacuum, temperature, and radiation of space (Jönnson et al. 2008). Cryptobiotic tardigrades are dispersed in the winds (Kinchin 1994) and thus are rained onto the canopies of the world, where they must find acceptable habitat to survive (Miller 2004). Despite our knowledge of tardigrade physiology, we know almost nothing about their ecology (Ramazzotti and Maucci 1983; Miller 1997). Mitchell et al. (2009) presented the first data of differential substrate and vertical distributional patterns for tardigrades. A canopy tardigrade hypothesis is that there are significant differences in the diversity and density of tardigrade populations at different elevations on different substrates (tree species) and that there exists more diversity in the higher reaches of the canopy (See Miller et al. this volume). We propose to engage mobility-limited students to address that hypothesis.

3.2 Herbivory

Tropical rain forest canopies clearly show that herbivory (Fig. 33.5) is lowest in the sun leaves of the upper canopy and that insect herbivores prefer to eat young leaves more than mature ones (Lowman 1985; Lowman and Heatwole 1992). Further,

Fig. 33.5 Herbivory

insect herbivory has been shown to correlate to leaf toughness more significantly than to leaf chemistry (Lowman and Box 1983). Thus, insect herbivory is another student-testable hypothesis in temperate forest canopies. Again vertical transects will be used to sample herbivory and test hypotheses; students will measure leaf attributes as well as spatial and temporal abundance and diversity of common herbivores (Lowman and Heatwole 1992; Lowman and Rinker 2004b).

Specifically, three projects—(1) long-term monitoring and discrete foliage sampling to quantify herbivory with respect to spatial and temporal factors (see Lowman 1983 for protocols), (2) enclosure experiments using mesh bags to assess the effect of insect herbivory on common genera of trees, and (3) laboratory analyses of leaf toughness and chemistry with respect to different ages and vertical distribution of foliage—can be performed using protocols developed for tropical forest canopy studies (see Lowman 1985; Lowman and Box 1983; Ernest et al. 2006).

3.3 *Insect Diversity*

A focus on insect diversity, density, and spatial distribution (Fig. 33.6) is possible because they are a major group of herbivores (Lowman and Selman 1983; Erwin 1982). The idea that there is a greater abundance of insects in the temperate forest understory than the canopy (Preisser et al. 1999) can be tested. Previous studies by Lowman (1985) in a Massachusetts forest suggested that temperate insect herbivores preferred feeding on young, shaded, understory foliage, as they do in the equatorial tropical forests.

Fig. 33.6 Herbivorous insect

3.4 *Leaf Chemistry*

Instrumented chemical composition analyses of samples of the stratified leaf/moss/lichen collections can be subjected to GC-ECD, Mass Spec, FID, and HPLC-UV testing. The chemical properties of the leaf/moss/lichen as food or habitat may be analyzed by time of year, damage, and stratification. Chromatographic techniques to test leaf toughness as well as chemical composition are now possible with the advent of solid-phase and rapid Turbo Vap® extractions.

Leaf extractions should follow the techniques of Li et al. (2009), Pereira et al. (2009), and Wang et al. (2009). Gas chromatography (GC) fitted with an electron capture (EC) and a mass selective (MS) detector may be used to confirm chemical identifications using an NIST data library, and unknown isolates can be subjected to HPLC for further separation and recovery. Water, methanol, ethanol, and other organic solvent extractions may be used to reveal plant allelochemicals. Homogenization (Ferreres et al. 2010), shaking (Chludil et al. 2009), grinding (Alves et al. 2010; Budryn et al. 2009), and/or sonicating will facilitate leaf extractions.

4 Discussion

Undergraduates with physical limitations compose 11 % of US student populations—yet in the life sciences and mathematics, they make up only 9 % (NSF 2009), and far fewer pursue careers in ecology or field biology (Ecological Society of America cites fewer than three attendees in 4,000 with wheelchairs). Opportunity for ambulatory-challenged students to consider life sciences, especially field biology, as a major course of study and ultimately as a career must be encouraged. By design, projects or classes should allow undergraduates to have prosthetics or be unable to ambulate without the assistance of crutches or a wheelchair; such rigors currently deter students with disabilities from seeking careers in field biology.

Upon graduation, the undergraduate canopy researchers should have studied a major forest type, conducted research via SRT/DRT climbing, and have used remote sensing systems, HPLC, GC, and SEM technologies to analyze the environment. Students will have discovered new relationships among the canopy trees and the epiphytes that grow upon them. They may describe new species, new habitats, and new chemical characteristics in the canopy environment of the temperate forests. They may have made poster and oral presentations, and many will have submitted a paper for publication.

We present these protocols in hopes that other researchers will consider the creation of protocols for mobility-limited students in their labs. Upon graduation, such students will be part of a growing research community of young scientists and canopy experts. All will have a core understanding of the achievements and challenges faced by colleagues with physical disabilities and be well prepared for graduate school in ecological sciences. *We boldly declare that canopy access and invertebrate ecology are disability-friendly arenas for rigorous scientific research.*

Acknowledgments This project was supported by NSF grant DBI: REU:115650.

References

Alves RC, Soares C, Casal S, Fernandes J, Oliveira MP (2010) Acrylamide in espresso coffee: influence of species, roast degree and brew length. Food Chem 119(3):929–934

Bergelson JM, Kitching RL, Lowman MD, McIntyre S (1993) The biodiversity of arthropods from Australian rainforest canopies: general introduction, methods, sites, and ordinal results. Aust J Ecol 18:181–191

Budryn G, Nebesny E, Podsedek A, Żyżelewicz D, Materska M, Jankowski S, Janda B (2009) Effects of different extraction methods on the recovery of chlorogenic acids, caffeine, and Maillard reaction products in coffee beans. Zeitschrift für Lebensmittel-Untersuchung Und-Forschung A, Eur Food Res Technol 228(6):913–922

Burgstahler S (2009) Application of Universal Design to a Science Lab. DO-IT, University of Washington. http://www.washington.edu/doit/. Accessed 24 Mar 2010

CBS (2002) http://www.cbsnews.com/stories/2002/10/18/sunday/main526192.shtml. Accessed 24 Mar 2010

Chludil HD, Vilarino MDP, Franco ML, Leicach SR (2009) Changes in *Lupinus albus* and *Lupinus angustifolius* alkaloid profiles in response to mechanical damage. J Agric Food Chem 57(14):5107–5113

Ernest KA, Lowman MD, Rinker HB, Shaw DC (2006) Stand-level Herbivory in an Old-Growth Conifer Forest Canopy. Western North Amer Nat 66(4):473–481

Erwin TL (1982) Tropical forests: their richness in Coleoptera and other arthropod species. Coleopt Bull 36:74–75

Ferreres F, Pereira DM, Valentao P, Oliveira JMA, Faria J, Gaspar L, Sottomayor M, Andrade PB (2010) Simple and reproducible HPLC-DAD-ESI-MS/MS analysis of alkaloids in *Catharanthus roseus* roots. J Pharm Biomed Anal 51(1):65–69

Gathright J, Yamada Y, Morita M (2008) Tree-assisted therapy: therapeutic and societal benefits from purpose-specific technical recreational tree-climbing programs. Arboric Urb For 34:222–229

Jönnson KI, Rabbow E, Schill RO, Harms-Ringdahl M, Rettberg P (2008) Tardigrades survive exposure to space in low earth orbit. Curr Biol 18(17):729–731

Kinchin IM (1994) The biology of tardigrades. Portland Press, London

Li B, Qi L, Wen X, Cao J, Li P, Zeng J, Yang H, Zhao J, Hu R (2009) Rapid and sensitive analysis of tannins and monotrepene glycosides in radix paeoniae alba products by HPLC-MS. J Liq Chrom and Rel Tec 32(15):2232–2245

Lowman MD. (1983*) Rain forest insects and their host plants. Biology in Action (Australian newsletter)

Lowman MD (1985) Temporal and spatial variability in insect grazing of the canopies of five Australian rain forest tree species. Aust J Ecol 10:7–24

Lowman MD (1999) Life in the treetops. Yale University Press, New Haven

Lowman MD, Box JD (1983) Variation in leaf toughness and phenolic content among five species of Australian rain forest trees. Aust J Ecol 8:17–25

Lowman MD, Heatwole HE (1992) Spatial and temporal variability in defoliation of Australian eucalypts and its consequence for the dieback syndrome. Ecology 73(1):129–142

Lowman MD, Rinker HB (eds) (2004a) Forest canopies. Elsevier Academic, San Diego

Lowman MD, Rinker HB (2004b) Insect herbivory in tropical forests. In: Lowman MD, Rinker HB (eds) Forest canopies. Elsevier Press, San Diego

Lowman MD, Selman BJ (1983) The biology and herbivory rates of *Novacastria nothofagi* Selman (Coleoptera: Chrysomelidae), a new genus and species on *Nothofagus moorei* in Australian temperate rain forests. Aust J Zool 31:179–191

Miller WR (1997) Tardigrades: bears of the moss. Kans Sch Nat 43:1–16

Miller WR (2004) Tardigrades: moss bears in the canopy. In: Lowman MD, Rinker HB (eds) Forest canopies. Elsevier Press, San Diego

Mitchell C, Miller WR, Davis B (2009) Tardigrades of North America: influence of substrate on habitat selection. J Pa Acad Sci 83(1):10–16

NSF (2009) Women, minorities and persons with disabilities in science and engineering. NSF, Arlington, 284 pp

Pereira D, Faria J, Gaspar L, Valentao P, Guedes de Pinho P, Andrade P (2009) Boerhaavia diffusa: Metabolite profiling of a medicinal plant from nyctaginaceae. Food Chem Toxicol 47(8): 2142–2149

Preisser E, Lowman MD, Smith D (1999) Canopy and ground level insect distribution in a temperate forest. Selbyana 19(2):141–146

Ramazzotti G, Maucci W (1983) Il Phylum Tardigrada (third Edition) Memorie dell'Istituto Italiano di Idrobiologia Marco Marchi, vol 41. p. 1–1012

Shaw DC, Ernest KA, Rinker HB, Lowman MD (2006) Stand-level herbivory in an old-growth conifer forest canopy. West North Am Nat 66(4):473–481

Wang Z, Li D, Zhou Z, Li B & Yang W (2009) A simple method for screening and quantification of ricinine in feed with HPLC and LC-MS. J Chrom Sci 47(7):585–588

Wilson EO (2005) The future of life. Harvard University Press, Cambridge

Chapter 34
Modeling Insect Outbreaks in Forest Canopies: Integration of Virtual Simulations with Hands-On Ecology for Undergraduates

Leon Kaganovskiy and Margaret Lowman

Keywords Forest canopy • Insect outbreaks • Virtual simulations

Bullet Points

1. Integrate virtual technology with hands-on ecology in a simulation exercise about insect defoliation of forest canopies.
2. Computational results show trends similar to extensive field measurements done by M. Lowman in different forest canopies.

Summary

Students increasingly utilize gaming and handheld PDAs in their everyday lives, leading to a "nature-deficit syndrome" where hands-on ecology is completely replaced by "indoor" STEM applications via computer screen. We integrate virtual technology with hands-on ecology in this simulation exercise about insect defoliation of forest canopies. In this project, we model the dynamics of tree canopies attacked by insects using the ideas similar to ecological models for fire and infectious decease propagation. We developed a stochastic model of a generic tree. A volume containing leaves is determined and broken into a large number of computational cells. Then the dynamics of the foliage including herbivory, aging, and other temporal and spatial factors is first evaluated within the computational cells and mapped back onto the tree crowns. The computational results show trends similar to extensive field measurements done by M. Lowman in different forest canopies.

L. Kaganovskiy (✉)
Touro College, New York, NY, USA
e-mail: leonkag@gmail.com

M. Lowman
North Carolina Museum of Natural Sciences, North Carolina State University, 121 W. Jones Street, Raleigh, NC 27603, USA
e-mail: canopymeg@gmail.com

M. Lowman et al. (eds.), *Treetops at Risk: Challenges of Global Canopy Ecology and Conservation*, DOI 10.1007/978-1-4614-7161-5_34,

1 Introduction

One of the new challenges facing STEM education is to integrate emerging technologies with inquiry-based environmental science (Lowman and Mourad 2010). On the one hand, ecological literacy of Americans is decreasing, perhaps in part due to "nature-deficit disorder" (Louv 2005). On the other hand, students are more apt to experience virtual ecosystems through gaming and computers than experience the outdoors firsthand (Balmford et al. 2002). We propose activities of modeling environmental issues to inspire students to seek careers in science and also to learn about their local ecosystem dynamics. This exercise requires some knowledge of ecology, math, computers, entomology, and climate change. By modeling insect outbreaks or other applied ecological issues, students have an interdisciplinary approach to problem-solving and also are motivated by using virtual and real systems. In this project, we develop an interdisciplinary approach to biology teaching. We created an activity that allowed students to change the parameters of insect outbreaks and the underlying canopy structure of a virtual tree to understand possible outbreak patterns. The learning curve is not steep and the basics of the model can be mastered within a single class with subsequent projects assigned to be completed in a week. This approach is of particular relevance to a biology teacher, since different outbreak scenarios can be considered with relatively short computer runs providing an outstanding supplement to the available historical data on forests and insect pests. In short, this is an interaction that can be observed in the school yard, garden, or at a national scale. From the modeling point of view, our approach is an extension of cellular automata models for fire and decease propagation but applied to tree structure. This approach gives our ecology/biology students a valuable exposure to contemporary agent-based modeling. First, we review some relevant work on the subject and then proceed to describing the algorithm and results.

The interface of ecology and math has facilitated more accurate modeling of biological events, including insect outbreaks (rev in Otto and Day 2007). Insect population fluctuations have long impacted the history, economy, mortality, and ecology of human civilization (rev in Lowman and Morrow 1998). Herbivores represent one of the most important economic groups of insect outbreak populations, where, for example, plague locusts represented one of the first-recorded environmental catastrophes, as reported in the Bible. But only in the last few decades have similar biological epidemics been mathematically quantified through ecological modeling, aiming to use such predictions to reduce their economic and ecological impacts (Green et al. 2003).

Early reports of insect outbreaks in forests were usually confined to casual observations by foresters or ecologists (rev. in Pimental 1986). Debate continues as to whether insect outbreaks should be regarded as agents of disturbance (Veblen et al. 1994) or simply as ecosystem responders to environmental change (Willig and Camilo 1991, 1995). Regardless of causality, insect herbivory can have catastrophic impacts to forest ecosystems, varying with the type, intensity, severity, scale, frequency, and regularity of the outbreaks. For example, outbreaks of Douglas-fir tussock moths (*Orgyia pseudotsugata*) in western North America occur at intervals of 9–10 years and generally last 2–3 years (Mason and Luck 1978), but scarab beetles (*Anoplognathus* sp.) have attacked gum trees in New South Wales, Australia, continuously throughout the last century, albeit with peak outbreaks at 100-year intervals (Lowman and Heatwole 1992). Similarly, many agricultural and forest insect pests consume moderate amounts of foliage every season, with intermittent outbreaks at longer-term temporal intervals.

Exhibiting yet a third temporal pattern, a chrysomelid beetle (*Novocastria nothofagi*) on Antarctic beech exhibits monthly temporal variability, with caterpillar populations peaking every October synchronously with the emergence of new leaf flushes. Insect outbreaks are increasingly important to quantify due to the expanding number of introduced species, both accidental as well as intentional (Dobson and May 1986). Such exotic populations usually exhibit epidemic population growth due to the absence of natural predators (Elton 1958; Simberloff 1986). Gypsy moths (*Lymantria dispar*) were introduced into North America in 1869 but exhibited lagging population outbreaks until 1900–1915 and then again in 1966–1990, at approximately 50-year intervals (Liebhold et al. 1992).

With the increasing economic importance of timber harvests, and the correlation between insect attacks and primary productivity, predicting insect outbreaks has become a critical component of forest conservation management. With the current trajectories for climate change, and with insect outbreaks often serving as an early warning event for drier and warmer conditions (Lowman 1992), the application of mathematical modeling to insect herbivory may be critical to understanding forest canopies.

Returning to modeling, Fisher (1937) first published a model of advantageous genes spreading through a population and Skellam (1951) produced the first application of diffusion models to the spread of invasive organisms. As a consequence of spending more than $194 million on gypsy moth outbreaks in the USA alone at the end of the twentieth century (Johnson et al. 2006), mathematical studies of the spatiotemporal variability in biological invasions expanded in scope. Since then, the mathematical modeling of ecological phenomena, especially for economically important events such as insect outbreaks, has become a critical component of both ecology and forestry. However, simple examples for teaching and understanding the dynamics of invasions are not easily available in the ecological literature, despite an obvious need to integrate math and ecology as a tool for environmental stewardship. In this chapter, we undertake three objectives:

1. To use long-term observations of M. Lowman on forest insect herbivory
2. To model expected outcomes applicable to forest health to create a useful case study for undergraduate students who are beginning to use mathematical modeling as a tool for conservation biology and ecology
3. To simulate one (of several) behavior of insect herbivores in forests based on real data to begin to predict the complex dynamics of tropical rain forest canopies

Accurate predictions of the insect outbreaks are critical, because these events are expected to scope in severity with global climate change. In this project, we use tree data structures to model the growth of tree canopies attacked by insects (Lowman 1985, 1987) using the cellular automata and percolation ideas similar to fire and infectious disease-spreading models (Beer 1990; Drossel and Schwab 1992; Malamud et al. 1998; Von Neumann and Burks 1966; Wolfram 2002; Fu and Milne 2004; Magdon-Maksymowicz 2004; Martins et al. 2000; Zorzenon dos Santos and Coutinho 2001; Morley Submitted; Situngkir 2004).

The chapter is organized as follows. First we discuss tree creation and underlying stochastic setup. Next, we describe five-state models based on five states identified in the work of Lowman (1985, 1987). Finally, we develop a different daily eating model and apply it to the data collected recently as a pilot study (Lowman 1985). We end with conclusions and directions for future research.

2 Tree Creation

The tree crown simulation idea draws upon one of the author's previous work in application of artificial trees to fluid dynamics simulations (Kaganovskiy 2007) as well as on computer science tree codes (Samet 1984; Samet 1990) and applications of data structures to artistic and scientific surface representations (Sederberg et al. 2003; Sederberg et al. 2004; Pauly et al. 2002).

In the tree creation part of the simulation, we start with the artificial "root" of the tree and give it user-defined length and radius (particular to each tree). Then we proceed recursively. At each step, we take a uniform distribution on [0,1] and user-defined probabilities of having two, three, or four branches—p_{2b}, p_{3b}, p_{4b}. Also starting from a user-defined level (in most cases we take it to be five), we allow for a possibility of a branch ending with leaves with probability p_L. Students can change these parameters to construct various trees.

Where branches need to be created, we randomly assign branch height angle ($\varphi_{min} < \varphi < \varphi_{max}$) measured from a vertical and plane angle ($0 < \theta < 2\pi$) measured from the x axis in the xy plane. In the pictures below, after some experimentation, $\varphi_{min} = \pi/6$ and $\varphi_{max} = 7\pi/24$. Then the length of each branch is taken to be current length divided by user-defined size ratio r_s, and its radius is parent's radius divided by r_s^2. The end of the first branch is always taken to be in (θ, φ) direction. Then depending upon how many other branches are present, the whole angle interval of 360° is broken into two, three, or four equal subintervals and the other branches are just rotated in θ versions of the first branch above. Finally, a user-defined radius of canopy (R_c) is used to stop branching procedure at some radius from the center of the tree creating treelike shapes of the branches and trees (Fig. 34.1). The colors in this figure represent the states of each leaf in terms of insect defoliation, which are discussed in the next section. Thus, size, branch structure, and leaf volume sizes can all be chosen by students to accommodate a specific tree.

Fig. 34.1 Generic tree

3 Five-State Model

Once such a tree structure is created, a volume containing leaves is determined and broken into a large number (>27 million) of equal size computational cells. From leaf average coordinates, we can immediately determine the corresponding computational cell and vice versa. Each cell and each tree leaf in a cell can be in one of the five states: young leaf (light green), mature leaf (green), slightly eaten by herbivores (light red), large percentage eaten by herbivores (red), and senescent leaf (black). At each step we determine the next state of a leaf given its current state and several user-defined probabilities as follows. First, each computational cell in light green or green state can be attacked by herbivores with user-defined probability p_{GtoE}. In addition, analogously to fire spreading models (Beer 1990; Drossel and Schwab 1992; Malamud et al. 1998), if any of the 26 neighbors of a given leaf is attacked by herbivores, we assign a probability (p_{spread}) of the spreading to current leaf. In Fig. 34.2, we show a simple analog of a current cell and its eight neighbors (marked 1–8) in 2-D case.

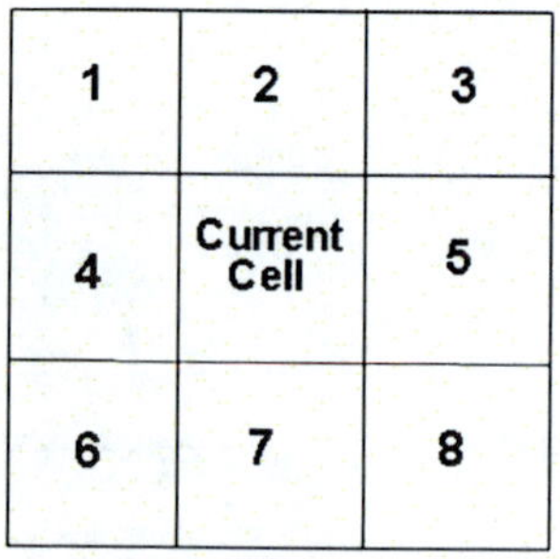

Fig. 34.2 Cell diagram

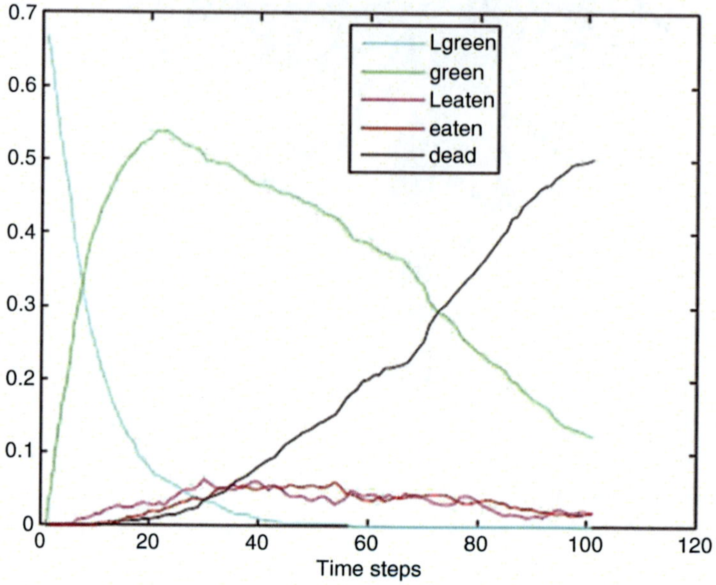

Fig. 34.3 Five states, no regrowth

Next, there are user-defined probabilities for going between the states in a leaf life:

p_{LGtoG}—probability of going from light green to green, as leaf matures
p_{LEtoE}—probability of going from light eaten to eaten
p_{EtoD}—probability of going from eaten to senescent
p_{AgeD}—probability of a leaf dying due to age

In addition, we allow for a possibility of "regrowth"—once a particular leaf has died, there is p_{regrow} probability of new foliage regrowing on the same branch. As a part of class and project activity, students have to experiment with these probabilities to see how they affect the final outcomes. Figure 34.3 shows one such run with the parameters indicated. No regrowth is assumed in this case. The percentages of the leaves eaten are similar to extensive results of Lowman (1985, 1987). On the

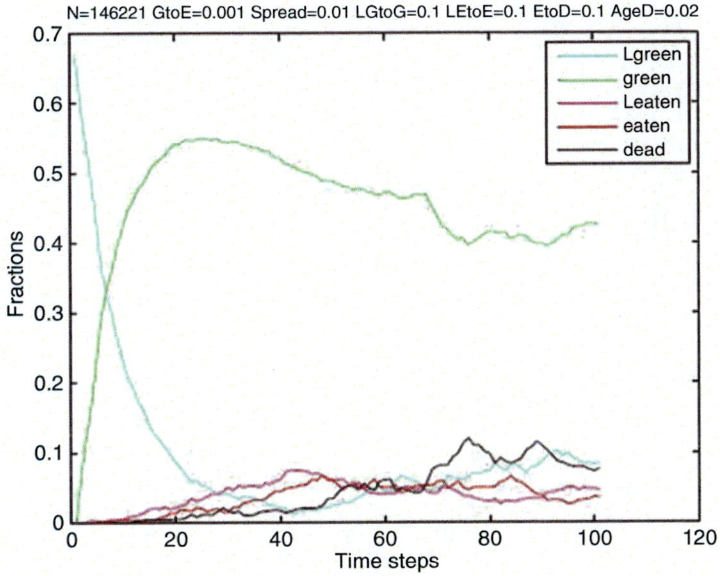

Fig. 34.4 Five states with regrowth

other hand, Fig. 34.4 illustrates small regrowth probability ($p_{regrow}=0.1$) and leads to a much smaller fraction of senescent leaves. We were also interested in a long-term behavior for some of the evergreen species including the genera Nothofagus, Ceratopetalum, and Eucalyptus, all important forest dominants (Lowman 1985, 1987). In this case, students set quite a high probability of spread of insect herbivores (as occurs with outbreaks) and state changes which leads to oscillations and the steady state for each of the states as shown in Fig. 34.5.

In addition, we can add effects of the variation of leaves eating due to spatial and temporal factors including age, light, and height. It is well known (Lowman 1985, 1987) that as any of the age, light, or height increases, the relative susceptibility of leaf to herbivores is reduced. In the case of age, herbivory is reduced to almost 0 % of leaf area eaten. To model these effects, say for height, we divide the total leaf volume height into three zones. Then we set a factor (<1) in each zone to reduce the user-defined probability of a leaf being eaten. A linear function is then fitted to these three values to give the height factor in the middle of each interval. The same is done for the light factor as a function of height and for the age factor as a function of age. Once all factors are determined, each probability of the process is multiplied by all these factors. The resulting run is shown in Fig. 34.6 and should be compared to Fig. 34.3, which has the same parameters except in the absence of specific spatial and temporal factors. As can be expected, this simulation reduces the fraction of dead leaves and slightly increases the proportion of green ones. Students were expected to investigate the defoliation patters with and without effects of light, height, etc.

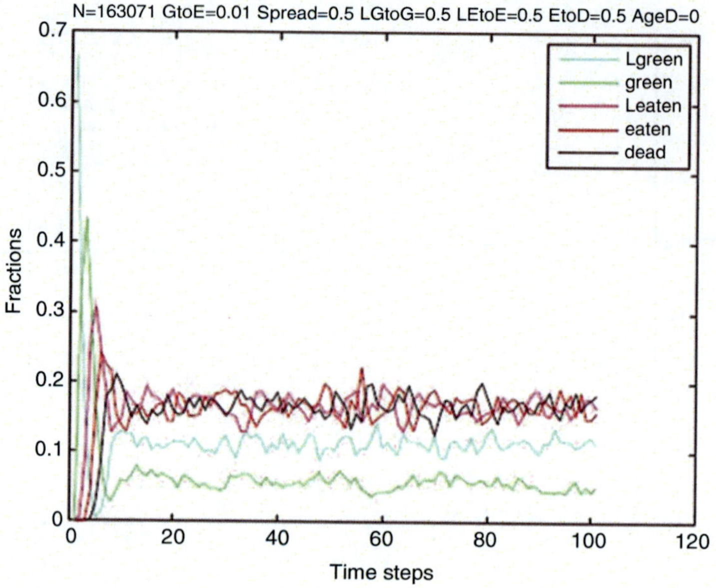

Fig. 34.5 Steady state

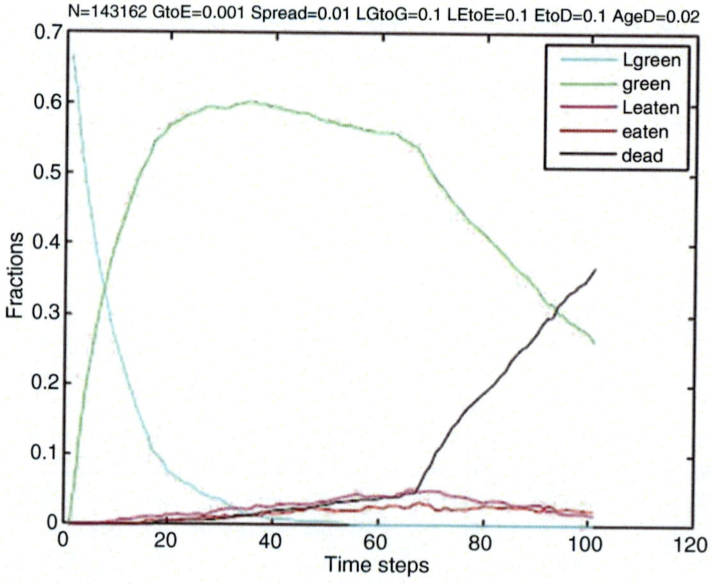

Fig. 34.6 Five states with height, light, and age factors

4 Daily Percentage Eating Model

This model is based on a slightly different idea. Instead of just five defined states, we introduce a user-defined parameter—d—percentage of a leaf eaten daily by herbivores. We start with green leaves—0 % eaten. Then, similar to five-state model of the previous section, there is probability (p_{eat}) for each leaf to be defoliated. Also, if any of the 26 neighbors of a given cell are being actively eaten, then with probability p_{spread}, this cell will get "infected." Once a leaf "enters" into the eaten phase, we set corresponding computational cell active eating toggle to true and at each step (day) increase the amount of the leaf eaten by d. In such a model, the percentage of the leaf eaten increases continually and once it reaches a user-defined fraction—max-alive—we consider this leaf dead. We also set a possibility for the regrowth with user-defined probability $p_{\_regrow}$ and allow for a possibility of herbivores to stop eating the leaf with probability $p_{\_stop}$. Similarly to the five-state model, students were expected to vary these probabilities and explain the ecological outcomes.

Figure 34.7 shows application of this model with relatively small probabilities of insect herbivore "attack" and spread of an outbreak between the computational cells. In this case, we do not have states and plot only average percentage of leaves eaten. The parameters have been chosen to correspond with data shown in Fig. 34.8 collected by our students recently as a pilot study (Lowman, 1985).

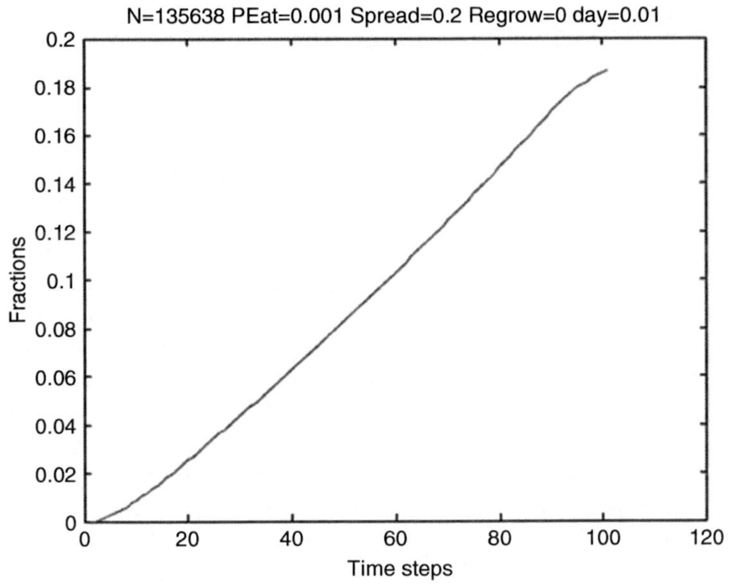

Fig. 34.7 Daily eating model

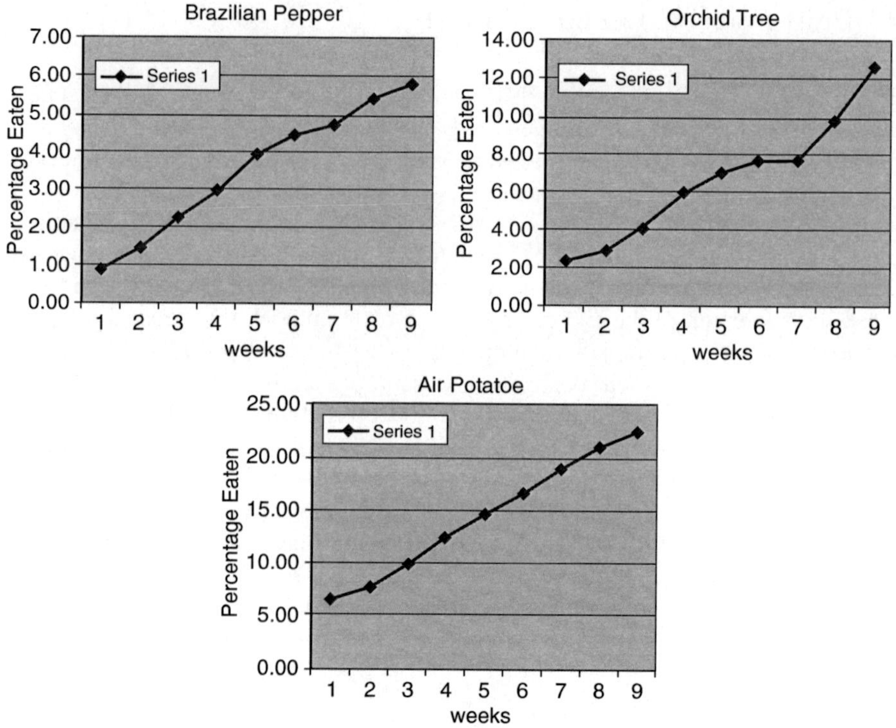

Fig. 34.8 Data collected at the New College of Florida

5 Conclusions

In this chapter, we discussed how to apply some of the tree-structure techniques to create a student activity investigating a model of stochastic tree and using percolation/cellular automata ideas to study insect outbreaks. Discrete five-state and continuous daily eating models have been investigated. The computational results show trends similar to extensive field measurements done by Lowman (1985) in different forest canopies. There are many avenues for additional investigations here, including tree age, upper versus inner canopy percentages, sun versus shade dependence, extending the results to entire forest, as well as specializing to particular types of trees and insect outbreaks. This modeling exercise sets the stage for additional case studies involving specific herbivore behavior in forest canopies with their estimated millions of insect species and equally large numbers of specific insect-plant interactions (Rinker and Lowman 2004). With inevitable increases in insect epidemics with the predicted hotter and drier climatic conditions (IPCC 2007), this convergence of mathematical modeling and ecology is critical for forest management. We also uphold that it provides essential training for undergraduates who will need the convergence of math and ecology in their training to achieve higher global competitiveness for American science education (National Academy of Sciences—Rising Above the Gathering Storm 2007).

References

Balmford A, Clegg L, Coulson T, Taylor J (2002) Why conservationists should heed Pokémon. Science 295:2367

Beer T (1990) Percolation theory and fire spread. Combust Sci Technol 72:297–304

Dobson AP, May RM (1986) Patterns of invasions by pathogens and parasites, vol 58. Springer, New York

Drossel B, Schwab F (1992) Self-organised critical forest fire model. Phys Rev Lett 69:1629–1632

Elton CS (1958) The ecology of invasions by animals and plants. Methuen, London

Fisher RA (1937) The wave of advance of advantageous genes. Ann Eugen 7:355–369

Fu SC, Milne G (2004) A flexible automata model for disease simulation. In: Proceedings of ACRI 2004, LNCS 3305, Springer, Berlin/Heidelberg

Green JL, Hastings A, Arzberger P, Ayala FJ, Cottinghan KL, Cuddington K, David F, Dunne JA, Fortin M, Gerber L, Neubert M (2003) Complexity in ecology and conservation: mathematical, statistical, and computational challenges. Bioscience 55:501–510

Johnson DM, Liebhold AM, Tobin PC, Bjornstad ON (2006) Allee effects and pulsed invasion by the gypsy moth. Nature 444:63

Kaganovskiy L (2007) Adaptive panel representation for 3d vortex ring motion and instability. Math Probl Eng 1. doi:10.1155/2007/668953

Liebhold AM, Halverson JA, Elmes GA (1992) Gypsy moth invasion in North America: a quantitative analysis. J Biogeogr 19:513–520

Louv R (2005) Last child in the woods. Algonquin Books, Chapel Hill, NC

Lowman MD (1985) Temporal and spatial variability in insect grazing of the canopies of five Australian rainforests tree species. Aust J Ecol 10:7–24

Lowman MD (1987) The biomass of New-England peppermint (Eucalyptus nova anglica) in relation to insect damage associated with rural dieback. Aust J Ecol 12:361–371

Lowman MD (1992) Heat futures in Australia's rain forests. Earthwatch Institute Newsletter, Boston, MA

Lowman MD, Heatwole H (1992) Spatial and temporal variability in defoliation of Australian eucalypts and its consequence for the dieback syndrome. Ecology 73(1):129–142

Lowman MD, Morrow PA (1998) Insects and their environment: plants, 4th edn, Textbook of entomology. Wm C. Brown, Dubuque

Lowman MD, Mourad T (2010) Bridging the divide between virtual and real nature. Front Ecol Environ 8:389

Magdon-Maksymowicz M (2004) Simulation of a horizontal and vertical disease spread in population. In: Proceedings of ICCS 2004, LNCS 3039, Springer, Berlin/Heidelberg

Malamud BD, Morein G, Turcotte DL (1998) Forest fires: an example of self organised critical behavior. Science 281:1840–1842

Martins ML, Ceotto G, Alves SG, Bufon CCB, Silva JM (2000) Cellular automata model for citrus variegated chlorosis. Phys Rev E 62(5):7024–7030

Mason RR, Luck RF (1978) Population growth and regulation. Technical report, The Douglas-fir tussock moth: a synthesis, USDA Forest Service Technical Bulletin 1585, USDA Forest Service, Washington, DC

Morley PD Critical behavior in a cellular automata animal disease transmission model. J Comput Phys (submitted)

National Academy of Sciences (2007) Rising above the gathering storm. The National Academies Press, Washington, DC
Otto SP, Day T (2007) A biologist's guide to mathematical modeling in ecology and evolution, 1st edn. Princeton University Press, Princeton
Pauly M, Gross M, Kobbelt LP (2002) Efficient high resolution wake modeling using vorticity transport equation. In: Proceedings of the conference on visualization, Boston, Massachusetts, pp. 163–170
Pimental D (1986) Biological invasions of plants and animals in agriculture and forestry, vol 55. Springer, New York
Rinker HB, Lowman MD (2004) Insect herbivory in tropical forests. In: Rinker HB, Lowman MD (eds) Forest canopies. Elsevier Press, San Diego
Samet H (1984) The quad-tree and related hierarchical data structures. Comput Surv 16:187–260
Samet H (1990) Applications of spatial data structures. Addison-Wesley, Reading
Zorzenon des Santos RM, Coutinho S (2001) Dynamics of HIV infection: a cellular automata approach. Phys Rev Lett 87(16):168102
Sederberg TW, Zheng J, Bakenov A, Nasri A (2003) T-splines and t-nurccs. ACM Trans Graph (TOG) 22:477–484
Sederberg TW, Cardon DL, Zheng J, Lyche T (2004) T-spline simplification and local refinement. ACM Trans Graph (TOG) 23(3):276–283
Simberloff D (1986) Introduced insects: a biogeographic and systematic perspective. In: Ecology of biological invasions of North American and Hawaii, vol 58, Ecological studies. Springer, New York
Situngkir H (2004) Epidemiology through cellular automata, case of study: avian influenza in Indonesia, Board of Science, Bandung Fe Institute. http://cogprints.org/3500/. Accessed 26 Nov 2012
Skellam JG (1951) Spatial and temporal variability in defoliation of Australian eucalypts and its consequence for the dieback syndrome. Biometrika 38:196–218
Veblen TT, Hadley KS, Nel EM, Kitzberger T, Reid M, Vilalba R (1994) Disturbance regime and disturbance interactions in a rocky mountain subalpine forest. J Ecol 82:125–135
Von Neumann J, Burks A (1966) Theory of self-reproduction automata. University of Illinois Press, Urbana
Willig MR, Camilo GR (1991) The effect of hurricane hugo on six invertebrate species in the luquillo experimental forest of Puerto Rico. Biotropica 23:455–461
Willig MR, Camilo GR (1995) Dynamics of a food chain model from an arthropod-dominated lotic community. Ecol Model 79:121–129
Wolfram S (2002) A new kind of science. Wolfram Media, Champaign

Chapter 35
Canopy Capture: Essay of a Photographer

Bhaskar Krishnamurthy

Keywords Photography • India • Forest canopies

A monkey's-eye view of the top of a tree is not nearly as good as a bird's-eye view, but to an Indian child out of school for the summer holidays, a mango tree laden with its golden fruit is akin to American ice cream on a hot day. It is also an invitation to plunder and risk grandmother's wrath. I remember running into the mango grove, clambering up the tree to pluck a "forbidden" fruit and gorge on it while feasting my eyes on the wondrous view of the world below through the foliage, perched on the fork of two obliging branches.

Venturing into photography much later in life, I had no intention to capture the vast expanse of any landscape, let alone mango canopies. As destiny changed the course, I got to wander around the forests and an appreciation for mother nature and an abiding interest in cameras, trapping her wonders. My enthusiasm thus kindled, I am no longer satisfied with what I see at eye level but dream of capturing the landscape from many vantages.

Naturally, for one who had monkeyed around in a mango canopy through many summers, the call of treetops is too persistent to be ignored. In most tropical countries, where coconut trees are common, humans have mastered the art of climbing to the top. In South India, where I grew up, men climb extraordinary heights with just a rope tied to their two feet, almost like an inchworm.

Being born in a tropical country covered with all kinds of vegetation—including Tropical Evergreen Rain Forests, Mangrove Forests, and Mountain Forests—climbing trees becomes essential to good nature photography. Yet, it is no joke climbing a tree with camera equipment hanging around the neck. Unlike other kinds of nature photography where one can comfortably photograph on the land from the comforts of a 4×4 vehicle, or load up a zodiac to capture the ocean denizens, it takes an

B. Krishnamurthy (✉)
TREE Foundation, Sarasota, FL, USA
e-mail: bhaskardk@yahoo.com

M. Lowman et al. (eds.), *Treetops at Risk: Challenges of Global Canopy Ecology and Conservation*, DOI 10.1007/978-1-4614-7161-5_35,

enormous effort to hoist big lenses and gear up a tree and then wait patiently for the action to unfold. There are no stairways. It is a ropes-and-harness trick. Once at the top, the early mornings and the late evenings are magical. Morning before sunrise is boisterous and the cacophony of animal calls is nothing short of enchantment.

Unlike other forests around the world, Indian forest canopies are the least scientifically explored. India's forests are probably also the least photographed, despite the unique flora and fauna living at the top. Through a camera lens, India's canopy represents unexplored territory and one that is filled with an extraordinary richness of natural history.

Chapter 36
Kids Can Save Forests

Lynne Cherry

Keywords Global Warming • Climate change • Rain forest • Youth solutions • Making a difference • Youth engagement • Environment • Rainforest • Youth empowerment • Youth action • Young voices • Nashua river • River clean-up • Pollution • Kids saving trees • Tree-planting • CO_2 mitigation • Empathy • Ecosystems • Local rivers

If young people have the opportunity to connect with the natural world, they often become greatly motivated to save the things they come to love. Their natural empathy for the animals and, by extension, the ecosystems they depend upon can make youth wonderful advocates for wildlife and their habitat—and places where they themselves can go to experience nature. From personal experiences, from the transformative experience of reading a book that reaches their hearts and souls or by experiencing firsthand a natural place that exudes the smells, sounds, and feeling of a living ecosystem, many young people go on to take action and clean up their local rivers and forests or distant rain forests.

There are many stories of the power of youth to create change. In the 1950's and in the early 1960's the Nashua River smelled terrible, and it ran a different color every day depending on what color the paper mills were dyeing paper. Rats could run across its surface on clumps of paper pulp. It was disgusting. Marion Stoddart and others who lived near the river were trying to prevent the local Fitchburg, Massachusetts, government from allowing further pollution to be discharged into the river. Children played an important role in cleaning up the river. They brought bottles of dirty river water to politicians and were instrumental in getting media attention for the citizen's cleanup efforts. Testifying at a local hearing, a child told the legislators, "We just want this river the way it was when you were kids. You could fish in it, you could swim in it, and you could drink it." The committee sat in

L. Cherry (✉)
Young Voices on Climate Change, PO Box 127,
Thurmont, Md. 21788, Lynne Cherry, Washington, DC, USA
e-mail: YoungVoicesforPlanet@gmail.com

M. Lowman et al. (eds.), *Treetops at Risk: Challenges of Global Canopy Ecology and Conservation*, DOI 10.1007/978-1-4614-7161-5_36,

Fig. 36.1 In *Lynne Cherry's classic children's book.* The Great Kapok Tree the rain forest creatures plead with the sleeping woodcutter to spare their tree

stunned silence, ashamed, and then voted against allowing increased dumping. The children also stood by the road along the river with signs: "Hold your nose; Nashua River ahead." This was the beginning of a cleanup effort that, in 10 years, changed the Nashua from a waste dump to a clean clear fishing stream. It was the precursor to the first state Clean Water Act passed by Massachusetts.

I have received hundreds of letters from children who had read my book *The Great Kapok Tree* and wanted to help save tropical rain forests like the one pictured in the book. In this dreamlike story, a man goes into a Brazilian rain forest with instructions to cut down a huge Kapok tree. The Kapok is the quintessential rain forest tree, an "emergent," poking up above the rain forest canopy, towering over the other trees. When in flower, their bright yellow bursts of color are visible for miles from a small plane. The woodcutter in the story is lulled to sleep by the heat and hum of the forest. As he sleeps, many rain forest creatures whisper in his ear, describing the important role the tree plays in their lives and how it supports an entire ecosystem. They plead with him to spare the tree and when he wakes up, he has an epiphany, seeing the overwhelming beauty of the forest as if for the first time. He drops his axe and walks out of the rain forest (Fig. 36.1).

A young scientist, Sharon Kinsman, spent many years in the Montverde rain forest in Costa Rica researching its ecology. Then, in 1987, she gave a slide show about the rain forest to students in Sweden and described to them how the rain forests were in danger. The students wished to help save this mysterious rain forest with its unique animals. A Swedish teacher established a nonprofit organization to collect funds to protect and preserve the forest. They named the forest and the project *Bosque Eterno de los Ninos*, The *Eternal Forest of the Children*, and raised contributions which Sharon brought to Monteverde to buy and protect the forest. By 1992, the Swedish children and their friends, with help from the Swedish International Development Agency (SIDA), had raised $2 million to purchase rain forest land. (B. Kern, *pers. comm.*). Sharon Kinsman told me that many of the young people who were interested in saving rain forest had been motivated after reading *The Great Kapok Tree* which has been included in many 3rd-grade textbooks around the world. Whereas Sharon's slideshow had described the science of the rain forest, the book had reached children's hearts.

To research *The Great Kapok Tree* in 1989, I traveled to the Brazilian rain forest with the help of the eminent scientist Dr. Thomas Lovejoy whose study site I visited and where I made preliminary sketches for the book illustrations. Subsequently over a 6-year period, I had the good fortune to be invited to teach an ecology course at the Amazon Conservatory for Tropical Studies (ACTS), in the Peruvian rainforest. Each summer, 60 educators, K-college, accompanied us in order to study the ornithology, entomology, geology, and ecology of the rain forest. But equally as important as learning the science, was their visceral experience of the rain forests. Standing at the topmost overlook on the canopy walkway and looking out at a sea of trees, unbroken rainforest for 360° from horizon to horizon, was a life-changing experience.

In Marquette, Michigan, Deb Vezetti, a 3rd-grade teacher regularly took her students out to the Estivant Pines forest on the Keweenaw Peninsula to study ecology. The students loved the forest and the many things that they discovered living there. When they found that the landowner was planning to cut down the forest for a development, they asked their teacher, "Is there anything that we can do?" They went to speak to the developer and asked him if he would consider saving the forest and calling it the *Eternal Forest of the Children*. The man told the students that he couldn't give them the forest, but he could sell it to them for $100,000. "We're only in 3rd grade. We don't have $100,000," they told him. He said he was sorry but that he could not just give them the land. Mrs. Vezetti created a math problem; the students went into the forest and counted the number of trees and divided the number of trees into $100,000. They found that each tree was worth $14. "Let's just try to save what we can, tree by tree!" they said. They wrote letters to the newspaper describing the beautiful forest and some of what they had learned during the years they had been studying it. In their op-eds and letters to the editor, the children asked the readers to contribute to a fund to try to purchase the forest. Then they were asked to do radio interviews. People started sending them checks. But when they spoke about their campaign to save the forest on TV, so many people saw them that checks began pouring in and in 6 weeks these 3rd graders had raised not $100,000—but $200,000. They bought the forest and named it the *Eternal Forest of the Children*.

Fig. 36.2 Students worldwide put on performances of The Great Kapok Tree to teach their communities about the importance of the rain forests and to raise funds to protect rain forest acres

Students worldwide put on performances of *The Great Kapok Tree* to teach their communities about the importance of the rain forests and to raise funds to protect rain forest acres (Fig. 36.2). Often teachers teach about the rain forest by creating crepe paper rain forests in their classrooms. But the best way to teach about the rain forests is to introduce students to a real live forest. The curriculum *Rain Forest/Your Forest* encourages teachers to teach about the rain forest comparatively. How is the rain forest similar to the forest in your own back yard? How is it different? Students come to understand that the real living breathing trees in their schoolyard—photosynthesizing, evaporating, transpiring, bringing deep water up to the surface, and providing sustenance to many living things and ecosystem services to humans—bear a much closer resemblance to the rain forest than do inert colored paper in the school library. There is great value in these creative endeavors for creating art and learning science, but children need to smell the sweet earth, feel the dampness of a misty morning, hear the singing of the birds whose home is in that forest, and experience a living ecosystem.

Young people have the power to change the world. The *Young Voices for the Planet* films document many youth successes including the story of 11-year-old Felix Finkbeiner from Germany. Felix wrote a report in school about the climate crisis.

He was very concerned about climate change but also learned that trees take up CO_2. He started a group *Plant for the Planet* and, to date, he has planted billions of trees! At the end of this short film (which you can view at YoungVoicesforthePlanet.com), Felix, sitting in a verdant forest, bathed in dappled light and surrounded by bird song, says, “You adults will only be around for maybe another 20 years but we will be around for another 80 and our children even longer. So we have to do it,” calling upon youth to speak out, find solutions, and lead the change.

Chapter 37
Forest Canopy Tourism: Analyzing a Flagship Attraction in the Ecotourism Arena from a Political Ecology Perspective

Markus Seibel

Keywords Canopy tourism • Rainforest • Development • Political Ecology • Actor-Network Theory

Summary

Carbon dioxide sequestration and high biodiversity are two distinctive features attributed to tropical forests. Whereas those features serve as justification for global forest conservation, their effects for the local communities are frequently neglected. Accordingly, projects that aim at conserving tropical forest areas are often associated with constraints on – or even the loss of – livelihoods for these actors. Against this background, ecotourism is a frequently discussed approach that combines the protection and use of tropical forests and as such is a potentially promising conservation strategy. One relatively new tourism approach is canopy tourism which allows access to the treetops. A Political Ecology analytical framework was used in two case studies in Rwanda and Malaysia to analyze the potential and consequences of canopy tourism.

1 Canopy Tourism: A Panacea in the Conservation Development Nexus?

Canopy tourism may be viewed as an intrusion into a hitherto untouched ecosystem, sometimes labeled the last biotic frontier on earth. This relatively new tourism approach might also be a way to create understanding of this ecosystem and hence contribute to its protection. Conservation projects are often associated with

M. Seibel (✉)
Geographisches Institut der Humboldt-Universität zu Berlin, Berlin, Germany
e-mail: seibel@forest-canopy.com

M. Lowman et al. (eds.), *Treetops at Risk: Challenges of Global Canopy Ecology and Conservation*, DOI 10.1007/978-1-4614-7161-5_37,

Fig. 37.1 Canopy walkway in Nyungwe National Park (Picture taken 2010)

constraints on – or even the loss of – livelihoods for local communities (Adams 2009). The ecotourism approach aims at working around this dilemma by trying to generate alternative income for these actors while protecting the environment (Ellenberg et al. 1997).

A standard work on canopy tourism as such does not exist (but see Lowman 2004). Likewise, it was impossible to determine how many canopy tourism facilities exist globally or how many tourists visit those annually. During the Fifth International Canopy Conference, the participants of the symposium "Conservation, Community Development, Rainforest Experience – the Environmental, Social and Economic Values of Forest Canopy based Ecotourism" pointed out difficulties to clearly distinguish between canopy tourism in particular and rainforest tourism in general. In this chapter, canopy tourism is defined as an ecotourism segment which allows access into the forest canopy via facilities like walkways, ziplines or observation towers – thus giving visitors the opportunity to experience and understand the forest ecosystem from a completely new perspective (Fig. 37.1) (Seibel 2012). As a flagship attraction, it could potentially generate significant revenue while minimizing ecological impact by providing access only to a defined part of the forest. By creating a special experience in the forest canopy, it could foster a better understanding of the need to protect this ecosystem. Two case studies were conducted in national parks in Malaysia and Rwanda, aimed at analyzing the potential and consequences of canopy tourism.

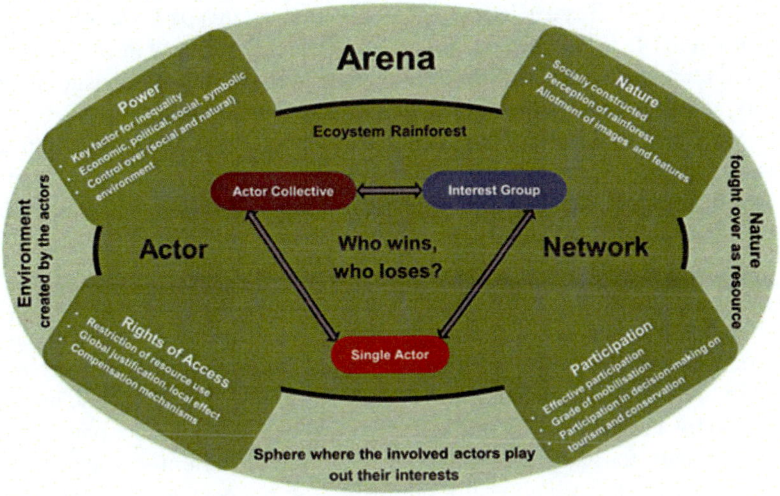

Fig. 37.2 Political Ecology analytical framework (Translated from Seibel 2012, 79)

2 Political Ecology Analytical Framework

Political Ecology analyzes such aspects of human-environmental relations by applying a rather static chain of explanation from micro to macro level (Robbins 2004). Tourism, conservation, and development, however, cannot be reduced to either of these spheres. Therefore, I combined Political Ecology with Actor-Network Theory to provide the theoretical base for the development of the analytical framework underpinning this study (Seibel 2012).

Within this arena, the forest becomes a natural resource which the involved actors fight over. By doing so, they are creating their own environment which in turn has an influence on their livelihoods and decision making (Fig. 37.2).

3 The Political Ecology of Canopy Tourism in Malaysia and Rwanda

Taman Negara National Park in Malaysia is one of the oldest lowland tropical rainforests in the world, with an exceptional biodiversity and a more or less undisturbed development for more than 130 million years. It is an integral part of Malaysia's rainforest tourism offer and is home to one of the longest canopy walkways, built in 1992 solely for tourism purposes.

Nyungwe National Park in Rwanda is located in a region of exceptionally high biodiversity and endemism (Plumptre et al. 2007). Furthermore, it comprises of the largest tropical mountainous forest in East Africa and is the most important freshwater source of the country. Besides its ecological importance, this park was chosen as the primary research site in Rwanda because the first canopy walkway in East Africa was inaugurated there in 2010 as part of a development project to increase visitor numbers to the park and generate additional revenue for local communities.

Both Malaysia and Rwanda are countries where the touristic use of tropical forest is of economic importance. On the macroeconomic level, however, there are significant differences. Whereas Malaysia is becoming increasingly industrialized, Rwanda remains one of the poorest countries of the world (Klugman 2011). Within and around the national parks, conservation and tourism lead to tensions and conflicts among the involved actors. A considerable amount of conflict was found in Taman Negara National Park which to a large extent arose between the national park management mainly responsible for conservation and those actors who are mainly involved in tourism development. The potential of the canopy walkway, in spite of being visited by a large majority of the tourists and being the number one reason for coming to the park for half of them, was not fully used. No rainforest interpretation took place at the walkway, and visitor numbers were not controlled. Queuing at the walkway, littering and noise were major issues leading to disappointment for many visitors who expected a serene and pure rainforest experience.

The lines of conflict in Nyungwe National Park were not as obvious, with the typical conflict between tourism development and conservation largely missing. Affected by the 1994 genocide, the protected area was gazetted a national park only in 2005. The newly opened canopy walkway received a lot of attention, as it is considered the flagship attraction of a USAID-funded ecotourism development project. Accordingly, interpretation of the forest and its canopy had high priority for the involved actors to position the forest experience of Nyungwe National Park in competition with the mountain gorilla tourism in the north of Rwanda.

In comparing both case studies, the central question is: why is there more conflict in the Malaysian actor network than in the Rwandan case and how does this reflect on canopy tourism as an ecotourism tool? Whereas Rwanda has a national ecotourism plan which is identical with the country's overall tourism strategy to a large extent, in Malaysia, ecotourism is a niche product in a market mainly geared towards mass tourism. This difference is reflected by a different understanding of nature. Whereas the involved actors in Malaysia mainly viewed forest conservation as a concept that should be balanced with economic development, in Rwanda, the tropical forest was largely understood as a threatened and valuable resource which should be conserved by all means.

4 Looking Beyond the Treetops

The origin of environmental conflicts revolving around canopy tourism could mainly be traced back to differences in perception of nature resulting in the way the involved actors created their environment. Canopy tourism has the potential to create a unique tropical forest experience and better understanding of the need to conserve this ecosystem. In order to fulfill this potential, it has to be embedded in an overall ecotourism concept, which follows a clear vision and includes all involved actors. Otherwise, canopy tourism will merely remain an adventure activity against the backdrop of the tropical forest.

References

Adams W (2009) Green development. Environment and sustainability in a developing world. Routledge, London

Ellenberg L, Beier B, Scholz M (1997) Ökotourismus. Reisen zwischen Ökonomie und Ökologie. Spektrum, Heidelberg

Klugman J (2011) Sustainability and equity. A better future for all. United Nations Development Programme (UNDP), New York

Lowman MD (2004) Ecotourism and the treetops. In: Lowman MD, Rinker B (eds) Forest canopies. Elsevier Academic Press, Amsterdam, pp 475–484

Plumptre A et al (2007) The biodiversity of the albertine rift. Biol Conserv 134(2):178–194

Robbins P (2004) Political ecology. A critical introduction. Blackwell, Malden

Seibel M (2012) Canopy Tourism in Tropenwäldern. Eine politisch-ökologische Analyse zum Tourismus in Baumkronen. Verlag F. Entwicklungspolitik, Saarbrücken

Part V
Ecosystem, Services and Sustainability

Introduction

In 2001, the World Research Institute formally launched the Millennium Ecosystem Services Assessment (MEA), and ever since, there have been many assessments of diverse habitats, from deserts to tropical forests and across continents. Although forest canopies were identified as a critical habitat and endorsed by the Convention on Biodiversity, to date there has been no comprehensive assessment exclusively on forest canopy. This section of the book integrates some selected research which directly and indirectly addresses the ecosystem services from the canopies. Two chapters – one on canopy ecotourism and one on non-forest timber products – delve into a political ecology framework. Such contributions have the potential to push back the frontiers of canopy ecology and offer a broader perspective to the field.

In contrast to this, another chapter addresses sustainability issues of harvesting from the canopies. This directly reflects the theme of Bangalore, India, 2009 canopy conference "Forest Canopies: Conservation, Climate Change and Sustainable Use" which was crafted to bring together scholars of both natural science and social sciences for the first time ever. Also, the conference achieved consensus from diverse attendees to apply interdisciplinary approaches to challenges in conservation. This becomes particularly important for canopy science, since over 40 % of terrestrial biodiversity inhabits this realm and harbors a sizeable number of bio-resources that are important for human well-being.

Chapter 38
Ancient Coastal Rainforest Canopies in Western Canada: Issues in Biodiversity and Conservation

Neville N. Winchester and Zoë Lindo

Keywords Structural complexity • Arboreal specificity • Canopy arthropods • Oribatid mites • Spatial scales • Coastal temperate rainforests

Summary

This chapter focuses on three interrelated themes that we have developed over the course of our canopy research: (1) structural complexity of canopy habitats, (2) biodiversity and arboreal specificity and (3) conservation challenges in coastal temperate rainforests of British Columbia, Canada. Habitat complexity in conifer canopies is enhanced through multiple spatial scales (within tree complexity and microhabitats, among different tree species, throughout a multitude of tree and forest stand ages and over gradients in elevation). Arboreal microhabitats include suspended soils, lichens and bark where microarthropod communities show arboreal specificity when compared to their ground counterparts. The most speciose group that illustrates these differences is the oribatid mites (Acari: Oribatida). We discuss conservation issues facing these unique rainforests that continue to be challenged by stressors such as forest harvesting and climate change.

N.N. Winchester (✉)
Department of Geography, University of Victoria, 3800 Finnerty Road, British Columbia, VIC V8P 5C2, Canada
e-mail: winchest@uvic.ca

Z. Lindo
Department of Biology, University of Western Ontario, London, ON N6A 5B7, Canada
e-mail: zlindo@uwo.ca

M. Lowman et al. (eds.), *Treetops at Risk: Challenges of Global Canopy Ecology and Conservation*, DOI 10.1007/978-1-4614-7161-5_38,

1 Introduction

As global biodiversity continues to decline, our understanding of species distributions, richness and diversity gradients in concert with broad-scale diversity patterns remain critical issues in conservation biogeography (Dinz-Filho et al. 2010). Evaluation of trends in diversity for insects in general, and soil arthropods in particular, is difficult since distribution patterns for most taxonomic groups remain unknown. However, local-scale diversity patterns have been documented for western coastal Canadian temperate canopy arthropods (e.g. Lindo and Winchester 2008, 2009; Winchester et al. 2008). This ecosystem has many of the features needed to test theories associated with large-scale biogeographical patterns and ecosystem-level conservation initiatives. For example, in combination with experimental approaches and observational studies, canopy arthropod communities may provide opportunities to test evolutionary influences on large-scale diversity patterns in response to climate change (Fig. 38.1). Additionally, factors that influence

Fig. 38.1 Artificial canopy habitats: Zoë Lindo conducting experiments in the canopy of an ancient Western red cedar tree in the Walbran Valley, Vancouver Island, British Columbia, Canada (Photo credit: Neville Winchester)

the geographic variation in soil fauna richness and abundance at different spatial scales can be compared and tested using the unique microhabitats found in the high canopy of intact temperate rainforests of British Columbia, Canada. In general, our research seeks to incorporate fundamental ideas discussed in soil community ecology and understand the factors that generate and maintain patterns of biodiversity.

In this chapter, we investigate coastal temperate arboreal ecosystems and address three major points. First, we investigate the structural complexity of ancient conifer forest canopies that provide a habitat template for diversity patterns that are most evident for canopy fauna (e.g. oribatid mites, Collembola). Numerous examples of the synergy between habitat complexity and species diversity at different scales exist, however few canopy studies attempt to quantify this relationship. The formation of arboreal community assemblages across a range of spatial scales as a function of structurally complex habitats that offer a mosaic of microhabitats can be described along three axes: heterogeneity, complexity and scale. An increase along any or all of these axes (e.g. suspended soil depth) results in greater species diversity. Structural diversity in conifer trees is further enhanced with age, because trunk and branch reiterations are common in older trees. This increase in architectural complexity promotes the development of microhabitats such as suspended soils (Lindo and Winchester 2006) that in turn result in greater microarthropod diversity (Lindo and Winchester 2007a).

Second, we document the unique community of arthropods associated with canopy microhabitats (e.g. suspended soils, canopy lichens) found in ancient temperate forest conifers. A large body of evidence suggests that age of conifers is positively correlated with increased arthropod diversity. Perhaps the most interesting and least explored habitat in conifer canopies is the 2–40 cm deep moss mat-debris accumulations that, over time, form a well-developed soil layer. Suspended high above the forest floor, this habitat is intimately associated with the architectural complexity of the tree that is determined by tree age, tree species, and geographic location. Until the early 1990s, this habitat was largely unexplored and arthropods using this habitat were unrecorded (Winchester 1997). Soil microarthropods dominate these suspended soils and are best represented by the oribatid mites. Significant progress on taxonomic resolution has been made (e.g. Lindo et al. 2010) leading to the conclusion that suspended soil communities are taxonomically distinct from the forest floor (Lindo and Winchester 2006).

Lastly, although tropical rainforests continue to receive much conservation attention and protection initiatives, coastal temperate rainforests have been largely overlooked on the global stage. Habitat loss, overexploitation, and habitat fragmentation are three main factors that continue to exert the heaviest pressures on coastal temperate rainforests of British Columbia, Canada. Notably, these ecosystems have many unique features, apart from being diverse and having many undescribed arthropod taxa; they also contain some of the largest and oldest conifer species on the planet. Most attractive is the potential for canopy studies to assist efforts in maintaining forest biodiversity and providing clarity in issues surrounding conservation of global forest biodiversity in the face of large-scale perturbations such as climate change.

2 Canopy Habitat Structural Complexity

Structural habitat complexities at local and regional scales are major determinants of arboreal oribatid mite species richness and this trend likely applies to most microarthropods (Lindo and Winchester 2007a). Spatial heterogeneity of the soil environment is one explanation for the higher diversity in soil communities and arboreal mites follow this pattern. Species richness increases with the structural complexity that is related to architectural complexity (e.g. suspended soil depth and development) and the spatial complexity that is associated with the distribution of suspended soils in space (Fig. 38.2).

In coastal temperate rainforests, habitat patches are relatively permanent and large with respect to the temporal-spatial scale of the organism; a species could persist by remaining within the patch without dispersal from external sources. How patch size, height in the canopy and moisture influence species richness and abundance in arboreal oribatid mites was investigated by Lindo et al. (2008a) who found a positive relation between species richness and abundance with patch size and moisture content. Communities became more similar (nested) with increased harshness of the habitat patch (smaller, drier, higher), suggesting that species-specific environmental tolerances combined with the differential dispersal abilities of species contributed to the nonrandom community patterns in these habitats. Simply stated, habitat complexity as measured by patch size (e.g. depth and area) results in a greater retention of moisture and amelioration of temperature fluctuations. This in turn acts as a niche-selection filter, eliminating species from the regional pool that cannot tolerate rapid fluctuations in environmental conditions.

3 Biodiversity and Arboreal Specificity

Standardized sampling and exploration of arboreal microhabitats (e.g. accumulation of organic debris, lichens, bark and branches) within conifer tree species (macrohabitats) has been investigated at several study sites on Vancouver Island, British Columbia, Canada (Winchester 1997; Winchester et al. 2008; Lindo and Winchester 2006, 2009; Lindo et al. 2010). These studies present strong evidence to support arboreal specificity in the coastal conifer rainforests of British Columbia, Canada. Although species richness in the oribatid mite fauna is greater on the forest floor than the canopy, several species are unique to the canopy microhabitats. Average abundance for most microarthropod taxa is significantly greater on the forest floor, but this trend is not apparent in the oribatid mites. Further oribatid mites in the canopy are numerically dominant as many other taxa are not present in high abundance (see Lindo and Winchester 2006). When similarity measures are considered, we have consistently noted that canopy sites are more similar to each other, as are forest floor sites, with the overall similarity between canopy and forest floor habitats ranging between 30 % and 40 %. In conclusion, the canopy community of oribatid mites is not a taxonomic subset of the forest floor assemblage and this observation

Fig. 38.2 Aspects of structural complexity in ancient forest conifers: (**a**) Structural complexity increases with age as vertical distribution of microhabitats increases. Examples include suspended soils (**b**) and lichens (**c**) (Photo credit: Neville Winchester, Zoë Lindo and Kevin Jordan)

is remarkably consistent over a number of studies (e.g. Behan-Pelletier and Winchester 1998; Lindo and Winchester 2006).

The discovery of several new species in ancient conifer forests is not surprising. Several of these species, mainly oribatid mites, however, are confined to microhabitats found only in the canopy. In particular, representatives of the oribatid mite families Camisiidae, Ceratozetidae, Scheloribatidae and Mycobatidae are usually numerically dominant in the canopy of old-growth Pacific Northwest forests. Recent species descriptions coupled with samples collected at multiple study sites and multiple microhabitats over a large spatial scale are starting to cumulate evidence to support the assumption that ancient forest canopies contain several canopy-specific species (Lindo 2011). For example, *Dendrozetes jordani* represents the first record of the genus *Dendrozetes* in North America (Lindo et al. 2010) and is numerically dominant on branch tips and arboreal lichens of *Abies amabilis* and *Tsuga heterophylla* where a new species of arboreal oribatid mite, a lichen associate in the family Achipteriidae, *Anachipteria geminus* also occurs (Lindo et al. 2008b). In addition, observed host tree specificity increases biodiversity in these forests. An example of this is found in the closely related oribatid mite species of *Jugatala*, where *Jugatala tuberosa* was only found on branch tips from *A. amabilis*, while *Jugatala* (undescribed) sp. was primarily found on branch tips from *T. heterophylla* within coniferous montane forests (Winchester et al. 2008).

Recent evidence suggests an environmental transition zone for ground-dwelling species occurs at the height of the understory (approximately 5 m above ground) where changes in moisture availability, rather than physical barriers, restrict distributions and dispersal, resulting in distinct canopy/ground microarthropod communities (Lindo and Winchester 2007b). As such, we conclude that the trunk of these ancient trees (e.g. often greater than 50 m) is not a dispersal corridor for ground species to colonize microhabitats (e.g. suspended soils) found in tree crowns. In essence suspended soils are true islands that are 'cut off' from the species inputs from the ground, and colonization events occur from within the canopy system. We suggest that these characteristic trends in community composition in ancient conifer forests may be indicative of all intact conifer rainforests where suspended soils and other microhabitats (e.g. lichens) are well developed.

4 Conservation Issues

In Canada some of the largest intact coastal rainforests remain within the province of British Columbia. Much of these forests are unprotected and are facing considerable threats due to a growing demand to extract natural resources (Inforain 2011). Leading international scientists continue to call for greater protection of temperate rainforests that are known to have more carbon-storing capacity than tropical rainforests and are equally threatened (DellaSala 2011). If conservation of these ecosystems is going to be successful, knowledge of biodiversity and ecosystem function is required; several initiatives that document the diversity, structure and function in these rainforests are currently underway.

A key component of biological diversity that is often identified in conservation initiatives are endemic and threatened species. While it is still too early to make explicit conclusions about endemic species in the arboreal system, our contribution towards categorizing the arthropod fauna of ancient coastal forests in British

Columbia continues and many species have features that fit into the theory surrounding extinction-prone and endemic species. Namely, several species appear to have specialized habitat requirements, are poor dispersers, have a low intrinsic rate of population growth and life histories associated with isolated habitats. Of particular importance to conservation is the recognition of those species that are specific to habitats that are only found in canopies of these forests. It is unlikely that habitat features of old trees will develop in second-growth forests that are in an 80–120 year rotation, as there is simply not enough time to develop the architectural complexity needed to support microhabitats that would contain these species. We suggest that canopy-specific species assemblages will be lost if these canopy habitats are not retained or allowed to develop in second-growth forests.

In conclusion, we suggest that ancient temperate rainforests in British Columbia, Canada, are repositories for biodiversity. In particular, the canopies in these ecosystems contain species that are specific to microhabitats found in large, old conifer trees. Our research has accounted for the confounding effects of spatial and temporal constraints on species composition and, coupled with investigations of scale-dependent diversity patterns, has provided the foundation to continue with the testing of hypotheses associated with diversity patterns and aspects of structural habitat complexity at local and regional scales. Most attractive is the potential for studies to assist efforts in maintaining forest canopy biodiversity and providing clarity in issues surrounding conservation of global forest biodiversity in the face of large-scale perturbations such as climate change.

References

Behan-Pelletier VM, Winchester NN (1998) Arboreal oribatid mite diversity: colonizing the canopy. Appl Soil Ecol 9:45–51

DellaSala D (2011) Temperate and boreal rainforests of the world: ecology and conservation. Island Press, Washington, DC

Dinz-Filho JAF et al (2010) Defying the curse of ignorance: perspectives in insect macroecology and conservation biogeography. Insect Conserv Divers 3:172–179

Inforain (2011) http://www.inforain.org/

Lindo Z (2011) Five new species of *Ceratoppia* (Acari: Oribatida: Peloppiidae) from western North America. Zootaxa 3036:1–25

Lindo Z, Winchester NN (2006) A comparison of microarthropod assemblages with emphasis on oribatid mites in canopy suspended soils and forest floors associated with ancient western red cedar trees. Pedobiologia 50:31–41

Lindo Z, Winchester NN (2007a) Local-regional boundary shifts in oribatid mite (Acari: Oribatida) communities: species-area relationships in arboreal habitat islands of a coastal temperate rain forest, Vancouver Island, Canada. J Biogeogr 34:1611–1621

Lindo Z, Winchester NN (2007b) Resident corticolous oribatid mites (Acari: Oribatida): decay in community similarity with vertical distance from the ground. Écoscience 14:223–229

Lindo Z, Winchester NN (2008) Scale dependent diversity patterns in arboreal and terrestrial oribatid mite (Acari: Oribatida) communities. Ecography 31:53–60

Lindo Z, Winchester NN (2009) Spatial and environmental factors contributing to patterns in arboreal and terrestrial oribatid mite diversity across spatial scales. Oecologia 160:817–825

Lindo Z, Winchester NN, Didham RK (2008a) Nested patterns of community assembly in the colonisation of artificial canopy habitats by oribatid mites. Oikos 117:1856–1864

Lindo Z, Clayton M, Behan-Pelletier VM (2008b) Systematics and ecology of *Anachipteria geminus* sp. nov. (Acari: Oribatida: Achipteriidae) from arboreal lichens in western North America. Can Entomol 140:539–556

Lindo Z, Clayton M, Behan-Pelletier VM (2010) Systematics and ecology of the genus *Dendrozetes* (Acari: Oribatida: Peloppiidae) from arboreal habitats in Western North America. Zootaxa 2403:10–22

Winchester NN (1997) Canopy arthropods of coastal Sitka spruce trees on Vancouver Island, British Columbia, Canada. In: Stork NE, Adis JA, Didham RK (eds) Canopy arthropods. Chapman and Hall, London

Winchester NN, Lindo Z, Behan-Pelletier VM (2008) Oribatid mite communities in the canopy of montane *Abies amabilis* and *Tsuga heterophylla* trees on Vancouver Island, British Columbia. Environ Entomol 37:464–471

Chapter 39
The Population Dynamics of Epiphytic Orchids: A Review and Methodological Guide

Demetria Mondragón and Daniela Dutra Elliott

Keywords Epiphyte • Demography • Methods • Orchidaceae • Metapopulation

Summary

Epiphytic orchids represent one of the most threatened plant groups in the world. Despite the urgent need for research on their population behavior, there is a dearth of in-depth studies on the topic. Population dynamic studies using matrix analysis are a widely used tool for the management and conservation of wild plant species, but their accuracy is contingent upon high-quality field data. Obtaining field data for epiphytic orchids presents challenges far beyond those encountered for terrestrial species, due to the challenges of access and also the multidimensional space of the canopy. The diversity of orchid habitat and population structure pose additional challenges for methodological design, which are further exacerbated by the scarcity of studies documenting appropriate methods for the study of epiphytic orchids. In this review, we summarize the methodologies from existing studies and compare them. In addition, we offer new ideas and key factors to insure more accurate conservation management plans.

D. Mondragón (✉)
Centro Interdisciplinario de Investigación para el Desarrollo Integral Regional (CIIDIR) unidad Oaxaca, Calle Hornos No. 1003, Santa Cruz Xoxocotlán, Oaxaca C.P. 71230, Mexico
e-mail: dmondragon@ipn.mx

D. Dutra Elliott
Botany Department, University of Hawaii at Manoa, 3190 Maile Way, Room 101, 96822 Honolulu, HI, USA
e-mail: ddutra@hawaii.edu

M. Lowman et al. (eds.), *Treetops at Risk: Challenges of Global Canopy Ecology and Conservation*, DOI 10.1007/978-1-4614-7161-5_39,

1 Introduction

The study of population dynamics using matrix analysis, which estimates population growth and determines the importance of each life stage for the destiny of a population, is a widely used tool for the management and conservation of wild plant species. In recent years, matrix analysis has been used as the basis for powerful methods to model population dynamics, such as population viability analysis (PVA), which estimates the probabilities of quasi-extinction and minimal population size; the life table response analysis (LTRE), which compares the differences between lambda values of populations in different circumstances or years produced by varying vital rates; and the integration projection model, which projects population growth rates using regression models of vital rates rather than by dividing populations into stage or size classes (Silvertown et al. 1996; Menges 2000; Crone et al. 2011). The commonality shared by these methods is a strong reliance on the quality of the field-collected data used to build the models.

Among vascular plants, the Orchidaceae are one of the most threatened families in the world and constitute the largest single group covered by the Convention on International Trade in Endangered Species (CITES) and in the International Union for Conservation of Nature (IUCN 2010) Red List. Multiple causes account for the threatened status of orchids: overexploitation of wild populations, habitat loss, loss of associated mycorrhizae, specificity of substratum, and limitation of resources and pollinators (Roberts and Wilcock 2002; Batty et al. 2004).

Even though more than 100 species of epiphytic orchids are on the Red List, the population dynamics have been studied for only nine of these species. One explanatory factor for the scarcity of studies on epiphytic orchid species is the challenge of canopy access, made easily available only during the last 20 years (Lowman and Rinker 2004). Due to the urgent need for further study of orchid population dynamics, we present guidelines to inspire future research.

2 Methodology

To identify existing studies on epiphytic orchid population dynamics, we searched in multiple databases (Web of Science, Google Scholar & WorldCat) for peer-reviewed journal studies in English and Spanish. We omitted theses and dissertations. From each study, we documented the methodology used and then augmented these results using input from our own field experience to develop this guide.

3 Results and Discussion

3.1 *Horizontal Distribution*

Epiphytic orchids are usually distributed in clumps, encountered as patches along a transect. Usually the patch with the highest density is chosen for a survey, to attain an adequate number of individuals in each stage and thereby diminish the sampling error. Munzbergova and Ehrlen (2005) suggest marking 50 individuals for each stage to obtain good estimates. For many epiphytic orchid species, few large patches exist and/or the density of individuals in each patch is low. In those cases, we suggest labeling and measuring all individuals following the methodology used by Tremblay et al. (2006) in their analysis of small populations of lithophytic orchids.

Another important consideration is that (although it is rare to encounter phorophyte (host tree) specificity) many orchids have a preference for certain tree species. This preference may be attributed to the effect of phorophyte characteristics on the demographic processes of epiphytic orchids. For example, Frei and Dobson (1972) reported that *Quercus magnoliaefolia* Nee, *Q. cyclophylla* Liebm, and *Q. peduncularis* Nee contain substances either toxic to orchid seeds and protocorms or inhibitory to the survival of orchid seedlings. A major cause of mortality in epiphytic individuals is the instability of the substrate (flaking bark, breaking branches, or fallen trees). Consequently, phorophytes with high peeling rates are likely to result in higher rates of mortality in epiphytes than those with stable barks and the same effect is likely in soft- versus hardwood trees. Additionally, fecundity is related to the visibility and density of the plant, so phorophytes with low density of orchids and abundant foliage are likely to present lower fecundity rates. However, this is not always an important variable (e.g., Flores-Palacios and García-Franco (2003) found no relationship between orchid density and the number of flowers produced in the fruit production of the epiphytic *Rhyncholaelia glauca* Lindl). Cardelús and Clark (2010) point to a possible link between the phorophyte and the mineral nutrition of epiphytic orchid via canopy soil, potentially important due to the demographic dependence on resource availability.

Another consideration is the existence of temporal and spatial variation in the demographic process. For example, Otero et al. (2007) compared the population structures of *Psychilis monensis* in two locations and found that the site with highest seedling density also had the lowest fruit and seed production, suggesting that the best sites for pollination and seedling establishment were not the same. Schödelbauerová et al. (2010) studied population dynamics of six populations of *Lepanthes rubripetala* and found differences in lambda values among populations, concurring with the findings of Zotz and Schmidt (2006) that the variation in annual rainfall significantly affects recruitment and growth rates of smaller orchid individuals of *Aspasia principipissa*. Since there is potential temporal and spatial variation, it is important to decide the length your study and also how many populations you will incorporate into the study. In this situation, more is better.

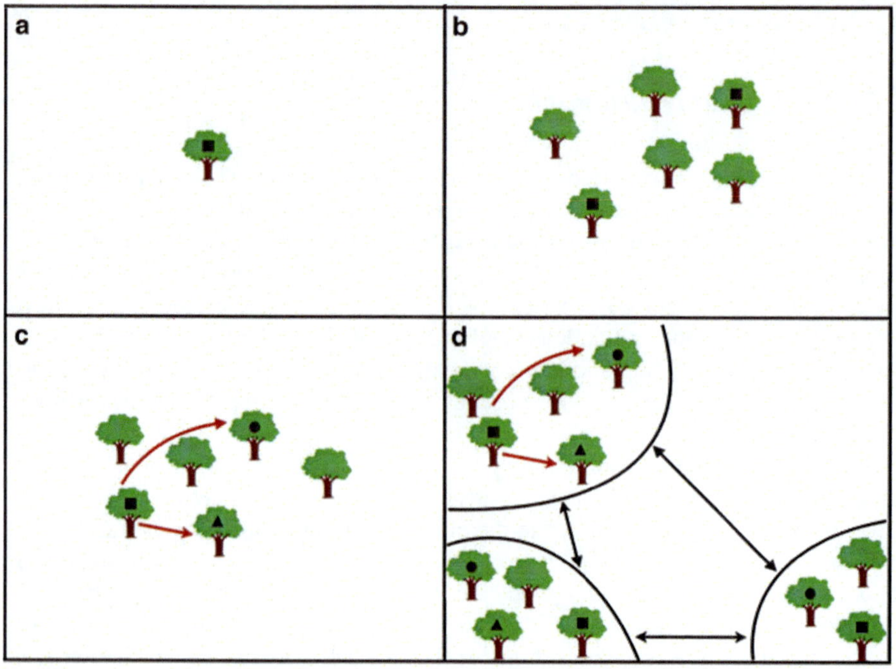

Fig. 39.1 *Squares* represent phorophytes found to have study species the first time the census was taken. *Triangles* and *circles* represent phorophytes with study species found the second and third times the census is taken, respectively. *Red arrows* represent seed dispersal within a patch and *black arrows* represent seed dispersal among patches. (**a**) A study conducted in a single phorophyte; in this case, the phorophyte is the population. (**b**) The study is conducted in a patch and all phorophytes with the study species are followed. Note the when the study was first set up, the researcher decided to only census the phorophytes that originally had the study species going back to the same phorophytes every time the census is taken. (**c**) The study is conducted in a patch using a metapopulation approach. (**d**) The study is conducted using a metapopulation approach with multiple patches

The dynamic of orchid populations can also be studied using a metapopulation approach; for groups of populations interconnected by seed dispersal, the dynamics of colonization and extinction of new patches are taken into consideration. For epiphytic orchids, this can be studied at two levels: at the patch level, where every patch is considered a population, or at the phorophyte level, where every phorophyte is considered a population (Fig. 39.1). Winkler et al. (2009) used the metapopulation approach to study the population dynamics of 3 epiphytic orchid species in Mexico. Plants growing in each phorophyte were treated as a subpopulation resulting in 1 orchid species having 5 subpopulations (5 photophytes) and 2 orchid species having 6 subpopulations each. In this study adjacent photophytes were not checked for new seedlings. Tremblay et al. (2006) also looked at the dynamics of an epiphytic/lithophytic orchid species using a metapopulation approach; however, this study looked at the occupied and unoccupied sites taking into consideration the dynamic among phorophytes (or bolders) in order to have an idea of the rate of incorporation of new phorophytes to be colonized or the loss of phorophytes in the system.

A number of questions must be answered in order for a metapopulation study for epiphytic orchids to be possible. For example, what defines a patch? What is the minimum number of individuals needed for a patch or a tree to be considered a population? How many species of phorophytes are necessary to conduct an accurate evaluation of the metapopulation process?

In order to apply the metapopulation approach at the patch level rather than at the phorophyte level, the kind of ecosystem being studied is an important consideration. When working in a forest where most of the canopy is interconnected, we suggest setting up a metapopulation study at the patch level. Alternately, when working in an ecosystem where phorophytes are separated by more than 5 m, we recommend setting up the metapopulation study at the phorophyte level.

3.2 *Vertical Distribution*

In some ecosystems, the microclimatic variation from the top to the bottom of the trees restricts the distribution of certain epiphytic species. Johanson (1974) identifies five vertical zones and subsequently researchers have fused or adapted these zones to account for the specificities of the ecosystem being studied. In these cases, it is important to delimit one part of the tree as a population for labeling. This step is not necessary in ecosystems where microclimatic differentiation is weak or when studying generalist species; in these cases, individuals throughout the entire tree must be labeled in order to capture the full range of variation. Labeling individuals throughout all areas of the tree is essential due to the potential effects of location on germination and seedling survival, as well as mortality due to dislodgement; different parts of the tree are more susceptible to lose bark and broken branches.

When working with tall trees, we recommend using climbing equipment and conducting the census and labeling from bottom to top; in smaller trees, a ladder can provide better accessibility. General safety precautions apply to all canopy work: avoid trees with wasp or bee hives, assess the weight bearing capacity of each species being climbed, and avoid diseased or dead trees.

3.3 *Individual Labeling*

The best surveying practice for labeling individuals is to work from bottom to top, following a branch. Individuals are frequently dislodged after the initial survey but can be identified easily in subsequent visits by interpolating from the anterior and posterior labels. The proper choice of materials greatly increases success in the difficult environment of the canopy. Aluminum or plastic labels offer the best resistance to weathering and the use of bright colors, such as red and blue, facilitates the surveying process, just as the use of green, brown, or black labels will hinder efforts to locate tags. An additional option is to use brightly colored wire to fix labels, in the event that brightly colored labels are not available.

3.4 How to Measure

In order to evaluate the demography of the population, it is necessary to follow the destiny of the individuals of the sample population. Many epiphytic orchid species have the ability to produce new independent individuals with the same genetic identity originated from the vegetative meristem (Harper 1977). You can follow the dynamic in two different ways (Harper and White 1974): (1) at the genet level, where the generation of each new ramet is considered a growth of the parent genet (Harper 1977; De Kroon and van Groenendael 1997), and/or (2) at the ramet level, where the generation of each new ramet is considered a new individual (Piquot et al. 1998).

When monitoring at the ramet level, the natality and mortality rate of the ramet only affects the size of the individual plant. On the other hand, the natality and mortality rate of genet has a direct influence on the effective size of the entire population having strong influence in the ecology and evolution of the species (Cook 1983).

When monitoring at the ramet level, each pseudobulb is considered a single individual. When monitoring at the genet level, each group of ramets is considered a genet. It can be discerned between genets since the connection among pseudobulbs can be easily broken or hidden under moss or other epiphytic plants, making the same genet appear as two different individuals. It is also possible for the genet to grow until it reaches a neighboring genet making it hard to tell individuals apart.

Once the individual unit being monitored is defined, estimators can be identified to measure the growth, fecundity, and survival of the individuals. For epiphytic orchids, the pseudobulbs are storage organs for water and minerals, and some also have photosynthetic functions (Yew and Hew 2000), so they are very important for the survival, growth, and fecundity of the individuals. In Table 39.1, we present the estimators that have been used in epiphytic orchid demographic studies. Most studies use the number and/or size of the pseudobulb, but in species that have no pseudobulbs, the number of leaves may be used. There are no studies for leafless species, like *Harrisella porrecta*. In this case, the lack of leaves and pseudobulbs leaves the option of measuring the individual stems.

Measuring plant survival can be straight forward; however, in some cases, even when the pseudobulbs look dead in 1 year, surprisingly a live plant can be found the next year the population is visited. If an apparently dead plant is found, wait at least 2 years to confirm it. In terms of fecundity, it is believed that epiphytes have a seed bank (Benzing 1990). Consequently, fecundity is estimated based on the number of protocorms produced by an adult individual, rather than on the number of seeds

Table 39.1 Structures and measures used in epiphytic orchid population dynamics studies

Species	Stages based on	Growth structure measured	Fecundity	
Aspasia principissa[a]	Length	Pseudobulb		F=# of seedling/# of ind. in reproductive categories
Erycina crista-galli[b]	Height and number and presence of rs	Pseudobulb	Fr	F=# of seedling/# of ind. in reproductive categories
Jacquiniella leucomelana[c]	Length of the longest	Ramet	Fr, Fl	F=# of seedling/# of reproductive plants
J. teretifoli[c]	Length of the longest	Ramet	Fr, Fl	F=# of seedling/# of reproductive plants
Guarianthe aurantiaca	Height and number, and presence of rs	Pseudobulb	Fr	F=# of seedling/# of ind. in reproductive categories
Lepanthes caritensis[d]	Height, and presence of rs	n/a	Fr, Fl	F=# of seedling/# of reproductive plants
L. eltoroensis[e]	Presence, and presence of rs	Lepanthiform sheet	Pollinaria, Fr	F=# of seedling/# of reproductive plants
L. rubripetal[f]	Presence, and presence of rs	Lepanthiform sheet		n.a.
Lycaste aromatica[c]	Volume and number, and presence of rs	Pseudobulb	Fr, Fl	F=# of seedling/# of reproductive plants

Fl flowers, *Fr* fruit, *rs* reproductive structures
[a]Zozt and Schmidt 2006
[b]Mondragón et al. 2007
[c]Winkler et al. 2009
[d]Tremblay 1997
[e]Schödelbauerová et al. 2010
[f]Tremblay and Hutchings 2003

being produced. This avoids creating a delay of a year in the matrix (Caswell 2001). It can be very hard to find protocorms because they are so small and can be hidden in cracks on the phorophyte stem or inside moss mats. It is easier to find seedlings because they are bigger, which is why most of the orchid studies evaluate fecundity based on number of seedlings found. When seedlings become visible, they are very delicate and can be easily detached, so it is important not to put a label on the seedling itself. Instead, it is more effective to attach a pin or tag on the stem or branch near the seedling. It is important to be consistent in the distance and orientation of the pin and to maintain detailed notes, to facilitate finding plants on subsequent visits for data collection. Also, you may find a large number of seedlings growing together; in this case, make an imaginary square with four pins surrounding the group of seedlings and count and measure them. Again, take good notes. If you have different adult categories, you will need to count the number of fruits that each adult produced, in order to give the proportion of seedlings that each adult category produced in function of the proportion of total fruit produced by that category.

Finally, when visiting the field site to take a new census, bring the measures from the previous time. This can be extremely helpful since you may need help locating a plant with a lost tag or accessing dead individuals.

4 Conclusion

Here we addressed key knowledge gaps in methods for epiphytic orchid research, which hinders research on population behavior critical to conservation management efforts of epiphytic orchid species. There is an urgent need of new research that can incorporate a metapopulation approach that takes into consideration spacial and temporal variations.

We suggest that it is necessary that more studies of epiphytic orchid species be conducted, specifically:

- The demography of species with different morphological characteristics (e.g., deciduous species, species without leaves, species without pseudobulbs)
- The demography of species in different types of ecosystems (e.g., temperate forest, dry forest, costal dune)
- The comparison of the population dynamics growing in different types of phorophytes
- Incorporation of the metapopulation concept at the patch level

References

Batty AL, Dixon KW, Brundrett MC, Sivasithamparam K (2004) Orchid conservation and mycorrhizal associations. In: Sivasithamparam K, Dixon KW, Barrest RL (eds) Microorganism in plant conservation and biodiversity. Springer, Netherlands

Benzing DH (1990) Vascular epiphytes. Cambridge University Press, New York

Cardelús CL, Mack MC (2010) The nutrient status of epiphytes and their host tree along an elevational gradient in Costa Rica. Plant Ecol 207:25–37

Caswell H (2001) Matrix population models: construction, analysis and interpretation. Sinauer, Sunderland

Cook RE (1983) Clonal plant populations: A knowledge of clonal structure can affect the interpretation of data in a broad range of ecological and evolutionary studies. American Scientist 71: 244–253

Crone EE, Menges ES, Ellis MM, Bell T, Bierzychudek P, Ehrlen J, Kaye TN, Knight TM, Lesica P, Morris WF, Oostermeijer G, Quintana-Ascencio PF, Stanley A, Ticktin T, Valverde T, William JL (2011) How do plant ecologist use matrix population models? Ecol Lett 14:1–8

Flores-Palacios A, García-Franco JG (2003) Effects of floral display and plant abundance on fruit production of *Ryncholaelia glauca* (Orchidaceae). Rev Biol Trop 51:71–78

Frei SJK, Dodson CH (1972) The chemical effect of certain bark substrates on the germination and early growth of epiphytic orchid. Bull Torrey Bot Club 99:301–307

Harper JL (1977) Population biology of plants. Blackburn Press, England. 922 pp

Harper J, White J (1974) The demography of plants. Annual Review of Ecology and Systematics 5:419–463

IUCN (2010) IUCN red list of threatened species. Version 2010.2. www.iucnredlist.org

Johanson D (1974) Ecology of vascular epiphytes in West African rain forest. Uppsala University

Kroon HD, Groenendael JV (1997) The ecology and evolution of clonal plants. Backhuys Publishers

Lowman MD, Rinker HB (2004) Forest canopies. Academic, San Diego

Menges E (2000) Population viability analyses in plants: challenges and opportunities. Trends Ecol Evol 15:51–56

Mondragón D, Maldonado C, Aguilar-Santelises R (2007) Life history and demography of a twig epiphyte: a case study of *Erycina crista–galli* (Orchidaceae). Selbyana 28:137–144

Munzbergova Z, Ehrlen J (2005) How best to collect demographic data for population viability analysis models. J Appl Ecol 42:1115–1120

Otero JT, Aragón S, Ackerman JD (2007) Site variation in spatial aggregation and phorophyte preference in Psychilis monensis (Orchidaceae). Biotropica 39:227–231

Piquot Y, Petit D, Valero M, Cuguen J, de Laguerie P, Vernet P (1998) Variation in sexual and asexual reproduction among young and old populations of the perennial macrophyte Sparganium erectum. Oikos 139–148

Roberts DL, Wilcock CC (2002) Fragmentation of tropical rainforests and its effect on orchid survival. In: Nair H, Arditti J (eds) Proceedings of the 17th world orchid conference "Sustaining Orchids for the Future", Natural History Publications, Borneo

Schödelbauerová I, Tremblay R, Kindlmann P (2010) Prediction vs. reality: can a PVA model predict population persistence 13 year later. Biodivers Conserv 19:637–650

Silvertown J, Franco M, Menges E (1996) Interpretation of elasticity matrices as an aid to the management of the plant population for conservation. Conserv Biol 10:591–596

Tremblay RL (1997) *Lepanthes caritensis*, an endangered orchid: no sex, no future? Selbyana 18:160–166

Tremblay RL, Hutching MJ (2003) Population dynamics in orchid conservation: a review of analytical methods, based on the rare species *Lepanthes eltoroensis*. In: Dixon KW, Kell SP, Barrett RL, Cribb PJ (eds) Orchid conservation. Natural History Publications (Borneo), Kota Kinabalu

Tremblay RL, Meléndez-Ackerman E, Kapan D (2006) Do epiphytic orchids behave as metapopulations? Evidence from colonization, extinction rates and asynchronous population dynamics. Biol Conserv 129:70–81

Winkler M, Hüber K, Hiez P (2009) Population dynamics of epiphytic orchid in a metapopulation context. Ann Bot 104:995–1004

Yew CK, Hew CS (2000) Orchid pseudobulbs – 'false' bulbs with a genuine importance in orchid growth and survival. Sci Hortic 83:165–172

Zotz G, Schmidt G (2006) Population decline in the epiphytic orchid *Aspasia principissa*. Biol Conserv 129:82–90

Chapter 40
Can Canopy-Dwelling Frogs Be Monitored from the Ground? A Case from Western Ghats of India

K.S. Seshadri and T. Ganesh

Keywords Amphibians • Long term monitoring • Automated sound recording • Diel activity patterns

Summary

Biodiversity in the canopies remains poorly explored in many parts of the world. Particularly, there exists a large knowledge gap in the tropics on the lesser known fauna inhabiting the canopies such as amphibians. Most canopy research programs require numerous human-hours, often in harsh environments which can pose limitations. But technological advancements of canopy access have assuaged this by opening up an entire new frontier in the topmost layer of the forests. In this chapter, we discuss the initial findings of a first-ever study to document the canopy- and ground-dwelling anurans by monitoring their vocalizations in the Western Ghats biodiversity hotspot of India, with a long-term objective to establish a monitoring program.

1 Introduction

In the larger context of global climate change, the forest canopy, with its direct interface with the atmosphere, stands to be most affected (Nadkarni and Solano 2002). Anurans in particular are unique organisms to model their impacts under climate change scenarios. Minor changes to the moisture availability or temperature have profound impacts on the breeding success of anurans (Gibbs and Breisch 2002). Further, unprecedented declines in amphibian populations were recently reported from the tropics, in part due to increasing incidences of fungal infections (Stuart

K.S. Seshadri (✉) • T. Ganesh
Ashoka Trust for Research in Ecology and the Environment (ATREE), Royal Enclave, Srirampura, Jakkur Post, Bangalore, KA 560064, India
e-mail: seshadri.ali@gmail.com; tganesh@atree.org

M. Lowman et al. (eds.), *Treetops at Risk: Challenges of Global Canopy Ecology and Conservation*, DOI 10.1007/978-1-4614-7161-5_40,

et al. 2004). Establishment of inventories and long-term monitoring for amphibians is advocated (Bickford et al. 2010). While monitoring anurans using their vocalizations has been widely practiced, no such programs exist in the canopy.

Anurans show temporal and spatial patterns in vocalizing. Often, manual acoustic surveys are challenging due to lack of human power, harsh weather conditions, and difficulty of identifying species with certainty. But automated sound recorders (ASR) which can be deployed in harsh field conditions for long durations are increasingly used to monitor wildlife (See Oseen and Wassersug 2002; Waddle et al. 2009). The system also provides permanent data in the form of recordings which can later be subject to multiple analyses by different observers or using complex algorithms on a computer program. In this chapter, we report on ASRs to monitor canopy-dwelling frogs.

The Western Ghats of India is considered a critical site for species radiation of amphibians (Roelants et al. 2007). So far, over 150 anurans have been documented from the forest matrix of the Western Ghats, including recent descriptions of new species (Seshadri et al. 2012). But studies on the natural history, ecology, and behavior have been limited. The landscape is undergoing rapid changes leading to forest loss, degradation, and fragmentation (Puyravaud et al. 2010). A protocol for long-term monitoring program for anurans was developed in the Kalakad Mundanthurai Tiger Reserve (KMTR), Southern Western Ghats. The protocol was designed and tested in the site for three seasons and is potentially applicable in different habitat matrices across the Western Ghats and elsewhere. Specifically, the objectives were to determine if vocalizing of anurans in the canopy is different from ground level and how such variation occurs across 24 h periods between the canopy and ground.

2 Study Area

Kalakad Mundanthurai Tiger Reserve (KMTR) is located in the southern Western Ghats, India. The area comprises of a habitat matrix of dry forest on the plains to wet evergreen forests in the higher reaches. The elevation ranges from 200 to 1,800 m msl. Mean annual rainfall is ~3,000 mm, received from two monsoon seasons—one during May–August and another between October and December (Ganesh and Davidar 1999). This study was carried out in the mid-elevation evergreen forests at about 1,500 m, with average canopy up to 35 m.

3 Methods

Initial reconnaissance was carried out on both ground and canopies at night (the later using Single Rope technique) to ascertain the presence of vocalizing anurans. Subsequently, commercially available automated sound recorders (Song Meter,

SM2, Wildlife acoustics Inc. USA) were rigged with climate data loggers (Kestrel Pocket weather tracker, NK Inc, USA) in the canopy (30 m) and on the ground (0 m). The sound recorders were programmed to record 60 s. every 29 min for a full 24 h. Subsequently, the recorder was shifted to another tree ~100 m away. This process was repeated for three locations within the forest stand for two seasons during 2009–2010. In 2009, only one sound recorder was available and it was placed in the canopy for 24 h and on the ground at the same location for a subsequent 24 h. In 2010, two sound recorders were placed simultaneously in the canopy and on the ground in each sample location. The recorders were enclosed inside a steel cage to prevent damage from primates. Intensive manual nocturnal surveys were conducted near each recorder. Vocalizing anurans were detected and their vocalizations recorded within 20 cm of the animal. The vocalizations from manual surveys were annotated and saved for reference. Anurans were identified based on photographs and in the sound recordings from reference recordings.

One observer listened to all recordings (576 min) over a period of a few months and the data were tabulated. For each minute of the recording, the presence of anuran vocalization (0=no vocalization, 1=vocalization) was noted, as well as temperature and humidity.

A total of 48 min of recordings were completed for each tree. Thus, three sampling locations over two seasons resulted in 288 min. of recordings for canopy and ground, respectively. The presence of anuran vocalization was compared between canopy and ground using Mau Tau's species accumulation curves followed by a Rich test (R package). Variation in the incidence of vocalizations over time between canopy and ground were compared using F and T tests. Difference in incidence of vocalization within a 24 h period was tested using one sample t test. Difference in species detections between canopy- and ground-based recorders was tested using a paired t test. All statistical tests except Rich test were carried out using PAST®.

4 Results and Discussion

Eleven species of anurans were detected in the sound recorders placed in the canopy and on the ground. Vocalizations of ten species were detected in the canopy and eleven at ground level. The species accumulation curves for the canopy- and ground-based recorders differed significantly (Fig. 40.1; Rich test: Canopy-Ground=1, $p=0.005$).

4.1 Incidence of Vocalization

A total of 576 min. of recordings over two seasons were analyzed. Anuran vocalizations were detected throughout the 24 h period (Fig. 40.2). The incidence of anuran vocalizations were marginally different between canopy- and ground-based

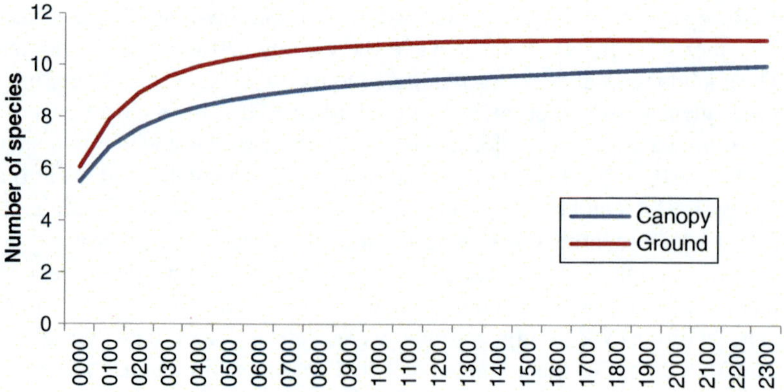

Fig. 40.1 Resampled species accumulation curves (Mao Tau) based on 288 min each (six recordings over 24 h period) for canopy- and ground-based recorder

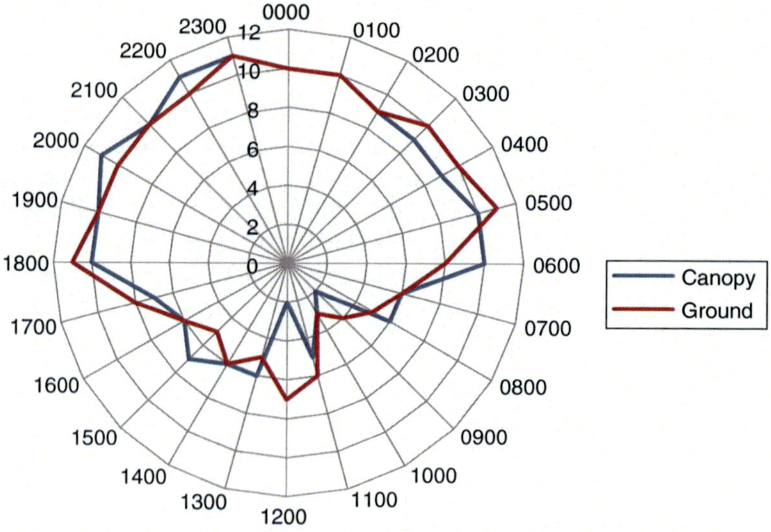

Fig. 40.2 Incidence of vocalization in canopy and ground recorders over 24 h period (y-axis = incidence of vocalization in 1 min summed for every half hour, recorded per hour over 24 h for six sample locations in canopy and ground, respectively)

recorders but not statistically significant (Fig. 40.2, paired t test: $t = -0.843$, $p > 0.05$, $n = 72$). Overall, the number of calling incidences peaked and lasted for 13 h between 1800 and 0600 h and 80–90 % of the species vocalized during this time. The vocalization activity was lower between 0600 and 1800 h, but a few vocalization peaks occurred during daytime with about 50 % of the species vocalizing. The period from 0800 to 1000 h had the lowest incidence of vocalization both in the canopy and

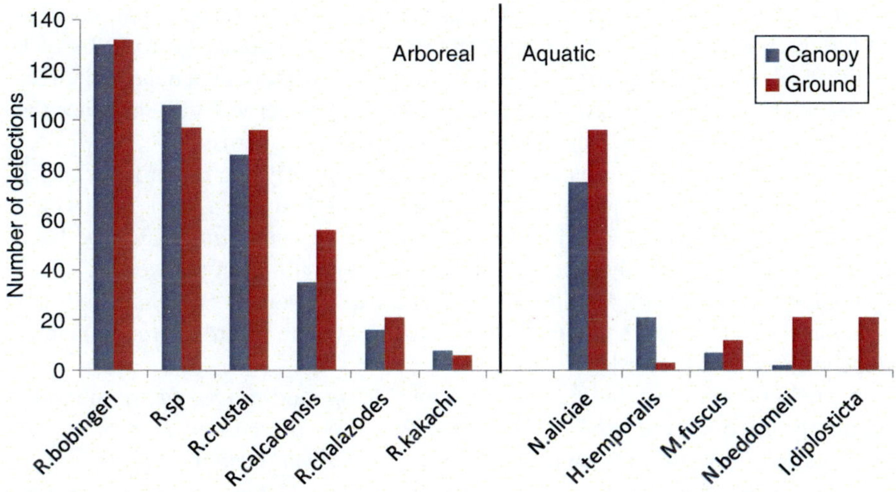

Fig. 40.3 Number of 1 min samples in which vocalization was detected in canopy- and ground-based recorders for arboreal and aquatic species

ground. The vocalization incidence could possibly correlate with favorable moisture and temperature requirements during daylight hours. There were intervals during both day and night in each tree when no anuran vocalized, but this is not evident when data from six trees were pooled. Though there were vocalization incidences during the day, only a few species like *Raorchestes bobingeri* and *Nyctibatrachus aliciae* vocalized but not the entire anuran community of 11 species.

4.2 Species Detections

Of the total species detected, six were arboreal and five were aquatic in habit (Fig. 40.3). Among the arboreal, *R. bobingeri* had the highest incidence of vocalization and *R. kakachi* the least. The other arboreal species included one unidentified anuran (*R.sp.*), *R. crustai, Rhacophorus calcadensis,* and *R. chalazodes*. Among aquatic anurans, *N. aliciae* had the highest incidence of vocalization and *Micrixalus fuscus* had the least. *Hylarana temporalis*, *Nyctibatrachus beddomii*, and *Indirana diplosticta* comprised the other terrestrial/aquatic anurans. All the anurans were detected in both canopy- and ground-based recorders, except *I. diplosticta* (terrestrial).

The variance and incidence of vocalization were compared between the canopy- and ground-based recorders at a species level using F and T tests. The variance was significant only for two arboreal species: *R. kakachi* and *R. calcadensis* ($F=3.215$ and $F=2.294$; $p<0.005$, respectively). Two aquatic species *H. temporalis* and *N. beddomii* had significantly different variance and means ($F=10.143$; $t=3.258$; $p<0.005$ and $F=7.989$; $t=-3.168$ $p<0.005$, respectively). The variation in the vocalizing incidence of these four species could be an attribute of their relative commonness and life history traits. *R. calcadensis* is relatively common compared to *R. kakachi* and breeds in slow-moving pools of water, while *R. kakachi* does not depend on free-flowing water to breed. The availability of free-flowing water will thus influence the occurrence of *R. calcadensis* and thereby its vocalization patterns.

Among the two aquatic species, *H. temporalis* is a large, common frog, which vocalizes from debris along streams; *N. beddomii* is a smaller, rarer frog which vocalizes from swampy areas. The vocalizations of *H. temporalis* may have been masked by the noise of adjacent streams and hence resulting in a lower incidence captured in ground recorders. Similarly, *N. beddomii* has a faint call which does not travel far into the canopy, and therefore, vocalizations were recorded predominantly on the ground-based recorder.

Our field observations indicate *R. bobingeri, R. sp,* and *R. crustai* are the three commonly encountered arboreal anurans. While *N. alicea* and *H. temporalis* are common along streams, the occurrence of *N. beddomii* becomes apparent once the faint vocalization of this species becomes familiar to the observer. However, vocalizations are only that of males, as none of the females have been observed to vocalize. When such measures are used in other areas, care should be taken to tease apart the fact that few individuals may vocalize more in a unit time and thus give the false notion of commonness.

This study demonstrates the potential for using an automated sound recording framework for long-term monitoring at a spatiotemporal scale, often in harsh environments where manual surveys are challenging. Some of the key outcomes of this study are discussed below. Most importantly, this study showed that arboreal species can be monitored by sound recorders placed on the ground. This could be due to the efficiency of the omnidirectional microphone, overall short stature of the forest trees (30–35 m), or presence of multiple layers of forests connecting understory to canopy through the midstory. The study also showed significant differences in the incidence of vocalization by the anuran assemblage over a 24 h period, revealing diel activity patterns among anurans.

The sound recorders detected 11 out of the 15 species known to be present in the forest stand indicating its efficiency. The automatic recorders are even more useful to detect the presence of cryptic species when deployed for long periods of time. The outputs from the recordings made in a spatial framework, when presented as simple bar graphs, could indicate the distribution of the species. This method could be used as a surrogate for the labor-intensive distance sampling or quadrat-based surveys. However, by using incidence of vocalization as a surrogate, it is likely that few individuals of particular species that have intense vocalization can lead to false positive information of abundance or commonness. This problem can be overcome by spatial replication of recorders in a particular forest stand which would help determine if the species was indeed common or if it was only a few individuals of that particular species having intense vocalizations in an area.

Considering the points discussed above, this framework of using automated sound recording systems in conjunction with manual surveys has immense potential in areas that require both rapid assessments as well as long-term monitoring of anuran presence.

One disadvantage of using sound recorders is that it only detects the vocalizing individuals. In most anuran species, only the male vocalizes, resulting in no information on the females. It is also possible that male individuals are present in an area but may not vocalize or may not be detected in the sampling units. Harsh weather can also affect this protocol. Although the recorders are moisture proof, moisture will likely affect the equipment in the tropics.

Because of their advantages, automated sound recorders are emerging as a useful tool for monitoring numerous communities including birds, insects, amphibians, and forest scapes. However, the importance of manual field surveys must be highlighted. Though automated sound recorders are proven efficient, they should not replace field observations. Many aspects of anuran natural history traits play a vital role in understanding the ecology of anurans; such information cannot be obtained through automated sound recording systems. More importantly, the sound recording systems are only a tool to execute better science and are not surrogates for being in the field and experiencing the joy of being in the canopy by ourselves.

Acknowledgments Financial support was provided the Conservation Leadership Program and Idea Wild. The Tamil Nadu Forest Department Provided research permits. Chian and John helped in the field. Chetan assisted with R package. We are thankful to all of them.

References

Bickford D, Howard SD, Ng DJ, Sheridan JA (2010) Impacts of climate change on the amphibians and reptiles of Southeast Asia. Biodivers Conserv 19(4):1043–1062

Ganesh T, Davidar P (1999) Fruit biomass and relative abundance of frugivores in a rain forest of southern Western Ghats, India. J Trop Ecol 15:399–413

Gibbs JP, Breisch AR (2002) Climate warming and calling phenology of frogs near Ithaca, New York, 1900–1999. Conserv Biol 15(4):1175–1178

Nadkarni NM, Solano R (2002) Potential effects of climate change on canopy communities in a tropical cloud forest: an experimental approach. Oecologia 131(4):580–586

Oseen KL, Wassersug RJ (2002) Environmental factors influencing calling in sympatric anurans. Oecologia 133(4):616–625

Puyravaud JP, Davidar P, Laurance WF (2010) Cryptic loss of India's native forests. Science 329:32–32

Roelants K, Gower DJ, Wilkinson M, Loader SP, Biju SD, Guillaume K, Bossuyt F (2007) Global patterns of diversification in the history of modern amphibians. Proc Natl Acad Sci 104(3):887–892

Seshadri KS, Gururaja KV, Aravind NA (2012) A new species of Raorchestes (Amphibia: Anura: Rhacophoridae) from mid-elevation evergreen forests of the southern Western Ghats, India. Zootaxa 3410:19–34

Stuart SN, Chanson JS, Cox NA, Young BE, Rodrigues AS, Fischman DL, Waller RW (2004) Status and trends of amphibian declines and extinctions worldwide. Science 306(5702):1783–1786

Waddle JH, Thigpen TF, Glorioso BM (2009) Efficacy of automatic vocalization recognition software for anuran monitoring. Herpetol Conserv Biol 4(3):384–388

Chapter 41
Just Harvest: Ecology and Politics of Forest Canopy Product Use in Protected Areas

Nitin D. Rai and Siddappa Setty

Keywords Rights-based conservation • Forest policy • Non-timber forest produce • Indigenous people • India

Bullet Points

1. State forest policy has retained control over forest products, denying people access and rights to livelihoods and management.
2. Ecological studies have shown that use of non-timber forest products (NTFPs) can be sustainable under conditions of local control and adaptive management.
3. Recent legislation giving people rights to forest use is providing opportunities for collaborative management of forests based on both science and local knowledge.

Summary

There exists a long history of state interest in non-timber forest products (NTFPs) harvest, as a result of which forest-dwelling communities have been "arborealized," i.e., made dependent on forest produce while being denied land rights (Li 2010). Recent conservation policies have added to this historic marginalization by further restricting a relatively low level of subsistence forest use. Such forest policies are mandated in spite of ecological studies that demonstrate that NTFPs can be sustainably harvested. The loss of forest access has resulted in adverse effects on livelihoods of local people. We present recent findings from ecological and socioeconomic studies in a protected area in India to argue that forest product harvest might be

N.D. Rai (✉) • S. Setty
Ashoka Trust for Research in Ecology and the Environment (ATREE), Royal Enclave, Sriramapura, Jakkur Post, Bangalore, KA 560064, India
e-mail: nitinrai@atree.org; siddappa@atree.org

M. Lowman et al. (eds.), *Treetops at Risk: Challenges of Global Canopy Ecology and Conservation*, DOI 10.1007/978-1-4614-7161-5_41,

supported even within protected areas subjected to a system of local and scientific knowledge and management. Using the case of *Phyllanthus* fruit collection and *Apis dorsata* honey harvest, we demonstrate how ecological, social-cultural, and economic aspects might be incorporated into equitable management regimes.

1 Introduction

India, as have most countries in Asia and Africa, has a long history of forest policies that have encouraged the use of forests for timber and non-timber products. The colonial enterprise was premised on extraction of resources to fuel the economies of European powers. While timber was used for larger infrastructural requirements, non-timber forest produce (NTFP) was targeted at a more local and regional consumer base. The state controlled resource extraction by taking over the management and access to forests from local people. This resulted in the denial of individual land titles to forest dwellers and instead limited them to harvesting and selling NTFPs to sustain themselves. The need for continued production from forests and the desire to stem degradation resulted in the adoption of scientific and data-driven forest management techniques that were not only focused on timber species but also on NTFPs. There has therefore been a tradition of scientific management of timber and NTFPs in India.

State control over forests and resources has continued even after independence from the British in 1947, with the result that forest access by local communities is limited, while their dependence on forest products (to offset the lack of income from agriculture and other sources) has been high. The resistance to giving people more control over forest produce harvest has a historical basis in colonial rule and in the received idea that local harvest of forest products inevitably leads to degradation. This is contrary to research findings that suggest that local management and improved access regimes to NTFP are ecologically sustainable and economically viable.

2 NTFP and Conservation Policy: A Troubled Co-existence

At least one-fourth of the world's poor depend partly or fully on forest products for subsistence and as significant sources of cash income. NTFPs play integral roles in the livelihoods and health of hundreds of millions of people across the globe. Yet there is very little appreciation of this in forest conservation efforts in India leading to increasingly restrictive legislations that prevent people from harvesting and selling NTFP in forest areas (especially protected sites). The most recent effort at such restriction was the amendment in 2002 to the Wildlife Protection Act that mandated a ban on the harvest of NTFP from India's wildlife sanctuaries. This was not based on any rigorous scientific evidence. On the contrary, results from research suggesting that harvest was not the cause for species declines were ignored in the drafting of these conservation laws. Lessons from long-term research in a wildlife sanctuary show that NTFP harvest is feasible and may have fewer adverse impacts than other local ecological factors such as invasive species and hemiparasites.

Soligas, who are an indigenous people, live in the Biligiri Rangaswamy Temple Wildlife Sanctuary (BRT) and earn an average of Rs 4 million per year from NTFPs to meet a proportion of their cash income. In BRT, harvesting of fruits from two *Phyllanthus* species (*Amla*) alone contributed approximately 6–11 % of the total cash income in a Soliga household, and up to 50 % of the cash income per household comes from extraction of NTFPs (Hegde et al. 1996). Aside from *Amla,* the most important commercial NTFP for Soligas is honey obtained from rock bee colonies (*Apis dorsata*), lichens, soapnut (*Acacia sinuata*), and soapberry (*Sapindus laurifolius*), all of which are removed from the forest canopy. Among these NTFPs, 15–18 t of honey is harvested per year making it one of the most important canopy forest resources in BRT (Setty et al. 2008).

BRT is located in the state of Karnataka between the Western and Eastern Ghats. The sanctuary has an area of 540 km^2 and consists of a number of hills with an average elevation of about 1,350 m and 1,400 mm annual rainfall. The major vegetation types are dry deciduous, scrub, evergreen, savanna, and shola. Around 16,000 Soligas live in and around BRT. Traditionally, Soligas were hunters and shifting cultivators and collected a wide range of NTFPs for their subsistence. When BRT was designated a wildlife sanctuary in 1974, shifting cultivation and hunting were banned and Soligas were allocated small pieces of land to practice settled agriculture but were not given titles to the land. However, the extraction of NTFPs continued with marketing assistance from a tribal cooperative society. BRT was one of the pilot sites in India where community were involved in monitoring forest resources. However, in 2002, a ban on NTFP collection from BRT prevented further harvesting, monitoring, processing, and marketing with adverse impacts on livelihoods. The protected-area approach has often resulted in adverse effects on the livelihoods of local people.

The fruits of *Amla* are collected from two species, namely, *Phyllanthus emblica* and *P. indofischeri*. *P. emblica* grows in dry deciduous forests as well as woodland savanna, while *P. indofischeri* grows in scrub forests and is mostly restricted to low elevations. *Amla* is an important medicinal plant, and the fruit, which is rich in vitamin C, is used in pickles, jams, preserves, jelly, and medicines. Fruits are also used for making certain dyes and in hair shampoo.

Honey from the rock bee is an important source of income for Soligas. Rock bees construct hives in open spaces such as rock cliffs and branches of tall trees in order to avoid predation. Such trees are often found along streams. It is believed that in addition to their relatively aggressive nature, nesting at great heights is one of the main defenses that these honey bees have against predators. Bee colonies are only found in BRT during the flowering season which lasts from April to June and a leaner season in November. Soligas collect honey from colonies that are on tall trees as well as from rock cliffs.

3 The Politics of Protected Area Management

Long-term ecological and socioeconomic research in BRT has revealed some interesting results. The ecological research has looked closely at what factors have influenced *Amla* population dynamics. Beginning with the publication of a set of papers in 1996 in the Journal of Economic Botany and continuing for the next 16 years, studies showed that harvest had little effect on population growth rates of *Amla*. Authors identified the presence of *Lantana camara* an invasive weed in the understory and a hemiparasite that infects adult trees (*Taxillus tomentosus*) that greatly affect the population ecology of *Amla*. *Lantana* prevents germinated seedlings from growing through the thick undergrowth, thus reducing recruitment into the sapling stage. A high rate of *Taxillus* infestation resulted in increased mortality of adult trees (54 % of infected trees died), which reduced population more than any anthropogenic factors.

These results should be discussed in the context of the history of forest management in BRT over the last 50 years. With the notification of the Wildlife Protection Act in 1972, protected areas were demarcated all over India. These areas were brought into a management system that suppressed local practices that were ongoing for several decades or longer. Thus in BRT, the use of cool season fires by Soligas was banned resulting in changes in vegetation and fire regimes. Soligas were illegally displaced from their shifting cultivation areas and settled into colonies where only about half of them were given small pieces of land for cultivation. These sudden changes in forest management resulted in changes to the forest. Soligas claim that the lack of fire increased *Lantana* cover and also increased the density of *Loranthus* which is susceptible to the heat generated by ground fires. Recent ecological work corroborates Soliga claims about the relationship between fire and *Lantana* showing that areas that experienced higher frequencies of fire showed a lower probability of invasion by *Lantana.* Thus, not only have studies and local observations shown that local harvest is not a driver of population change, but also that the cessation of long-term local practices dramatically changed the forest.

Long-term monitoring of rock bee colonies reveals an equally complex story. The highest densities of bee colonies were recorded in evergreen forests, followed by dry deciduous forests. The level of extraction was high in dry deciduous forest, followed by evergreen, and least in cliffs due to accessibility. Our results show that colony numbers fluctuate widely over the entire period and that there is no clear pattern to the fluctuations. There was a gradual decrease in bee colony numbers from 1995 to 2011. Honey harvest data did not seem to reduce bee colony numbers, and it is still unclear what is influencing these numbers. Our conversations with Soligas has thrown up the possibility that the use of pesticides in the surrounding agricultural landscape, into which bees travel after the main flowering season in BRT, might be responsible for the decline in bee colony numbers. This resonates with studies elsewhere in the world where pesticides are linked to massive bee population declines.

Regional processes that occur far from boundaries of protected areas have large impacts on forest function. Pollinators play an important role in maintaining forest structure, composition, and function. By focusing on harvest and other local factors, managers have failed to see the influence of distant factors such as agricultural intensification in the surrounding landscape. Forests are part of a complex and integrated landscape and not a closed system as is currently perceived by forest managers in India. That the Soligas were able to identify the factors affecting their bee population decline is testament to their understanding of the forest. Using this case we argue for a more nuanced understanding of the role of situated knowledge, history and non-local processes in conservation.

4 The "Right" Way Forward

A recent and parallel legislative process has attempted to repair some of the long-term erosion of rights of forest dwellers. In 2008, the Forest Rights Act was notified by the central government, returning land and forest use rights to forest dwellers. This has however been met with resistance from the forest administration with fears that giving rights to people will erode its control over forests. The politics of forest control is therefore being played out across the country; subsequently, forests are not being managed in ways that are adaptive and cognizant of history, while local people continue to be marginalized by forest management policies.

In BRT, the FRA has been successfully implemented with >1,500 households getting rights to cultivable land and 32 villages gaining community rights to harvest, add value, and sell NTFP. This is the first example in India of such comprehensive rights given in a protected area. The exercise of these rights is however another matter as little has happened since these rights were given by the state administration. Citing the provisions of the WLPA, the forest administration has strengthened its protectionist position.

Forest canopy products such as *Amla* and bee colonies are important as systems for study by ecologists. They also provide subsistence level incomes to forest-dwelling people while serving as ecological indicators for forest management that could be collaboratively monitored by scientists, managers, and local communities. We suggest that current conservation policy pays heed not only to scientific evidence that indicates forest produce can be used sustainably but also to the history of forest use and local knowledge to evolve a collaborative approach to the conservation of forest canopy resources.

References

Hegde R, Suryaprakash S, Achoth L, Bawa KS (1996) Extraction of nontimber forest products in the forest of Biligirirangan Hills, India. 1. Contribution to rural income. Econ Bot 50:243–251

Li TM (2010) Indigeneity, capitalism, and the management of dispossession. Curr Anthropol 51:385–414

Setty SR, Bawa KS, Ticktin T, Madegowda C (2008) Evaluation of a participatory resource monitoring system for non-timber forest products: the case of *Amla* (*Phyllanthus* spp.) fruit harvest by Soligas in South India. Ecol Soc 13(2):19

Chapter 42
Tropical Rainforests of Africa: Can Conservation Projects Reconcile Forest Conservation and Development of Forest-Dependent Communities?

Bernard-Aloys Nkongmeneck, Rose Caspa, and Evariste Fongnzossie Fedoung

Keywords Cameroon • Rainforests • Protected Areas • Local communities

Bullet Points

1. The rainforests of Africa in the Cameroon and Congo basin comprise unexplored canopies and have come under severe threat due to resource extraction. To counter this, a series of protected areas have been established.
2. Most of the development goals of these projects are not embedded within the socioculture of local communities, and hence, conservation success is marginalized.

1 Introduction

Most of Africa's remaining rainforests are found in the Congo River Basin and in Cameroon. Parts of these forests still comprise intact canopy but are relatively unstudied compared to other continents. Local communities derive a bulk of resources in the form non-timber forest products from the forest canopies of these areas (Amrose-OJI 2003). To protect these forests, protected areas have been created, followed by

B.-A. Nkongmeneck (✉)
University of Yaoundé I, Yaoundé, Cameroon
e-mail: bnkongme@yahoo.fr

R. Caspa
Agronomic and Development Research Institute, Yaoundé, Cameroon
e-mail: rosegusua@yahoo.com

E.F. Fedoung
University of Douala, Douala, Cameroon
e-mail: fong_nzossie@yahoo.com

M. Lowman et al. (eds.), *Treetops at Risk: Challenges of Global Canopy Ecology and Conservation*, DOI 10.1007/978-1-4614-7161-5_42,

a number of restrictions vis-à-vis the local population. Many local people believe that they have been alienated and would like to exercise their traditional rights to forests as they have always done in the past. Also there is a general frustration among these communities because these protected areas were established without consideration of their requirements. There have been an increasing number of pilot or demonstration biodiversity conservation projects which now involve local people. Led by the conservation NGOs (national and international), these projects have mostly been based on innovative land use strategies, including protected areas, multiple-use conservation areas, buffer zones on protected area boundaries, community forestry, and a variety of other approaches. All these projects aim to conserve biodiversity with local participation. Despite numerous NGO projects during recent years, successful and convincing examples with local needs reconciled with biodiversity conservation remain difficult to find. There is increasing evidence that national parks and other protected areas alone are insufficient for the conservation of Africa's biodiversity. Similarly, sustainable community development in buffer zones around protected areas has proved elusive for adjacent rural dwellers. So the question remains if we need an alternate conservation paradigm for Cameroon.

2 Overview of Current Approaches

Conservationists recognize that local participation is crucial for the sustainability of protected areas in developing countries (Wells and McSchane 2004). Several approaches were implemented in Cameroon, whereby community forestry was developed under the forestry law of 1994 as a means of ensuring decentralization of forest management and increasing the participation of local people. Seventeen years after, the results from this process are still subject to severe criticism from various stakeholders. Yet another approach was initiated via biodiversity observatories and the "science-policy" platform (Jürgens et al. 2012). Science-policy platforms call for scientists to make the results of their research available to policy makers to enable them to make informed decisions.

In the third approach, the private sector was engaged in conservation under the Congo Basin Forest Partnership (CBFP 2006; Wasseige et al. 2008), a trans-boundary collaboration was developed between Central Africa, Europe, and North America in Congo basin countries. Given the demographic growth in most Congo basin countries and the importance of timber reserves and minerals among the richest in the world, the nations of Central Africa face a challenge of reconciling ecosystem conservation and human use (Kamdem-Toham et al. 2003). The landscape approach identifies and involves all stakeholders in a landscape or region (regional and local authorities, private sector, logging companies, etc.). This approach takes into account the concerns of all components of a landscape, which can lead to a sustainable conservation combining ecological, economic, and social aspects.

Local government budgets for conservation in Cameroon are inadequate, and the consequences are insufficient staff who lack motivation and technical capacities to effectively implement conservation. Conservation is a long-term goal and requires long-term investments. But unfortunately, the main donors are engaged in short-term planning, causing a cyclic appearance and disappearance of conservation projects. Many programs are intermittent, which allow poachers to gain entry.

3 Case Studies in Cameroon

In the Dja Biosphere Reserve in Cameroon, local communities were offered training in intensive agriculture and animal husbandry. Also, goods such as corrugated iron sheets and blankets were distributed. These solutions were not successful simply because they ignored the culture of the Baka, Badjoué, Kaka, Nzimé, and Fang people living in the region. It seems that people who received the goods gave them to their families who live in cities. Attempts such as small-scale fish pond development projects also failed, as it is not easy to convince communities who usually fish in the river to create fish ponds. White and Edwards (2001) have rightly proposed that "it is imperative that the protected area staff establish real relationships with local communities and encourage their participation from designing to implementation, monitoring and evaluation. Neglecting this aspect could have a more profound influence on the success as poor biological data collection." Alternate approaches using traditional or indigenous knowledge may be the way forward (Wit and Zagt 2009; Nkongmeneck 1998, 2003).

A transition from "top-down" to "bottom-up" approach in the design of conservation projects is more likely to succeed in the African rainforests. Designing the biodiversity conservation projects with the participation of local residents allows them to take ownership of the project from the outset. Monitoring can also be done with relatively low financial means by local people themselves (Njatang 2001).

Improving knowledge and skills of local people through science and technology access inevitably helps improve their living conditions. For example, beekeeping projects work very well as income-generating activities, as it fits perfectly into the habits of the region. Another project that might work instead of fishponds is that of organized river fishing, with good training and good equipment of artisanal fishermen. Such activity could provide sufficient fish for local consumption and supply other parts of the country. So it is really critical to identify activities that lie within the culture of the communities and then blend them with current technology.

A final example is the harvesting of wild fruits from forest canopies, with the possibility of developing agroforestry systems to reduce the biological cycle of the species. In developing this type of enterprise, it will be important (1) to consider on-site processing of certain fruit into butter, oil adjuvant, food supplements, etc., and (2) to shift from traditional to more sophisticated processes and create profitable jobs.

Conservation and sustainable management of biodiversity through protected areas requires cooperation of all stakeholders, donors, governments, national and international NGOs, local people and communities, municipalities, and politicians.

Management and monitoring should be coordinated and funding must be provided by donors or governments concerned and must be sustainable. A challenging question remains: what will be the first country in Central Africa that will succeed in establishing a conservation and sustainable management of biodiversity?

Biodiversity conservation and local livelihoods are integrated, and therefore, conservation and development should be mutually supportive. As suggested by Torri and Hermann (2010), research efforts, rather than trying to provide a quick and universal solution to problems of poverty and biodiversity loss, need to focus on the contextual details that make particular outcomes more or less likely. Further, for research to be policy-relevant, new studies need to focus on the dynamics of the relationship between various measures of poverty and biodiversity and on how these dynamics are affected by macrosocial and political variables.

References

Amrose-OJI B (2003) The contributors of the livelihoods of the 'forest poor': evidence from tropical forest zone of south-west Cameroon. Int For Rev 5(2):106–117

CBFP (2006) Etat des forêts du Bassin du Congo

Jürgens N, Schmiedel U, Haarmeyer DH, Dengler J, Finckh M, Goetze D, et al (2012) The BIOTA Biodiversity Observatories in Africa–a standardized framework for large-scale environmental monitoring. Environ Monit Assess 184(2):655–678

Kamdem-Toham A, Adeleke W, Burgess ND, Carroll R, D'Amico J, Dinerstein E et al (2003) A vision for forest conservation in the Congo Basin. Science 299:346.

MINFOF (2006) Bilan des acquis de la foresterie communautaire au Cameroun et définition de nouvelles orientations. Rapport CARFAD

Njatang U (2001) Un exemple de gestion communautaire des ressources naturelles par des femmes, la réserve naturelle de Popenguine au Sénégal. www.ecoloinfo.com/2009/08/16 odysseO2-et-long-brésilenne-ecologica. 2009

Nkongmeneck BA (1998) Les populations des zones forestières africaines et les projets de conservation des écosystèmes et des ressources naturelles: exemple du Cameroun. Com. Propos. au sém. FORAFRI, Libreville

Nkongmeneck BA (2003) Le contexte socioculturel dans la gestion durable des produits forestiers non-ligneux au Cameroun. Séminaire GTZ sur la gestion durable des PFNL, Edéa, pp 25–26

Torri MC, Herrmann TM (2010) Biodiversity Conservation versus Rural Development: What Kind of Possible Harmonization? The Case Study of Alwar District, Rajasthan, India. J Hum Ecol 31(2):93–101

Wasseige C, Devers D, deMarcen P, Eba'a Atyi R, Nasi R. Mayaux P (ed) (2009) Congo Basin Forests-State of Forests 2008. Office des publications de l'Union Europeenne, pp. 209–221

Wells MP, McShane TO (2004) Integrating protected area management with local needs and aspirations. Ambio 33:513–519

White L, Edwards A (2001) Conservation en forêt pluviale africaine. Méthodes de recherche

Wit M, Zagt R (2009) Making knowledge for forests and people, annual report. Tropenbos International

Chapter 43
Sacred Groves as Sanctuaries for Mistletoe Conservation in Kathmandu Valley

Mohan P. Devkota

Keywords Sacred groves • Mistletoe conservation • Kathmandu • Nepal himalayas

Bullet Points

1. Sacred groves have centuries-old tradition of conserving forest in the valley.
2. Erosion in religious beliefs and tradition values has threatened biodiversity of groves.
3. Rich diversity of groves constitutes 45 % of mistletoe species of Nepal.
4. Mistletoes play an important role in balancing the natural ecosystems.
5. A separate management plan is proposed to conserve and develop groves as mistletoe sanctuary.

Summary

This chapter discusses the potentiality of sacred groves of Kathmandu Valley, Nepal, in developing them as sanctuaries for mistletoe conservation and also recommends measures for that. Deeply rooted religious and cultural beliefs in Nepal have centuries-old traditional history of conserving patches of forests around the religious sites and have provided refuge to a variety of local flora and fauna, which may otherwise have become locally extinct. These groves constitute 45 % of mistletoe species recorded from the country in the remaining patches of old growth forest of the valley. Due to negligence in management practices from the last two decades, these groves are in the verge of extinction and mistletoes are facing threat of extinction due to an alarming rate of environmental degradation. A number of pitfalls have been identified that need to be taken into account for the proper management and

M.P. Devkota (✉)
Department of Botany, Tribhuvan University, Amrit Campus,
PO Box-7342, Kathmandu, Nepal
e-mail: mistldevkota@yahoo.com

M. Lowman et al. (eds.), *Treetops at Risk: Challenges of Global Canopy Ecology and Conservation*, DOI 10.1007/978-1-4614-7161-5_43,

conservation of sacred groves. Existing conservation strategies of Nepal are limited, with no specific plans implemented in the Nepal Biodiversity Strategy Implementation Plan (2006–2010) despite its commitments to the management of the country's sacred groves. All sacred groves are currently under threat, because of cultural breakdown, pressures on land and resources, and poor governance that permit deleterious use. The spiritual values of sacred groves need to be considered seriously while planning conservation, and conservationists should take a lead role to deal effectively with sacred groves and the people for whom they are sacred. Thoughtfully planned and implemented conservation strategy can contribute in the conservation of sacred groves of the valley and will help Nepal to fulfill its commitments made to IUCN and Convention on Biological Diversity.

1 Religion and Conservation in Nepal

Establishment of small patches of forest in the form of sacred groves is a universal human phenomenon not associated with any specific religion or worldview but with strong religious context influenced by traditional local beliefs. In many regions of the world, sacred groves have major effects on conservation due to taboos and restrictions associated with them. It has been believed that sacred virgin forests date back to several thousands of years when human society was in the primitive state and all forms of vegetation in the sacred groves were supposed to be under the protection of the reigning deity of that grove, and the removal of even a small twig was taboo. Long before the existence of officially protected areas as we now know them, people were protecting sacred lands. Sacred groves are probably the oldest method of habitat protection on the planet and still form a large and mainly unrecognized network of sanctuaries around the world. Thus the role of sacred groves in the conservation of biodiversity has long been recognized (reviewed in Bhagwat and Rutte 2006; Khan et al. 2008).

Nepal is no exception to this practice which has centuries-old traditions of establishing sacred groves devoted to a particular deity. Sacred groves provide important sites for spiritual and cultural value and have significant role in local, regional, or even international level. Therefore, they are important in biodiversity conservation associated with deeply rooted religious and cultural beliefs related to Hindu and Buddhist religions. The influence of religious beliefs and rituals on forest conservation in Nepal is significant because they give rise to appropriate institutions and organizations for cooperative resource management and provide symbols of forest conservation in Nepalese society. In addition, sacred groves contribute in providing refugia to local species that may otherwise become locally extinct, in a changing landscape throughout Nepal. Nepal is rich in its ethnic diversity so there are diverse human societies with different religious customs, myths, and beliefs. These societies are intricately interconnected through their religion and customs and have been practicing religious traditions by establishing sacred groves devoted for their deities. Much of the social life in rural communities of Nepal occurs in and

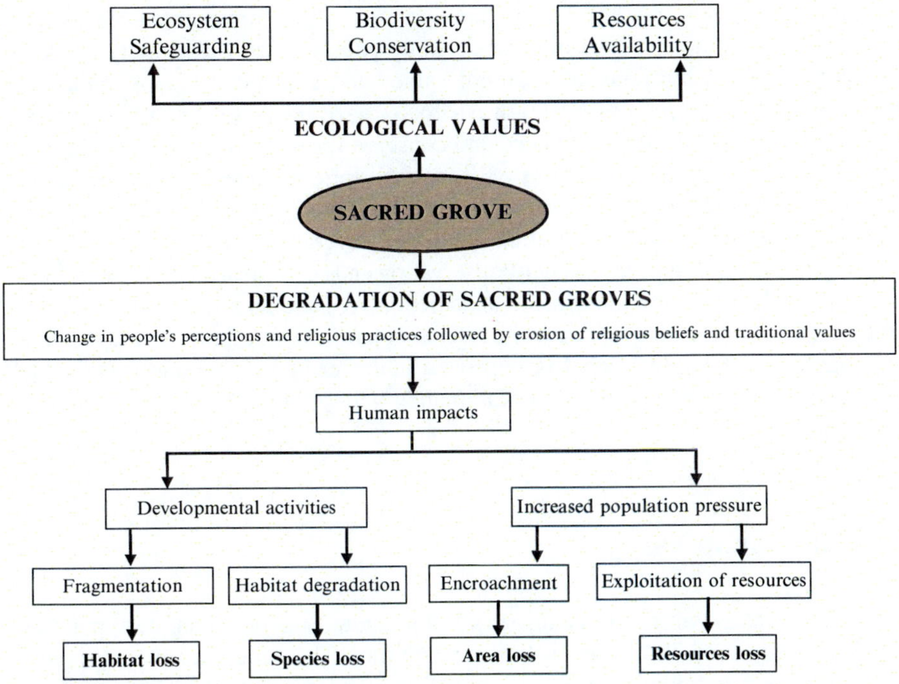

Fig. 43.1 Ecological value of sacred grove and their relationship with eroding cultural values and impacts

around sacred groves; even urban communities visit groves during various religious festivals. In Nepal, sacred groves serve to conserve small patches of forest from a sociological, cultural, and religious perspective. It has been argued that religious beliefs and practices affect the way forests are perceived and managed in Nepal. Ingles (1995) describes the influence of religious beliefs and rituals on forest conservation in Nepal by examining how the dominant social philosophy of the Indian subcontinent influences the way Nepalese communities perceive their environment and how religious beliefs and practices affect the way forests are managed. These groves are probably the oldest method of habitat protection and still form a large and mainly unrecognized network of conserved areas in Nepal. Long before the existence of officially protected areas as we now know them, Nepalese people were protecting their sacred lands thus contributing to biodiversity conservation. The government of Nepal, under "Forest Act 1992," has given legal provision to establish and use sacred groves.

Human pressure, as resources dependency on groves, may possibly lead to decline in the species diversity and change floristic composition (Fig. 43.1) of the groves. In Nepal, sacred groves are hailed as models of community-based conservation and storehouse of valuable biodiversity. Their importance to forest conservation becomes even more significant given the current rate of deforestation in Nepal. Kathmandu Valley contains centuries-old sacred groves which have been the places of birth of strong religious and sociocultural practices; therefore, a greater focus and understanding of these groves in plant diversity conservation is definitely needed. The deeply rooted religious and cultural values and associated taboos related to Hindu and Buddha religions have always encouraged ancient rulers of the valley and communities to establish, protect, and conserve forests surrounding a religious site. Written evidence, in the form of stone inscriptions, dates back to the sixth century, describing the strict conservation rules of sacred groves.

Anthropological contributions that challenge several common perceptions on sacred groves are often overlooked in conservation and ecological studies. Therefore, this chapter discusses the importance of sacred groves and its potentiality in developing them as future sanctuaries for mistletoe conservation (assuming the existing national forest policy of Nepal is improved and renewed to include sustainable management of sacred groves of the Kathmandu Valley, which has been overlooked in the past).

2 Kathmandu Valley

Kathmandu Valley, with its three districts Kathmandu, Bhaktapur, and Lalitpur, lies at 1,350 m elevation and covers an area of 665 km^2. The valley lies in subtropical zone, but the surrounding mountains have forests of upper cool temperate zone. Climate of the valley is characterized by typical monsoon seasons, with rainy summers receiving approximately 80 % of rain, and dry winters. Summer is hot where temperatures reach up to 35 °C, and winter is cold where temperatures occasionally drop below 0 °C when surrounding mountain tops receive snow.

The valley has 20,945 ha of forest, about 32.7 % of the total area of the valley (ICIMOD 2007). The natural vegetation, except in few protected areas, has been under intense pressure. The area under natural forest cover, excluding shrubs, is 9,580 ha (45.7 % of the total forest land), of which only about 22 % has good forest cover with more than 50 % of crown coverage (ICIMOD 2007). Mature hardwood forests are now confined to protected areas such as Gokarna, Phulchowki, and Shivapuri-Nagarjun National Park and sacred groves such as Pashupatinath, Swayambhu, Bajrabarahi, and Dakshinkali. The major vegetation types in the valley are *Schima-Castanopsis* on the valley floor and hill slopes, *Pinus roxburghii* on lower drier hill slopes, *Quercus lanata* and *Q. semecarpifolia* dominating upper hill slopes of surrounding mountains, Oak-Laurel at middle elevation, *Quercus lanata* and *Pinus wallichiana* at higher elevation, and *Rhododendron arboreum* on the reaches of the surrounding mountains of the valley. About 1,500 vascular plant species, which is about 21 % of the Nepal Flora, have been reported from the valley, showing its rich biodiversity. At the same time, Kathmandu Valley is losing some 18 orchids and 6 fern species due to urbanization, deforestation, environmental degradation, and other human impacts. Joshi (2000) has reported that 7.3 % of the Kathmandu Valley flora is under various levels of threats. There are 12 sacred groves covering an area of 226.18 ha within three districts of the valley, but only five large sacred groves were chosen for the study.

Table 43.1 Major groves of mistletoe inventory

Name of the grove	Location (district)	Area (ha)	Dominant vegetation type	Status
Pashupatinath	Kathmandu	90	Deciduous mixed hardwood forest	UNESCO World Heritage site
Syambhunath	Kathmandu	32	Deciduous mixed hardwood forest	UNESCO World Heritage site
Bajrabarahi	Lalitpur	19	Deciduous mixed hardwood forest	Locally managed
Dakshinkali	Lalitpur	56	Deciduous mixed hardwood forest	Locally managed
Suryabinayak	Bhaktapur	18	Deciduous mixed hardwood forest	Locally managed

3 Studied Groves and Mistletoe Inventory

Five sacred groves were explored in three districts of the valley for mistletoe inventory (Table 43.1). Necessary information on each grove was acquired from District Forest Offices and local communities. All sacred groves under study have an open structure, both in the overstory and understory, with an absence of a well-established pool of regeneration. These groves are believed to be the remnants of the forest which had once dominated the valley but over time had gone through the cycle of decline and regrowth. They had been degraded to a significant amount in the past and had regenerated to closed forests after the strong protection measures or reforestation activities were initiated.

Each grove was visited twice in different times of the year for mistletoe inventory between 2007 and 2009. Walk-on survey was the main method of inventory, and every part of each grove was visited to collect information on the mistletoe, their host species, and condition of the groves. Most of the groves contained man-made walking trails so it was comparatively easy to carry out the inventory. Mistletoe was observed with a pair of binoculars, collected by using pole pruners, and identified in the field and herbariums deposited at KATH.

4 Results

A total of nine mistletoe species, seven belonging to five genera in the family Loranthaceae and two species in one genus in the family Viscaceae, were recorded from five sacred groves of the valley (Table 43.2). Genus *Scurrula* was largest with three species followed by two species in genus *Viscum*. Three mistletoe species *Helixanthera ligustrina*, *Dendrophthoe pentandra*, and *Viscum articulatum* are reported from all five sacred groves; similarly Suryabinayak sacred grove had the highest number of mistletoe species. The groves contribute 45 % of total mistletoe species recorded from Nepal compared to Devkota (2009) and Hara et al. (1982)

Table 43.2 Mistletoe record in each sacred grove

Mistletoe species	Sacred groves				
	Pashupatinath	Syambhunath	Bajrabarahi	Dakshinkali	Suryabinayak
Helixanthera ligustrina	✓	✓	✓	✓	✓
Loranthus odoratus	–	–	–	–	✓
Macrosolen cochinchinensis	–	–	✓	✓	✓
Dendrophthoe pentandra	✓	✓	✓	✓	✓
Scurrula elata	–	–	✓	–	–
Scurrula parasitica	✓	–	–	–	✓
Scurrula pulverulenta	–	✓	✓	✓	✓
Viscum album	✓	✓	–	–	–
Viscum articulatum	✓	✓	✓	✓	✓

who reported 20 mistletoe species in different times from Nepal. Nine mistletoe species occurring in the valley is a clear indication of high mistletoe diversity in these groves.

5 Conservation Challenges of Sacred Groves in Nepal

Modernization followed by liberalization, privatization, and globalization is leading to erosion of cultural values that are vital for the conservation of sacred groves in Nepal. As a result, the groves are losing cultural values and appear to be in the verge of depletion. Ecological values and impacts facing sacred groves in the valley due to various human activities are summarized in Fig. 43.1. The unavailability of official statistics on the number and area covered by sacred groves indicates that sacred groves in Nepal are still unrecognized, their roles in conserving biodiversity are less understood, and they have remained overlooked by the government of Nepal. The major problems responsible for the loss of ecological and cultural values of the groves that severely affect the biodiversity of the groves include:

- Loss of biodiversity due to habitat fragmentation, habitat loss, and over dependency for resources
- Replacement of local plant species due to the introduction of exotic species
- Soil erosion and poor regeneration
- Fire due to human negligence
- Increasing number of visitors leading excess garbage production
- Encroachment and erection of new structures

- Changing perceptions of people towards religious and cultural values
- Poor management and conservation practices and lack of awareness

Nepal had developed and endorsed its long-awaited Biodiversity Strategy in 2002, in the process of implementing Convention on Biological Diversity and meeting its three major objectives of conservation, sustainable use, and fare and equitable sharing of benefits of biodiversity. The strategy had identified and prioritized various programs to be included in future conservation activities. Management of religious forests of the country is one of the identified conservation activities prioritized in forest sector under sectoral strategies of the government. Regardless of this, no specific programs for the management of religious forests were planned to be included in five years program of Nepal Biodiversity Strategy Implementation Plan (2006–2010). The country's national programs for nature conservation have been deficient in scale and have been characterized by inadequate planning and sluggish progress in implementation, thus overlooking the importance of sacred groves in biodiversity conservation. The National Conservation Strategy for Nepal (1988) had identified the introduction of western culture and lifestyles as a constraint in conserving Nepal's cultural heritage, which includes sacred sites and religious forests.

6 Dilemma in Sacred Grove Conservation

The presence of sacred groves in the landscape provides a reminder of conservation messages in Hindu and Buddha religions in Nepal. However, the effectiveness of this symbolism in promoting conservation ethics is under threat from the increasing influence of western culture on the Nepalese population. In many parts of the world, sacred groves play a significant role in conserving local biodiversity and environment protection. But is the conservation and management of sacred groves important to biodiversity conservation in Nepal? This question can be addressed by considering the following questions:

- How do sacred groves contribute to the biodiversity and resources conservation?
- How can the local system of managing sacred groves be adopted in rural development programs by conserving the resources?
- How do attitudes towards sacred groves differ among different ethnic groups and contribute in biodiversity conservation?

The potential role of sacred groves in conserving biodiversity has been greatly studied, but has anyone actually proved the links? There has been no comprehensive global survey into the conservation effectiveness of sacred groves (neither has there been a survey of protected areas). The question also relates in many cases more generally to whether or not indigenous and local people are effective in conserving biodiversity. Generalizations are dangerous, and while there is good evidence that some local communities have been very effective in conserving biodiversity and managing natural resources, others have caused degradation; periods of rapid change in cultures are particularly dangerous in creating the conditions where mismanagement can occur. Differing perspectives on sacred groves influence management; some people regard sacred groves as wonderful places suitable for praise, while others regard them as fearful or dangerous places. Today, as space and resources become limited, many local communities choose immediate necessities over long-term sustainable use. In these conditions sacred groves are particularly important from a biodiversity perspective, both because they are likely to be among the last areas disturbed and also because they provide a focus and an additional reason for encouraging conservation management.

The limited quantitative evidence suggests that sites protected by faiths for their spiritual values can also perform a valuable function in protecting wild nature. Sacred groves are generally highly protected (although limited harvesting may be acceptable in some cultures) and also often long established, maintaining sanctuaries of natural or seminatural areas where the original ecosystem has been widely disturbed or destroyed.

7 Scared Groves as Future Mistletoe Sanctuaries

The importance of mistletoe in various types of plant communities has been recognized by many workers in the past. Watson (2001) proposed mistletoes to function as key stone resources in forests and woodlands throughout the world, having a disproportionately large effect on faunal community composition and structure as they provide abundant nutritional and nesting resources for a diverse range of birds, mammals, and insects, and changes in mistletoe number can lead to concomitant changes in community structure. Many studies in the past have shown a positive relationship between mistletoe density and avian richness, such that areas with more mistletoe plants have higher species richness. Thus, mistletoe helps to promote biodiversity, regardless of their parasitic nature, and has an important natural role in "balanced" forests, but the benefits of mistletoe are lost in severely infected "out-of-balance" forests.

Until very recently, mistletoes of Nepal Himalayas have been overlooked despite a large number of past botanical explorations. Although, mistletoes were never recognized as an important component of Nepalese biodiversity, the recent addition of five mistletoe species to Nepal clearly indicates their diversity value. Their role in rich Himalayan diversity is unexplored and it is unknown how they affect the diversity pattern. Mistletoe faces the threat of extinction not only by deforestation, land degradation, and habitat fragmentation but also due to threats to the mistletoe bird population which plays a significant role in their life cycle. Now is the time to develop mistletoe conservation strategies to protect them before they disappear from Nepal Himalayas. Mistletoe cannot be controlled in natural plant communities since they are dispersed by a special group of highly adopted mistletoe bird species. Conservation of mistletoe can also direct in developing sacred groves as mistletoe bird sanctuaries as they complement each other in forests (Devkota 2009). This may contribute significantly in developing ecotourism in the country. But where mistletoes are seen as a threat, strategies should be carefully considered and researched, taking into account all factors that have contributed to the occurrence of a particular mistletoe species.

The social values of sacred groves, their contribution to biodiversity conservation, and the social arrangements for managing them can tell us about the practical influence of religious beliefs and rituals on nature conservation in general. Obviously, sacred groves of the valley are part of the cultural heritage of Nepal and provide important sites for the conservation of its valuable biodiversity. They represent the ancient forest of the valley and have been established for a long time, often modified greatly by local inhabitants, could constitute an important reserve of the natural vegetation, and play an important role in the conservation of local mistletoe species.

Following are the recommendations which need to be applied, as soon as possible, by the government of Nepal to protect the sacred groves of the valley:

- Carry out a detail inventory of all sacred groves which could provide important information on their role in in situ conservation.
- In order to promote them for community-based conservation of biodiversity, the government should declare sacred groves as preservation sites and incorporate them into modern conservation systems.
- Help communities find alternative ways to meet their needs for fuel and other forest resources in order to diminish the pressure on sacred groves.
- Educate and aware local communities.
- Include such sacred groves into National Biodiversity management priority list.

After protecting the sacred groves from further environmental degradation, the following are the guidelines proposed for their management and conservation to develop them as future mistletoe sanctuaries within the valley:

- Develop a management plan for sacred groves as committed in the Nepal Biodiversity Strategy (2002).
- Implement a separate conservation action plan for mistletoe in sacred groves and provide legal protection to mistletoe and mistletoe birds.
- Restore the ecology of sacred groves and reforest with indigenous plant species.
- Restore and maintain the healthy population of potential host species as healthy mistletoe population requires healthy host population and translocate mistletoe.
- Involve various stakeholders (faith groups) in planning, designing, and implementing the conservation plan for sacred groves.
- Each faith group should be involved in decision-making roles for conservation of each grove, and conservation organizations should work much more closely with faith groups to identify ways of collaboration.

The management of mistletoes for conservation and preservation has recently become more of an issue in worldwide plant conservation. The International Union for the Conservation of Nature (IUCN) has listed four species of Loranthaceae and 15 species of Viscaceae on the official IUCN Red List of Threatened Species (2006). Worldwide, rare parasitic plants, like mistletoes, in general have emerged as an important and specialized issue for conservation and need ecosystem management approach to sustain a healthy mistletoe population.

Mistletoe is an important component of tree canopy. Tree density, which directly influences the canopy structure of a forest, also influences mistletoe population dynamics. In turn, mistletoe population dynamics is greatly influenced by management activities because they obviously influence stand density, species composition, and forest canopy structure. Sacred groves of Kathmandu Valley have great potential and can prove as future mistletoe sanctuaries provided thoughtful management plans are designed and implemented timely by the government of Nepal.

References

Bhagwat SA, Rutte C (2006) Sacred groves: potential for biodiversity management. Front Ecol Environ 4(10):519–524

Devkota MP (2009) Mistletoe diversity and their status in Nepal Himalaya. In: Hartmann M, Weipert J (eds) The proceedings of biodiversity and natural heritage of Nepal Himalaya III, Natural History Museum, Erfurt

Hara H, Chater AO, Williams LHJ (1982) An enumeration of flowering plants of Nepal, vol III. British Museum (Natural History), London

Ingles AW (1995) Religious beliefs and rituals in Nepal: Their influence on forest conservation. In: Hallaway P, Gilmour DA (eds) Conserving biodiversity outside protected areas: The role of traditional agro-ecosystems. The World Conservation Union, Gland, pp 205–224

Rana G, Murray AB, Maharjan DR, Thaku AK, ICIMOD (eds) (2007) International centre for integrated mountain development. Kathmandu Valley Environment Outlook, Kathmandu

IUCN (2006) IUCN red list of threatened species. www.iucnredlist.org. Last accessed on 26 July 2011

Joshi DP (2000) Rare, threatened and endangered vascular plants of Kathmandu valley. In: Biotechnology applications for reforestation and biodiversity conservation. In: Proceedings of the 8th international workshop of BIO-REFOR, Kathmandu

Khan ML, Khumbongmayum AD, Tripathi RS (2008) The scared groves and their significance in conserving biodiversity: a review. Int J Ecol Environ Sci 34(3):277–291

Nepal Biodiversity Strategy (2002) Ministry of forest and soil conservation. HMGN, Kathmandu

Watson DM (2001) Mistletoe–a keystone resource in forests and woodlands worldwide. Annu Rev Ecol Syst 32:219–249

Chapter 44
Nutrient Recycling Starts in the Canopy: The Secretive Action of Termites

Yves Roisin, T. Bourguignon, and Maurice Leponce

Keywords Isoptera • IBISCA • Panama • Feeding ecology • Vertical stratification • Xylophagy

Bullet Points

1. Termites are abundant decomposers at the ground level in tropical rainforests, but their presence and action in the canopy are easily overlooked.
2. In a Panamanian rainforest, 7 termite species out of 45 were exclusively or preferentially found in the canopy.
3. In the canopy, species with small- or medium-sized colonies are entirely accommodated within dead branches on which they feed. By contrast, large-colony Termitidae (e.g., *Nasutitermes* spp.) build arboreal carton nests and foraging galleries to reach scattered sources of dead wood.
4. Limitations in the amount of wood available favor the acquisition of nitrogen by atmospheric N_2 fixation in small-colony canopy dwellers. By contrast, large-colony central place foragers may afford to extract N from wood, which they consume in large quantities.
5. Termites may exert a substantial influence on the C and N balance of the forest before dead wood reaches the forest floor, but the magnitude of their action remains to be investigated.

Y. Roisin (✉)
Evolutionary Biology and Ecology, Université Libre de Bruxelles, Brussels, Belgium
e-mail: yroisin@ulb.ac.be

T. Bourguignon
Department of Biological Sciences, National University of Singapore, Singapore
e-mail: dbstmdb@nus.edu.sg

M. Leponce
Biological Assessment Section, Royal Belgian Institute of Natural Sciences, Brussels, Belgium
e-mail: Maurice.Leponce@naturalsciences.be

M. Lowman et al. (eds.), *Treetops at Risk: Challenges of Global Canopy Ecology and Conservation*, DOI 10.1007/978-1-4614-7161-5_44,

1 Introduction

The role of termites as decomposers is undisputed at the level of the tropical forest floor (Martius 1994; Bignell and Eggleton 2000). The family Termitidae is especially diverse, with many taxa, collectively considered as soil feeders, feeding at a definite level in the humification gradient. Yet, the fundamental source of food for termites is wood, which can be exploited in several ways. Two extreme strategies are illustrated by small-colony dry-wood termites (family Kalotermitidae) and large-colony *Nasutitermes* (family Termitidae).

Most Kalotermitidae colonies live within a single piece of wood, which they consume. Colony size ranges between 10^2 and 10^3 individuals, including a pair of reproductives, a few permanent, sterile soldiers, and many polyvalent immatures (Fig. 44.1a). Termites being hemimetabolous, these immatures are active: they are the ones that dig galleries, chew at the wood and digest it, and may transfer food to dependent individuals (i.e., the youngest instars, reproductives, and soldiers). As the piece of wood approaches its end, the immatures will proceed along the classical hemimetabolous development, reach the winged stage, and fly away to colonize other sites and found new colonies.

Nasutitermes species reach huge colony sizes (>10^5 individuals) and function as central place foragers: a central nest (Fig. 44.1b) houses the reproductives and brood, whereas specialized workers, protected by accompanying soldiers (Fig. 44.1b, inset), forage along covered runways and bring food back to the nest in their gut. Sexual brood (wing-budded nymphs) mostly remain in the nest and, like other dependent stages, receive salivary secretions from workers.

In forest canopies, either strategy may be applicable, but the presence of termites, their abundance, and their potential role are difficult to assess. In addition, "suspended soils" that form in large epiphytes or within hollow trunks may be accessible to humus or litter feeders. Because termites spend most of their life inside galleries in the wood or under shelter when foraging, they can only scantily be collected by classical sampling methods such as fogging, beating, or activity traps (Eggleton and Bignell 1995). Flying sexuals can be collected by flight interception traps or, for night-flying species, at lights, but these methods require long-lasting monitoring and are subject to large fluctuations. Here, we synthesize results of intensive manual collecting in the canopy of a Panamanian rainforest, suspended soil sampling, and alate trapping.

2 Methods

Field work took place in the San Lorenzo Protected Area (Colón Province, Republic of Panama), within the framework of IBISCA-Panama, an international project focused on investigating the biodiversity of arthropods at all levels of a neotropical rainforest (Basset et al. 2012). Ground-level fauna was sampled according to standard protocols (Jones and Eggleton 2000; Roisin and Leponce 2004), along eight

Fig. 44.1 Representative termites of a Panamanian forest canopy. (**a**) *Calcaritermes brevicollis* (Kalotermitidae). Dead branch housing the colony was split to expose termites and their galleries. Immatures surround one soldier (*center*), able to block a gallery with its pluglike head (photo by M. Leponce). (**b**) *Nasutitermes nigriceps* (Termitidae). Nest in canopy seen from the San Lorenzo crane gondola. Covered runways connect the nest with dead branches, such as the one in the background, on which the termites feed. Inset: two workers (*middle*) and soldiers (with *pointed head*) on a runway just exposed by removing the wood carton cover (still visible on the *left side*) (photos by Y. Roisin)

transects of 20 successive 10 m^2 quadrats in 2003 and along one transect of 40 quadrats of 5 m^2, placed every 10 m, in 2004. Canopy fauna (defined as residing >10 m above the ground) was manually sampled from trees along the ground transect lines: termites were collected from arboreal nests and galleries or from dead limbs cut down by professional tree climbers (Roisin et al. 2006). Suspended accumulations of debris were sampled from 33 trees (8 core samples of 3×5 cm per tree). An equal number of soil cores were taken at the base of the same trees. These samples were extracted for 48 h in a Berlese-Tullgren apparatus. Additionally, light traps and long-lasting flight interception traps were set up to collect flying dispersers (alates) (Bourguignon et al. 2009; see also Basset et al. 2007 for detailed methodology). One occurrence was defined as the presence of one species (no matter how many individuals) in one sampling unit (ground quadrat, climbed tree, soil core, light trap, or flight interception trap).

3 Results

Manual collecting in the canopy yielded 77 termite occurrences, representing 10 species, all of which were wood feeders (Table 44.1). Four of these were not found at ground level, although 550 occurrences, representing 41 species, were recorded from ground transects. Specific components of the canopy fauna include several dry-wood species (Kalotermitidae) forming small colonies within dead branches or stubs and *Termes hispaniolae*, a wood-feeding species from a genus generally classified as soil-feeding. The rhinotermitid *Prorhinotermes molinoi* was also exclusively found in the canopy (two occurrences). Among common wood-feeding

Table 44.1 Number of termite species occurrences resulting from manual collecting and Berlese sampling of ground and canopy in the San Lorenzo forest

	Manual	Berlese	Manual	Berlese
	Ground	Ground	Canopy	Canopy
Kalotermitidae				
Glyptotermes angustus (Snyder)	1	–	3	–
Neotermes holmgreni Banks	–	–	3	–
Rugitermes panamae (Snyder)	–	–	4	–
Calcaritermes brevicollis (Banks)	3	–	15	–
Rhinotermitidae				
Prorhinotermes molinoi Snyder	–	–	2	–
Termitidae: Termitinae				
Microcerotermes arboreus Emerson	37	–	12	1
Termes hispaniolae (Banks)	–	–	7	–
Termitidae: Nasutitermitinae				
Nasutitermes nigriceps (Haldeman)	2	1	22	5
Nasutitermes guayanae (Holmgren)	23	1	5	2
Nasutitermes banksi (Emerson)	11	–	4	3
Nasutitermes corniger (Motschulsky)	5	–	–	3
34 other species	468	2	–	–
Total occurrences	550	4	77	14
Species	41	4	10	5
Total species: 45	Ground: 41		Canopy: 11	

species, *Microcerotermes arboreus* and *Nasutitermes banksi* were found at all heights. All the other species showed a definite preference either for the canopy or for the ground. For instance, *Calcaritermes brevicollis*, *Termes hispaniolae*, and *Nasutitermes nigriceps* were preferentially found in the canopy, whereas *Heterotermes convexinotatus*, *Cylindrotermes macrognathus*, and *Amitermes beaumonti* were only collected by ground transects. *Nasutitermes guayanae* was also more common at ground level (Roisin et al. 2006).

Suspended debris accumulations were poor in termites (14 occurrences out of 264 cores). Only the gallery-building, wood-feeding genera *Nasutitermes* (4 species) and *Microcerotermes* (1 species) were represented (Table 44.1).

Alate trapping yielded 1630 individuals representing 36 species in 551 occurrences: 34 species (478 occurrences) for flight interception traps and 16 species (73 occurrences) for light traps (Bourguignon et al. 2009). Only one species (*Dolichorhinotermes* sp.) caught as alates was not represented in samples collected by other methods. Overall, flight interception traps were much less efficient per unit time than light traps but could be left operating day and night for long periods. The vertical distribution of flight interception captures indicates that alates were globally more frequent in the canopy than in the understory or near the ground. This occurred because species that preferentially nest or forage in the canopy also flew more in the upper strata. In particular, dry-wood termites (Kalotermitidae) showed a clear preference for flying in the canopy. By contrast, species that preferentially nest or forage on or near the ground were caught at all heights (Bourguignon et al. 2009).

4 Discussion

Termite presence in the forest canopy was expected under three situations. First, dead branches could be colonized by small-colony termites, spending their whole life cycle within them. Second, species with large arboreal-nesting colonies could forage for food at all levels of the forest through a network of trails or wood carton tunnels. Third, humus feeders might colonize "suspended soils" formed in large epiphytes or within holes in trunks or branches.

Our results from manual collecting demonstrate that the canopy significantly contributes to the species richness of the termite assemblage. Here, on a total of 45 recorded species, 11 were found above 10 m, and 7 of them showed a clear preference for the canopy. We were able to confirm the common presence in the canopy of small-colony termites of the Kalotermitidae family, whose colonies entirely reside within dead branches. Large arboreal-nesting species were also represented by species of *Nasutitermes* and *Microcerotermes* (Termitidae), which forage for dead wood in the canopy along covered runways. No true soil feeder was found to reach suspended soils. As these species live on an energetic knife-edge and require high amounts of food (Eggleton and Tayasu 2001), the exploitation of small and far-between pockets of soil up in trees probably lies below the sustainability threshold for them.

Findings of *Prorhinotermes molinoi* and *Termes hispaniolae* in the canopy were unexpected. Both genera form rather populous colonies, but do not build externally visible nest structures. *Prorhinotermes* species are island or coastline specialists. *Termes hispaniolae* is also present on several Antillean islands, where it was routinely collected in wood at ground level (Scheffrahn et al. 1994). It is possible that in species-rich mainland rainforests, the canopy constitutes a marginal habitat, out of the reach of most competitors, in which large dead branches allow these species to subsist.

Sampling protocols based on ground transects easily overlook small-colony canopy-dwelling species. Whether the scarcity of Kalotermitidae in tropical rainforest termite inventories (e.g., Davies et al. 2003) is genuine or results from sampling bias deserves further investigation. Specific search for canopy termites by professional tree climbers is efficient but costs much time and money. A more balanced view of the local termite assemblage can be provided by alate trapping, especially if flight interception traps are hauled into the canopy. Drawbacks of this method are its low output, requiring long-term trapping, its seasonal fluctuations, and, for many areas, the lack of keys and reference samples for alate identification.

The presence of specific niches for termites in the canopy raises the question of their ecological function. The C/N ratio of wood is very high, making the acquisition of nitrogen a critical issue for wood-feeding termites. Termites living in dead branches have access to limited C resources and are therefore likely to obtain nitrogen by symbiont-mediated fixation of atmospheric N_2. By contrast, large-colony central place foragers able to exploit widely distributed food sources through foraging runways can afford adopting a C-burning strategy, i.e., consuming wood in excess of their energy needs to fulfill their nitrogen requirements (Eggleton and Tayasu 2001). In either case, termites may substantially alter the amounts of C or N present in the wood before it falls down to the floor –where most forest ecologists wishing to collect it for quantitative studies are waiting. The actual importance of canopy termites in the C and N cycles deserves further investigation.

Acknowledgments Special thanks are due to the organizers of the IBISCA-Panama project (Yves Basset, Bruno Corbara, Héctor Barrios), the supporting institutions (Pro-Natura International, Océan Vert, l'Université Blaise Pascal, the Smithsonian Tropical Research Institute (STRI), la Universidad de Panamá), and fund providers (Solvin-Solvay S.A., STRI, the United Nations Environment Programme, the Smithsonian Institution (Walcott fund), the European Science Foundation, and the Global Canopy Programme). Authors received additional support from the Belgian National Fund for Scientific Research (F.R.S.–FNRS), through personal grants to YR and a predoctoral fellowship to TB.

References

Basset Y, Corbara B, Barrios H, Cuénoud P, Leponce M, Aberlenc H-P, Bail J, Bito D, Bridle JR, Castaño-Meneses G, Cizek L, Cornejo A, Curletti G, Delabie JHC, Dejean A, Didham RK, Dufrêne M, Fagan LL, Floren A, Frame DM, Hallé F, Hardy OJ, Hernández A, Kitching RL, Lewinsohn TM, Lewis OT, Manumbor M, Medianero E, Missa O, Mitchell AW, Mogia M,

Novotny V, Ødegaard F, Oliveira EG, Orivel J, Ozanne CMP, Pascal O, Pinzón S, Rapp M, Ribeiro SP, Roisin Y, Roslin T, Roubik DW, Samaniego M, Schmidl J, Sørensen LL, Tishechkin A, Van Osselaer C, Winchester NN (2007) IBISCA–Panama, a large-scale study of arthropod beta-diversity and vertical stratification in a lowland rainforest: rationale, description of study sites and field methodology. Bull Inst R Sci Nat Belg (Entomol) 77:39–69

Basset Y, Cizek L, Cuénoud P, Didham RK, Guilhaumon F, Missa O, Novotny V, Ødegaard F, Roslin T, Schmidl J, Tishechkin AK, Winchester NN, Roubik DW, Aberlenc H-P, Bail J, Barrios H, Bridle JR, Castaño-Meneses G, Corbara B, Curletti G, da Rocha WD, De Bakker D, Delabie JHC, Dejean A, Fagan LL, Floren A, Kitching RL, Medianero E, Miller SE, Oliveira EG, Orivel J, Pollet M, Rapp M, Ribeiro SP, Roisin Y, Schmidt JB, Sørensen L, Leponce M (2012) Arthropod diversity in a tropical forest. Science 338:1481–1484

Bignell DE, Eggleton P (2000) Termites in ecosystems. In: Abe T, Bignell DE, Higashi M (eds) Termites: evolution, sociality, symbioses, ecology. Kluwer, Dordrecht

Bourguignon T, Leponce M, Roisin Y (2009) Insights into the termite assemblage of a neotropical rainforest from the spatio-temporal distribution of flying alates. Insect Conserv Divers 2:153–162

Davies RG, Eggleton P, Jones DT, Gathorne-Hardy FJ, Hernández LM (2003) Evolution of termite functional diversity: analysis and synthesis of local ecological and regional influences on local species richness. J Biogeogr 30:847–877

Eggleton P, Bignell DE (1995) Monitoring the response of tropical insects to changes in the environment: troubles with termites. In: Harrington R, Stork NE (eds) Insects in a changing environment. Academic, London

Eggleton P, Tayasu I (2001) Feeding groups, lifetypes and the global ecology of termites. Ecol Res 16:941–960

Jones DT, Eggleton P (2000) Sampling termite assemblages in tropical forests: testing a rapid biodiversity assessment protocol. J Appl Ecol 37:191–203

Martius C (1994) Diversity and ecology of termites in Amazonian forests. Pedobiologia 38:407–428

Roisin Y, Leponce M (2004) Characterizing termite assemblages in fragmented forests: a test case in the Argentinian Chaco. Aust Ecol 29:637–646

Roisin Y, Dejean A, Corbara B, Orivel J, Samaniego M, Leponce M (2006) Vertical stratification of the termite assemblage in a neotropical rainforest. Oecologia 149:301–311

Scheffrahn RH, Darlington JPEC, Collins MS, Křeček J, Su N-Y (1994) Termites (Isoptera: Kalotermitidae, Rhinotermitidae, Termitidae) of the West Indies. Sociobiology 24:213–238

Chapter 45
Valuing Ecosystem Services Flowing from the Indian Himalayan States for Incorporation into National Accounting

S.P. Singh and Rajesh Thadani

Keywords Ecosystem services indicators • Valuation • Payment

1 Introduction

The degradation of ecosystems has resulted in adverse impacts on 60 % of ecosystems services globally (MEA 2005). This is a matter of great concern as humans heavily depend on these life supporting services (Costanza et al. 1997). While not generally recognized, India is unique among developing countries in having incorporated the value of forest ecosystem services of its states in national accounting. As per the recommendations of the 12th and 13th Finance Commissions of the country, grants are transferred to forest-rich states in amounts corresponding to their forest covers (Singh 2007). However, funds allocated based on forest cover are not an accurate indicator of ecosystem services flowing from these states.

Nevertheless, the recognition by the government that ecosystem services are important for sustainable development is a major advancement in the field of conservation and for issues relating to equity. If payments are to reflect the true value of the services provided, then these services need to be both measured and assigned unit values, requiring two separate types of analysis. The necessary biophysical measurements of ecosystem services based on easily verifiable and objective methods remain a problem (Boyd and Banzhaf 2006). We present here an effort to suggest a system that could be used to valorize ecosystem services in the Himalayan states of India in a way that is consistent with the national mission on Himalayan ecosystems. The rationale used is that ecosystem services of a state benefitting the

S.P. Singh (✉) • R. Thadani
Centre for Ecology, Development and Research (CEDAR),
41/1 Vasant Vihar, Dehradun, Uttarakhand 248006, India
e-mail: surps@yahoo.com; thadani_rajesh@hotmail.com

M. Lowman et al. (eds.), *Treetops at Risk: Challenges of Global Canopy Ecology and Conservation*, DOI 10.1007/978-1-4614-7161-5_45,

Table 45.1 Criteria for selecting ecosystem services for incorporation into national accounting

Consider only the services of national (e.g., water) and international (e.g., carbon sequestration) significance
Focus on services which are easily measurable in a reliable and valid way in biophysical terms and are verifiable independently, for example, forest carbon sequestration
Use surrogates or proxies where possible that in reality cover a broad range of services, for example, biodiversity should not be considered an independent service; rather, it is a contributor to several ecosystem goods and services. Thus, species diversity, because it is precisely measurable, may be considered an indicator of a bundle of relevant biological services including biodiversity. Forest carbon may also be a measure of a variety of other forest variables

rest of the country should be valued and that these values should be incorporated into national accounting.

States or regions that are the principal providers of ecosystem services often need to curtail or restrict development activities. For example, they may not be allowed to construct a road because it restricts the movements of some charismatic megafauna such as elephants, or a hydropower project is stopped due of the threat to biodiversity. To address these restrictions, alternatives might need to be considered – for example, constructing an elevated railway track on columns that allows elephants to move freely. Some alternative forms of clean energy may be required to fulfill the electricity deficit caused by the loss of the hydropower. Such alternatives have additional costs. Hence, we have proposed that the central government should give credit to states which provide high levels of ecosystem services, in the form of finance or resources required to maintain the integrity of ecosystems while supporting economic growth.

We first consider how to estimate or quantify the flow of ecoservices in biophysical terms, and then we rank the Himalayan states with regard to the amount of ecosystem services they are producing for the benefit of the country. A policy paper has already been submitted to the Government of India for use as a basis for deciding on rates of payment for the Himalayan states. Precedence exists where the Indian Central Government has approved the transfer of a block of funds to forest-rich states, although no criteria on dividing the funds among the state were evolved.

Our main argument is that since the market does not perform money transfer from regions that benefit from ecosystem services to regions which produce them, the central government should perform this transfer. However, to implement this policy, the central government would need some reasonable methods to compare the production by different ecosystem services of national and international importance. This chapter is an attempt to support this idea. Table 45.1 lists some of the criteria we used for choosing ecosystem services. In the next Sect. 2, we deal with the potential for measuring ecosystem service delivery, and Sect. 3 addresses ways in which these services can be attributed value.

It is also clear that the stakeholders who are actually managing the lands that provide the ecosystem services are local communities. Hence, a mechanism is needed to ensure that these communities are the major beneficiaries of the money transfer from center to states. Mechanisms are needed to determine the type of benefits they should receive, and a fair way of distributing the funds would have to be developed. For example, local communities could be given access to clean cooking energy (benefits in kind). Additionally, we also see scope for the involvement of the local communities in the measurement of ecosystem services through participatory monitoring and associated with this the possibility of direct financial payments for their survey work.

2 Measuring the Delivery of Ecosystem Services

A critical area that needs to be looked at is how to measure ecosystem services. Several different methodologies exist and each uses its own set of assumptions and criteria. We focus here on developing a methodology that is easy to replicate and simple and leads to relatively consistent values. Optimally the variables should be measured by government agencies or better still be based on criteria for which relatively accurate values exist.

2.1 Identifying Ecosystem Services for Payment

A large number of ecosystem services are identified in the literature. In conjunction with the criteria listed in Table 45.1, we propose three sets of ecosystem services as the most suitable for determining payment to states. These are forest carbon sequestration, biodiversity-related services, and services that accompany mountain rivers flowing to adjacent plains. How the selected ecosystem services relate to traditionally described services is explained in Table 45.2. In what follows, we suggest how these three services could be verifiably measured in ways that are based on scientific, biophysical methods, involving local people, giving scores for each state on each indicator.

2.2 Biophysical Measurement of Carbon Sequestration

Carbon sequestration is a global service with a national responsibility. Standard methods of measuring biomass increase over time in forest trees can be found in forestry manuals, and the IPCC has approved versions of these for estimating changes in carbon stock (IPCC 2003). India is a strong supporter of international policy on reduced emissions from REDD+, under which reduced emissions of carbon from deforestation and degradation may be credited. However, deforestation is prohibited by law (through a ban on green felling) in many parts of the Himalayas. Although degradation is undoubtedly taking place because of population pressure and the continued dependence of local people on forest products, reduced degradation is very difficult to estimate since this requires a baseline (past rates of degradation). For this reason, we focus rather on enhancement of forest carbon between two different dates (i.e., on additionality), fully recognizing that this will give a

Table 45.2 Types of ecosystem services (ESs) mentioned in literature, their spatial scales, and our categories in which they have been included

Ecosystem service	Spatial scale	Service providers within an ecosystem	ESs types under which we consider
Aesthetic/cultural	Local/global	All biodiversity	Biodiversity
Ecosystem goods	Local/global	Diverse species	Biodiversity, generally paid
Air purification	Regional/global	Microbes, plants	Carbon and biodiversity
Climatic stability	Local/global	Vegetation	Carbon, biodiversity
Flood and drought mitigation	Local-regional	Vegetation	Mountain vegetation reduces flood intensity in adjacent plains (covered under forest stock and river flow); drought generally has local relevance, hence not considered
Pollination	Local	Insects, birds, mammals	Local significance, not applicable
Pest control	Local	Invertebrates, parasitoids, predators	Local significance, not applicable
Water purification	Largely local but also regional	Vegetation, soil, microbes, aquatic invertebrates	Partly included under river flow and forest stock; water in rivers is cleaner in mountainous terrain than plains
Detoxification and decomposition of wastes	Local-regional	Vegetation, microbes, aquatic invertebrates	River flows and forest stock
Soil generation and soil fertility	Local, but transported to plains through rivers, it becomes regional	Litter, soil fauna, soil microbes, N_2fixing plants, plant production	Forest stock and river flows
Seed dispersal	Local	Ants, birds, mammals	Not applicable

conservative estimate of carbon savings, since any increase in stock would imply also reversal of earlier degradation processes. However, REDD+also includes conservation of the forest stock. Even when brought under sustainable management, forests may not always succeed in sequestering additional carbon, since poor mountain people often depend on forest biomass for various aspects of their livelihoods. If however people take adequate preventive measures, such as cutting only large branches, rather than twigs with next year's buds; avoiding damaging or cutting young trees; or collecting primarily fallen wood pieces, the annual loss of biomass may be averted. In other words, although it is very difficult to quantify the reduction of degradation, it may be possible to make qualitative judgments that show that an area that had been degrading is now no longer degrading. This could then be assigned a conservation value. Thus in total, two different measures of carbon will be included: stock increases and area conserved/saved from further degradation.

Forest carbon is a national and international service (criterion 1) and easily measurable (criterion 2) and moreover is a fairly reliable indicator of the several other co-ecosystem services such as forest goods, litter fall and nutrient cycling, hydrologic regulation, and protection of soil against erosion. Hence, this indicator will carry with it several other important variables (criterion 3). Forest carbon is already being measured. Deforestation may be assessed through remote sensing technologies; the Forest Survey of India (FSI) measures forest cover regularly at an interval of 2 years (FSI 2011). Moreover, India, being a supporter of international policy on REDD+, has in any case to prepare itself at the national level to reduce forest degradation and maximize carbon credits that accrue. As per existing policy, India will heavily depend on its states to manage REDD+, and the states, in turn, will depend on local communities. The measurement of changes of carbon in a forest that is being degraded is challenging as remote sensing methods cannot pick up the small annual stock changes typical of areas which are used by communities. Nonetheless, a large work force is available in the form of local communities in Indian Himalayan states to measure forest carbon through periodic forest inventories at local level. Following some training, large amounts of data could be collected. "The Kyoto: Think Global Act Local (KTGL)" (Skutsch et al. 2009; Skutsch 2011) project which was carried out in the Himalayas and African countries showed that local communities can carry out the main steps of measuring forest carbon, and this would be quite handy for cases of forest which are already under community management, most of which are quite degraded.

In terms of interventions which could be used to encourage communities to preserve carbon stocks in their forests, several technological interventions are possible which not only reduce forest degradation but also improve livelihoods. For example, grass-based fodder development can result in the decrease of tree leaf cutting for fodder as well as dairy development. Access to clean cooking energy would drastically reduce the use of firewood. These interventions would speed up restoration of degraded forests and enhance their capacity to sequester carbon. Since forest fires, almost all of which are human induced, are an important degrading agent in these mountain states, the issue of fire also needs to be addressed. Schemes like the "no-fire bonus" (Tacconi et al. 2010) as applied in Philippines could work in conjunction with other rewards for carbon services.

In addition to carbon sequestration in forests, we should also include the carbon stocks in soils, particularly in the Alpine pasture systems. Maintenance of these reserves is important and should be considered as a separate item in the analysis.

For the case of the Himalayan region, forest productivity would potentially increase from west to east among the Himalayan states, but anthropogenic pressure can clearly modify the pattern. In the alpine belt, however, difference from east to west is likely to be small, because low temperature brings about uniformity in high altitudinal areas. All other conditions being equal, the potential rate of forest carbon sequestration is expected to be the highest in Arunachal Pradesh and lowest in Kashmir.

2.3 Biophysical Measurement of Biodiversity-Related Services

We have elected to use species richness as the basic indicator for a range of values relating to biodiversity. For this, we consider the number of plant species in the concerned state and the presence of old-growth forests. Species diversity has implications as regards several ecosystem services and goods. While carbon sequestration is a purely global ecosystem service, biodiversity has local, regional, and global significance (criterion 1) and also has intergenerational dimension; it can be readily quantified (criterion 2) and it stands as a surrogate for a variety of other biological values (criterion 3). Old-growth forest area is taken as a second variable in this area, and the potential asylum value of ecosystems as a third.

(a) *Species diversity of forests and alpine grasslands* is an easily measurable attribute. Species diversity (species richness data are available for several forest and grassland sites) contributes to ecosystem functioning and several ecosystem services. The loss of biodiversity is a matter of concern throughout the world. The number of plant species occurring in a state could be used as its biodiversity contribution, but a more complex index could be constructed in which different values are given to endemism and easily measurable groups like birds. Here community registers for biodiversity could be useful to fill the gaps, with the involvement of local communities as "spotters." Other organisms could be added, as our knowledge expands.

(b) *Old-growth forests make another biodiversity attribute*. They store a large amount of carbon, but more importantly they have considerable habitat and regulatory values (criterion 3). Somehow, old-growth forests have remained neglected in Indian forestry and conservation programs. They still exist in higher ranges, need to be conserved, and people need to be made aware of their significance. An old-growth forest provides habitat to many species of great conservation values. By counting the number of trees of above certain stem diameter, one can give an estimate of old-growth forests for particular ecosystems (criterion 2). Old-growth forests primarily are of regional and global importance (criterion 1), local people generally get little from them in terms of goods but often give considerable cultural value to them. Many so-called sacred groves (protected for religious/cultural purposes) fall in this category.

(c) *Asylum value for migrating species under global warming*. The Himalayas are the only major geographical area in south Asia where species would be able to escape from warming condition by moving uphill. Under the influence of warming, species would not only migrate toward high altitude areas, they would also show range shift from east to west. The asylum value of a state would therefore depend on (i) its position along the east-to-west arc within the Himalayas, (ii) extent of alpine areas within it, (iii) upper altitudinal limit of areas that support biota, and (iv) scale of fragmentation of vegetation and presence of corridors and stepping stones. All these indicators can be readily measured, and in terms of management, and fulfill criteria 1, 2, and 3. Humans can assist species migration by reducing fragmentation and creating corridors between isolated fragments. Such efforts would need to be encouraged to conserve the species.

Despite marked differences in biodiversity of forests, the dominant genera do not differ much across the Indian Himalayan states. *Shorea robusta* in foothills, *Pinus* spp. in lower outer flanks, *Quercus* and *Castanopsis* in mid-elevations, *Alnus nepalensis* on landslips, and firs (*Abies* spp.), birches (*Betula* spp.), rhododendrons (*Rhododendron* spp.), and junipers (*Juniperus*) in subalpine and alpine belts occur in much of the east-to-west arc of the Himalayas. The similarities of genera make it easier to correlate forest productivity to climate (precipitation and temperatures decreasing from east to west).

The number of plant species per unit area generally decreases from east to west as the aridity rises (Singh and Singh 1992). However, the number of species in any state would depend on several other factors including geographical area, altitudinal range, heterogeneity of soils, and habitats. The number of recorded flowering plants, as well as endemism, is relatively high in Arunachal Pradesh, Uttarakhand, and Jammu and Kashmir. Having a large area above 3,000 m altitude, the Kashmir region is particularly important with regard to asylum values for species migrating as a result of global warming.

2.4 Biophysical Measurement of River Flow Services

The expanse of mountain rivers can be used to give an estimate of transport of soil and water to adjacent plains for various economic activities, particularly agriculture. The catchment areas of rivers in adjacent plains and the population that they support would be the criteria for assessing their services to nation. Mountain rivers contribute considerably to water tables in the plains. For example, the heavy rains of September 17–18, 2010, in Uttarakhand are reported to have raised the water table in parts of the Gangetic plain by several meters. Forest hydrology plays a vital role in regulating water quantity as well as quality of mountain rivers and their contribution to adjoining lowland areas. River flows detoxify the environment and decompose wastes. Mountain forests and grasslands purify water and thus limit the pollution load of water reaching lowland areas low.

In terms of the flow of services through rivers, Uttarakhand state stands out among the Himalayan states, as the river Ganga which serves India's major northern states, namely, UP, Bihar, and Bengal, originates here. However, the entire water that flows from the river Ganga and silt it deposits in the plains are not contributions of Uttarakhand state. This should be kept in view while deciding on the amount of service flow. The area and population living in river basins should also be considered during ecosystem service estimation.

3 Valuing the Ecosystem Services

The measurement of ecosystem services is a necessary step in the process, but it is not the same as their valuation, which depends among other things on the weight which society gives to different services as well as the opportunity costs involved. We focus here on weighting, that is, assigning a relative value to the service.

As concerns the weighting for the different services, this is a process which might be considered more political than scientific, since it requires the comparison of relative importance to society and survival, of the different elements among the ecosystem services. Setting such weights will always require consultation and participation of different groups in society (including scientists) in order to reach an agreement or compromise. In Table 45.3, we propose a set of weights (W), subjectively derived on the basis of our understanding of the robustness of the measures in reflecting overall ecosystem health: carbon services get the highest weights because these are the most easily and reliably measured but also because they correlate with other services and can be seen as a partial measure of other services too. However, these weights should be considered as illustrative only.

It should also be understood that in any weighting exercise, a common base is required for comparison between indicators of different types. One cannot, for example, directly compare "tons of carbon" with "number of species present" or "liters of water flowing to plains" because they are measured in different, and to some extent, arbitrary units (tons rather than number, liters rather than tons; using a different unit in each case would grossly distort the final value assigned). The measured level of each ecoservice first needs to be normalized before weights can be applied. This is usually done by calculating the position of each state on a scale for each indicator and then weighting their relative positions. The simplest way to do this is to rank the states, but this obscures a lot of important differences (on forest

Table 45.3 Proposed distribution of weights between ecosystem services flowing from Indian mountain states to the rest of the country and the world

Sl	Set of services	Weight to be applied to normalized score (W)
1.	**Carbon services**	**50**
1.1	Forest carbon sequestration (increases in stock over time, measured annually)	25
1.2.1	Forest carbon conservation (size of forest area maintained)	20
1.2.2	Alpine meadows soil carbon stock (area maintained)	5
2.	**Biodiversity-related services**	**20**
2.1.1	Number of plant species	10
2.1.2	Number of endemic species[a]	02
2.2	Area under old-growth forests	05
2.3	Asylum value of the region (measured in terms of the relative availability of territory at higher elevations)	03
3	**River flow** services (measured in terms of volume of water flowing from each state into states in the plains)	**30**

[a]Data for angiosperms only have been used

Table 45.4 Example of normalizing score of states as regards biodiversity service indicators

	Number of flowering plant species				Number of endemic flowering plant species			
State	Number of species	Rank of state	Normalized index (unit-less)	State score (max. 10)	Number of species	Rank of state	Normalized index (unit-less)	Score of state (max. 2)
J&K	4,252	3	13.3	1.33	224	2	25.3	0.51
Himachal	3,047	5	9.5	0.95	81	4	9.2	0.18
Uttarakhand	4,700	1	14.7	1.47	120	3	13.6	0.27
Sikkim	4,500	2	14.1	1.41	58	7	6.6	0.13
Arunachal	3,974	4	12.4	1.24	239	1	27	0.54
Nagaland	2,431	8	7.6	0.76	14	9	1.6	0.32
Tripura	1,546	10	4.8	0.48				
Meghalaya	3,000	6	9.4	0.94	65	6	7.3	0.14
Manipur	2,376	7	7.4	0.74	74	5	8.4	0.17
Mizoram	2,141	9	6.7	0.67	27	8	3	0.06
Total	**31,967**		**99.9**	**9.99**	902	**45**	**102**	

cover, for e.g., Arunchal Pradesh has three times as much as the next state, while the differences between the second, third, and fourth states are small), so we present an alternative way of normalizing on this indicator, based on percentage of the total.

An illustration is shown in Table 45.4. Normalized indexes for the other indicators could be calculated in a similar way. Note that the average cannot be used as a point of reference, since this would automatically result in negative values for some of the states.

The score that a state would receive would then be calculated by multiplying each of the normalized biophysical measurement score (BM) by the relevant weight (W) for each service.

As concerns opportunity cost, it is clear that some states have foregone greater opportunities than others in conserving ecosystems – for example, a protected area created in foothills, such as Jim Corbett National Park in Uttarakhand, has a far higher opportunity cost than a biosphere reserve in high mountains, such as Great Himalayan National Park, H. P. The creation of Jim Corbett National Park means that this area cannot be used, as is much of surrounding land, for highly productive agricultural and industrial activities. Indeed, Uttarakhand has dedicated 13.8 % geographical area to protected areas, which is among the highest in Indian states. Some of them, namely, Jim Corbett and Raja ji parks with combined area, 1,341 km^2, are located in a highly productive region. Their combined opportunity cost is estimated as per some Indian Forest Service internal documents at roughly Rs. 50,000 million, when agricultural values of the land are considered.

In a fair system for ecosystem payments, and particularly in a system which is designed to encourage the maintenance of such services, this should be reflected in the size of payments. This is a different issue from the relative importance of the ecosystem service as such and needs to be handled independently and additionally to the weighting.

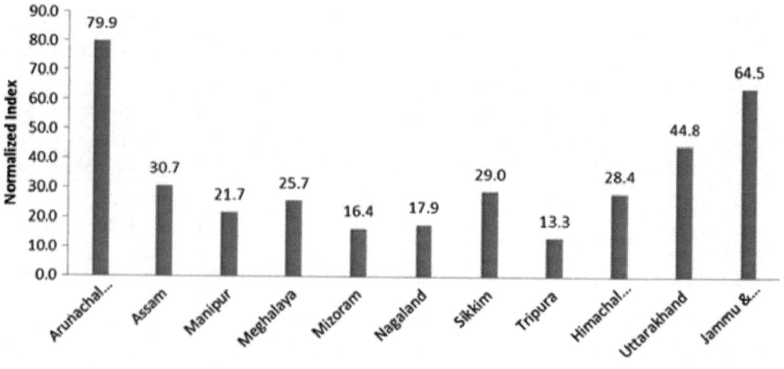

Fig. 45.1 Normalized index (unit-less) with regard to ecosystem services flowing from Indian Himalayan states. It is sum of all services: (i) carbon services which include carbon stock as well as carbon sequestration by forests, (ii) biodiversity services which include plant species number in a state and endemic species number, and (iii) asylum value, based on area of above 3,000 m altitude which supports forest and meadows. Species would migrate there under the influence of climate change

In reality, opportunity cost will most likely be related to particular projects, such as the creation of reserves as mentioned above. But if it is to be evaluated at state level, a fair surrogate measure could be taken as the net profit from agricultural yield from similar lands. This will vary between states, and the final payment to each state could be adjusted to take these differences into account.

Figure 45.1 below gives the sum of normalized indices based on a few ecosystem services of the Himalayan states, namely, biodiversity (which we have based on both the number of flowering plant species as well as the number of endemic species), asylum value for migrating species under the influence of global warming (which is based on the area of alpine meadows and areas above 3,000 m in altitude), and carbon services (including carbon stock and carbon sequestration). While we have tried to use the available data, the numbers given here are estimates and at best approximations as there exist many gaps in the data.

Nonetheless, the normalized values presented in Fig. 45.1 do give a fairly good idea of relative positions of the states with regard to their ecosystem services. Arunachal Pradesh, which clearly stands out among the Himalayan states, is known for its relatively pristine forest areas with dense tree canopy. Jammu and Kashmir scores well due to its large geographic area, high endemism, and high asylum value.

4 How Much Should the States Be Paid?

The payments to the states would not represent market prices as a market does not operate for ecosystem services. Rather than trying to determine a market value, we propose that the payment should be sufficient to help the state enable people to address issues related to poverty which exacerbate the degradation of natural resources. An element that we feel is particularly important is access to an appropriate clean cooking energy, such as LPG, biogas, hydroelectricity, and solar power in place of fuelwood which is a primitive form of energy and associated with forest degradation, women's drudgery, and health issues related both to collection and burning. The advantages of this kind of support are many: reduction in women's drudgery, more time for child care (women often travel miles to collect firewood), reduction in kitchen air pollution, which is an important cause of early mortality, lowered black carbon emissions, and increased carbon sink value of forests with the release of pressure on them from fuelwood collection. The payment could also be used to promote livelihoods with a linkage to conservation. For example, by developing grass-based fodder production, the harvest of tree leaves could be reduced. Apart from increasing milk yield, grass fodder would increase forest carbon sequestration and reduce physical stress on women. It could also be used to lengthen the slash and burn cycle or replace it with more economically rewarding activities.

Some mountain states are committed to organic form of agriculture, which has helped keep the environment free from toxics and eutrophication of water bodies. Funds transferred from the center could be used to strengthen organic agriculture and market the products of these states.

What is essential is that states receiving funds as payments for their ecosystem services should be required to provide clear plans for how the money is to be spent, showing that the lion's share is indeed going to the stakeholders who are responsible for the delivery of the ecosystem services and that they should be audited on this before subsequent payments are made. Otherwise, there is every risk that the money will be spent on general expenditure and not be applied to programs that are actively designed to support and maintain ecosystem services.

5 Conclusion

What is advocated here is the development of a transparent methodology that can help determine the distribution of funds released by the Indian Government for ecosystem services provided by various hill states.

The methodology can be best described as a work in progress and there is certainly scope for improvement. Values initially calculated may be approximations based on available data. Some of the estimates used in the calculations are also based on numbers that would best be further verified and validated. However, we see this as a positive start and of great utility given that the process of payment for ecosystem services has already been initiated by the central government in India. This also can be an example that can be replicated in other countries where some regions contribute a disproportionate share of ecosystem services in a bid to incentivize those regions.

We propose in this chapter that the payment should be used for activities which contribute to flow of ecosystem services, as well as have social equity. By enabling the economically deprived to have access to clean cooking energy and better quality fodder, poverty will be reduced while at the same time advancing conservation and ESs flow. States would need to provide necessary data using standard methods. Their claims would need to be certified by a proper authority or agency. As long as a state is unable to produce proper data, it can be given a minimum base value. The allocation of funds by the center in itself would be a stimulus for better research on biodiversity, forest carbon, and river and ecosystem flows and encourage the creation of better data.

Acknowledgments The authors gratefully acknowledge the extensive comments, suggestions, and inputs of Prof Margaret Skutsch of the University of Twente.

References

Boyd J, Banzhaf S (2006) What are ecosystem services? The need for standardized environmental accounting Units. Resources for the Future, Washington

Costanza R, d'Arge R, de Groot R, Farberk S, Grasso M, Hannon B, Limburg K, Naeem S, O'Neill RV, Paruelo J, Raskin RG, Suttonkk P, van den Belt M (1997) The value of the world's ecosystem services and natural capital. Nature 387:253–260

Forest Survey of India (FSI) (2011) India state of forest report. FSI, Dehradun

Intergovernmental Panel on Climate Change (IPCC) (2003) Good practice guidance for LULUCF. IPCC, c/o Institute for Global Environmental Strategies, Japan

Millennium Ecosystem Assessment (MEA) (2005) Ecosystems and human well-being: synthesis. Island Press, Washington, DC

Singh SP (2007) Himalayan forest ecosystem services: incorporating in national accounting. CHEA, Nainital

Singh JS, Singh SP (1992) Forests of the Himalaya. Gyanodaya Prakashan, Nainital

Skutsch M (ed) (2011) Community forest monitoring for the carbon market: opportunities under REDD. Earthscan, London

Skutsch MM, McCall MK, Karky B, Zahabu E, Peters-Gurain G (2009) Case studies on measuring and assessing forest degradation: community measurement of carbon stock change for REDD. FAO, Rome, 2009. Forest resources assessment working paper 156

Tacconi L, Mahanty S, Suich H (2010) Payments for environmental services, forest conservation and climate change: livelihoods in the REDD? Edward Elgar, Cheltenham/Northampton

Chapter 46
Epiphytic Plants as NTFPs from the Forest Canopies: Priorities for Management and Conservation

Daniela Dutra Elliott and Tamara Ticktin

Keywords Epiphyte • Harvest • Trade • Market • Orchidaceae • Bromeliaceae • Ferns • Moss

1 Introduction

The harvest of wild plants for cultural and economic reasons is highly important in the lives of hundreds of millions of people worldwide (FAO 1995; Cunningham 2001). Non-timber forest products (NTFPs) are used for food, shelter, medicine, and as a source of income (FAO 1995; Cunningham 2001). NTFPs may also play an important part in other cultural and religious practices (FAO 1991; Ticktin et al. 2006, 2007). Some of the most important but understudied NTFPs include canopy species of cultural and economic value that have been used extensively by people (Bennett 1992; Nadkarni 1992). Indeed, a wide diversity of canopy species have been documented to be used across the globe (Alcorn 1984; Acharya and Rokaya 2010; Bennett 1995; Bulpitt et al. 2007; Haeckel 2008; Hornung-Leoni 2011; Lawlor 1984; Ossenbach-Sauter 2009; Upreti et al. 2005).

Epiphytic plants are characteristic and distinctive components of forests, contributing significantly to total biomass, species diversity, and nutrient cycling in these ecosystems (Gentry and Dodson 1987; Nadkarni and Matelson 1992). The potential for over-harvesting has generated concerns over the conservation of epiphytic species in many parts of the world (Koopowitz 2001; Flores-Palacios and Valencia-Diaz 2007; Mondragón 2009). In this chapter, we define canopy as "the above ground plant organs within a community" (Moffett 2000) and consider NTFPs from the canopy to be epiphytic plants. Although fruits, bark, and resin can be considered canopy NTFPs under this definition, they are not the focus of this chapter.

D. Dutra Elliott (✉) • T. Ticktin
Botany Department, University of Hawaii at Manoa, Honolulu, HI 96822 USA
e-mail: ddutra@hawaii.edu; ticktin@hawaii.edu

M. Lowman et al. (eds.), *Treetops at Risk: Challenges of Global Canopy Ecology and Conservation*, DOI 10.1007/978-1-4614-7161-5_46,

We know very little about how harvest impacts populations of epiphytic plants, the species that depend on them on the canopy, and their ecosystems. Wolf and Konings (2001) were the first to develop theoretical recommendations for epiphytic species harvest. Since then, there have been only a few published studies to evaluate the ecological effects of harvesting vascular (Mondragón 2009; Mondragon and Ticktin 2011) and nonvascular (Peck and Christy 2006; Peck and Frelich 2008; Peck et al. 2008; Molleman et al. 2011) epiphytes. Moreover, understanding the effects of epiphyte harvest goes beyond biological data and requires information on cultural and socioeconomic patterns of epiphyte use and trade. However, very few studies have focused on epiphyte or moss markets (Flores-Palacios and Valencia-Diaz 2007; Peck and Muir 2007; Mondragón 2008). Below we review what we know to date about harvesting from the forest canopy, focusing on the four main categories of epiphytic plants harvested from the canopy: orchids, bromeliads, ferns, and nonvascular plants. Based on this review, we then present some research priorities.

2 Orchids

A very high percentage of harvested epiphytes belong to the Orchidaceae. Epiphytic orchids are especially threatened with extinction caused by the high demand generated by their horticultural and cultural uses, habitat destruction (Soto et al. 2007), and their vulnerability to harvest. The latter is due to their life history characteristics, including their requirements for mycorrhizal associations, breeding systems, and nutrient limitations (Ackerman et al. 1996; Tremblay 1997; Otero et al. 2005). Orchids are one of the few plant groups that receive blanket protection internationally (CITES) due to conservation concerns generated by the high-volume trade (Koopowitz 2001; Koopowitz et al. 2003).

Epiphytic orchid use has been recorded historically in different parts of the globe. In Australia the pseudobulbs of epiphytic orchid species have been used as binder for paintings (Boustead 1966) and medicinally to treat pediatric cases (Pearn 2005). In pre-Columbian times, the Mayans of Central America used many species of orchids (Ossenbach-Sauter 2009), and Aztec Mexico orchid pseudobulbs were used as a source of gums (Berdan et al. 2009). Today, the demand for orchids is diverse and is generated by specialized international collectors, traditional cultural uses at the local level, and widespread commercialization as ornamentals.

The impact of orchid collecting, which usually involves harvest of the whole plant, depends on the life history of the species and the type of collecting being conducted (Cribb et al. 2003). A detailed understanding of population biology, structure, and dynamics of populations is therefore needed to fill knowledge gaps on tropical orchid species as well as to provide accurate metrics for the development of sustainable management plans in areas where overharvest is likely to occur. Unfortunately, such information for tropical orchid taxa is rare (Koopowitz 2001).

For example, we know that many epiphytic orchid species are harvested in China for medicinal use (Zhenga and Xing 2009; Ghorbani et al. 2011; Bulpitt et al. 2007), but there is little information on the impacts of harvest on populations. Rajbhandari et al. (2000) and Koirala et al. (2010) state that in certain areas of Nepal, orchid species are disappearing due to over collection. Among 92 species of orchids used medicinally in Nepal, 43 are epiphytes (Acharya and Rokaya 2010). However, very little is known (or is currently published as peer review publications) about the harvest of these species. Koirala et al. (2010) assessed of the current stock of traded orchid species in the Rolpa district of Western Nepal and documented the species being harvested and traded. This is an important step towards the understanding of trade in the area; however, further research is needed to understand the impacts of harvesting these epiphytic orchid species.

We are aware of only one published study to date on the ecological impacts of harvesting epiphytic orchids, which was carried out on a commonly harvested species, *Guarianthe aurantiaca,* in Mexico (Mondragón 2009). This study shows that populations of this species can withstand only very low rates (<5 %) of harvest.

Information on the intensity and frequency of harvest of epiphytic orchid species is also lacking, but all evidence indicates that harvest is extensive and unsustainable for orchid populations worldwide. The severity of harvest impacts at the population level varies according to the decisions local harvesters make, specifically on harvest practices such as plant selection, intensity, and frequency. The impacts of harvest can be mitigated by practices informed by local or traditional ecological knowledge (Ticktin 2004).

Finally, very little is known about the trade of orchids. Only one study has been published on the market sales of orchid species. Conducted in Veracruz, Mexico, this study illustrated that 81 % of epiphytic plant species being traded in markets were orchids (Flores-Palacios and Valencia-Diaz 2007). This illustrates how strong harvest pressure can be for orchid species and how little is known about this issue. Unless we understand how and why people make decisions to harvest, we cannot design locally appropriate plans for management or conservation. Clearly research is urgently needed on the decision making and environmental behavior driving orchid collecting and trade.

As a response to heavy harvest and trade, propagation protocols for native epiphytic orchid species have been developed in recent years (Åvila-Diaz et al. 2009; Dutra et al. 2009; Santos-Hernandez et al. 2005). Although many orchid conservation efforts in the tropics have involved ex situ micro-propagation labs to relieve the pressure on rare species, these techniques are expensive and require expertise in vitro propagation. They therefore limit the power of local communities to be involved in conservation efforts. In areas where epiphytic orchids are used as decorations and then discarded (e.g., Mexico), programs that foster the reestablishment of these home gardens where they can be reused in following years are extremely valuable. Damon et al. (2005) investigated ways to develop rustic cultivation techniques for seven epiphytic orchid species using plants that had been propagated in an ex situ lab. Community members established rustic orchid galleries in their home gardens and coffee plantations. However, propagating orchids from seeds can be time consuming, since most species take many years to bloom depending on the species and cultivation conditions. Reusing mature plants or harvesting only parts to be used for propagation from the forest can speed up the process of flowering and give incentive to people to propagate plants instead of only relying on harvesting new plants from the forest.

3 Bromeliads

The Bromeliaceae family is comprised of over 3,000 species spread across the Neotropics, with one species found outside the Americas. Epiphytic bromeliads play important roles in nutrient cycling (Nadkarni and Matelson 1992) and in supporting animal communities in the canopy. Among some of the animals that epiphytic bromeliads house are crabs (Diesel 1989), ants (Blüthgen et al. 2000), endemic ciliates (Foissner 2003), and even earthworms (Fragoso and Rojas-Fernández 1996).

The harvest of bromeliads is common in the Neotropics (Bennett 1990; Aceby et al. 2010; Haeckel 2008; Negrelle et al. 2011) and especially for use as ornamental plants and for religious decorations (Mondragón and Villa-Guzmán 2008). Throughout the year and especially during the Christmas and Holy week holidays, thousands of plants are collected to sell in urban markets (Rees 1976; Mondragón 2008) and to decorate churches (Haeckel 2008) and nativity scenes in Southern Mexico. Flores-Palacios and Valencia-Diaz's 2007 market study in Veracruz demonstrated the high diversity and volume of bromeliad species that are sold. However, very little is known about the effects harvest has on epiphytic bromeliad populations and the surrounding ecosystems. Mondragon and Ticktin (2011) used matrix population models to study the demography of two species of bromeliads (*Tillandsia macdougallii* and *T. violacea*) in Oaxaca, Mexico, and the potential to sustainably harvest their populations. They found that the populations for these two bromeliads are unlikely to support even low levels of harvest. This is because these species, like the other (non-harvested) epiphytic bromeliads studied to date (Mondragón et al. 2004; Zotz et al. 2005; Winkler et al. 2007; Valverde and Bernal 2010), are most sensitive to changes in survival of large or reproductive individuals, and these are precisely the stages that are harvested.

However, throughout the year high volumes of bromeliads fall naturally to the forest floor, where they are unable to survive. Mondragon and Ticktin (2011) illustrated that harvest of these fallen bromeliads from the forest floor can be a both ecological and economically sustainable alternative to harvest from the canopy. This has recently been shown to be true also for other bromeliad species in Veracruz, Mexico (Toledo & Hernandez, unpublished results). The harvest of fallen bromeliads may offer a promising alternative for epiphytic bromeliads elsewhere and needs further investigation.

4 Ferns

Ferns and their allies have been used all over the world for thousands of years: their rhizomes, leaves, and spores have been used as medicine, their fibers to construct baskets and other artifacts (May 1978; Thomas 1999), and their fronds for cultural traditions such as adornment in hula in Hawaii (Ticktin et al. 2006, 2007). Although there is literature on the effects of harvest on terrestrial ferns, very little has been published on epiphytic fern harvest. Porembski and Biedinger (2001) showed that the commercial harvest of *Platycerium stemaria* (Polypodiaceae) in the Ivory Coast caused a large local decline in the population size of this epiphytic fern.

The harvest of epiphytic ferns is likely misrepresented in the literature since most studies in the ethnobotanical literature do not mention the life form, and it is difficult to know if species are being harvested from the canopy or not. However, much can be learned from the literature pertaining to terrestrial fern harvest. For example, many tree fern species (Cyatheales) are critically endangered due to harvest of their trunks to be used as orchid medium in the ornamental trade. This type of harvest affects many different epiphytic fern species (and other vascular and nonvascular epiphytes) that are growing on tree fern trunks. Therefore, the harvest of terrestrial ferns can have an indirect impact on the demography of other species. Roberts et al. (2005) assessed the diversity of epiphytic fern species found on Tasmanian tree ferns and found 97 fern and bryophyte species on one species of tree fern (*Dicksonia antarctica*) and 64 species on the trunks of a second, *Cyathea cunninghamii*. Similar studies on the diversity of epiphytic ferns on the trunks of tree ferns have been conducted in other places including Brazil and Argentina (Schmitt and Windisch 2010; Marquez and Yanez 2012), and they show a high diversity of epiphytic ferns growing on tree fern trunks.

5 Nonvascular Plants

Moss harvest has been defined as the collection of either moss or a mixture of mosses, liverworts, and lichen and some vascular plants (Peck and Muir 2007). Epiphytic moss harvest has been studied mainly in North America. Commercial moss harvest of epiphytic species takes place in the Pacific Northwest (Peck and Christy 2006) and Appalachians (Muir et al. 2006) and has been going on for decades. Three million kilograms of mosses are harvested per year (Muir et al. 2006). Large mats are peeled off from tree trunks and branches and sold in the national and international markets (Peck 2006a). Harvested moss is mainly used in the floriculture and horticulture industry as decorations, packing material, and as a medium for growing epiphytes such as orchids.

Moss harvest mainly takes place in National forests (Peck 2006b) and permits are required. We know more about moss harvest in North America than we know about harvest of any other epiphyte. Species involved in the commercial trade have been identified (Muir 2004; Peck and Muir 2007), the epiphyte communities removed by harvest have been described (Peck 1997), and harvest impacts such as biomass accumulation and changes in relative species composition over time have been assessed (Peck and Frelich 2008). The effects of different harvest practices on species abundance, richness, and composition have also been quantified (Peck and Christy 2006). Based on this research, several harvest guidelines have been established and implemented for the sustainable management of moss. For example, the method of harvest (partial vs. total stripping) can have enormous impacts on how epiphytic communities recover after harvest in terms of species diversity, abundance, and composition and how long recovery takes (Peck and Frelich 2008). However, there are still many gaps of knowledge on moss harvest research (Peck 2006a). Moss is harvested from areas other than the United States, for example, in Mexico, moss is heavily harvest for nativity decorations during the Christmas season.

In India, epiphytic lichens are very important NTFPs. They are harvested by indigenous communities as a form of income (Molleman et al. 2011) and are sold as spices, in the perfumery and medical industries, and as a raw material for paint (Kumar and Upreti 2001; Upreti et al. 2005). Harvesters scrape the branches to remove the lichens and large volumes of lichens are harvested. In 1997 alone, 1,000 metric tons was harvested nationally (Shah 1997). To date, only one quantitative lichen harvest study has been published: Molleman et al. (2011) showed that 7 years after harvesting, the lichen community showed significant regeneration capacity. They also demonstrated that the harvest method makes a difference: shallow harvesting resulted in a quicker recovery of species abundance and richness compared with deep harvesting. This highlights the importance of studies that investigate how people make decisions to harvest and their effects on species composition and population structure.

6 Research Priorities

To gain a better understanding of epiphyte harvest and its ecological consequences, the following research is needed:

1. More research, and especially integrated research, on the population dynamics, market dynamics, local management, and ecosystem impacts of epiphyte harvest is needed. The information is essential for designing sustainable management plans.
2. Traditional ecological knowledge should be taken into account in harvest studies. There is little information available on how local people make decisions on the harvest or management of wild epiphyte populations, which are often common pool resources. This knowledge is critical for understanding patterns

and intensity of use and their ecological implications. Imposing management policies and regulations that ignore preexisting management practices, and related social institutions, may deteriorate existing land and resource management capacities (Alcorn and Toledo 1998; Agrawal and Ostrom 2001).

3. National and international studies on the trade of epiphytes are especially important. We know, for example, that orchids are being heavily harvested in many parts of the world; however, they are greatly underreported in CITES (Phelps et al. 2010). In parts of Brazil, for example, orchids are harvested from natural areas and sold on sides of roads, but there are no publications that we know of the internal trade of orchids in that country.
4. Research on the effects of heavy epiphyte harvest, on the organisms that depend on them, is needed. No studies to date have examined this for epiphytes, but, for example, a study on *acai* palms showed that heavy harvest of canopy fruit can lead to changes in the abundance and composition of frugivores (Moegenburg and Levey 2003).
5. Community-based propagation of heavily harvested epiphytes may reduce pressure on wild populations and provide income to local communities, and more research is needed in this area.
6. More research on a broader spectrum of epiphytes is needed. Not only do we have little or no information on many of the most commonly harvested species of epiphytes, but there are also entire plant families of epiphytes for which no harvest studies are available. For example, Flores-Palacios and Valencia-Diaz (2007) report that epiphytic species belonging to the Araceae, Cactaceae, and Lycopodiaceae are harvested in Veracruz, Mexico; however, nothing is known about their harvest or of its ecological consequences.

References

Aceby A, Krömer T, Maass BL, Kessler M (2010) Ecoregional distribution of potentially useful species of Araceae and Bromeliaceae as non-timber forest products in Bolivia. Biol Conserv 19:637–650

Acharya K, Rokaya M (2010) Medicinal orchids of Nepal: are they well protected? Our Nat 8(1):82–91

Ackerman JD, Sabat A, Zimmerman JK (1996) Seedling establishment in an epiphytic orchid: an experimental study of seed limitation. Oecologia 106:192–198

Agrawal A, Ostrom E (2001) Collective action, property rights, and decentralization in resource use in India and Nepal. Polit Soc 29:485

Alcorn JB (1984) Huastec Mayan ethnobotany. University of Texas Press, Austin Texas

Alcorn JB, Toledo VM (1998) Resilient resource management in Mexico's forest ecosystems: the contribution of property rights. In: Linking social and ecological systems: management practices and social mechanisms for building resilience. Cambridge University Press, Cambridge, pp 216–249

Åvila-Diaz I, Oyama K, Gomez-Alonso C, Salgado-Garciglia R (2009) In vitro propagation of the endangered orchid Laelia speciosa. Plant Cell Tissue Organ Cult 99:335–343

Bennett BC (1990) Ethnobotany of bromeliads: indigenous uses of tillandsias in the southern Andes of Peru. J Bromeliad Soc 40(2):64–69

Bennett B (1992) Use of epiphytes, lianas and parasites by the Shuar people of Amazonian Ecuador. Selbyana 13:99–114

Bennett BC (1995) Ethnobotany and economic botany of epiphytes, lianas, and other host-dependent plants: an overview. In: Lowman MD, Nadkarni NK (eds) Forest canopies. Academic, San Diego

Berdan FF, Stark EA, Sahagún JD (2009) Production and use of orchid adhesives in Aztec Mexico: the domestic context. Archeol Pap Am Anthropol Assoc 19(1):148–156

Blüthgen N, Verhaagh M, Goitía W (2000) Ant nests in tank bromeliads–an example of non-specific interaction. Insect Soc 47:313–316

Boustead W (1966) Conservation of Australian Aboriginal bark paintings with a note on the restoration of a New Ireland wood carving. Stud Conserv 11(4):197–204

Bulpitt CJ, Li Y, Bulpitt PF, Wang J (2007) The use of orchids in Chinese medicine. J R Soc Med 100(12):558–563

Cribb PJ, Kell SP, Dixon KW, Barrett RL (2003) Orchid conservation: a global perspective. Orchid conservation. Natural History Publications (Borneo), Kota Kinabala, pp 1–24

Cunningham A (ed) (2001) Applied ethnobotany: people, wild plant use & conservation. Earthscan, London

Damon A, Perez Soriano M, del Lourdes RM (2005) Substrates and fertilization for the rustic cultivation of in vitro propagated native orchids in Soconusco, Chiapas. Renew Agric Food Syst 20:214–222

Diesel R (1989) Parental care in an unusual environment: metopaulias depressus (Decapoda: Grapsidae), a crab that lives in epiphytic bromeliads. Anim Behav 38:561–575

Dutra D, Kane ME, Richardson L (2009) Asymbiotic seed germination and in vitro seedling development of *Cyrtopodium punctatum*: a propagation protocol for an endangered Florida native orchid. Plant Cell Tissue Organ Cult 96(3):235–243

FAO (1991) Non-wood forest products: the way ahead. Food and Agriculture Organization of the United Nations, Rome

FAO (1995) Non-wood forest products for rural income and sustainable development. Food and Agriculture Organization, Rome

Flores-Palacios A, Valencia-Diaz S (2007) Local illegal trade reveals unknown diversity and involves a high species richness of wild vascular epiphytes. Biol Conserv 136:372–387

Foissner W (2003) Morphology and ontogenesis of Bromeliophrya brasiliensis gen. n., sp. n., a new ciliate (Protozoa: Ciliophora) from Brazilian tank bromeliads (Bromeliaceae). Acta Protozool 42:55–70

Fragoso C, Rojas-Fernández P (1996) Earthworms inhabiting bromeliads in Mexican tropical rainforests: ecological and historical determinants. J Trop Ecol 12:729–734

Gentry AH, Dodson C (1987) Contribution of nontrees to species richness of a tropical rain forest. Biotropica 19:149–156

Ghorbani A, Langenberger G, Feng L, Sauerborn J (2011) Ethnobotanical study of medicinal plants utilised by Hani ethnicity in Naban River Watershed National Nature Reserve, Yunnan, China. J Ethnopharmacol 134(3):651–667

Haeckel IB (2008) The "arco floral": ethnobotany of Tillandsia and Dasylirion spp. in a Mexican religious adornment. Econ Bot 62(1):90–95

Hornung-Leoni CT (2011) Bromeliads: traditional plant food in Latin America since prehispanic times. Polibotánica 32:219–229

Koirala PN, Pyakurel D, Gurung K (2010) Orchids in Rolpa district of Western Nepal: documentation, stock, trade and conservation. Banko Jankari 20(2):3–13

Koopowitz H (2001) Orchids and their conservation. Timber Press, Portland

Koopowitz H, Lavarack PS, Dixon KW (2003) The nature of threats to orchid conservation. In: Dixon KW, Kell SP, Barrett RL, Cribb PJ (eds) Orchid conservation. Natural History Publications, Borneo

Kumar K, Upreti D (2001) Parmelia spp. (lichens) in ancient medicinal plant lore of India. Econ Bot 55:458–459

Lawlor LJ (1984) Ethnobotany of Orchidaceae. In: Arditti J (ed) Orchid biology: reviews and perspectives. Cornell University Press (Comstock), Ithaca
Marquez GJ, Yanez A (2012) Helechos epifitos de Alsophila setosa (Cyatheaceae, Pteridophyta) en la provincia de Misiones, Argentina. Boletin de la Sociedad Argentina de Botanica 47(3–4):435–442
May LW (1978) The economic uses and associated Folklore of Ferns and Fern Allies. Bot Rev 44(4):491–528
Moegenburg SM, Levey DJ (2003) Do frugivores respond to fruit harvest? An experimental study of short-term responses. Ecology 84:2600–2612
Moffett MW (2000) What's "up"? A critical look at the basic terms of canopy biology. Biotropica 32:569–596
Molleman L, Boeve S, Wolf J, Oostermeijer G, Devy S, Ganesan R (2011) Commercial harvesting and regeneration of epiphytic macrolichen communities in the Western Ghats, India. Environ Conserv 38:334–341
Mondragón D (2008) La comercializaon navidena de bromelias epifitas en la ciudad de Oaxaca, Mexico. Etnobiología 6:24–28
Mondragón D (2009) Population viability analysis for Guarianthe aurantiaca, an ornamental epiphytic orchid harvested in Southeast Mexico. Plant Species Biol 24:35–41
Mondragon D, Ticktin T (2011) Demographic effects of harvesting epiphytic bromeliads and an alternative approach to collection. Conserv Biol 25:797–807
Mondragón D, Villa-Guzmán DM (2008) Estudio etnobotánico de las bromelias epífitas en la comunidad de Santa Catarina Ixtepeji, Oaxaca, México. Polibotánica 26:175–191
Mondragón D, Durán R, Valverde T (2004) Temporal variation in the demography of the clonal epiphyte Tillandsia brachycaulos (Bromeliaceae) in the Yucatán Peninsula, Mexico. J Trop Ecol 20:189–200
Muir PS (2004) An assessment of commercial "moss" harvesting from forested lands in the Pacific Northwestern and Appalachian regions of the United States: how much moss is harvested and sold domestically and internationally and which species are involved. Department of Botany and Plant Pathology, Oregon State University
Muir P, Norman K, Sikes K (2006) Quantity and value of commercial moss harvest from forests of the Pacific Northwest and Appalachian regions of the US. Bryol 109:197–214
Nadkarni N (1992) The conservation of epiphytes and their habitats: summary of a discussion at the international symposium on the biology and conservation of epiphytes. Selbyana 13:140–142
Nadkarni NM, Matelson TJ (1992) Biomass and nutrient dynamics of epiphytic litterfall in a neotropical montane forest, Costa Rica. Biotropica 24:24–30
Negrelle RRB, Anacleto A, Mitchell D (2011) Bromeliad ornamental species: conservation issues and challenges related to commercialization-doi: 10.4025/actascibiolsci. v34i1. 7314. Acta Sci Biol Sci 34(1):91–100
Ossenbach-Sauter C (2009) Orchids and orchidology in Central America: 500 years of history. Lankesteriana 9(1/2):1–268
Otero JT, Bayman P, Ackerman JD (2005) Variation in mycorrhizal performance in the epiphytic orchid *Tolumnia variegata* in vitro: the potential for natural selection. Evol Ecol 19:29–43
Pearn J (2005) The world's longest surviving paediatric practices: some themes of Aboriginal medical ethnobotany in Australia. J Paediatr Child Health 41:284–290
Peck JE (1997) Commercial moss harvest in northwestern Oregon: describing the epiphyte communities. Northwest Sci 71:186–195
Peck JE (2006a) Towards sustainable commercial moss harvest in the Pacific Northwest of North America. Biol Conserv 128:289–297
Peck JE (2006b) Regrowth of understory epiphytic bryophytes 10 years after simulated commercial moss harvest. Can J Forest Res 36:1749–1757
Peck JE, Christy JA (2006) Putting the stewardship concept into practice: commercial moss harvest in Northwestern Oregon, USA. Forest Ecol Manag 225:225–233

Peck JE, Frelich LE (2008) Moss harvest truncates the successional development of epiphytic bryophytes in the Pacific Northwest. Ecol Appl 18(1):146–158

Peck JE, Muir PS (2007) Conservation management of the mixed species nontimber forest product of "moss" – are they harvesting what we think they're harvesting? Biodivers Conserv 16:2031–2043

Peck JE, Hoganson H, Muir P, Ek A, Frelich L (2008) Using inventory projections to evaluate management options for the Nontimber forest product of epiphytic moss. Forest Sci 54(2):185–194

Phelps J, Webb EL et al (2010) Boosting CITES. Science 330(6012):1752

Porembski S, Biedinger N (2001) Epiphytic ferns for sale: influence of commercial plant collection on the frequency of *Platycerium stemaria* (Polypodiaceae) in coconut plantations on the southeastern Ivory Coast. Plant Biol 3(1):72–76

Rajbhandari KR, Bhattarai S, Joshi R (2000) Orchid diversity of Nepal and their conservation need. In: Bista MS, Joshi RB, Amatya SM, Parajuli AV, Adhikari MK, Saiju HK, Thakur R, Suzuki K, Ishii K (eds) Proceedings of 8th international workshop of BIO-REFOR: biotechnology applications for reforestation and biodiversity conservation, Kathmandu, 28 Nov–2 Dec 1999. BIO-REFOR, IUFRO/SPDC, Tokyo

Rees J (1976) The oaxaca christmas plant market. J Brom Soc 26(6):223–232

Roberts NR, Dalton PJ, Jordan GD (2005) Epiphytic ferns and bryophytes of Tasmanian tree-ferns: a comparison of diversity and composition between two host species. Austral Ecol 30(2):146–154

Santos-Hernandez L, Martinez-Garcia M, Campos JE, Aguirre-Leon E (2005) In vitro propagation of Laelia albida (Orchidaceae) for conservation and ornamental purposes in Mexico. HortScience 40:439–442

Schmitt J, Windisch P (2010) Biodiversity and spatial distribution of epiphytic ferns on Alsophila setosa Kaulf. (Cyatheaceae) caudices in Rio Grande do Sul, Brazil. Braz J Biol 70:521–528

Shah N (1997) Lichens of economic importance from the hills of Uttar Pradesh, India. J Herbs Spices Med Plants (USA) 5:69–76

Soto M, Solano R, Hàgsater E (2007) Risk of extinction and patterns of diversity loss in Mexican orchids. Lankesteriana 7:114–121

Thomas BA (1999) Some commercial uses of pteridophytes in Central America. Am Fern J 89(2):101–105

Ticktin T (2004) The ecological implications of harvesting non-timber forest products. J Appl Ecol 41:11–21

Ticktin T, Whitehead AN, Fraiola H (2006) Traditional gathering of native hula plants in alien-invaded Hawaiian forests: adaptive practices, impacts on alien invasive species and conservation implications. Environ Conserv 33:185–194

Ticktin T, Fraiola H, Whitehead AN (2007) Non-timber forest product harvesting in alien-dominated forests: effects of frond-harvest and rainfall on the demography of two native Hawaiian ferns. Biodivers Conserv 16:1633–1651

Tremblay R (1997) Lepanthes caritensis, an endangered orchid: no sex, no future. Selbyana 18:160–166

Upreti DK, Divakar PK, Nayaka S (2005) Commercial and ethnic use of lichens in India. Econ Bot 59:269–273

Valverde T, Bernal R (2010) Hay asincronia demografica entre poblaciones locales de Tillandsia recurvata?: Evidencias de su funcionamiento metapoblacional. Boletin de la Sociedad Botanica de Mexico 86:23–36

Winkler M, Hulber K, Hietz P (2007) Population dynamics of epiphytic bromeliads: life strategies and the role of host branches. Basic Appl Ecol 8:183–196

Wolf JHD, Konings CJF (2001) Toward the sustainable harvesting of epiphytic bromeliads: a pilot study from the highlands of Chiapas, Mexico. Biol Conserv 101(1):23–31

Zheng X, Xing F (2009) Ethnobotanical study on medicinal plants around Mt. Yinggeling, Hainan Island, China. J Ethnopharmacol 124(2):197–210

Zotz G, Laube S, Schmidt G (2005) Long-term population dynamics of the epiphytic bromeliad, Werauhia sanguinolenta. Ecography 28:806–814